~ to be returned on ~
the last d~~

Facility Manager's Operation and Maintenance Handbook

Other McGraw-Hill Handbooks of Interest

Facility Manager's Operation and Maintenance Handbook

Bernard T. Lewis

McGraw-Hill

New York   San Francisco   Washington, D.C.   Auckland   Bogotá
Caracas   Lisbon   London   Madrid   Mexico City   Milan
Montreal   New Delhi   San Juan   Singapore
Sydney   Tokyo   Toronto

Library of Congress Cataloging-in-Publication Data

Lewis, Bernard T.
   Facility manager's operation and maintenance
handbook / Bernard T. Lewis.
     p.   cm.
   Includes index.
   ISBN 0-07-040048-2 (alk. paper)
   1. Factory management. 2. Plant engineering.
3. Plant maintenance. I. Title.
TS155.L3678 1999
658.2—dc21
                                   98-34425
                                    CIP

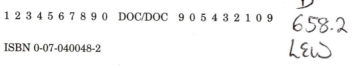

McGraw-Hill

A Division of The *McGraw-Hill Companies*

1 2 3 4 5 6 7 8 9 0 DOC/DOC 9 0 5 4 3 2 1 0 9

ISBN 0-07-040048-2

*The sponsoring editor for this book was Linda Ludewig, the editing
supervisor was Frank Kotowski, Jr., and the production supervisor was
Tina Cameron. It was set in Century Schoolbook by Terry Leaden of
McGraw-Hill's Professional Book Group composition unit.*

Printed and bound by R. R. Donnelley & Sons Company.

McGraw-Hill books are available at special quantity discounts to use
as premiums and sales promotions, or for use in corporate training programs. For more information, please write to the Director of Special
Sales, McGraw-Hill, 11 West 19th Street, New York, NY 10011. Or contact your local bookstore.

To my wife Ruth, she withstood and understood; and to my grandsons, Michael, Jeffrey, Neil, Joseph, and Mark, and great-grandson Seth.

Contents

Part 1 Organizing for Maintenance Operations

Chapter 5. Preventive Maintenance (PM) Plan   5.1

Chapter 6. Predictive Maintenance Plan   6.1

Chapter 7. Indoor Air Quality (IAQ) Plan   7.1

Part 3 Equipment and Systems Operations and Maintenance Procedures

Chapter 10. Mechanical Equipment and Systems Maintenance Procedures

Chapter 11. Electrical Equipment and System Maintenance Procedures

Foreword

The people whose responsibility it is to maintain a nation's building stock and constructed infrastructure have always been challenged. Balancing the conflicting imperatives imposed by the effects of time and nature against a sometimes indifferent source of necessary resources has been a problem since we began building buildings. Constructing roads, bridges, and buildings that we cannot afford to maintain properly is not a new concept. Examples exist throughout recorded history and lay all about us today, although this now represents a challenge that is more intense than experienced by those before us for several reasons.

First, our building stock is aging at a rate much faster than it is being replaced. This means that there are more, older buildings each year, that buildings and their components are expected to last longer, perhaps longer than they were intended, and many buildings are or will be pressed into services for which they were not designed. All of these and other factors as well will increasingly stress our constructed environment, and place even greater demands on facility management professionals to deliver increasing value for the repair and maintenance dollar.

The second factor, which will become more apparent over the midterm future, is the outcome of the high-pressure development era of the 1970s and 1980s. During this time, the success of a development project was often measured by how inexpensively a building could be constructed and how quickly it could be occupied. Spurred by pressures to reduce construction cost and time, the building industry responded with a flurry of new construction methods and materials which have yet to withstand the test of time. This era also attracted developers whose only objective was to get the fastest and highest return on their investment, and to dispose of the property at the first profitable opportunity. Worries about the quality of construction and life cycle value were left to those who followed. Fortunately, many of these influences are gone now. Unfortunately, their legacy will remain with us for decades to come.

An outgrowth of the hyperactive development era was the inevitable depression of real estate values, which in turn led to the transition of many companies from lessee to owner. While these decisions were made for good and sound business reasons, they also left some of our buildings in the hands of owners who are inexperienced and unsophisticated in matters of asset preservation. While in most cases there is a high level of interest among these owners in preserving the asset, the expertise to do this is usually far removed from their core business.

Another, broader influence is the strong trend today among U.S. industries to become more productive and competitive in world markets. The competition within these organizations for capital and expense dollars is being more heavily weighted toward productivity and competitiveness and less towards maintaining constructed environments, unless or until deterioration has advanced to the point where it threatens the core business. Although maintenance of the constructed environment has never been a high priority in most industries, it is becoming even less so today. In a time when every dollar has to produce a return, preventive and predictive maintenance expenditures sink to a low priority.

Having adequate maintenance funding is only useful if the money is spent wisely. Our industry is replete with examples of bad maintenance decisions, from solving the wrong problem to ignoring a seemingly small detail that later results in massive building damage. Some of these decisions arise because buildings and their systems are regarded as simple assemblies whose problems and solutions are believed to be obvious and intuitive when in fact they are not. For example, a high-rise apartment building in the mid-Atlantic region underwent a multimillion dollar masonry restoration simply because a defect in a relatively minor flashing detail went uncorrected.

In this new arena, the best weapon any facility manager can have is knowledge: knowledge of the processes by which buildings and their systems interact, skillful use of the replace/repair model, knowing when and how to apply predictive maintenance, and most of all avoiding inappropriate measures because the problem was misunderstood.

Few individuals will possess the breadth of knowledge needed to make the right decision in all cases at all times. The judicious use of expert advice is indispensable to the process of making consistently sound maintenance decisions. Advice should be sought from recognized experts who have a record of cost-effective solutions and who are willing to look beneath the superficial to determine the problem's true cause. Every problem will bring a flurry of well-intentioned but anecdotal advice. While courtesy demands that we listen, one needs to resist the temptation to apply a quick fix to an offhand diagnosis.

It would be easy to predict catastrophe if the foregoing processes were allowed to reach their conclusion. However, catastrophe is not an acceptable outcome and so ways must be found to counter the trends we see happening in our commercial environment. Doing more with less has become the marketplace battle cry in every aspect of American commerce, and facility management will be no exception.

Gary G. Briggs
Senior Vice President / Chief Operating Officer
Consolidated Engineering Services, Inc.

Preface

This handbook was developed to provide for the tremendous and apparent need for a comprehensive information source on facilities' equipment and systems operations and maintenance requirements. Facilities operations and maintenance budgets are spiraling today. Facilities' operating costs must be lowered in the face of facility occupants' constant demands for increased services. With the ever-increasing use of HVAC equipment and systems that are state-of-the-art in technology, proper quantity and quality equipment and systems operations and maintenance support becomes essential. Because these equipment and systems are becoming widely utilized, facility managers, facility engineers, and operating engineers require instruction and guidance in their use and care.

These personnel are expected to be experts in practically every phase of facility operations and maintenance—from HVAC operations to electrical maintenance. With the constant influx of new equipment and systems in recent years, they must have continuing updated knowledge of the methods for making the physical changes required to adapt this equipment and systems to facilities; and to modify, accordingly, existing methods and procedures for operations and maintenance tasks. This handbook presents a program for improving equipment and systems reliability to provide required services (minimize breakdowns) at the least overall "cradle-to-grave" cost.

Delivery of facility operation and maintenance services, in today's economy, will be shaped by the same challenges facing every property owner: how to keep pace with spiraling regulatory, equipment, systems, and facility occupant demands on budgets limited by competitive organization needs. This handbook proposes three broad objectives in this regard:

- *Continuity in all services* to avoid equipment, systems, and facility downtime and consequent loss of occupants' comfort, safety, and productivity.

- *A professionally managed property* in terms of the extent of use of state-of-the-art management, information, and specialized technical services, with backup support available.

- Achieving each of the above objectives at an affordable cost.

Experience has shown that the following functional program areas are of special interest and need for facility managers in the next millenium. They will be covered in detail in the handbook.

Technical Improvements Plan

A comprehensive technical improvements plan will reduce annual operation and maintenance costs. Services will include the following examples:

- Total Quality Management (TQM) Program to develop cost/benefit plans which when implemented will provide cost-effective long-term solutions to management operations and maintenance problems.

- Value Improvement Program to develop successful value engineering approaches that will add value to operations and maintenance design, engineering, and installation efforts.

- Predictive Maintenance Plan that will describe predictive maintenance technologies, the predictive process, how to start a program, and how to measure the program's effectiveness. Predictive maintenance techniques' usage will enhance, and make more effective the Preventive Maintenance Program, as well as improve the operational efficiency of equipment and systems.

Quality Control

Performance of quality control inspections, aided by trend analyses. Facility department managers and supervisors will conduct routine, but random, inspections. Results will be recorded and tracked in logs and in the Facility Management Department's Computerized Maintenance Management System ("CMMS"). Inadequate performance will be quickly addressed with responsible individuals. Contractors' managers will perform quality control inspections for their own staffs, thus reinforcing the Facility Manager's Quality Assurance inspections of their work.

Indoor Air Quality (IAQ) Services

The explosive growth of sealed building design has spawned growing concern over indoor air quality (IAQ) as attested by EPA and OSHA issuance of IAQ regulations. This potential problem area must be managed by development and use of a comprehensive IAQ program to minimize property deficiencies that affect occupant productivity and add to maintenance costs.

Computerized Maintenance Management Systems (CMMS)

The operations and maintenance routines must be guided by computerized software (modules) to cover service calls, preventive and predictive maintenance, work orders, inventory control, personnel data, performance standards, warranties, financial and accounting systems, and reports (both production and financial).

Outsourcing Considerations

Outsourcing operations and maintenance functions/tasks must be guided by the need for use of specialty contractors to reduce capital expenditures, to minimize downtime, to provide flexibility in tasks performance, and to reduce overall operations and maintenance costs. Essentially, outsourcing should be used to avoid in-house staffing of highly skilled, specialized trades personnel with their incumbent costs and work-management requirements.

I am sure that use of this handbook as a reference source will assist facility manager practitioners, both in the public and private sectors, in providing high-quality, least-cost property operation and maintenance support by describing proven, useful procedures and techniques.

Bernard T. Lewis, P.E., C.P.E.

Contributors

Gary G. Briggs *Senior Vice President/Chief Operating Officer, Consolidated Engineering Services, Inc.* (FOREWORD)

Michael D. Burns *President/CEO, Engineering Management Consultants, Inc.* (CHAP. 2)

Arthur J. Freedman *President, Arthur Freedman Associates, Inc.* (CHAP. 16)

Tony Graf *Client Services Division, Microwest Software Systems, Inc.* (CHAP. 5)

Dana Green *General Manager, Substation Test Company* (CHAP. 11)

Joseph E. Humphrey *Diaganostic Test Foreman, Potomac Electric Power Company* (CHAP. 6)

Joel Levitt *President, Springfield Resources* (SEC. 2, CHAP. 4)

Dr. Bernard T. Lewis, P.E., C.P.E. *Facilities Management Consultant* (SEC. 1, CHAP. 1; SEC. 5, CHAP. 3; CHAP. 10; SEC. 1, CHAP. 12; APPENDIX)

Paul S. Lewis Attorney at Law (SEC. 2, CHAP. 12)

Dennis Mulgrew *Control Center Manager, Consolidated Engineering Services, Inc.* (CHAP. 9)

Richard S. Nietubicz *Manager, Maintenance Contracts Services, Metropolitan Washington Airports Authority* (SEC. 2, CHAP. 1)

Richard Payant *Director, Facilities Management Department, Georgetown University* (SECS. 1–4, CHAP. 3)

Alan Pearlman *Vice President, Electronic Systems Associates* (SEC. 3, CHAP. 8)

A. J. Ruta *President, Microwest Software Systems, Inc.* (CHAP. 5)

Steven B. Sadler *Vice President (New England), Pritchard Industries of New England, Inc.* (CHAP. 13)

Ronald D. Schloss   *Executive Director, SeeC, L.L.C.* (CHAP. 15)

Clive Shearer, P.E., C.Eng.   *Management Consultant* (SEC.1, CHAP. 8)

Richard M. Silberman   *Manager of Business Development, Technical Project Manager, Healthy Buildings International* (CHAP. 7)

Harlen Edmund Smith, P.E., C.V.S.   *Manager of Value Engineering Services (for the Parsons Brinckerhoff family of firms) and Assistant Vice President of Parsons Brinckerhoff Construction Services, Inc.* (SEC. 2, CHAP. 8)

William G. Suter, Jr.   *Director, Physical Plant Operations, American University* (SEC. 1, CHAP. 4; CHAP. 17)

Carleen Wood-Thomas   *Landscape Architect, Consolidated Engineering Services, Inc.* (CHAP. 14)

Paul Zeski   *Engineering and Quality Control Manager (Texas), Pritchard Industries of New England* (CHAP. 13)

Organizing for Maintenance Operations

Program Organizational Structure

Richard S. Nietubicz

Maintenance Contract Services
Metropolitan Airports Authority
Alexandria, VA

Bernard T. Lewis

Facilities Management Consultant
Potomac, MD

1.1 Responsibilities and Authority

The facilities management function is a logical outgrowth of competitive markets that characterize the 1990s business environment. Acceptance of facilities operations and maintenance costs as fixed expense items in the budgeted cost of a product or the furnishing of a service has given way to the realization that these costs can be minimized by the application of technical and management effort in the planning, design, and operational life of a facility.

To derive maximum financial benefit from such efforts, the grouping of all facilities functions involving an interrelationship or interdependence should be achieved. Thus site planning should be consistent with the state of the design art and the dollar limitations of the capital-investment capability of the cognizant public or private sector organization. The design itself should incorporate lowest *overall total* cost features—not lowest initial cost or lowest maintenance cost. The maintenance program should relate the cost of breakdowns to the investment in a preventive/predictive maintenance program. The heating, ventilation, and air-conditioning (HVAC) energy and utilities support function should be objectively analyzed and evaluated in terms of "produce-buy?" and demand peaks and production costs. The payoff on the replacement of obsolescent or deteriorated facilities and equipment should be an integral part of the facilities equation that spells lowest *overall total* cost.

The grouping of these closely interlocked functions is an essential element in the achievement of a lowest overall total cost posture for the public or private sector organization. Recognition of the need to establish a single responsible organization to achieve this result has been the key to success in meeting the competitive challenges that face the public or private sector facility manager. By taking advantage of specialization in the facilities function, the public or private sector facility manager acquires the same degree of concentrated expertise that has historically been devoted to the areas of production, marketing, sales, finance, etc. In this sense, the facility function is the last frontier of industrial management.

Facility managers can exert a significant or a minor influence on the capital and expense costs of providing essential facilities and equipment in support of the provided product or service. Their influence will be significant if they are assigned the responsibility for melding the technically diverse facilities functions into a coordinated, integrated "support machine" which will ensure lowest *overall total* cost. Concurrently, their authority would provide for decision making in the area of responsibility.

The influence of facility managers will be minor in terms of cost control if their assigned responsibilities are fragmented and they have no direct control over all the cost factors involved in providing adequate space, HVAC support, systems, and equipment support so that they can be grouped into a facility "package."

1.2 The Facility Manager

The facility manager is the single responsible individual in a public or private sector organization to whom top management looks to coordinate and control the efforts of all employees engaged in

- Facility planning
- Engineering and design of new facilities
- Engineering and design of modifications to facilities

- Engineering and design in support of maintenance and repair functions

- Construction of facilities and installation of equipment

- Maintenance and repair of facilities, systems, and equipment

- Generation, purchase, and maintenance of HVAC systems

- Evaluation of proposals for replacement of facilities and equipment which are based in whole or in part on maintenance, energy, or utilities savings

1.3 Facility Planning

Facility planning is the process of determining the characteristics and costs of the land and physical facilities which will satisfy the operational and functional requirements of the user. A facility master plan establishes both the carrying capacity and a logical strategy for the development of a site. It should provide for future site development. It should be applied to the development of raw, new, as well as existing sites. Master plans should be updated regularly to reflect changes in site and/or strategic business conditions. Each improvement on the site should be developed consistent with this plan. In addition, a facility plan should be prepared for each improvement. The facility plan should provide directions, cost estimates, and focus for its design, construction, and operations. The facility planning function provides an objective analysis and evaluation of land usage, site selection, buildings, structures, and utility and transportation infrastructure as well as on-site equipment and systems required to accomplish specified business objectives. It includes the base mapping of the site, traffic and transportation studies, and local and regional development plans and identifies important aspects of the site (i.e., building zones, main egress, planting buffers, etc.) Planning for individual sites includes the preliminary layout of facilities, systems, and equipment and a determination of the engineering and construction feasibility of a project.

Stated in somewhat different terms, it can be said that facility planning is the development of factual quantitative data and the formulation of recommendations based thereon for the provision of the physical facility and its production and support equipment and systems necessary to satisfy top management's stated requirements.

The decision to build a new facility or to expand or modernize an old facility can be based upon several factors, some or all of which may be interrelated in a particular decision-making situation, as in the following:

- The need to expand production to keep pace with sales growth
- The need to provide production capacity for a new product line or support service
- The need to lower distribution costs
- The need to lower labor costs
- The need to lower facilities operations and maintenance costs
- The need to lower energy costs

In the first two cases, the sales and marketing organizations will assist in developing specifications for required facilities, systems, and equipment expansions in conjunction with the production organization. In the third situation, the distribution organization will specify the requirement. The reduction of labor costs is an issue that transcends the entire organization. The last factors, operations, maintenance, and energy, are the functional responsibility of the facility organization.

After review and approval of the requirements for a change in the capacity, location, or size of a facility, and its equipment and systems, by top management, it is the responsibility of the facility manager to develop a series of engineering evaluations which will specify the feasibility and cost of each of the possible alternate courses of action. Based upon these evaluations, a single recommended solution to the problem of satisfying the stated requirements should be developed to provide quantitative data relative to:

- The cost of the site purchase within the specified geographical area
- The configuration of the facilities so as to provide for least overall construction cost consistent with production layout and aesthetic considerations
- Lost production time in the case of a modification to an existing facility
- Availability of transportation and communications links with distribution centers and raw-material sources
- Availability of adequate and reliable utilities and energy sources
- The estimated construction and installation costs of the proposed expansion or modification
- The estimated relocation costs of personnel, systems, and equipment
- The estimated purchase price of required new equipment and systems

The feasibility studies of various combinations of the above factors will be integrated into parallel studies of the purchasing organization relative to raw-materials supplies, the personnel organization relative to labor supplies and wage rates, the financial organization relative to capability of the organization to finance the total cost of the project and the relative tax advantages of alternate sites, and the distribution organization relative to warehousing and delivery costs.

Each of these parallel studies will have a significant impact upon the evaluations of various sites and the facilities configuration. It is therefore essential that the facility manager be provided initially with the best-defined and specifically detailed statement of requirements that is possible. It is equally important that any change in the initial planning information be immediately conveyed to the facility manager so as to avoid needless expenditure of planning worker-hours on unnecessary evaluations.

The end result of the facility planning effort will be a documented statement of the feasibility and cost of providing the facilities, systems, and equipment which will meet the requirements of the various functional managers. This statement should be in the form of a single best solution or, in the event that conflicting requirements of the functional managers cannot be resolved, a series of alternate proposals.

In either event, the facility planning studies should provide detailed breakdowns of estimates on:

Site-development costs:

1. Access facilities (roadway, waterfront, air transportation, and/or railway)

2. Clearing and grading

3. Extension of water, sewerage, and electrical utility systems

4. Fencing and other security features

5. Intraplant roads, walks, parking areas

6. Exterior lighting

7. Landscaping

Costruction costs:

1. Foundation and utilities

2. Structures

3. Building systems

4. Materials layout and storage yards

5. Materials-handling systems

6. Production equipment installations
7. HVAC generating facilities
8. Ground structures, towers, test facilities
9. Storage tanks
10. Waste-processing systems
11. Crane trackage and cranes
12. Installed support equipment
13. Other identifiable facilities, systems, and equipment

Operating costs:

1. Maintenance and repair of facilities
2. Maintenance and repair of production and support equipment
3. Operation of support equipment
4. Energy and utility usage
5. Operation of facility support equipment
6. Trash and recycling collection and disposal
7. Custodial services
8. Operation and maintenance of motor vehicles and equipment, construction, and materials-handling equipment
9. Security and fire protection
10. Insurance rates

Support equipment costs:

1. Office furnishings and miscellaneous installed equipment
2. Maintenance shop equipment

Other costs:

1. Facilities engineering and architectural design
2. Inspection and administration of construction and maintenance contracts
3. Real estate costs and fees
4. Relocation costs
5. "Lost production" costs

In developing the solutions for the provision of facilities and equipment to satisfy the requirements of the various functional managers, the facility planner should ensure that a full and complete analysis is made of alternates to construction or modification. The merits of the leasing of real property and rental of support equipment should be fully explored and compared with the costs of ownership. A thorough space and equipment utilization study should be a required part of the planning program so as to avoid the construction or leasing of new facilities or equipment when existing facilities or equipment are underutilized.

In considering the degree of planning that is required on a particular project, the facility manager should evaluate the influence of the planning effort on the cost of the project. For a multimillion dollar facility proposal, the full spectrum of the planning function is usually justified. For an extension to an existing facility, a much reduced level of effort is required.

In either situation, it should be stressed that facilities planning should be based upon the criteria and requirements of the various functional managers. The facility planner cannot work in isolation. The realities of sales and marketing problems, labor availability, production materials, supplies, etc., must be an integral part of the site-selection, preliminary engineering, and economic feasibility aspects of the planning for a new project. The requirements of these functions must be translated into practical solutions if the facility planning function is to make a constructive contribution to the industrial process.

1.4   Facility Engineering and Design

Facility engineering and design is the conversion of concepts into solutions, working drawings, and specifications for the construction of facilities, and the procurement and installation of equipment and systems. It includes the study of facilities problems and the development of criteria and standards for the economical solution of such problems.

In discussing the role of engineering and design, the principal tasks that constitute this function in terms of facility management should include:

1. The evaluation of facilities, systems, and equipment
2. The translation of requirements into specific solutions

In each of these areas of functional interest, the engineering and design organization is responsible for:

1. The design of new facilities

2. The design of modifications and major repairs to existing facilities

3. The furnishing of engineering support to the repair and maintenance shops

4. The development of design criteria

5. The furnishing of specialized facilities engineering consulting services

6. The development of operations, maintenance, and repair criteria

The functional role of the facility manager in the design of new facilities includes all aspects except the production equipment and the process flow. These designs are logically the province of the production and/or industrial engineering organizations. But the housing of the production equipment, the furnishing of the HVAC services, and the configuration of the various facility elements in support of the production process are the responsibility of the facility engineering and design organization.

The integration of the production equipment and the process flow into the final facilities design requires a high degree of teamwork and cooperation with the production and layout staffs. On the one hand, the layout should not be developed without a clear understanding of the impact that each plan will have in the cost of constructing the facilities and providing the HVAC services. Also to be considered are the specialized requirements of the production equipment for precise humidity control, vibration-damping foundations, acoustical attenuation, radiation shielding, etc. On the other hand, the facilities design should be a lowest *overall total* cost design incorporating a balanced treatment of (1) construction costs, (2) installation costs, and (3) operation and maintenance costs in terms of the *facilities function* and (1) production costs and (2) handling, warehousing, and delivery costs in terms of the *production function.*

The responsibility of the facility organization should be clear-cut and explicit in providing this balanced treatment. In no case should facility managers presume to be expert in production and layout matters. Nevertheless, they should provide objective and cost-oriented recommendations on the proposals of the manufacturing groups relative to location and layout of equipment and systems, and the design features required to support such equipment and systems.

Once a decision has been reached by top management and the generalized concepts of the production and facilities planners have been established as the approved approach, the facility designer establishes a means of providing for the desired end product by preparing plans and specifications for construction and installation.

Although the responsibility for the preparation of plans and specifications is a function of the facility management organization, the facility manager has options on how he or she should discharge this responsibility in the design of new facilities.

The first option is to establish a design group with all the technical capabilities required by the type of facilities that need to be supported. In exercising this option, the facility manager must recognize that the workload characteristics of facilities design are rarely in a steady-state condition. In most cases, they are unpredictable as to both time-frame and worker-hour requirements. However, if the new facilities design workload can be predicted and the volume is sufficient to support a full-time staff, the facility manager should provide all necessary design service with an in-house engineering group. This is the most cost-effective approach.

The situation most frequently encountered by the facility manager is that the requirements for the design of new facilities are either sporadic or so volatile as to be unpredictable within an arbitrarily established time frame. In either case, facility managers are faced with the problem that they cannot hire and fire engineering talent as the workload peaks or falls and still maintain any semblance of quality in the plans and specifications for which they are responsible.

The options in this case are to perform all the design for new facilities by means of contract arrangements with consulting architect-engineering firms or to contract for some of the workload and perform the remainder with an in-house staff.

If the latter solution is selected, it must be accepted that at certain times the facility engineering and design staff will be unable to perform any other function by reason of the volume and time limitations involved in producing plans and specifications for construction. The impact of this problem upon the furnishing of design services for modifications to existing facilities, and engineering support to the facility shops, is such that it will only be the occasional facility manager who can afford the prolonged interruptions to the day-to-day facilities engineering requirements.

The alternate course of action is usually taken by most facility managers. Recognizing the economic impracticality of staffing at a level which would permit accomplishment of an unpredictable design workload for new facilities concurrently with the day-to-day facilities engineering workload, they select qualified consultants to accomplish the design workload for new facilities and monitor and guide their performance with in-house staff.

In contracting for such services, the facility manager retains the responsibility for the end product. The consultant can provide quality architectural and engineering services only when parameters of the design task are established. In some cases, a clear definition of scope can be established initially. In other cases, the consultant may have to explore several alternate solutions before definition of scope is established. The solutions to problems posed by production and layout requirements must be consistent with the many other overlapping and coincident problems that surround a new facilities design task. Therefore, it is incumbent on the facility manager, through the facility engineering and design organization, not only to provide thorough guidance but also to review thoroughly the consultant's work as it progresses, so as to ensure both technical adequacy and functional consistency.

1.4.1 Modifications or major repairs

Construction or installation cost	Normal fee multiplier
$ 0–$10,000	2.5
$ 10,000–$25,000	2.0
$ 25,000–$75,000	1.8
$ 75,000–$200,000	1.6
$200,000–$500,000	1.4
$500,000–$1,000,000	1.2

These factors are, of course, only indicative of the relative range of "average" jobs. A consultant involved in a complex design for a heating, ventilating, and air-conditioning modification could reasonably expect higher fee multipliers to be used and should alternately expect a lower multiplier for a parking lot–sealing project.

1.5 Engineering Support to the Maintenance and Operations Functions

The furnishing of engineering support to the maintenance and operations functions should be accomplished by an in-house staff. This includes the provision of sketches and plans for repair and maintenance projects requiring engineering determinations, such as the layout and sizing of piping and ducts, the design of concrete foundations and floors, the design of switching and instrumentation installations, the design of minor modifications and alterations to facilities and support equipment—these and the thousands of similar tasks that require the services of a qualified facility engineer during the day-to-day operation of a facility should be performed as part of the facility engineering function by facility department employees. Where the workload volume is small, facility managers can choose to employ a "general" engineer or devote a portion of their own time to this vital facility engineering function.

1.5.1 Specialized technical services

The furnishing of specialized technical services is a day-to-day requirement for the normal facility engineering organization. Where such services are a full-time workload, a qualified facility (graduate)engineer should be employed as an assistant to the supervisor of the engineering section. This could occur in almost any industry or government agency, and examples are almost endless. The need for a sanitary engineer to monitor, control, and advise on the liquid-waste problems of the chemical industry is a common instance. Another is the services of a qualified entomologist in the food-processing industry.

If the workload is not sufficient to warrant a full-time employee, the facility manager has the option of part-time assignment of a qualified facility engineer from the organization's engineering staff or contracting for intermittent services from qualified consultants. If consulting organizations are utilized, it is advisable to engage the required services on a "retainer" basis so as to ensure immediate responsiveness when such is required. This would involve the payment of a fee for the availability of the consultant when called and an additional fee for specific services required above a minimum scope specified in the "retainer" agreement. In this way, possibly costly delays can be avoided by the prior commitment of the consultant to provide responsive service.

If the part-time services of an in-house facility engineer are used for specialized technical assignments, rigid control must be exercised to ensure that his or her routine assignments do not delay or prevent providing responsive service. The nature of these specialized technical services is such that extensive delay in accomplishment will usually result in needless expense or legal difficulties. Thus the task of studying the peak demand and energy use of a facility which is delayed by the unavailability of an electrical specialist can result in considerable excessive electrical cost until it is accomplished and all possible reduction schemes are implemented. In many cases, facilities have experienced a peak demand for only one day in a year and have had to pay for this peak for the other 364 days because of the demand clauses in their energy contracts. The identification of demand problems and their correction, if possible, is a classic example of a specialized technical assignment, which should not be deferred, since every day of deferment adds to the existing loss.

Legal difficulties requiring facilities engineering study are most commonly encountered in waste disposal, asbestos abatement, noise abatement, and damage to adjacent property problems. In these cases the continuance of the cause of the problem can have adverse results requiring costly litigation and incurring loss of governmental and public goodwill. Again, the technical service should be made available by the facility manager as soon as the need arises and for as long a period as will be required to develop a solution.

1.5.2   Cost reduction

The development of maintenance and repair criteria is of paramount importance in developing maximum long-range cost reduction alternatives in the facilities management function. However, it is the most consistently ignored area for application of effort that is encountered in a production or service organization. The advantages of governmental and private research and the experience gained through local experience with products and techniques should be consistently applied to the programmed tasks that are being specified for the maintenance craftspeople, the operating engineers, and the construction contractor. The need for major improvement in this area of the facility management and engineering function cannot be overstated. The available literature is full of examples of designed and then built-in maintenance problems and costs that could have been avoided by a consistent operations and maintenance engineering effort and application of the criteria developed from this effort.

In one instance some 320,000 square feet of roofing was applied to a number of flat concrete slabs in a hot, dry environment subjected to sudden heavy rainfalls. Despite all the technical reasons for avoiding a built-up roof structure in this situation, the designers specified a four-ply tar-and-gravel installation. The end result was a constant maintenance job after less than 2 years of use. The expansion of the concrete slabs during the heat of the day and the rapid contraction at night produced a differential movement of the joints, and any rain trapped in the resultant cracks was turned to steam by the sun's heat. Repeated bubbling of the steam pockets created a ruptured surface, and the entire installation was a continuous maintenance problem. The development of maintenance criteria for this area would have avoided the design errors that were translated into maintenance costs.

In developing maintenance and repair criteria it is essential that a feedback be established between the design engineer and the cognizant maintenance foremen. The repetition of past errors is a common experience in facilities maintenance and operations, and it is only by informing the designer of the problems created by a flashing detail, a piping intersection, an undersized motor, a cramped working space, and similar day-to-day expense items that the designer can ensure future avoidance of the same problems. A second essential facet of the development of maintenance and repair criteria is a constant searching of the technical literature by the design staff to ensure that the experience and research of others are applied in the local facilities environment.

1.6 Construction of Facilities and Installation of Equipment and Systems

Construction and installation are the translations of design concepts into physical reality. The plans and specifications of the engineers become working units of a facility only through construction and installation contracts or by work accomplished by the maintenance and repair shops. In either case, it is necessary to monitor the performance of the construction or installation crew to ensure that the designer's plans and in some cases even the specified techniques are faithfully followed.

The facility manager, in almost all cases involving major construction, installation, or modification projects, will utilize the services of a qualified contractor. For minor work of this nature the maintenance shops are usually utilized. The distinction between major and minor is a relative one; but in facilities with fewer than 100 personnel in the facility management organization, it is recommended that all individual jobs requiring more than 500 worker-hours be performed by contractor forces.

The maintenance and repair staff should be devoted to the preservation of the facility and its equipment and systems. It is a common and costly error to staff at a level sufficient to support assignment of personnel to construction or installation tasks beyond a minimum level of effort. At times, need for a 500-worker-hour job will arise. In a facility staffed with a 100-person facilities organization there will be a considerable number of such jobs during each year. Thus the diversion of their efforts to a nonrepetitive workload creates planning, scheduling, and accomplishment problems which can be avoided in almost every case by the hiring of a construction or installation contractor.

The argument of economy is usually put forward by the proponents of the accomplishment of major projects by the facility maintenance shops or the honing and use of technical skills by in-house workforces on larger jobs. Such arguments usually do not include the various factors that go to make up the true cost of such accomplishments. This true cost in many cases is hidden in the accounting system of the facility or is the unidentifiable cost of deferment of needed repairs or preventive- and predictive-maintenance tasks in order to accomplish a major construction or modification job.

In the larger facility organizations (i.e., over 100 personnel) a progressively increasing scale of maximum worker-hours that shop forces could expend on single construction or modification projects should be established. The following rules of thumb are cited in this regard.

1.6.1 Personnel in facility organizations—Single-project worker-hour limit

No of personnel	Single-project worker-hour limit
0–100	500
100–300	1000
300–500	1700
Over 500	2500

There are, of course, individual projects which, because of governmental security requirements, product secrecy, or an expensive contractor learning process, can be more economically accomplished by shop forces, irrespective of these cited recommended limits. In these cases a careful record of deferred maintenance and repair tasks should be kept and a catch-up program established so as to ensure that the programmed maintenance and repair workload is accomplished as planned. The costs of any overtime or extra-material costs involved in such catch-up work should be estimated as part of the true cost of accomplishment of major project work by the maintenance and repair shops.

Where construction, modification, and equipment and systems installation contracts are used as the method of accomplishment of required projects, the responsibility of the facility manager can be divided into two distinct phases:

1. On-site inspection and acceptance of completed work
2. Contract administration

On-site inspection and acceptance of completed work is necessary to ensure that the work performed is in conformance with the designer's specific intent. If changes in the designer's detail are required because of unforeseen conditions, it is necessary that such changes be made known to and approved by the facility manager or his or her designated representative.

Inspection should be just sufficient to produce the desired end product. It is important that design details be followed and that material specifications be met. However, overinspection can be a costly and aggravating error. Blind adherence to plans and specifications is not inspection. It results in claims, counterclaims, and litigation which neither the owner nor the contractor can truly turn into profit. The rule of reason must be applied in the area of inspection as well as in all other functional areas.

A typical case history involved the specification by a designer of welded tees in accordance with the standards of the American Society for Testing and Materials. The installation of these tees in a steam main for various lateral connections was questioned by the contractor, who made a suggestion to use "Weldolets" as an alternate and at a reduction in contract price. The inspector refused to consider the substitution, and the contractor, after very carefully reading the debatable language of the specification, installed the welded tees under protest and instituted litigation for reimbursement of extra costs beyond the terms of the contract. He ultimately received extra compensation for his work instead of a reduction in his contract price.

In this case, overinspection was costly. No design is perfect, nor is any designer. Errors of omission and commission are built-in mementos of design failings. Intelligent and competent on-site inspection and understanding of a contractor's work and problems can save time and money.

On the other hand, underinspection is also costly. The failure of a contractor to meet machine performance specifications or to furnish the quality of material specified can lead to costly failure or maintenance problems throughout the life of the equipment or facility.

It is the task of the inspector to:

- Ensure that the terms of the contract are met by the contractor

- Review and accept or reject the work of a contractor in the name of the owner

- Evaluate, make recommendations on, or approve changes in materials or procedures that are beneficial to the owner's interests or required by unforeseen conditions

- Certify the validity of progress payment vouchers submitted by the contractor

The inspection function can be accomplished by employment of a consulting engineering firm or by members of the facility management staff. Normal practice is for a consulting firm that has been retained to develop contract plans and specifications for a project to inspect the construction or installation during the project-implementation phase.

If a consulting firm is engaged to inspect a contractor's work, a clear understanding should be reached on the limit of its authority and responsibility. It is recommended that specific approval of the facility manager be required for all changes involving extra costs or variation from intent. The firm's responsibility is to ensure that the design is faithfully carried out, and although it is not a party to the construction or installation contract it must assure the interest of the owner in all contractual matters.

If the facility organization performs its own inspection functions, the responsibilities are the same as those of a consulting firm, and the authority of the inspection staff should be limited to the same degree as is that of the consultant.

The definitions of the limits to the authority of a consultant firm's or in-house inspection staff cannot be specifically detailed. Where the design detail and the specification are critical to the success or failure of the end product to perform as intended, all proposed changes should be referred back to the designer for approval or disapproval. Where the failure tolerance is less stringent, more or less latitude can be authorized to the inspector in determining whether to accept or reject proposed changes or completed work.

It is good practice to hold preconstruction conferences to define the limits of authority of the inspection staff and to alert the contractor to the degree of inspection tolerance to be expected in the progress of the work.

A first meeting should include the facility manager, representatives of the facility engineering group, or the consulting firm, which prepared the plans and specifications, representatives of other organizational components of the facility which would be affected by the construction or installation (e.g., production, warehousing, industrial engineering, security, etc.) and representatives of the legal department or counsel retained by the company and the inspection staff. Specific requirements of the project should be discussed in depth, problems of interference delays, shutdown of production or other operations should be reviewed, and critical inspection tolerances should be identified and a clear understanding reached as to the intent of various elements of the contractual documents.

A second meeting between the contractor and the inspection staff should reflect the decisions reached at the first meeting and should clearly define the working arrangements for utility shutdowns, overtime work, storage areas, safety requirements, material deliveries, and similar day-to-day operational matters. Any interpretations of vague or nonspecific details in the plans or clauses in the specifications should be made at this time and an understanding should be reached on the method of payment and the procedure for requesting changes to the requirements of the plans or specifications. All legal questions should be resolved and if required by the contract terms, performance bonds, insurance, wage agreements, and proposed subcontractors should be certified as acceptable or modified to meet the contract requirements.

Contract administration should be clearly understood to be a technical and not a legal function. It involves the monitoring of a contractor's performance so as to ensure compliance with the terms of the contract as observed and reported by the construction inspectors. The determination of whether a contractor has so violated the terms of the contract as to warrant litigation is the province of the legal department after all the facts have been presented by the facility manager.

In its essence the contract administration function can be divided into six main areas:

1. Arranging for and participating in the bidding or negotiating process

2. Evaluating the bids or negotiated price offers

3. Selecting a contractor

4. Negotiating change orders and claims

5. Processing progress vouchers for payment

6. Supervising the inspection process

The ever-present danger that must be avoided is that the contract administrator will unwittingly become involved in decisions that are properly the responsibility of the legal or accounting departments.

In order to ensure that a contractual document is legally binding and includes all applicable organization, state, federal, municipal, and regulatory agency requirements (e.g., liability, insurance coverage, citation of wage and hour laws, etc.), it is recommended that the tender documents be separated into two sections:

1. Legal requirements or "boiler plate"

2. Technical requirements or statement of work

In this way standard formats for the legal requirements can be developed which will be applicable to the various types of work required by an organization. These formats can then be added to the technical specifications prepared by the facility manager, and a satisfactory contract format will be achieved.

In terms of overlap with the accounting function it must be accepted by facility managers that their role is to certify that required construction and/or installation tasks have been accomplished by the contractor in accordance with the terms of the contract and that the contractor is entitled to payment for such work.

The facility manager should avoid the temptation to set up a bookkeeping and payment group. It is a waste to duplicate the organization and functions of the accounting or other organizational functional groups who already are in existence and staffed to process payment vouchers.

1.7   Maintenance of Plant Facilities and Equipment

The term *maintenance* includes all work relating to the economical preservation of facilities, equipment, and systems at a level satisfactory to perform their designed functions. In more general terms, maintenance includes:

1. Preventive and predictive maintenance

2. Routine maintenance and minor repairs

3. Major repairs

4. Emergency repairs

5. Alterations and improvements

6. Housekeeping

The subjective terms *major* and *minor* can be defined only in relation to the type of workload of the facility organization and staffing of the maintenance shops. For a very large shop (say, 300 craftspeople) which is fully engaged in maintenance and repair work, even a $1000 alteration or improvement work request might be considered a major project because craftspeople are not available. Conversely, a small shop (35 to 40 craftspeople) might be able to handle a $5000 alteration or improvement project during a period of fluctuating workload and thus consider a $1000 project as minor.

The facility manager's maintenance responsibilities include:

1. Planning—all administrative and technical effort necessary to ensure that each craftsperson and laborer will be working at *maximum efficiency* on a *necessary task*

2. Estimating—establishing material, supplies, equipment, and worker-hour budgets for maintenance work performance

3. Scheduling—allocation of labor power, materials, and equip-

ment, at specific times and locations, for accomplishment of pre-scribed maintenance work

4. Evaluation—appraisal of craftspeople's and supervisors' perfor-mance by means of a computerized maintenance management system (CMMS) for reporting

5. Action—initiation of positive measures to improve performance or correct deficiencies

Detailed treatment of these responsibilities and specific techniques for their accomplishment will be found in Chaps. 3–6, 10, 11, and 13–15.

The facility manager should be the sole judge, within preestab-lished policy limits, of what operation and maintenance work should be performed on facilities, equipment, and systems. A basic manage-ment error that regularly recurs is to fragmentize responsibility and authority for the operation and maintenance program through the medium of profit center accountability. Profit center accountability for maintenance cost control is a subject beyond the scope of this hand-book. However, the very real and very human tendency of any profit center manager to reduce total costs can work to the disadvantage of *total* product cost if responsibility for authorization of maintenance and repair is not separated from profit-center responsibility.

The concept of *lowest overall total cost* must be established as the keystone of an effective facility management program. The deferment of a roof leak repair, or the refusal to stop production to correct a mis-aligned shaft, are decisions that must be made only within this con-text. As a service organization the facility management department must recognize that the operation and maintenance requirements of facilities, equipment, and systems must be equated in terms of the cost of deferment and the cost of breakdown. As a service organization it is subjected to the checks and balances of the production organiza-tion's demands and needs on continuous operation. However, a profit center manager with authority to authorize or defer maintenance work can make arbitrary decisions on such matters without reference to lowest overall total cost but rather in terms of lowest immediate manufacturing or service cost. The nature of the maintenance func-tion is such that uneconomical deferment in some cases can be achieved over several years before a need for a major capital invest-ment identifies the lack of lowest overall total cost decisions.

For this reason it is essential that facility managers be the sole decision makers on operation and maintenance work performance. In making such decisions they should be provided with, and work with-in, specific policy guidelines set by top management. They should also be required to secure the concurrence of an affected profit center before authorizing work which would be charged to a profit center. If such concurrence is not secured, and the facility manager can justify a recommended course of action, the facts should be provided to top management for a decision.

The simplest solution to this problem is to prorate the total cost of operation and maintenance among the various profit centers on the basis of floor space, number of machines, cubic footage, and/or other similar quantitative measures. However, this method can vitiate the concept of profit center accountability by distorting the true cost of a profit center's operation. Thus a profit center manager could retain high-maintenance-cost machinery in production operations, knowing that the real cost for maintenance is more than that charged on a prorated basis.

The best solution is to treat operation and maintenance as a separate cost for the facility as a whole in the same sense as the sales organization or the distribution group. In this way the facility manager can provide an expert input to the lowest overall total cost equation without interference from or disapproval by a profit-center manager who might be overly prejudiced in favor of immediate short-term manufacturing costs rather than lowest overall total cost over the marketable life of a product or service.

The basic principle to be recognized is that accounting techniques should never be allowed to distort or interfere with the facility manager's responsibilities for ensuring that the operation and maintenance of facilities, equipment, and systems is achieved at lowest overall total cost.

Operation and maintenance planning includes all administrative and technical effort necessary to ensure that each operation and maintenance craftsperson or laborer will be working at *maximum efficiency* on a *necessary task*. It is important to recognize that both these factors must be satisfied or the operation and maintenance planning function will have failed.

In order to achieve maximum efficiency from each worker, the facility manager will need to establish the workload generated by facilities and equipment and determine the most economical method of accomplishing it. The maintenance workload consists of:

1. Recurring workload

2. Nonrecurring workload

The recurring workload can be established by facilities control inspections and the development of preventive and predictive maintenance programs (see Chap. 6 for a discussion of predictive maintenance programs). The nonrecurring workload can be identified by extrapolation of prior maintenance history or by engineering evaluation. Detailed treatment of specific techniques for these methods can be found in Chap. 9.

Once the workload is identified, each job, both recurring and nonrecurring, should be estimated in terms of required worker hour, materials, and equipment expense by shop or by craft. These estimates should be prepared from the planning input, which will have established a most economical method for accomplishment of each job. Estimating methods are covered in Chap. 4.

All work given to the shops for accomplishment should be scheduled in terms of specific worker-power allocations and availability, equipment and materials availability, assist trades support, job site availability, and the time frames for performance and completion. Schedules should be kept flexible enough to accommodate unscheduled necessary changes, but variation should be limited to the true emergency. Short- and long-range scheduling procedures are included in Chap. 3.

Evaluation of craftspeople's and supervisors' performance can be accomplished by a combination of:

1. Day-to-day observation of job performance

2. Management reports analysis and data use

Reliance on either of these methods, to the exclusion of the other, is a dangerous practice except in the very small facility organizations (say, fewer than 35 personnel) where management reports should be minimized almost to the point of elimination.

There is no better way to evaluate performance than to supplement the cold facts of expense vs. estimates with unscheduled field and shop inspections by the facility manager. In this way, the facility manager can objectively determine if the planning and estimating staff is performing its role satisfactorily and whether the shops are operating at a satisfactory efficiency and effectiveness.

Action by the facility manager is required whenever the performance of the shops does not meet a reasonable cost or time standard. It is important, in this regard, to distinguish between recrimination and constructive action. Faultfinding is an essential part of the management process, but the reason for finding a fault is to correct the situation that caused it, not to invoke punishment or recrimination.

The effective facility manager must be a skilled practitioner of the art of human relations when it is necessary to take corrective action. It is not sufficient to order improvement—the ways and the means of improving must be provided to the shop foremen, and they must be convinced that corrective action is warranted in the first place. The use of objective performance statistics and subjective personal observation will normally be sufficient to effect an improvement, *provided* that performance targets have been established and that the foremen know that they are being measured against such targets.

Action, to be effective, must be timely. This requires that the evaluation procedure, on which the action depends, must be timely. This, in turn, requires that data for evaluation and action must be provided to the facility manager within time frames that will permit meaningful action—normally within 3 days or less from time of project(s) completion and reporting. To find fault with a shop's performance as of 3 months ago is useless. The circumstances surrounding such performance are dimmed by time, and excuses are usually convincing. It is of utmost importance that management action be initiated immediately after the occurrence of an act or the identification of a trend that requires correction.

In order to achieve a reasonable order of immediacy, the facility manager must ensure that all necessary data, both statistical and observational, be provided for evaluation on a scheduled basis. If the CMMS does not provide timely reports, the facility manager should direct top management attention to this reason for an inability to take *effective* action. If the facility management department does not identify a faulty equipment installation until the production line has started up, the facility manager should take action not only with the shop foreman but, perhaps even more importantly, also with the engineering supervisor to ensure that timely deficiency reports are made.

1.8   Preventive Maintenance

Preventive maintenance is defined as that prebreakdown work performed on a facility's equipment and systems to eliminate failures and/or breakdowns, or to keep such failures and/or breakdowns within predetermined economic limits. It is an obvious fact, but one which is often ignored, that a preventive maintenance program should be established only on the basis of a cost-benefit analysis. Preventive maintenance is neither good nor bad per se. What is good for one facility may be bad for another. It is the responsibility of the facility manager to evaluate the *degree* to which a preventive maintenance program should be initiated so as to ensure maximum economy for the facility as a production entity. A predictive maintenance program is a beneficial adjunct to a preventive maintenance program's efficiency and cost-effectiveness.

It is important that the cost of insurance (preventive maintenance) does not exceed the probable loss (breakdown cost) for a particular set of production and maintenance cost variables. Too often a facility manager will be overly proud of a record of "no breakdowns." Yet this is not the real lowest overall total operational cost. The factors to be considered in such an evaluation are:

1. Cost of production downtime
2. Cost of repairs without preventive and predictive maintenance programs
3. Cost of repairs with preventive and predictive maintenance programs balanced against (*a*) cost of "no breakdowns" preventive and predictive maintenance programs and (*b*) cost of "limited breakdowns" preventive and predictive maintenance programs

The result of such an analysis will be reflected in realistic rather than idealistic preventive and predictive maintenance programs. The key factor to be stressed in any consideration of preventive and predictive maintenance programs is that they must pay for themselves—not in terms of shiny machinery but in terms of cost. It is no disgrace to *justify* the acceptance of breakdowns. It is intolerable to accept unnecessary cost. The effective facility manager will balance the total cost of preventive and predictive maintenance services against the total cost of "breakdown" maintenance and will establish economical programs of preventive and predictive maintenance consistent with the benefits derived from them.

Although some facility managers may claim that preventive and predictive maintenance is a waste and that a responsive repair group is more economical, it is almost axiomatic that some degree of preventive and predictive maintenance will provide a major contribution to the establishment of lowest overall total cost. Guidelines for the establishment of preventive and predictive maintenance programs are included in Chaps. 5 and 6.

1.9   Routine Maintenance

Routine maintenance is the day-to-day upkeep of facilities and equipment which will ensure their capability to perform their designed functions. Although definitions in the area of facility management are varied and sometimes incompatible, the range of work generally accepted as under routine maintenance includes:

1. Upkeep and minor repairs to equipment
2. Maintenance painting
3. Upkeep and minor repairs to HVAC distribution systems
4. Upkeep and minor structural repairs to buildings and structures

5. Upkeep and minor repairs to pavements

6. Upkeep and minor repairs to roofs

The reason for establishing a definition for routine maintenance is the need to identify what is susceptible to accomplishment by contract and what should be accomplished by facility employees. Routine maintenance should usually be accomplished by facility employees except where overriding economic or labor-relation reasons dictate contract performance. The reason for this statement is the familiarity and detailed experience with the facility and its equipment and systems which is required in the day-to-day effort that the routine maintenance function encompasses.

1.10   Major Repairs

Major repair is that work required to restore a seriously deteriorated or broken-down facility, equipment, or system to a state of usability for its designed function.

The distinction between routine maintenance and major repairs is essentially one of degree. In most facility management organizations the distinction is established by an arbitrary standard of dollars or worker-hours. Thus, in one facility all repair work involving over $1000 in labor and material could be classified as major repair. In another the breakpoint between routine maintenance and major repair could be $2000 or perhaps $100.

Most major repairs can be more economically accomplished by contract except where they involve:

1. Emergency stoppages of the production process

2. Specialized equipment

Even for specialized equipment it is often necessary to seek specialized contractual assistance because of the small frequency of breakdowns and the resultant inecomony of staffing for such rare occurrences.

The use of contractual services for major repairs is recommended as a means of leveling the shop workload so as to make possible a "steady-state" worker power staffing situation (i.e., continuing workload exactly equal to shop capability). The "steady-state" condition is, of course, not normally achieved, but it can be closely approximated by using contract services for the accomplishment of those major repairs which can be defined and scheduled without interfering with the production process.

For purposes of definition, major repairs would include cyclic replacement or refurbishing of major components of a facility or its equipment and systems. Thus the intermittent patching of a 40,000 square foot roof should be included under routine maintenance, but the cyclic *replacement* of 40,000 square feet of roofing, on a biennial schedule, should be considered as a major repair. As a major repair the facility manager should evaluate whether the cyclic-replacement program can be accomplished at the lowest total cost by the facility's personnel or by contract.

1.11   Alterations and Improvements

An *alteration* is that work required to transform a facility, equipment, or system so that it may perform a different function from that for which it was originally designed.

An *improvement* is that work required to increase the functional or productive level of performance of a facility, equipment, or system.

The facility manager must be continually alert to the workload impact of alterations and improvement requests. If they reach a level of 10 percent of the total maintenance and repair workload (including the alteration and improvement work), there is serious requirement for top management attention to the *need* for such work. The need should be clearly established as necessary to produce a lowest overall total cost product or service. However, many facility managers will find themselves on the defensive and subjected to criticism as "unresponsive" and "uncooperative" when they question or disapprove requests for alterations and improvements.

In such cases it is essential that the facility manager have established with top management a modus vivendi. It is not in the best interest of the organization to have continuing bickering over the cost-benefit ratio of changing the color of the halls in the research department. Only top management can resolve such a dilemma: Will the change in the environmental surroundings of the research department result in lowest total cost? Or to express it in a different way: Will the change in the color of the halls result in increased payoff from research efforts?

Obviously, facility managers are in no way qualified to pass judgment on this type of problem. Yet, their budgets must bear the brunt of such expenditures. If these expenditures are significant (i.e., more than 10 percent of the total operations and maintenance budget), the facility manager will necessarily be required to forgo necessary maintenance and repair requirements, thus building into the facility an accelerated capital-investment expense for replacement or major repair. For these reasons it is incumbent on top management to establish guideline procedures and controls for the approval and limitation of alterations and improvements. Within such parameters, the facility manager can then reach lowest overall total cost decisions with the operating staff relative to the timing, scope, and necessity for the requested alterations and improvements.

1.12   Housekeeping

The housekeeping function includes:

1. Janitorial services

2. Maintenance of grounds

3. Operating services

Janitorial services usually include the cleaning, dusting, and waxing of general usage and office spaces and the furnishing of expendable toilet and personnel-service supplies. It may also include cleanup of storage and production spaces where such cleanup is not included in this category of housekeeping. Chapter 13 covers the areas of custodial services, pest control, trash and debris removal, and recycling.

It is important for the facility manager to stress the cost of grounds maintenance in any review of plans for new construction or requests for alterations or improvements. A lowest total cost decision will in many cases involve additional capital investment for low-maintenance perennial plantings as opposed to annual bedding flowers. In some cases lawn areas should be minimized in favor of natural-cover landscaping or crushed-stone effects. Such decisions made in the planning stage can render a significant return in reduced costs of grounds maintenance over the operating life of a plant. Again, the *lowest overall total cost* equation must be solved in order to achieve lowest product or service cost. Chapter 14 covers landscaping services.

1.13   Principles of Organization

1.13.1   Definition and scope

In developing an organization to operate the facility management function, there is no "best" organization that will fit all needs. The organization must be tailored to fit the requirements of the company or agency. In reviewing these needs, remember that the prerequisite for the organization is an attitude of *teamwork*. An organization could be nothing more than a group of people coming together hopeful for a common objective. "To organize" is to establish either formally or informally the relationships, the lines of authority, and the responsibilities of these people. The thread that ties all this together is teamwork. With good teamwork a poorly organized group can function well. With poor teamwork a "properly" organized group will fail. Therefore, the attitude of teamwork must be recognized as the most important aspect of the organization. All the other elements (charts, job descriptions, lines of authority, spans of control, etc.) are only aids and guides in developing an attitude of working together—teamwork.

1.14 Facilities Management Defined

For a clear understanding of facility management, definition is important. Throughout the public and private sectors, facility management was usually looked upon as a function similar to plant engineering. The difference was usually one of emphasis. In government, municipal, and college facilities, buildings and grounds are paramount. Thus, for these facilities, facility management is the term often used. In the industrial complex (public sector), where the emphasis is on process equipment and machines, and service providers; facility management, not plant engineering, now denotes the function. Therefore, facility management is now defined as "the effort expended to provide complete operations and maintenance service support so that a physical facility (buildings, equipment, machinery, system, and grounds) may operate at an optimum *lowest overall total cost.*"

1.14.1 Facility management functions

The facility management effort normally includes the engineering functions of design, construction, maintenance, and operations. Design engineering is that function where new processes or new developments are incorporated into completed drawings and specifications so that facilities erection and equipment and systems placement can take place. Construction engineering uses the drawings and specifications to construct buildings, set equipment and systems, and provide services. Maintenance engineering is concerned with the everyday problems of keeping the equipment, systems, and physical facilities in good operating condition and repair. HVAC utilities are the segments of facility services that control steam, air, water, and electric generation, distribution, and disposal.

1.14.2 Organizational prerequisites

Facility managers must first be professional managers. This means being people-oriented as well as results-oriented. Today, the facility manager's multimillion-dollar budget pays for *labor* as well as equipment, materials, system, services, etc. More money will not solve today's and tomorrow's problems. A new philosophy of improved planning and scheduling, adequate application of techniques, and motivation of the personnel doing the work must be developed. In view of employee and community requirements, the motivation of personnel will take a more important place in the facility manager's life. Today's engineers and scientists have exhibited their ability to solve technical problems. This ability must now be extended to social problems. These problems are known to be difficult and demanding...but they must be coped with and eventually solved.

Several basic elements should be considered when establishing an effective organization. Those responsible for structuring the organization must develop a clear understanding of the organizational objectives. The giant inarticulate bureaucracy, or sprawling octopus, organization must be avoided. The following organizational parameters will assist those who are revising an existing organization or structuring a new one.

1.14.2.1 Philosophy. All too often facility management organizations appear to have no discernible management philosophy or management policies. Their attitude often seems to be that professionally trained engineers do not need this type of direction or that it does not apply to these types of people. Nothing is further from the truth. Lack of direction does not accomplish worthwhile results. Everyone wants to know the stance of the facility manager and all related supervisory and management personnel. Philosophy and policy must fit into the company stance; otherwise continual conflict will be in evidence. Every facility manager must develop a philosophy and related policies. For example, one may subscribe to Douglas McGregor's theory X (people hate work; thus they have to be forced to do it). On the other hand, a more meaningful philosophy is represented by McGregor's theory Y (create conditions so organization members can make their own efforts toward the goals and success of the company).[1] Whatever the philosophy, it must be reflected by everyone in the organization. The philosophy must be sincere and honest.

1.14.2.2 Policy. From the philosophy, policies are developed to operate the organization. Policies are immensely important to an organization to the extent that they are tangible. They guide the organization in a consistent direction. Without policies the organization would wander with conflicting decisions, inequitable administration of matters, and inequitable information and guidance.

Policies may be formal or informal. Informal policies tend to be unwritten; thus they become stories that are passed from one person to another. Formal policies are written but may become obsolete from lack of updating. Informal policies are difficult to disseminate, whereas formal policies can be read and more easily understood. Some policies may be proprietary or confidential so it may be advisable not to write them. In the long run, formal policies accomplish a more meaningful result; therefore, they are recommended.

Setting policies for the facility management organization is the responsibility of the facility manager. Policies should be related to the organization and what is expected of the person nel. A general list of policies that normally apply to the facility management function may include but are not limited to:

Hours of work	Product quality
Attendance	Grievances
Employee benefits	Employee publications
Company rules	Medical examinations
Safety	Equal opportunity
Security and plant protection	Employee transfer and promotions
Overtime	Wage and salary administration
Military service	Maintenance of building equipment and machines
Leave of absence	Value engineering
Vacation	Cost improvement
Sickness	Continuing education (tuition refund, etc.)
Associations (credit union, etc.)	Reduction in force
Acceptance of gratuities	Jury duty
Travel	

Even though policy statements are available from textbooks, consultants, etc., care should be exercised when these statements are prepared so that they fit the organization in question.

1.14.2.3 Span of control. There are a variety of opinions, studies, and writings relating to spans of control. The conclusions relating to span of control appear to be twofold: (1) there is a limit to the number of workers a supervisor can effectively supervise, and (2) the optimum number of workers supervised depends primarily on the given supervision and ability. Figures to support virtually any claim can be found. The important point is that too few subordinates supervised is a waste of supervisory time and talent, and too many may result in wasted employee effort.

In developing an organization, considerable thought should be given to the amount of supervision necessary. One-on-one supervision (one supervisor with one subordinate) must be closely watched and usually avoided. Even situations where one person supervises only two or three persons should be questioned. The deciding factor in the number of subordinates managed or supervised is the amount of detailed direction and control exercised plus the variation in the organization or work supervised. For example, a facility manager who manages an organization containing operations and maintenance services, and engineering, manages a highly varied and complex field. Consequently, two or three superintendents may be very realistic. The operations manager may need a powerhouse foreman just to oversee the generating problems, while the operations manager handles the HVAC technical functions. Thus care should always be exercised when following standards such as one supervisor should supervise between three and six subordinates, or sometimes ten to twelve if the work is generally the same. These can be satisfactory ratios for craft foremen or an engineering supervisor, but not for the facility manager.

1.14.2.4 Personalities involved. The organization must fit the personalities involved. In theory we should develop the optimum organization and adapt the people to its requirements. However, actual practice works out as a blending of the optimum organization fitted to the personalities present. Personalities must be recognized and dealt with accordingly. A person who is an excellent engineer may or may not be a good supervisor. The organization should be structured so that the good engineer who cannot successfully make the transformation to supervision and management can have position, status, and salary as an engineer.

1.14.2.5 Developing subordinates. It is important that the organization structure provide for employee development. In some successful organizations, key executive and supervisory personnel are given large numbers of subordinates so that it is impossible for them to exercise too close supervision over the subordinates' activities. This type of organization structure places subordinates largely on their own. This tends to improve the growth and development of the employee and will weed out those who lack self-confidence and personal ability. On the other hand, in organizations characterized by close supervisory control (one supervising two subordinates or one supervising three subordinates) subordinates may have little opportunity to develop ingenuity and capacity and will often leave the organization for areas in which they can develop.

1.14.2.6 Levels of organization. The tendency is to have an assistant, an assistant to, or a staff assistant. Care should be taken not to overdo the assistant category. This only adds to confusion, especially in the area of overlapping duties. When assistant, assistant to, etc., are used, a clear definition of their functions must be made. Watch for overlapping functions among the various levels of the organization. If these overlapping functions exist, there may be too many levels in the organization.

An organization with few layers of supervision, and a minimum of formal controls, places a premium on ability to stimulate and lead plus a premium on creative and self-directing subordinates. The "driver" type of manager, or supervisor, who achieves objectives through constant pressure and uses fear will not operate as effectively in the organization with few levels (called a flat organization). The driver type operates better in the "tall" organization (many middle management layers in the organization).

1.14.2.7 Individual facility requirements. Inasmuch as no two facilities are the same, the organization must be developed according to the specific facility operations, situations, and local requirements.

1.14.2.8 Skill level of workforce. In a large industrial area, where there is considerable competition for the skills needed, the organization may have to be structured to handle a lower skill level. This may mean more and closer supervision.

1.14.2.9 Training of workforce. The organization is as effective as its training. Where the skills needed in the facility are not available locally, intensive training will be required. If so, there will be a twofold effect on the organization. First, provision will have to be made to train new employees and/or retrain existing employees. The organization may reflect this effect in terms of personnel administering apprentice programs, performing training duties, etc. Second, newly trained employees require closer supervision than experienced employees. Additional supervision may be necessary to provide this closer control. However, after a period of time, supervision should be adjusted to reflect the experience of the workforce.

1.14.2.10 Operating schedules. A 5-day, single-shift operation and a 7-day, three-shift schedule will affect the size and structure of the organization differently. When working a 7-day, three-shift schedule, how will the twenty-first shift be covered? How much authority will the night supervision have for overall facility operation? How will coordination between shifts be handled? These are all valid points imposed by the operating schedule. For example, on a 7-day, three-shift schedule, a senior shift supervisor should be present on every shift. Through a shift log, and communication with the incoming shift, continuity of the operation is maintained.

1.14.2.11 Size of operation. The size of the facility has a great deal to do with the number of supervisory personnel needed. The number of supervisory personnel has a direct bearing on the number of levels required to provide effective management. The number of supervisory and staff personnel relates directly to the number in the wage-roll workforce. Also, the number of staff personnel needed will depend upon the size of the operation.

There are no mathematical rules for determining the ratios of supervisors or staff to subordinates. However, the smaller the workforce, the higher the ratio of supervisors and staff to wage-roll personnel usually is. The one supervisor to one subordinate situation must always be challenged. This situation can be minimized by planning the organization.

1.14.2.12   Type of operation.   Normally the facility management organization is slanted toward the dominant feature of the operation (e.g., buildings, process piping, pumping equipment, HVAC systems, electrical systems, etc.). Therefore, this dominance in the operation will be reflected within the organization in terms of the types of technical worker power required, types of supervisors needed, and the wage-roll skills demanded. Staff assistants can also play an important part in such an organization, especially where a particular field predominates. In fact, it is better to have a staff assistant who is an authority in a given field than a manager who has the authority but is weak technically. To convert a strong technical background or skill level to a strong managerial or supervisory function is often very difficult. This difficulty must be recognized. There are no guarantees.

An organization can very quickly become stilted and narrow if all those in it have identical backgrounds, skills, and training, especially among the supervisory, technical, and staff personnel. However, the principle could apply anywhere and must be considered in every organization.

1.15   Analyzing the Needs

An interesting thing happens when the facility manager prepares an organization statement. The statement usually reflects the way he or she views the organization. A very important point is how others view the organization. It must be remembered that the primary purpose of the facility organization is to provide a service. It is a necessary service but still a service.

Therefore, the facility manager and his or her staff should take time to review how others view the organization. What is the image of the facility management organization? What are they (users of the service) saying about us (the organization)? Such comments as "Maintenance is not concerned about breakdowns," "Maintenance never fixes anything right the first time," "Nothing those engineers design ever works" indicate how others view the organization. This review of the organization's image, although very important, can be painful at times, but every facility manager should do this task periodically and methodically and include both the good and bad points.

The result of such an analysis of the organization's image can be converted into the needs of the organization served. Those directing, creating, or changing organizational structures must have a clear understanding of the needs of those they serve. Facility managers *tend to think* they know these needs. Often they do not. They do not take the time and effort to determine what these needs are because they often conflict with their preconceived ideas. Also, when making this analysis, facility managers usually find things they do not want to hear and would rather not be reminded of. Anything displeasing will be avoided if at all possible. Consequently, the facility management organization ends up trying to fill needs which may or may not be the needs of those being served. Thus the statement, "That department is a necessary evil." Those being served, whether outside customers or other facility organizations, end up having to make the best of the situation. Facility managers must realize that when planning for the organization, the needs of those they serve must come first. Then and only then can an effective organization be structured.

1.16 Establishing Goals and Objectives

Goals and objectives are terms that have become well worn over the past decades. Often these are just two words with little or no actual impact on the organization. The reason they have much impact on the organization is that almost everybody agrees that goals and objectives are necessary, but when it comes to establishing and applying them, almost everybody becomes very vague and hazy regarding their application. This is not a criticism of those who are trying to establish goals and objectives for their organization and subordinates but a fact that points to the need for a systematic approach to establishing them. However, before a systematic approach can be applied, a clear understanding of setting goals and establishing objectives is needed. Once these elements are understood, the organizational planning process can begin.

1.17 Administering the Organization

A crucial part of the facility manager's function includes not only the setting of objectives and goals and organizing the department but also the day-to-day, hour-to-hour supervision and guidance called administration. The prime function of administering the organization is to make the best use of human resources. Effective utilization of human resources ultimately contributes to effective utilization of facilities, materials, supplies, and equipment. Therefore, administration is defined as providing effective leadership, which in turn guides the organization in its function.

It is not unusual to find lack of leadership in the facility management organization. The lack of leadership stems from the attitude held by many managers that engineers and skilled craftspeople do not need close supervision and detailed direction. Such an attitude is one of construing detailed supervision and direction to mean normal effort. The personnel in facility management want information for guidance, but they object to detailed direction. They want communication but object to having someone looking over their shoulders. Most of the leadership shortcomings within the organization are largely within the scope of the facility manager's responsibility.

1.18 Organization vs. People

Organizations do not turn people "on or off"; people do it to each other. Therefore, leaders cannot hide behind the facade of an organization and expect those in it to be motivated. They will not be motivated, but demotivated. People have more than the basic needs of foods, clothes, and shelter. They want to be recognized. They want to be well thought of. They must have their social needs met. The chart on the wall will not meet these higher needs of security, recognition, and achievement. This is an intangible item that must be supplied by the leaders of the organization.

The successful organizations in the facility management realm are those organizations which have recognized the fact that there can be problems such as the organization vs. the people. They acknowledge the fact that things like titles are important, that the organization is the people, and that people have basically the same needs regardless of their vocations. Thus, no matter how well facility managers do their planning, organizational analysis, etc., they will face failure unless they recognize the needs of people. Too often, we think of engineers and skilled personnel as members of a group or organization rather than as individual employees. The major concern should be to see that personal satisfactions are met and that each employee receives proper recognition for contributions to the achievement of the organization's goals.

Caring for the needs of one's employees does not suggest being permissive. This is a common misunderstanding among many organization leaders. Being firm, fair, and friendly is the attitude that must be generated.

1.19   Using Titles

The use of titles in the organization may vary from using virtually no titles to using titles that require an extra line to write. The relative meaning of titles also varies tremendously from one company or installation to another. Excellent examples are the title of facility manager. A facility manager's title in one company may depict the overall responsibility of operations and maintenance support, and design engineering, while in another it may be concerned only with the maintenance effort.

The primary consideration relating to titles is that they are relative within the activity of the organization. For example, the person in production who heads a production unit may be titled an area supervisor. If so, the person heading the maintenance department should be designated "maintenance supervisor." This is especially true if the overall facility has an organizational structure such as that shown in Fig. 1.1.

Consideration should be given to the qualifications of the person who fills the position. For example, in maintenance, the first-line supervision (the person to whom the craftspeople report) is generally called a foreman. However, if this position is normally filled by a technically trained engineer, the title of "engineer" may be more desirable.

The four most frequent titles used are manager, superintendent, supervisor, and foreman. However, titles change just as fashions change. Titles have prestige; so this must be considered when estab-

```
                          Manager
                         _____


                      Assistant Manager
                     _____

                                                    Operating
    Accounting                                      superintendent
    superintendent        Facility Manager
   _____      _____       _____


    Accounting           Maintenance
    supervisor           supervisor             Area supervisor
   _____        _____          _____


                           HVAC                   Operating
                         supervisor               foreman
                        _____            _____
```

Figure 1.1   Overall facility organization.

lishing and assigning titles. Certain titles should be reserved for certain levels within the organization. The title "director of engineering" is not normally one to be assigned to a facility-level engineering supervisor. Such a title suggests a corporate-level position. When titles are indiscriminately handed out, they usually become an embarrassment to the employee and the company. This problem is illustrated by the many jokes perpetrated on the janitor whose title has been upgraded to "sanitation engineer."

Selecting titles for staff positions often is a source of concern. If the titles pertain to the activity, no trouble should be experienced. Many technical engineers in the production department who serve as staff troubleshooters are designated "process engineers." By the same token, technical engineers in operations or maintenance could be titled "facility engineers."

Most problems relating to titles can be eliminated if a basic structure is established and followed which relates to the facility activity. The total facility organization can be viewed as different levels of responsibility illustrated in Fig. 1.2.

1.20   Performance Measurement

Measuring performance of the organization relates directly to personnel performance in the organization. Therefore, to say that the goals and objectives of the organization are being met may not be enough. The organization's goals and objectives may not be met in the future unless the performance of the individual employee is maintained. Hence an accurate review of each person's performance within the organization is in order.

Before performance can be measured, criteria must be established for measurement purposes. In establishing these criteria, the facilities manager will be faced with establishing goals and measuring performance over a variety of groups and persons. The criteria for the goals and measurement will likewise have to vary. For example, standard times on each maintenance work order may be used to establish goals of performance and also be compared with actual times as one method of measuring performance. The completion dates on an engineering project are indicators that can be used to measure the performance of the engineer. These goals and measurements must be as objective as possible. Also, a system for accumulating actual time and costs by job for the craftspeople, engineers, and other job- or project-oriented personnel should be initiated. This is one of the important steps in providing data for performance measurement. Without a knowledge of costs to do specific tasks, the controlling and measuring of performance are impossible.

Manager
———

Total facility responsibility and direction.
May have an assistant manager

Superintendent
————

Major operation responsibility. Such titles
As Production Superintendent, Facility Manager,
Accounting Superintendent are used

Supervisor
————

Department responsibility. Such titles as
Area Supervisor, Maintenance and Operations Supervisor,
Engineering Supervisor are used

Foreman
————

Group responsibility. Such titles as Foreman
And Engineer are used (i.e., Instrument
Engineer, Labor Foreman, Pipefitter Foreman,
Mechanical Engineer, etc.)

Wage roll

Figure 1.2   Facility organization levels of responsibility.

Finally, when performance-measurement goals and indicators are established, they first must relate to the overall organizational objectives and second must be fully communicated to those persons involved in the organization. A possible method of establishing communication is to have each group assist in developing measurement criteria. Everyone realizes that measurement is important. It is the methods that will be criticized. Thus subordinates outline the standards for their own jobs and the leader insists on realism. Subordinates develop measures for results and the supervision questions them as to what the measures portray (i.e., do they measure costs by job?). It is a total process of getting those in the organization to suggest the goals and methods of measuring performance. When the leadership obtains commitments and agreement to the goals and measuring systems, the end result is a *practical* system for measuring performance of the individual and the organization.

1.21   Organizing Considerations for Outsourcing Functions

Top management must recognize the value of managing facilities as an asset.[2] As soon as a facility is constructed it begins to deteriorate. Clearly, funds that must be used for operations and maintenance could otherwise be directed to increase an organization's growth and/or profit. It is the responsibility of an organization's facility maintenance organization to preserve an organization's capital assets.[2]

Facility management is a diverse complex function. To satisfactorily accomplish all the required tasks internally would take a large diverse highly trained staff. Very small facilities cannot afford to provide this required level of staffing. Even most large facilities usually cannot justify staffing in-house specialists that are required only occasionally. Outsourcing allows an organization to expand its capabilities and resources without the need to expand its workforce. Outsourcing means purchasing services by contract from vendors. Outsourcing involves contracting for a contractor's time and effort rather than for a specific end product. These contracts are typically referred to as service contracts. Common service contracts in facility maintenance include:

Operation, maintenance, repair, improvement, alteration, and salvage for supplies, systems, and equipment

Operation, maintenance, repair, improvement, alteration and salvage of real property

Architect and/or engineering services

Expert and consultant services

Virtually all or part of a facility management organization's functions can be outsourced. In addition, several of these functions can be provided by a single vendor.

The term outsourcing has become a catchall for a wide variety of arrangements for procuring operations and maintenance functions contractually. These range from the contracting out of individual services to hiring a vendor to take on a wide variety of facility management operations, maintenance, and support functions. Outsourcing became an accepted tool along with total quality management, and business reengineering for dealing with changes in the facility management environment. It is being used to achieve better customer service, competitive advantage, and lower costs.

According to a recent survey by the Association for Facilitation Engineering (AFE), nearly 70 percent of facilities professionals believe that downsizing and outsourcing will continue to increase.

AFE expects the market for outsourced facilities services to surge dramatically in the next 5 years. The key benefits of outsourcing facility operations and maintenance functions, according to Robert Dickhaus of Johnson Controls World Services, Cape Canaveral, FL, include:

- Time to focus on core business

- Accelerated reengineering

- Technology infusion

- Shared risks

- Extended technical and management resources

For outsourcing to be successful, management must have clear goals and objectives. When a decision is made to outsource existing functions in the organization, management must be sensitive to and address the anxiety this may create among its workers. Management should make a good faith effort to assist its employees who may be displaced by outsourcing in finding other employment. Frequently, the outsource provider will provide employment opportunities for these individuals. However, if the main goal of outsourcing is a reduction in staff, management must prepare itself for the gloom which is like to permeate the workforce.

Outsourcing offers many potential benefits if the associate contract is structured and managed properly. Outsourced services may cost less because the contractor may be able to pay its employees 15 percent more or less in wages than what the current in-house staff is being paid. Vendors usually have a large pool of skilled labor and supervisors so that they are able to satisfy overtime, turnover, and vacancy requirements. The organization can also avoid the costs that are associated with hiring, training, scheduling, and disciplining in-house personnel. Similarly the vendor can reduce the facility's exposure to Worker's Compensation. In addition, vendors, because of their experience and size, are often able to obtain better insurance rates and to absorb services for on-the-job accidents and injuries. Large vendors often can provide their skilled employees with training in professional development programs and career paths that are not possible in-house. Therefore, they are more likely to be better able to attract and retain the better employees.

The facility maintenance organization needs to pay only for the worker power and services it requires. In-house personnel are more likely than vendors to be distracted by other jobs or influenced by organizational policies and tradition. The productivity of in-house personnel is often diminished because they tend to stick with comfortable work routines and avoid change. Outsourcing reduces an organization's need and costs to maintain inventories of supplies, materials, parts, tools, and equipment. Furthermore, it is likely that, owing to the economy of scale, a vendor may be able to purchase these items at substantially lower costs than can otherwise be done in-house.

When considering outsourcing, organizations should not overlook the possible disadvantages. Since the organization must pay for the profit and overhead of the vendor's employees, the vendor's hourly cost to the organization may be greater than in-house costs. Outsourcing removes the organization's direct control over the individual who actually performs the work. The goals and priorities of the vendor may not be the same as those of the organization. On-site supervision provided by vendors may change more frequently than that provided in-house. Outsourcing will require the organization to expend allocation resources to procure and manage the services which are outsourced.

The outsourcing processing often causes an organization to evaluate the cost and benefit for these services and realistically assess their actual needs. In addition, the facility has the ability to use the vendor's expertise as a consultant.

The facility functions that are frequently outsourced include:

Facility maintenance functions such as plumbing, painting, paving, electrical, grounds, etc.

Facility engineering functions such as architectural and/or drawings

Janitorial

Security

Trash removal and recycling

Abatement of safety and/or environmental hazards

Outsourcing is an important part of many flexible staffing strategies. According to Professor Michael Bur of the Harvard Business School, "Expertise and excellence come from specialization."

Despite the potential benefits, facility managers are often concerned about the potential loss of control over services which are outsourced. Hiring outsiders to perform important functions requires facility managers to assume additional responsibilities. These responsibilities include the development of specifications, solicitation of services, preparation and award of the contract, and control of the contracted work while in process and on completion. In addition, the provider's performance must be measured and evaluated. Clearly outsourcing is not a simple solution.

The preferred way to procure outsourced services is by a competitive solicitation process. This process requires a written solicitation package, a vendor list, and bid evaluations. The solicitation package should include a detailed written statement of work. A statement of work is the "specifications" used to describe the requirements of the services. The form of the statement of work may be either term or completion. A term form expresses a level-of-effort requirement during a period of time. A completion form describes providing a completed product or service.

It is important to write clear statements of work for the services that are to be outsourced. The courts usually interpret ambiguities in a specification against the drafter. A good specification defines the minimum requirements, not the personal preferences of the users. The specification should also include provisions to establish that the supplier can meet all the user requirements. Some work like landscape maintenance, elevator maintenance, roof repairs, and custodial services is easily definable, and a very definitive "do it this way" statement of work is appropriate. However, other work such as architectural, engineering, and consulting services is more abstract and difficult to define. In these situations, a performance or "this is what we think we want to accomplish" statement of work is usually most appropriate.

When dealing with large complex issues such as facility planning or construction, it is usually beneficial to talk with prospective vendors prior to preparing the statement of work to learn more about the range of the services they offer. It is also helpful to break larger projects into smaller more manageable phases each with its own statement of work.

A detailed description or statement of work is the specifications used to describe the requirements for contract services. There are two types of specification, term and completion. For example, a term topic provides a requirement to provide a boiler operator for the month of December. A completion statement of work describes the actual work the user requires. The contractor must provide all the resources (i.e., worker power, supervision, supplies, tools, and equipment) to accomplish the work. A poorly written statement of work can result in unreasonable prices, failure to obtain offers, failure to obtain the denied level of effort from the contractor, etc. Words that are either vague such as "reasonable," "clean," "functional," and/or," should not be used. Words such as "including" and "similar" and phrases such as "good workmanship" and "neatly finished" should be avoided.

The user must identify the services that are desired to be procured. These requirements must be described in a written document called a "specification." The user then must determine the type of contract and method of procurement of these services.

There are basically two methods of procurement. These are sealed bidding (IFB) and the negotiated method (RFP). Sealed bidding is generally the preferred method of contracting. This method generally affords the maximum level of competition and thus results in the lowest prices. The main goal of sealed bidding is to give all qualified sources an opportunity to bid competitively on an equal and fair basis. Under this process the offeror must solicit bids (i.e., prices or price-related factors that are responsive to the solicitation). This type of solicitation is usually called an invitation for bids (IFB).

An offeror's submission is nonresponsive if it does not comply with the terms of the solicitation in a manner which could affect price, quantity, quality, and/or delivery. Generally the bids are opened publicly and remain effective for a set period of time. Offerors may not change or withdraw their bids after they are opened except under certain conditions. To be viable, the following conditions must exist:

Sufficient time is provided to complete the process

Adequate competition is expected

Specifications are satisfactory

Price and price-related factors are sufficient for determining the award

If the above are not met, the negotiated method of procurement should be pursued. Unlike sealed bidding, the negotiated methods proposals do not have to be entirely responsive to the solicitation, and they may or may not be opened in public.

These proposals may be withdrawn at any time an award is made. This type of solicitation is usually called a request for proposals (RFP).

The goal of negotiation is competition. Negotiation allows greater flexibility through the use of discussion and the opportunity for prospective contractors to modify their offers. If discussions are held with offerors, they are usually encouraged to submit the "best and final" offer.

The award of an RFP is generally more time-consuming and complicated than award of an IFB. RFPs should be evaluated by a rational set of predetermined technical and cost-evaluation criteria. The solicitation should give the offerors reasonable notice of the relative importance of the solicitation criteria. However, it is not necessary to identify the weights of each criterion in the solicitation. Technical criteria can be of the threshold type. These are mandatory requirements which when applied will result in either a "yes" or "no" response from the prospective offeror such as "Does the offeror have at least 5 years of relevant experience?" The other type of criterion is in "variable" criteria. This type of criterion is scored according to the degree to which it fulfills this requirement. An example of a question which should be evaluated by the criteria is the management plan the contractor proposes to implement, or the key personnel that the offeror proposes to use, to fulfill the requirements of the specification. Frequently the evaluation of technical proposals is performed by a group or panel rather than a single individual.

On occasion it may be necessary to procure services through other than a competitive process in order:

To maintain continuity of services

To maintain an existing business partnership

To acquire contractors with unique qualifications and experience

Other procurement methods may be developed using various elements of the sealed bid and negotiation.

Communication of the user's requirements to prospective vendors can be facilitated by using a uniform contract and specification format. A conference and/or tour of the job site should be scheduled prior to the recruitment of bids or proposals. A conference should be scheduled to provide prospective bidders an opportunity to discuss and clarify potential discrepancies. A job site visit allows prospective offerors opportunity to observe conditions which may hinder or facilitate ability to fulfill the requirements of the specification. As soon as the solicitation process is completed, a contract needs to be executed and administered with the successful vendor. In addition, the unsuccessful vendors need to be notified of the outcome of the solicitation.

When outsourcing is employed in the facility management organization there is a need to designate someone in-house to be responsible to administer development of specifications and the solicitation and award process and to inspect the work of the contractor as well as administer and ensure compliance of the contract after it is awarded.

In small organizations, a single individual may be responsible to fulfill all these duties. Consultants may be used as the circumstances warrant to develop the specification. It is prudent to use a lawyer experienced in contract law and facility operations and maintenance issues to develop the required contract documents. For example, the contract should include provisions which protect the interests of the facility management organization. Such provisions include but are not limited to the ability of the facility management organization to amend, extend, or terminate the contract for the convenience of the facility management organization. In addition, the facility management should designate an individual with assigned specific authority and responsibility to enter into and administer contracts for outsourced services on its behalf. The various duties necessary to administer contracts include:

Work initiation conference

Schedule and coordinate the contractor's activities

Implement an effective quality assurance program

Ensure compliance with all the terms and conditions of the contract specification, terms, and conditions

Verify and approve invoices for payment

Pay invoices in a timely manner for work that the contractor completed pursuant to the terms of the contract

Maintain an accountability for all work performed and costs that are incurred

Ensure work is performed within budget allocation

Service contracts may be amended to reduce or increase the scope of work during contract performance.

References

1. *Douglas, McGregor,* The Human Side of the Enterprise, McGraw-Hill, New York, 1960.
2. *Donn W. Brown,* Facilities Maintenance. The Managers Practical Guide and Handbook, AMACOM, New York, 1996.

Program Operations

Michael D. Burns

Engineering Management Consultants, Inc.
Houston, TX

2.1 Designing the Maintenance Management Information System

2.1.1 What kind of information do you need?

Everyone has a need to maintain their equipment. The information needed to schedule and perform the maintenance is reasonably self-evident. The real problem is "who besides the maintenance personnel needs information from the maintenance effort and what information is needed?"

To answer the information question requires an examination of the organization for which the equipment must be maintained, the structure of the organization, and the goals of that organization.

Hardware and software selections for the organization will depend on the nature and composition of the organization.

2.1.1.1 Small, localized companies. Small organizations with limited reporting needs may work with hardware that consists of an individual PC or a PC networked through a local area network (LAN). In this case the need to share information is less, and the quality goals and objectives are usually reflected by personal directives and actions.

2.1.1.2 Larger, dispersed companies. Medium-sized to large businesses usually have an "intranet" that may consist of mini or mainframe computers acting in a client-server relationship which may branch into a wide area network (WAN). Quality considerations are usually driven by formal policies. Such organizations may be committed to the current hardware they possess, having invested a great deal of money in expensive equipment which cannot be changed without major capital investment.

2.1.1.3   Companies with special needs.   Considering the diversity of companies, some companies may have special needs. A good example is a company whose business is dependent upon a fleet of vehicles or a fleet of marine vessels. The company may have a very small corporate or administrative staff, and the maintenance activity may be very widespread. Compiling and sharing information in this case may require special techniques.

Other companies may be driven by special needs. Power generation is an industry whose scheduling for availability or concern for downtime may drive the maintenance effort.

2.1.1.4   Companies with special requirements.   Some companies have special accounting or quality requirements. If the company is trying to integrate a total quality management system (TQM), there may be special reporting requirements.

Other companies may wish to link their maintenance system to a purchasing system or even a total accounting system. (This is generally referred to as the "general ledger.")

Companies who are involved in implementing environmental management systems such as EPA RMS or ISO 14000 may have special reporting requirements for hazardous materials (HAZMAT).

2.1.2   Common information needs

Despite the diversity of information needs within a company, some elements associated with the maintenance effort are common to all of the efforts. Maintenance activities must be planned. Therefore, what must be done, when it must be done, and by whom it must be done have to be entered into the maintenance system. Any required parts and special tools must be obtained in a timely fashion to be sure that maintenance can be performed within the scheduled period.

The basic information required for maintenance planning and scheduling is shown in Table 2.1.

2.1.3   Special information needs

Before beginning, determine any special information needs of the business. Although it is not possible to cover all organizations, it is worthwhile to discuss a few topics to see what types of special information may be required by different types of organizations.

TABLE 2.1 **Basic Information Required**

Facility information	A facility is a logical grouping of equipment. It may be a building, a plant, a vehicle, or any other logical arrangement of information which is consistent with the operation of the business
Equipment information	The equipment consists of the items to be maintained. However, some care needs to be made in defining the equipment. A generator set, for example, is certainly a logical piece of equipment, but it may have important components or subassemblies
Maintenance procedures	Maintenance procedures tell what must be done to equipment at a given time. Considerable thought should be given to how procedures are expressed. What they say depends upon the needs of the organization and the skills of the personnel
Calendar information	Each set of maintenance procedures should be tagged with a frequency, i.e., daily, monthly, semiannually. Then there must be a scheduling routine which notes what has been done and when the next repetition must occur
Parts information	The maintenance procedures often require parts to be inspected and changed. For maintenance to be efficient, the planning of parts logistics must be assessed

2.1.3.1 Facilities driven by availability. To some extent, all facilities are driven by the availability of equipment, but some especially so. A good example is power generation. Certainly, some equipment in a power generation plant, whether it be fossil fuel, nuclear, or cogeneration, is absolutely critical to the timely generation of product of the plant—electrical power.

Facilities driven in this manner have a very difficult time scheduling planned maintenance for critical equipment. The critical equipment can only be maintained when the plant is shut down. Of course, major failure will shut down the plant and "corrective" maintenance must be performed. This is not conducive to efficient production. Therefore, all such plants have "scheduled shutdowns." The suspension of production of part or all of the plant is stopped intentionally to prevent failure in the future.

When the plant is intentionally shut down, maintenance must be performed efficiently and quickly to allow return to operation. This involves a higher degree of planning than does routine maintenance of equipment which does not shut down the facility. Personnel and parts must be mobilized before the planned shutdown to ensure this efficiency.

This situation also makes a growing field, "predictive maintenance," greatly desirable. If the proper conditions within the facility are monitored and recorded and if the characteristics of the equipment are well known, the schedule of shutdowns may be more efficiently managed to ensure that planned shutdown time is minimized.

2.1.3.2 Facilities encompassing environmental requirements. Facilities such as refineries and chemical process plants may possibly fall under three current standards in the United States: OSHA 1910.119, EPA RMS, and ISO 14000. These environmental management systems require that critical equipment and systems which are used to safely manufacture a set of defined hazardous materials do not fail while in operation. Failure of this equipment could cause an unacceptable risk to the environment or people. Moreover, these standards require the traceability of any hazardous materials generated or used in the process.

2.1.3.3 Facilities encompassing safety requirements. Special safety requirements are mandatory in most countries. In the United States the general extant standard is OSHA 1910.119, but other systems such as the IMO International Ships' Management code (ISM) will apply to marine and eventually to offshore oil vessels worldwide. The maintenance program is a good arena for keeping records which show the proper maintenance of equipment and route of hazardous materials, issues which impact safety.

2.1.4 Hardware and software requirements

2.1.4.1 Hardware options. There is probably little that can be done in most organizations by those charged with developing the maintenance program in terms of selecting any of the major hardware or operating systems. Usually, these are in place, a good deal of money has been spent in acquiring them, and there is little practical hope of changing them before another major capital spending period. In the next purchase of capital equipment, the maintenance team will be only one of many voices expressing a need for basic hardware and software configurations. The tool for planning maintenance and keeping records is the CMMS (computerized maintenance management system). In recent times, CMMS packages frequently come with the ability to span operating systems. The primary question is whether the effort may require one person to access the software or several persons. If only one person is necessary, a "standalone" PC may be adequate. If several people need to access the program or information, then the software should be installed on the corporate system whether it be LAN, WAN, or Intranet.

- *Mainframe.* Should be utilized only if the company already has such a system in place or if the operation is so large that a mini or

PC-based system will not be sufficient. Mainframes are becoming less common, and the term itself is becoming an obsolete nomenclature. Most "big" systems now run on what is called a client-server basis where the server may be a UNIX operating system and the computer a mini and the clients usually connect from a PC using some form of "Windows."

- *Mini.*   Should be utilized for large centralized operations where all system components are in one location and remote facilities are networked to the central location. Typically this will be used in a WAN or Intranet unless the company has an investment in a mainframe computer. Larger organizations have a WAN (wide area network) and smaller organizations may operate from a LAN (local area network). Currently there is a movement toward an Intranet, a company-based system using some software common to the Internet and usually linked with the Internet.

- *PC-Based.*   This should be sufficient for most smaller maintenance efforts and consists of an individual PC which is accessed by only one person at any time.

2.1.4.2 Software requirements.   The software should be selected with the maintenance personnel and their primary mission in mind. If the basic purpose of maintaining the equipment is ignored, the software selected will have no value. Characteristics required in all CMMS software consist of the following:

- Easy to use
- Logically formatted
- Fast operating
- Clear screens
- User-friendly
- Able to grow and update with the equipment
- Available reports useful to the audience of the report
- Capable of sorting information in many different ways
- Able to generate custom reports
- Able to operate on any of the above listed hardware options

The program should keep track of the following, at a minimum:

- Capital equipment records (the major equipment itself)
- Scheduling
- Recording corrective maintenance

- Recording maintenance and equipment history
- Cost factors
- Running-hour tracking
- Parts management

The selected software should use a separate database for maintenance procedures which is linked to the equipment. This reduces the need of having to reenter maintenance procedures for each piece of equipment. As discussed in the hardware section above, the database language in which the software is written is usually determined by the company operating system. Those who are not maintenance technicians but may be involved in the maintenance effort from a planning or quality management standpoint may require some of the following features:

- Skill code tracking of the maintenance technicians
- Scheduling capability for the various skills (if the workforce is large)
- Labor content estimates (closely allied with the cost and scheduling features)
- Inventory management
- Reporting functions to determine cost performance over time (management information)
- Reporting functions to determine vendor and part performance (used to isolate problem parts or vendors)
- Integration with the purchasing function of the company to supplement inventory management (reduce duplication of effort)

2.2   Functions of the System

The most important function of the system is to supply the "hands-on" personnel with proper materials and information to conduct the physical maintenance. Beyond this, the functions are driven by corporate goals. Integration of purchasing and inventory modules may eliminate replication of effort and save money. Measurements of downtime, cost of maintenance, reduction in maintenance cost, and management of vendors are all associated with the management of quality and costs within the company.

It is not possible to generically state total system requirements, but it is essential for each organization to make a list of what is desired to be accomplished in the system. Normally, software will be purchased from a vendor who deals in such software. Rarely is it beneficial or cost-effective for companies to write their own software, even though the "off-the-shelf" software may not have all of the features desired. Because of recent advances in database connectivity most "off-the-shelf" software can be customized to meet specific needs at a cost-effective price by most software vendors.

2.2.1 Maintenance planning and scheduling

2.2.1.1 Facilities with a maintenance cycle. Fundamental to all planned and preventive maintenance programs is the job of generating, inputting, assigning, and scheduling maintenance. This "pre-work" must be done before the planned and preventive maintenance program can function properly.

The period of a maintenance cycle should be determined by the user. It can be daily, weekly, monthly, etc. The cycle should be able to be altered at the users' discretion. Most commonly the cycle is monthly and consists of issuing instructions, or "work orders," to maintenance technicians, performing the work orders, and then closing the completed work orders, making ready to begin the cycle again.

2.2.1.2 Facilities that are downtime dependent. In facilities where downtime is crucial, such as power generation, the goal of a computerized maintenance system is to increase the time between maintenance shutdowns and reduce the amount of time a facility will need to be down for maintenance. Doing this safely, without unscheduled maintenance and accidents, is best accomplished by using both predictive monitoring and planned maintenance procedures. In these situations the maintenance system should evaluate when the cost of unplanned maintenance outweighs the savings of longer running times.

2.2.1.3 Maintenance procedures. Maintenance actions and instructions need to be collected. Maintenance actions are service procedures recommended by manufacturers and operators of the equipment. Maintenance instructions describe how the maintenance actions are to be performed. These instructions may be extremely detailed or in the form of a brief outline depending on the skills of the maintenance technicians and their familiarity with the specific facility needs.

2.2.1.3.1 Abbreviated procedures. Procedures should be concise and clearly written and identify any special tools or requirements that will be needed to perform the maintenance.

2.2.1.3.2 Complete instructions. Complete "how-to" information should be used only when the person performing the maintenance may be unfamiliar with the piece of equipment or when the particular maintenance is performed so infrequently that details on how to do it will be helpful. *There is never any substitute for skilled technicians.*

2.2.1.3.3 Supplementary information. Although there is no substitute for maintenance technicians familiar with the facility, some facilities which are dependent upon availability may have to mobilize personnel during planned shutdowns for maintenance. In cases of this nature, it may be useful to store drawings and diagrams concerning the maintenance effort which will assist persons who are skilled but are not familiar with the site.

It is possible to store such drawings and information as image files which can then be viewed or printed before the maintenance begins. However, image files can be large in size and may necessitate more expensive computing equipment if they are to be employed.

2.2.1.3.4 Feedback, the overlooked element. The maintenance system should not only keep track of the maintenance effort, it should allow the moderated feedback of maintenance technicians. If constructed properly, the maintenance program should be a "living" system, continuously modified by good comments from the personnel who must deal with the hands-on aspect of the system.

The comments of field personnel should be encouraged, captured, and reviewed on a regular basis. Then the system should be amended (usually by amending maintenance procedures) to make the system match the specific characteristics of the facility.

2.2.2 Corrective maintenance

Corrective maintenance is the repair or replacement of equipment that has failed. Planned maintenance attempts to avoid this. However, some maintenance should be corrective in nature. For example, preventive maintenance is not effective on light bulbs and other such consumable parts and equipment.

2.2.3 "House calls"

House calls are often used in the maintenance of facilities like buildings. They are not, strictly speaking, maintenance. They are requests from tenants or clients to do numerous small tasks such as the replacement of light bulbs that are out. The tasks are mundane, but it is difficult to keep track of their receipt, to record that they have been done, and to ensure that the tasks are performed in a timely manner.

2.2.4 Projects

Some planned maintenance software includes a "projects" option. Projects actually consist of modifications to the facility and are not truly maintenance. Most software which includes this function addresses the maintenance of buildings or structures where tenants need to rearrange space.

Building space modifications typically involve the same materials and procedures for "building out" areas of the facility. Therefore, it may be convenient to have areas where materials can be ordered and common procedures can be stored for the change of the layout of a building.

Such an option in the maintenance software should be incorporated with caution, and the nature of the organization's business should be carefully examined to see if this is the appropriate place for such information. It is very possible that the maintenance software is not the appropriate tool for improvements or modifications. Refineries and petrochemical plants, for example, will have modifications which are extremely complex. Such modifications are a major engineering effort. In this case, it is unlikely that the maintenance software is the proper planning and scheduling tool.

2.2.5 Parts availability

Having the parts available and being able to find the parts in a timely manner can make a vast difference in how long it takes to do maintenance. Other "parts" which must be available are any special tools to perform the maintenance. Some types of equipment such as pumps and prime movers (engines) may have massive requirements for both special parts and special tools. Therefore, it is necessary to "outplan" types of maintenance having these heavy parts and tools requirements. Maintenance planners must be aware of upcoming maintenance well beforehand.

2.2.5.1 Parts inventory vs. parts management. Parts management is the timely receipt of the parts and tools needed for the performance of maintenance. Parts inventory is a general accounting function which includes purchase, receipt, storage, and physical inventory of the parts. The parts management must always be done if the maintenance is to be performed correctly. However, the extent of the parts inventory and management is dependent upon the size and nature of

the overall organization. The time and expense for large organizations to purchase, store, and issue such parts will depend on how closely the maintenance software is integrated with the purchasing system.

2.2.5.2 Integration with purchasing and accounting. Parts management is the actions of maintenance personnel to ensure that they have the necessary parts and tools for the performance of the maintenance. Parts inventory is a broader term. What parts does one have on hand? This could be a simple database that just lists the parts in stock or an extremely sophisticated maintenance software that "hooks" to the purchasing and inventory function of the accounting package of the company. A good parts management system is capable of recording what parts are in stock, adding the parts which come in, and removing parts which are used. If the maintenance scope is large enough geographically, it is also necessary to know where the parts are currently located. Most CMMS software does not initially interface with a general ledger package and does not generate purchase orders through an overall system to keep track of assets. There are too many extant accounting programs to make "beforehand" integration possible.

2.2.6 Information system integration

There are a few general ledger programs which have a "maintenance module." However, such programs are generally written by programmers concerned with accounting who are not familiar with the practical aspects of performing maintenance. Integrating a proper maintenance function with the accounting function was once a formidable, expensive problem. However, the state of the art in computing has advanced to include the ability to transfer information between disparate databases. Microsoft, one of the largest software vendors, terms this "open database connectivity" (ODBC). Other vendors have similar terms. Database connectivity makes it possible to take programs which are specially written for a particular function such as maintenance and integrate them into overall programs such as accounting programs.

A company's accounting program usually governs all activity because it is the method by which management judges the state of business. Whether you love or hate this, all organizations will be driven by management's need to assess the financial health of the business. However, all accounting programs are actually databases. If the accounting program is written in a modern database language with open database connectivity, activities such as purchasing and inventory management may be linked to the accounting functions of the organization by transferring information between the programs.

This is not necessarily easy, but it is becoming more common and easier as the database connectivity becomes easier.

2.2.7 Use of preventive and predictive information

Maintenance can be performed on a scheduled basis or when a piece of equipment begins to show signs that maintenance must be performed. Most large facilities benefit by using a combination of both of these methods.

2.2.7.1 Planned maintenance. Planned maintenance is scheduled by time (either regular periods or running hours). Schedules are determined by past performance, manufacturer's recommendations, and suggestions from technicians and employees. This type of maintenance requires an efficient, dependable, and functional software program.

2.2.7.2 Predictive maintenance. Predictive maintenance requirements are identified by and performed when empirical data, collected and reviewed, indicate that maintenance is required. It should be noted that this method could be very costly and result in some unnecessary parts replacement if applied across the board. The method has its best results when applied to equipment the loss of which would cause expensive downtime or result in a major safety hazard. Proper use of predictive maintenance maximizes intervals between scheduled planned maintenance activities, increasing the overall time the production process is available.

Predictive maintenance can also be used to keep key pieces of equipment operating at their peak, therefore making the facility more productive. This would be true in most operations, but it is especially true for chemical processes.

Applying predictive techniques is similar to a doctor giving a patient an EKG or blood test or gathering other diagnostic information.

Continuous study is ongoing for methods, processes, and equipment to be used to improve predictive maintenance. It is not possible to list all of the various methods which can be used to analyze the state of equipment because they are constantly changing. However, some of the major methods are listed as follows:

Method	Description
Visual inspection	The oldest method known. While very subjective, an experienced technician can often find indications of problems with equipment by observing the equipment and its performance
Oil analysis	One of the oldest quantitative methods of prediction. Oil, used to lubricate any kind of rotating or moving machinery, can often give an indication of excessive wear or impending problems
Vibration analysis	Any type of rotating machinery has some inherent vibration. Measurements of this vibration can give important information to knowledgeable persons. However, it is of most use when a large amount of empirical data has been gathered and analyzed by experts working with specific equipment
Thermography	Infrared imaging demonstrating the heat distribution in equipment can give important clues of pending failure
Process parameters	Drops in yield or efficiency in a process will often be future predictors of the equipment which is used in the process. However, the meaning of variations from the normal process requires personnel who are expert in the exact process
Electromechanical testing	Regularly recorded information as simple as voltages measured across points in electric circuits or sound levels around machinery will give information to specialists. Usually, a simple record of such parameters as they vary over the history of operation will allow trending analysis that may be used to predict time to failure of equipment

2.2.8 Warranties information

If proper maintenance records are kept, the risk of vendors voiding warranties because of lack of proof of maintenance on their equipment is greatly reduced. Expenses associated with the equipment can also be greatly reduced by keeping careful records on warranties and their expiration dates. The CMMS is a good mechanism for recording and tracking this information.

2.2.9 Production and financial report

The CMMS should provide information concerning the costs of parts, downtime, labor, and overhead expenses associated with the maintenance effort and the cost of lost production at the facility. This information should be able to be accessed in a usable manner. Reports should be able to highlight problems with specific portions of the facility. This will allow a company to see what part of the maintenance effort has excessive cost and remedy the situation. Sometimes this may mean evaluating the quality of a vendor's parts or replacing a piece of machinery because it is ineffective.

2.2.10   Personnel data

2.2.10.1   System administrator.   The administrator of the maintenance system should not be confused with the computing system administrator. The computing system administrator should ensure that the software is properly installed on the organization's computing system and that the information is properly "backed up."

The maintenance system administrator is an expert in maintaining equipment. This person is responsible for ensuring that the maintenance steps and schedule are properly entered, that parts are available, and that field comments are properly incorporated. This person is also responsible for maintenance and evaluation of records concerning predictive maintenance, if used in the facility.

2.2.10.2   Maintenance personnel.   Maintenance technicians should be familiar with all of the equipment and, if they will be doing data entry, with the computer system. The supervisors should be familiar not only with the equipment but with downtime scheduling and worker loading procedures.

Typically, the classifications of the required maintenance skills fall into the following broad categories:

- Electrical
- Mechanical
- Electronic
- Vendor specialists (e.g., prime movers, special process equipment)

Depending upon the organization, these personnel may be further subdivided into specialties or mastery of the skills. Vendor specialists may be categorized with respect to the type of equipment. The exact definitions depend on the equipment used in the facility.

2.2.10.3   Data entry personnel.   Data entry personnel should be computer literate and be familiar with the engineering language being used. Usually this task may be performed by the maintenance technicians or supervisors if the software is well designed to be "user-friendly" and express the entry fields in terms to which maintenance personnel can relate.

2.2.11   Quality control and quality assurance information

Keeping the following points in mind when selecting the maintenance program will assure that the plan also meets a company's effort to improve quality.

- Identify the *requirements*.
- Select the *process* to satisfy the requirements.
- Select the proper *personnel* to accomplish the process.
- Provide the proper *tools* to accomplish the process.
- *Listen* to the people for ways to *improve* the process.

2.3   Annual Maintenance Operational Report

Having the information collected in the maintenance software will permit the generation of reports which will tell an important story to those who manage the maintenance effort and the organization. The cost of maintenance, in both dollars and time, can be weighed against the downtime which has resulted from failure within the system. Vendor performance can be analyzed. Problems with specific parts of the facility or the processes in the facility can be isolated and plans can be made to improve the efficiency of the operation.

The annual operation report should be constructed with the information gathered within the system. It should then be used with the maintenance outplanning document to act as the "road map" of the maintenance effort for the upcoming time period, usually the coming year.

2.3.1   Elements of a successful system

The report(s) should include, at a minimum, the cost of the maintenance effort; the amount of downtime experienced with the equipment; and some assessment of the cost of the downtime to the operation of the organization. Suggestions should be made from the analysis of the data as to the most profitable areas of improvement. Where will the least effort produce the biggest cost savings or increase production yield? These are questions vital to the operation of the business which justify keeping numerous records in the maintenance software and the expense of proper overall maintenance.

2.3.2   Maintenance history
vs. machinery history

The information kept by the CMMS should include two types of information concerning the equipment. It should provide an audit path to prove that the planned maintenance has been performed, and it should keep a record of all the failures and problems associated with the equipment.

Maintenance history consists of demonstrating that work orders have been issued for the equipment for all of the desired maintenance and should confirm the maintenance has been performed. It may also include information concerning any parts used, the time which was required, and any special problems encountered. This is also the area where comments or "feedback" may be entered by the maintenance personnel.

The record of problems, breakdown, and repair is termed machinery history. At a minimum, it should include the date of the problem, the date of repair, and a description of the nature of the problem. Depending on the size of the facility and the maintenance effort, it may be worthwhile to provide specific fields where cost, lost time, required parts, and repair labor may be kept for future analysis of the performance of the equipment.

Careful records of machinery history are a key to determining such things as replacement of capital equipment and management of vendors. Maintenance history is important for historical cost analysis and providing clues in examining why equipment is failing or not performing to the desired standards.

2.3.3 Production and financial reporting

Planning, scheduling, budgets, and historical cost analyses are important tasks for which the CMMS must supply information. These topics are discussed in Sec. 2.3.5. Although maintenance outplanning is presented as a planning function, the types of reports and analyses may be used at any time to gauge performance and to isolate and solve problems.

2.3.4 Quality management impact

The CMMS and associated activities are an essential part of any quality system. Whether the organization calls the effort TQM, ISO 9000, or just keeping up with business, the CMMS will provide a tool to manage the effort. However, it is not an end to itself. It must be used in conjunction with some overall program which is instituted and desired by the management of the organization.

All quality systems seek "continuous improvement" in the practices of the business. Although this term is normally used in total quality management (TQM), it should be the goal of the organization to isolate problems and to solve them in such a way as to prevent their recurrence. What the quality system is called has little to do with the success of this goal. The diligence with which information is defined, kept, and used is the key to any of these processes.

Any organization must define their goals for running the business, create procedures to accomplish those goals, implement the procedures, and constantly monitor the progress toward the goals, amending the procedures where they can be improved.

The means used by an organization is dependent upon the nature of the organization. Individuals in business may keep everything in their heads. Large organizations may have a staff which continuously publishes procedures and conducts training. The best method of accomplishing the goal is to take time to give dedicated thought to what is to be accomplished, decide for the organization the best means of recording the goals and processes, and review them on a regular basis.

2.3.5 Maintenance "outplanning"

To ensure that the planned maintenance effort is conducted efficiently, it is necessary to review projected maintenance on a regular basis. This review should occur at least on an annual basis, at the time that budgets and business planning occur. To assist this planning effort, the software selected for the CMMS should have features that address budgeting, logistics, personnel, and vendor issues.

2.3.5.1 Budget planning. Budget planning is usually the first item accomplished because it has the most impetus from management. To be effective, the CMMS should be able to keep cost information from a parts, consumable, and labor standpoint.

The ease with which this information is produced for a selected period of time is linked directly to how much time, money, and effort will be spent in the budgeting process.

In evaluating any CMMS, care should be taken that the CMMS has the right information and can produce the information in some usable form. Although software having complex statistical manipulation packages is impressive, the information is useless if it is wrong, is poorly organized in the database schema, or cannot be exported into some other software which can be effectively used by the personnel doing the financial analysis. If the correct information is put into a simple spreadsheet program, a great deal of direct, meaningful analysis can be done on the budget by a knowledgeable person.

As in all cases with computers of any description, "garbage in" produces "garbage out." A great deal of effort should be exerted to ensure that correct and complete information has been entered into the CMMS or any other program which stores useful information.

2.3.5.2 Logistics planning. Any "outplanning" process over a period of time is a golden opportunity to review the schedule for maintenance which involves parts, personnel, or tools which are not easily obtained on short notice. Failure to use the crystal ball of "outplanning" on a regular basis can produce a situation similar to that of parents trying to assemble a child's' bicycle at midnight on Christmas Eve, only to discover they are missing wing nut 11B. This is the opportunity to avoid the Christmas Eve syndrome.

2.3.5.3 Personnel planning. One of the least obvious advantages of outplanning is personnel management. A review of the history of the maintenance effort and a knowledge of any capital equipment which may be added could affect the implementation of training programs or the addition of personnel with different skills.

The worker power loading may indicate to a business that there are certain peak periods for which outside labor is needed. It is possible that failure records may show that it is cost-effective to outsource to factory technicians for complicated equipment, something which may have previously been deemed too expensive.

2.3.5.4 Vendor management. What does the last period data say about failures in equipment due to vendor parts? Impressive improvements have been made by finding vendors and vendor's parts which fail too often. Corrective actions for these problems include working with vendors to correct the failure rate in the part or refusing to do business with them.

The information from the CMMS arranged in an effective manner can allow purchasing personnel considerable leverage in dealing with vendors. Some vendors may even be thankful for bringing a problem that they were unaware of to light so that it can be corrected. In this case, the information is beneficial to everyone.

2.4   Maintenance Management Manual

A maintenance management manual should be created which describes the policy and procedures to be used in performing the maintenance effort. The organization and layout of the manual must be adapted to the organization. Those organizations having a formal, written quality management system will follow the formats prescribed by the overall documentation system. Other organizations may create their own formats so long as they are concise and understandable.

2.4.1   Scope of the manual

The maintenance manual written for the organization's entire maintenance effort should be all-inclusive. It should begin with a statement of the objective of the organization and follow through the entire set of policies and procedures which define how the effort will be conducted, documented, and monitored.

2.4.2   Maintenance objective

The organization should state precisely what it hopes to accomplish with the maintenance effort. The statement should include any measurable parameters that can be used as a yardstick to determine the success of the effort. For example, it could include the objective that "available time for main throughput streams shall be increased by an amount of 5 percent per annum until achieving a 95 percent availability, which shall be held thereafter."

2.4.3   Lines of authority

Regardless of the complexity of the maintenance effort, every person involved must have a clear assignment of responsibility, must understand the responsibility, and must have the authority and resources to implement that portion of the effort.

Note that this does not reference the actual individual maintenance actions but covers the broader issues of types of personnel involved in the effort and their overall responsibilities. A very effective method of defining the responsibility and authority is through the use of job descriptions. Job descriptions should be written with specific tasks in mind rather than specific individuals already in the organization.

2.4.4   Task responsibility

The assignment of overall duties should be covered in the lines of authority. The specific maintenance tasks should be assigned from the software. This is usually done by the assignment of "skill codes." As discussed previously, the nature of the maintenance effort and the size of the organization will often determine this, but skill codes corresponding to personnel descriptions must be available for any organization that has more than three to five individuals performing the maintenance. Even in small organizations, some distinction must be made for maintenance specialists such as original equipment manufacturers (OEM) representatives who may be needed for complex overhauls or special maintenance of equipment. A skill code should always be assigned to any given work order generated by the CMMS.

Skills also often determine the labor cost. It is the most effective way of inputting costs from labor estimates in the outplanning exercise. Linking an average cost to the skill code will permit changing cost estimates with little effort and open up the possibility of "what if" planning for the maintenance budget.

2.4.5   Corrective action

The manual should include provisions for corrective action. This is not corrective maintenance on machinery but a mechanism to analyze procedures within the maintenance effort to isolate problems and amend procedures to prevent their recurrence. Any human effort may expect human failure which may cause a problem. However, if the problem is caused by the procedures themselves, the problems will repeat with great regularity. Therefore, the results of the maintenance effort must be continuously monitored and corrected.

Some mechanism must be conceived and implemented to detect when the maintenance effort has gone wrong and institute procedures to correct the problems. Empowering organization personnel to raise flags is a good real-time method. Review of measurable quantities such as downtime, cost of maintenance, and implementation of schedule are other methods.

Whatever means are conceived to alert the organization to problems, they must ensure that the problem does not lie in the way maintenance is conducted. If the problem is caused by the system, it will repeat itself with deadly regularity. Change the system before it can happen again.

2.4.6   Regular review of feedback

Those who actually perform the hands-on maintenance often have valuable things to contribute to the maintenance effort. The system should be structured to garner their comments, review them, and incorporate them into the system. Although the subject of the feedback may cover a broad range of topics, the feedback is usually concerned with the maintenance procedures themselves. Therefore, the manual should include a provision for modifying the maintenance actions using the moderated feedback of the maintenance technicians.

2.4.7   Document control

If there is not a standard organization-mandated documentation policy, the manual must address which documents must be kept, where they must be kept, hold points for keeping the documents, and for how long they must be kept.

In a quality management system, such records are called *quality records*. They are information which either provides proof that something has been done or keeps information concerning what has happened. Typically, in the CMMS, the record of what has been done is called the *maintenance history* and the record of what has happened is called the *machinery history*. The CMMS should keep these records automatically.

Hold points concern holding documents for review to ensure they are correct or to make some use of them. A good example is the review of feedback which can be used to generate modified, improved maintenance actions. If the organization performed maintenance by assigning work orders to maintenance technicians, then recording the maintenance from the returned orders, the supervisor reviewing and entering the data would be a hold point.

Because records, even electronic records, can become too massive after a time, the decision of how long to keep the records before archiving, and how long to keep the archives must be made. If the decision is not made in the beginning, considerable problems can result after a period of time.

2.4.8   Regular Review of Entire System

Periodic reviews of the system, no less than annually, should be scheduled to determine the overall accomplishment of the maintenance system of the organization. At this time a careful analysis needs be made to determine if any overall changes would improve the system. All managers of the organization should be privy to this review, and everyone should understand the progress, accomplishments, and necessary changes to the system.

2.5   Other Considerations

The maintenance system must be tailored to the needs of the organization. This encompasses not only the selection of the CMMS but the entire structure of how the organization plans to deal with the maintenance effort. Each organization must examine what it hopes to accomplish and carefully structure its system in light of its needs.

The only consideration which can be specifically addressed is the needed commitment of the management of the organization to a proper, efficient, and effective maintenance effort within the organization. *Without this commitment, any maintenance system is doomed to failure.*

Facility Operations
and Maintenance Plans

3

Operations Plans

Richard P. Payant

Facilities Management
Georgetown University
Washington, DC

Bernard T. Lewis

Facilities Management
Potomac, MD

Facility operations and maintenance (O&M) is the heart of facilities management. Why? Because O&M is big business and as such must be well organized, closely managed, and run like a business. An effective O&M program has the end result of extending the useful life of any facility. Since the 1970s all facilities management organizations have been forced to become "leaner and meaner." Consequently, these organizations have strived to become more productive, quality-conscious, and cost-effective. The key to success today, for a facilities manager, is being able to plan and schedule effectively, being proactive in anticipating potential problems, knowing how and when to leverage help to the organization through the use of contractors, using measurement techniques to help diagnose problems and improve maintenance, being prepared to respond to emergencies, and understanding people in general and the customer in specific. The intent of this chapter, then, is to define and describe methods that may be used to control and manage the maintenance effort.

3.1   Management Operational Plan

3.1.1   Work control methods and procedures

The facility manager's primary goal is to manage resources wisely by providing responsive, high-quality maintenance and repair services to all entities being supported. To accomplish this mission the facility manager must establish well-defined procedures and the organizational structure to fulfill this work. The procedures involve coordination and planning to ensure that all the elements of skills, tools, equipment, and materials are synchronized at the right time and in the right mix to produce the desired result of satisfying the customer while concurrently controlling costs. The organization must be such that all components are orchestrated to function as a smoothly running team. Work control methods allow work requirements to be identified, screened and evaluated, planned and scheduled, checked and inspected, closed out and cost accounted, results recorded, analyzed, and measured; and finally, feedback given to the customer. A typical work control cycle is shown in Fig. 3.1. Finally, it is important

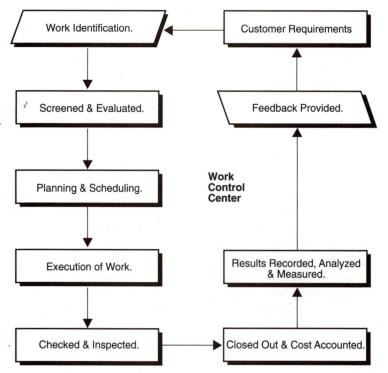

Figure 3.1 Management operational plan.

to point out that there is no standard model to follow. One method would be to establish a work control center capable of controlling all work received in the facilities department. Work control centers can range in size from a one-person operation, as an added duty, to an entire department supporting a large municipality or billion dollar corporation.

3.1.2   Work control center

The work control center is the "heartbeat" of any facilities organization. This is the central point where all work requirements are funneled, then coordinated, planned, costed, scheduled, and measured. As the primary interface between the customer and the organization the work control center can significantly influence the facility management image. Continuous coordination with workers and "closing the loop" or providing feedback to customers is essential to develop the professional reputation of the department both internally within the facilities management department and externally with customers. Figure 3.2 depicts a large work control center.

Work Control Center

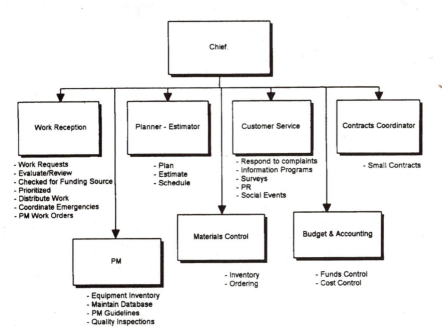

Figure 3.2   Work control center.

3.1.2.1   Responsibilities

- Centralized communications. One central telephone number for customers to call. The work control center manages the department communications systems.
- Receive and record all requests for work.
- Operate a work management system either manually or computerized using a computerized maintenance management system (CMMS).
- Coordinate all work with workers and customers.
- Coordinate and assign work priorities.
- Coordinate small contract work.
- Develop work estimates.
- Plan and schedule work.
- Operate preventive and predictive maintenance program and oversee quality control inspections.
- Assess productivity.
- Manage emergencies and special support work such as snow removal, utility outages, and special events.
- Track work backlog.
- Respond to customer complaints.
- Control supplies, parts, and materials inventory.
- Operate cost accounting and budgeting systems.
- Market "best practices" and continuous improvement.
- Measure performance of work against established goals and standards.

3.1.2.2   Benefits

- Facilities workforce is more productive.
- Management decision process is simplified because information is readily available.
- Utilities and services are more reliable because equipment is better maintained.
- Supplies, parts, and materials are better controlled, which translates into savings.
- Fewer equipment breakdowns as a result of planned and scheduled maintenance.

- Increased customer knowledge and understanding, resulting in fewer complaints and greater customer satisfaction.

- Facilities and equipment will economically attain their anticipated life expectancy.

- Facilities personnel will have improved morale due to clarification of operations and maintenance duties and a planned uniform workload.

- Less capital is needed to be invested because facilities will be used to their full capability and life expectancy.

- Professional image and reputation of the department is enhanced.

3.1.2.3 Staffing and location

- The work control center should be accessible and convenient to all departmental components. Each of its elements should have access to data and information such as historical facility information, O&M manuals, "as-built" drawings, and warranties.

- Ideally, the work control center area should be bright and open and central to the department. It should be the main hub of communications for the department and should include:

    One central telephone number with multiple work stations.

    Sufficient work receptionists to respond to customer requests and input those requests into the work management system.

    Access to read the energy management computer screens in order to triage "too hot" or "too cold" calls.

    Radio communications with mechanics in the field in the event of emergency calls.

    Facility and utility maps in order to pinpoint facility and utility emergency locations and for special support requirements.

    Policies, procedures, and operations plans.

    Access to contractor telephone numbers, on a 24-hour basis, in the event of emergencies.

    Personnel who have been trained in customer service techniques, are polite, and are willing to assist the person or group being served.

3.1.2.4 Elements

3.1.2.4.1 Work reception. Work requirements are generated from various means: customers, either verbally or through the completion and submission of a work request form, and operations and maintenance personnel as a result of preventive and prediction maintenance tasks, tours through facilities, and facility inspection programs. These requirements are received telephonically, from walk-in customers, in writing, and in some cases through electronic mail. Upon receipt, requests are evaluated for validity, reviewed for compliance with regulatory requirements and feasibility, checked for funding source, prioritized, assigned a project number, and entered into the work control system, which can be manual or automated. If additional information is needed, the caller or initiator of the request is contacted and asked to provide additional detail. Service requests are then generated and forwarded (electronically or manually) to the appropriate shop or zone for accomplishment. In the event of emergencies, calls are made via radio or other expeditious means to field mechanics.

Work reception entities vary in size and complexity depending on the size of the organization. In small companies, customers can contact the "maintenance man" or, for example, if in a hotel, the front desk, or the front office secretary in small companies. In larger organizations such as large companies, government agencies, or large universities, users would contact the customer service center or "trouble desk." The point is that regardless of size, every facilities organization should have a formal method for customers to request assistance.

3.1.2.4.2 Planner-estimator. The normal shortage of facility management resources dictates that work be carefully planned and estimated to ensure efficient accomplishment.The planner-estimator must be an experienced and highly competent individual who can generate work that is easily understood by facilities workers. This individual is responsible for planning and estimating in-house work which could involve only one trade to more complex projects involving multiple trades and requirements. To ensure accurate estimating, realistic scheduling, and timely materials coordination, it is important that the work plan be accurate, understandable, and thorough. Sketches and drawings should be provided describing technical characteristics, sizes, and dimensions of the job. Sources of information for use in planning, scheduling, and measuring the efficiency of work performance should include engineered performance standards (EPS) from recognized estimating handbooks and computer software such as F. W. Dodge and R. S. Means. A bill of materials specifying quantities of materials and items of equipment required for the job is also developed and provided to shop supervisors and the materials coordinator.

3.1.2.4.3 Materials control. The materials control section stores parts and supplies to support planned maintenance requirements, preventive maintenance tasks, and other repair and new work activities. Its main purpose is to provide timely supplies and materials while concurrently controlling the availability of the inventory which ultimately impacts the organization's budget. Stated in simpler terms...to provide materials, parts, and supplies at the lowest possible cost to the right place and at the right time. In many facilities departments space for storage or warehousing of parts, supplies, and materials is insufficient. One alternative is to partner with a local supplier who is willing to store materials, parts, and supplies, at no extra cost, and guarantee 24-hour delivery, in return for preferential treatment of purchases. Large-volume items such as filters and light bulbs lend themselves very nicely to this cooperative type of agreement. In addition to providing storage of parts and supplies, most suppliers will also give discounts of up to 20 percent.

3.1.2.4.4 Budget and accounting. The work control center is responsible for developing the budget and performing cost accounting and fund control for the facilities department. In large facilities organizations these functions are accomplished by the budget and accounting section. This section is responsible for planning, supervising, and coordinating the preparation and analysis of the facilities management budget. Once the budget has been formulated and approved, the facility manager is responsible for its proper execution. The execution phase is then divided into two categories: funds control and cost control.

- *Funds Control*   The status of various facility management accounts must be closely monitored to ensure that the budget is not overobligated. One method that is commonly used is a monthly review of the operating budget to analyze and compare it to the year-to-date actual expenses and encumbrances. This is then compared to the overall budget and gives the facility manager a real-world view of expenditures against budgeted funds.

- *Cost Control*   Requires good tracking and record keeping, and without it the facility manager will not have good job control. A good cost control system provides for tracking by budget item, work classification, and job function. While not totally exact, it will give the facility manager a good approximation of cost.

3.1.2.4.5 Contracts. With shrinking budgets, and the impact on the workforce, facility managers must find ways to leverage the way they do work. Some facilities departments have established, within the work control center, a small contracts office to control contracts up to a specified dollar amount. A popular limit used in universities is $50,000. There is no concrete way of determining the method of work assignment; however, guidelines should be established to help in making the decision whether to accomplish work in-house or go contract. Work requirements that exceed the available capacity of the in-house workforce are scrutinized by the work reception center as to complexity and customer necessities and can be contracted out. Examples of work that is normally contracted include highly skilled work requiring specialized knowledge, facilities, or tools; painting (the governing factor is that to do all the required painting would require a large in-house painting crew); renovations and alterations; and minor repairs. Work falling under the established contract limit can be contracted in a relatively simple manner using small-purchase procedures. There are several variations to these procedures that can simplify the process.

- *Indefinite Delivery Contracts (IDT)*   Essentially, there are two types that are useful: indefinite quantity contracts and requirements contracts. Figure 3.3 provides examples of work normally accomplished using an IDT contract.
- *Indefinite Quantity Contract (IQC)*   Under this type of contract the facility manager can order an indefinite service or supply quantity within a fixed dollar limit. Various contractors can be used. As an example, if the contract is for carpet repair and restretching, the contract coordinator probably will not know how many carpets will require repair or stretching in a given year and therefore writes a delivery order against the contract for each carpet repair requirement. Funds are obligated by each delivery order with a contract minimum agreed to up front.

    Requirements Contract   These are the same as indefinite quantity contracts except that this contract is with one contractor who provides all the services and supplies. This type of contract reduces your flexibility. It does have the benefit that the contractor is encouraged to maintain larger inventories and a larger workforce

---

Examples of work under indefinite delivery contracts:

- Repair walls, ceilings, and millwork (maximum number of square feet)
- Repair cabinets and countertops
- Repair resilient tile and sheet flooring (maximum number of square feet)
- Repair carpeting and restretch loose carpet (maximum number of square feet)
- Repair or replace appliances
- Repair broken glass
- Repair concrete (maximum number of square feet)
- Tree pruning and removal
- Painting

Figure 3.3   Indefinite delivery contracts.

capability, which means that lead time for supply procurement is reduced and service is faster.

- *Job-Order Contracts (JOC)* This type of contract is very similar to the indefinite delivery requirements type of contract. It is used primarily for work that exceeds a specified dollar amount. Its major feature is its unit price book. This book contains a listing of preestablished, and agreed upon, prices for defined tasks. When a task is generated, the quantity of work units and price are jointly established between the contractor and the facility management contracts coordinator, and the order is placed. The advantage of this type of contract is that it allows for quick response to customer needs.

- *Individual Contracts* Formal contracts for individual work are required when the cost is estimated to exceed the small contract limitation. Typical services that can be contracted include professional consulting services, construction services, management services, and maintenance and operations services.

3.1.2.5 Preventive Maintenance (PM). The preventive maintenance manager operates the PM program for the department. This can be done manually or with a more sophisticated computerized maintenance management system (CMMS). The latter is necessary in order to manage large electrical and mechanical systems consisting of thousands of items. A key component of a good PM program is being able to recognize which major equipment components, units, and other items should be included. Consideration must be given to:

- *Equipment* Specifically, is it critical to the operation, severe impacts if it fails, is it under warranty, how long will it take to repair if breakdown occurs, and does the cost of PM exceed the cost of repair.

- *Personnel* Focusing on the level of knowledge of each mechanic and his or her ability to provide feedback on their observations, findings, and recommendations while conducting the PM.

- *Manufacturer's Recommendations* This is a primary source for identification of preventive maintenance requirements. Additionally, the PM section, in the work control center, should have its own quality assurance system. This can be accomplished through random inspections of less critical items and a 100 percent inspection of all critical equipment. Preventive maintenance guides also must be developed for mechanics to follow. These guides, along with the PM work order, also provide a good mechanism for the PM manager to receive feedback as to whether the PM was completed, when it was completed, who completed it, and possible deficiencies or problems that could result in the future that would require further evaluation and examination. The PM section should also maintain certain records, publications, and specialized tools to help with diagnosing and predicting potential equipment defects. Records include PM inspection lists and schedules, access to "as-built" drawings, either paper copies or through the computer assisted digital data (CADD) system, and repair histories of equipment and components (organizations that are computerized can have this information located in their CMMS). Publications include parts, service, and operations manuals, and other engineering data. Specialized tools common to predictive maintenance include infrared imaging cameras, vibration analysis collectors, borescopes, temperature measuring devices, ultrasonic testing devices, listening devices such as a stethoscope, and other high dollar equipment. All of the above can be used by various entities throughout the facilities management department when it is necessary to determine the repair history of a system or component, repair parts reference, and asset value, and for planning and scheduling information.

3.1.2.6   Customer service.   While everyone in the facilities department must be customer service–oriented it is beneficial to have a section dedicated to this function and also focused on improving the image of the department and marketing its positive contributions. Excellent customer service is a prerequisite for success. It exists when the facilities department meets and exceeds customer expectations of service. In order for this to happen, the department must have a program in place which allows it first to provide excellent service and second to find ways of reminding the customer of the great service it does give. What customers perceive is reality to them. This section, in the work control center, then, has the goal of shaping perceptions. Specific functions of this section include:

- Developing information programs in order to educate the customer about facilities management services.
- Creating, continually refining, conducting, and analyzing internal and external surveys to determine trends, strengths, and weaknesses.
- Coordinating with agency, organization, or company media relation outlets to ensure accurate public relations coverage is provided to the facilities organization.
- Brainstorming, planning, coordinating, and conducting a variety of social events throughout the year in order to improve employee morale.

3.1.2.6.1 Work identification. Work for the facilities management department is generated when a customer identifies a requirement and asks that it be accomplished. Requests can come from externally supported customers or facilities staff personnel through the normal discharge of their duties. Facilities personnel generate routine work requests, for example, in the performance of preventive maintenance, through scheduled inspections such as monthly fire extinguisher inspections, and by simply walking and looking at their facilities on a daily basis. Normally, changes occur in the physical condition of the facility from age, environment, and use; by changes in regulatory requirements; and by requirements for construction, including alteration, due to expansion and changes in mission and operational needs.

The work receptionist, located in the work control center, is the primary point of contact for all customers. In a facilities management organization the work receptionist has the responsibility for receiving all requests for work and entering them into the work management system. A formal method must be in place for receiving these requests. The work request form is the tool established for just this reason. It is an official document used to record a description of the work to be done; who requested it (and how to contact); when it must be completed by; a budget number if the work is reimbursable; and finally, the signature of an approving individual who is authorized to obligate funds for that particular department. Work requests can be generated in several different ways: fax, telephonic, electronic mail, walk-up, and written. Figure 3.4 depicts a sample written work request form and Fig. 3.5 illustrates an electronic form. The initiator of the request normally completes the form. If the work is called in telephonically, the work receptionist will fill in the form. Written forms normally consist of several carbon (manifold) sheets in order to track the work through completion. One of the manifold copies is provided customers for their record.

Once the work request is received and entered into the work control cycle it becomes a generic work order and is assigned a work order number. The work control center then tracks each work order as to status: waiting scheduling, waiting materials, ongoing, or completed. A typical work order flow is shown in Fig. 3.6.

3.1.2.6.2 Work screening and evaluation. The work receptionist then classifies each request as either maintenance, repair, or new work. See Fig. 3.7. Maintenance work is defined as work performed to keep the facility operating and prevent breakdown. Repair is work necessary to fix something that has already failed. Finally, new work is work that is being added to expand, enhance, or reconfigure a facility.

Facility Work Request

Section - 1

Control Number	Time:	Date: ___ / ___ / ___
Title		

Section - 2

Facility Name:	Location:

Description of Work:

Justification:

Request for Cost Estimate Only ? Yes ☐ No ☐   Sketch Attached? Yes ☐ No ☐

Requested Completion Date: ___ / ___ / ___

Requesting Organization:	Cost Center:
Requestor Contact:	Telephone:

Requestor Authorized Signature:

Name: _____   Date: ___ / ___ / ___

Section - 3

Type of Work:	Classification:
Cost Estimate: $	Date: ___ / ___ / ___

Funding:	Cost Center	Amount	Percentage
Source 1	_____	$	_____ %
Source 2	_____	$	_____ %

Division of Facilities Authorization:	Date: ___ / ___ / ___
Division of Facilities Contact:	Telephone:
Division of Facilities Comments:	

Figure 3.4 Sample written work request.

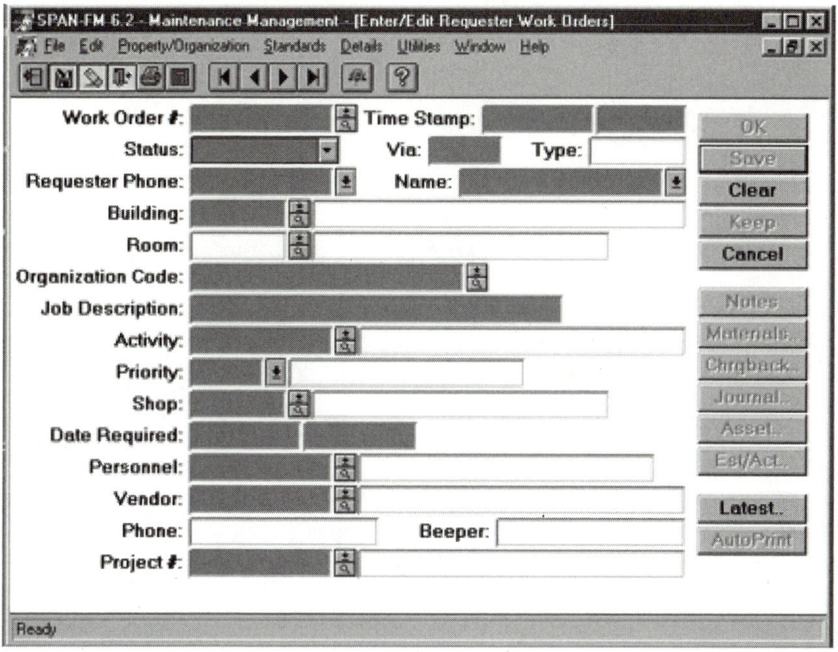

Figure 3.5 Electronic work request form. (*Copyright © 1998, Innovative Tech Systems, Inc.*)

Once work has been classified, it is categorized into one of four categories: service orders, work orders, standing operating orders, and preventive maintenance.

3.1.3   Service orders (SO)

Service orders are small, service-type maintenance jobs that need immediate attention and cannot be deferred. Normally, there are time and cost limitations established for this type of work, for example, $500 and 8 worker hours of effort. Planning, estimating, and scheduling are not performed. Examples of typical service orders include leaking roof, minor electrical and plumbing work, and broken window repairs. There should be continuous review and analysis of effort expended on service orders. This should be done by shop and/or by zone, depending on how the department is organized, and by similar jobs to determine workload and efficiency. Productivity can also be measured by comparing work effort to established performance standards such as F. W. Dodge and R. S. Means mentioned earlier.

The number of service orders can be reduced by strictly enforcing work priorities, forcing customers to pay for damages caused as a

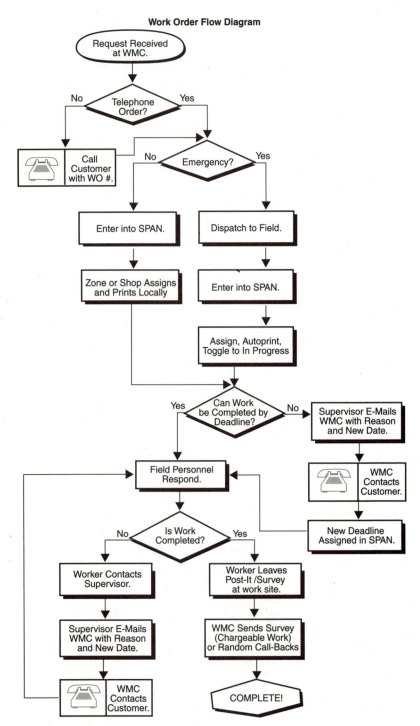

Figure 3.6 Work order flow diagram.

Maintenance:	Painting, filter changes, pothole repairs
Repair:	Repair of plumbing leaks, broken glass, torn carpet
New work:	Installing new electrical outlet, constructing a new addition to a building

Figure 3.7 Examples of work classifications.

result of vandalism or for repairs other than fair wear and tear, adhering to established performance standards, and applying strong preventive predictive maintenance programs.

3.1.4 Work orders (WO)

Work orders can include maintenance, repair, and new work. They are formally estimated, planned, and scheduled and can be considered project-type work. This type of work can be accomplished by the in-house workforce or by contract, depending on limitations that have been established. Work done in-house is estimated, planned, and scheduled by the planner-estimator. Work done by contract is sent to the contracts coordinator. In either case, upon completion of the job, documents are returned to the budget and accounts section for appropriate action.

3.1.5 Standing operating orders (SOO)

Standing operating orders are special work orders where the specific work and manpower requirements are relatively constant and predictable. They include operations, services, and routine maintenance where requirements are repetitive and can be planned annually. Examples include fire protection inspections, custodial services, utility plant operations, refuse collection and disposal, grass cutting, etc.

Work is planned, estimated, and scheduled prior to the start of the fiscal year.

Careful analysis of all standing operations. Scheduling must be complete and detailed. There should be periodic inspections and ongoing work sampling for each standing operation to determine the quality and quantity of the provided service.

3.1.6 Preventive maintenance (PM)

Preventive maintenance is routine, recurring work such as regular servicing of HVAC systems. PM reduces emergency responses, allows work to be efficiently scheduled, and ensures that replacement parts are available when needed.

PM work orders are generated similar to normal work orders.

3.1.7   Work Priority System

After the work is classified and categorized, it is prioritized. Each facilities management department should establish its own work priority system. Normally, there are three standard levels of work priorities: emergency, urgent, and routine. These priorities should be carefully and clearly defined and distributed to all customers for their understanding. By doing this, the job of the work receptionist should be made easier and become a little more pleasant.

3.1.8   Emergency priority

This takes priority over all other work and requires immediate action. Work continues until the emergency is corrected. Normal response time is within 1 hour. Examples include lock-ins, gas leaks, broken water and steam lines, utility failures, hazardous and toxic spills, elevator trap calls, and stopped-up commodes (when only one is available).

3.1.9   Urgent priority

This priority involves correcting a condition that could become an emergency. Normal response times range from immediate up to 72 hours depending on availability of the workforce. Once started, the work should continue until completed but is dependent on availability of materials. Examples include garbage disposal problems, lights that are out, pest control, lock changes, and appliance (refrigerator, range) problems.

3.1.10   Routine priority

This is work that is an inconvenience to the customer. The completion of this work is normally accomplished within 7 days. Examples include dripping faucets, screen door replacement, countertop repair, and floor replacement.

3.1.11   Work approval

The chief of the work control center is responsible for work approval. This is normally delegated. Experienced work receptionists can usually estimate the cost of service order work fairly accurately and then, using established criteria as guides, assign the work to the workforce. For more complex and costly work, such as work orders, the chief of the work control center should be the approval authority. This work should be examined from the standpoint of work assignment, i.e., whether it is reimbursable work or not. If reimbursable, for example, new work, a budget number should be provided by the customer. Repetitive work, standard operating orders, and PM work, is planned on an annual basis and does not require individual work approval.

3.1.12 Planning

Once work orders are approved, work reception forwards them to the planner-estimator. This individual then develops the detail of the scope of work, estimates the total cost of the job, which includes labor, material, length of time, and equipment required, and then prepares plans, sketches, and specifications to do the work at minimal cost and time while concurrently achieving the highest quality of work possible. To be effective at job planning the planner-estimator needs to query the initiator of the work for specific detail, then visit the site where the work is requested to verify information compiled to date and examine specific site conditions that will impact cost, and finally determine if permits are required. In larger, complex jobs involving multiple crafts it would be beneficial for the planner to use a planning worksheet.

3.1.13 Scheduling

Scheduling is putting what has been planned into action. This is where the planner juggles the availability of labor hours; work site availability; special customer requirements; materials, parts, and equipment availability; weather conditions and the relative priorities of the job. A key factor that must be considered and expected is the need for flexibility. No job runs exactly as planned and scheduled. There are always disruptions, delays, or changes that must be anticipated and planned for. One objective that every facility manager should have is to accomplish the largest portion of maintenance work on a scheduled basis. This is possible with the help of a good PM program. The facility manager will then see that as the amount of scheduled work increases so will workforce productivity. This is due to a decrease in unscheduled and emergency work.

The Association of Higher Education Facilities Managers (APPA) defines three intervals for scheduling work. These intervals are printed with permission from APPA.

3.1.13.1 Long-term scheduling. Usually, 3 months or greater, based on the PM master schedule, work order backlog, priority of the work to be done, availability of equipment and materials, time of the year (weather), and availability of worker power. These schedules must be continuously reviewed and revised as necessary. Characteristics of a typical long-term schedule include:

- Only large jobs or projects are scheduled. Typically, large means anything with more than 24 to 32 work hours of labor. No attempt is made to schedule emergency work.

- A separate master schedule work order should be maintained for each shop so that individual skill or trade bottlenecks can be seen.

- Care is taken to not exceed the capacity of each shop. To do this, the capacity of each shop must be calculated and updated regularly.

- All available capacity is not scheduled. Typically, only 50 to 80 percent of available work hours are scheduled. The remaining is reserved for absenteeism, nonproductive work, and emergencies.

- In well-run master planning scheduling systems, the schedule becomes frozen 1 to 2 months out; changes are allowed only in the third month and beyond. This greatly stabilizes the planning and scheduling process and contributes to high productivity and schedule reliability. Figure 3.8 is an example of a long-term schedule.

3.1.13.2 Weekly scheduling. This interval is for the coming week. It is developed from the long-term schedule and is a detailed commitment of resources. Characteristics of this schedule include:

- While master work order schedules cover only 50 to 80 percent of shop capacity, weekly schedules attempt to cover 80 to 90 percent.

- Jobs under 24 to 32 work hours duration are not included on the weekly schedule.

- Weekly schedules issued for each shop reflect known absenteeism in the shop capacity figure for a given week.

- The weekly schedule shows simply the number of work hours that should be spent by each shop on each job for every day of the work week. Figure 3.9 is an example of a typical weekly schedule.

- Jobs are not scheduled until major items of material have been checked for availability.

- Weekly meetings are held to communicate actual versus scheduled performance.

Typical Master Schedule

Shop __Carpenter__   Date __3-14__

Time Period	1st Half April	2nd Half April	May	June	July	Aug	Sep	Oct
Capacity (workhours/work day)	24	24	24	24	32	32	32	32
No. of work days in period	10	9	20	21	21	22	20	22
Planned workhours available	240	216	480	504	672	704	640	704
Total load (workhours)	200	220	470	450	370	340	340	240
Net workhours available	40	-4	10	54	302	364	300	464

Scheduled Workhours

Jobs or Projects	1st Half April	2nd Half April	May	June	July	Aug	Sep	Oct
# 6387 ABC Hall	180	10						
# 8522 XYZ Dorm		150						
# 3701 Bldg. #8			160	160				
# 2232 Bldg. #10 - Room 201	20	20						
# 6757 Bldg. #11 - Room 20		40						
# 8700 DEF hall			100	50	50			
New addition - XYZ Hall				80		120		
Dean Doke's office			180					
Alumni Bldg. roof			30	160		180	180	
Admin. Bldg. renovation					160			
Stairway West Parking					160			
Bldg. # 8- Rms. 101, 102, 103						40	160	
Bldg. # 12								240

Figure 3.8 Long-term schedule. (From Rex O. Dillow, ed., *Facilities Management: A Manual for Plant Administration*, 2d ed., Association of Physical Plant Administrators of Universities and Colleges, Alexandria, VA, p. 1011.)

TYPICAL WEEKLY SCHEDULE

DATE _____   Gross Workhours Available _____

SHOP _____   Minus Leave Time _____

Net Work Hours Available _____

Job Number	Location	Daily Workhours Scheduled					Total Hours Scheduled	Percent Completed	Comments
		Mon	Tues	Wed	Thu	Fri			

Figure 3.9 Typical weekly schedule. (From Rex O. Dillow, ed., *Facilities Management: A Manual for Plant Administration,* 2d ed., Association of Physical Plant Administrators of Universities and Colleges, Alexandria, VA, p. 1011.)

3.1.13.3 Daily scheduling. The daily interval of scheduling is developed from the weekly schedule. It results when there are changes that interrupt the weekly schedule such as excessive absenteeism, materials that have not arrived, emergencies, and last-minute special customer requirements. Daily schedules are normally developed by the immediate supervisor.

3.1.14 Execution of work

This is the actual conduct of maintenance and repair work in response to a work request. It can be accomplished in one of two ways: in-house workforce or contract. It is the point at which all the planning and scheduling pay off. Many activities (such as ordering materials and equipment; requesting and supplying support services, for example, scaffolding, land surveys, utility marking and identification; safety inspections; and developing trade and individual work schedules) are synchronized in order to achieve the common purpose of completing the work within the estimated cost and time frame.

An in-house workforce should record labor, and material, supplies, parts, and equipment usage on a daily employee activity sheet (see Fig. 3.10). This document provides a record of worker activities for that day. It is completed, daily, by the worker and turned in to the supervisor, who then checks the time associated with each work request. Supplies and materials usage can be annotated to this form or included on the work request.

For contracted work an inspector should be assigned to periodically check and inspect as the work progresses. Nothing should be allowed to be closed in or covered unless it has been inspected.

Emergency work is handled on a case-by-case basis and requires special attention. Emergencies called into the work receptionist should be verified by requesting someone (either a supervisor or

Employee Activity Sheet

Name:	Date:	ID #:

S.O. Number	Bldg	Location	Time spent	Complete?	Tag#	Customer Requestor/Contact
Problem:						
Description/Parts/Notes						

S.O. Number	Bldg	Location	Time spent	Complete?	Tag#	Customer Requestor/Contact
Problem:						
Description/Parts/Notes						

S.O. Number	Bldg	Location	Time spent	Complete?	Tag#	Customer Requestor/Contact
Problem:						
Description/Parts/Notes						

S.O. Number	Bldg	Location	Time spent	Complete?	Tag#	Customer Requestor/Contact
Problem:						
Description/Parts/Notes						

Page ___ of ___

Signature: ____________

Figure 3.10 Employee activity sheet.

craftsperson) to verify the emergency. Once verified, requests to shops or zones should be made via radio or pager for quick response. Emergency work takes priority over all other work.

3.1.15   Checked and inspected

Good facility managers check! check! check! Even with continuous quality improvement programs wherein quality is built into everything that is done there still exists a need for inspection. Everyone in a facilities department, whether they believe it or not, is part of the inspection program. Inspection is a continuous undertaking...it never ceases! Craftspersons should be trained in their trade and pride instilled in them to produce the best quality work they can. They should not be satisfied with mediocre results. Supervisors and managers should be trained the same way—to strive to continuously improve upon everything they do. As maintenance deficiencies are identified they are reported to the work reception center, where they are entered into the work control system.

Inspections are also an important component of the PM program. It is important to note here that as PM inspections are properly accomplished there will be a decrease in reactive maintenance and a corresponding increase in scheduled maintenance. A mechanic conducts PM of a specific piece of equipment and annotates any unusual information on the PM work ticket so that it can be entered into the CMMS PM module database. In most cases the PM inspector will verify the information in the field. If severe enough, repair can be planned and scheduled without the need for an emergency outage. As equipment ages and deteriorates, the importance of PM inspections is obvious...serious problems resulting from breakdowns can be reduced. Through planning and scheduling, downtime is reduced.

3.1.16   Closed and cost accounted

Work is properly closed out when excess materials, parts, and supplies are returned to materials control; punch list items are completed; required tests, demonstrations, or verifications have been completed; operation and maintenance manuals have been collected, reviewed, and accepted; "as-built" drawings have been submitted; lessons learned are documented; labor, material, and equipment used have been assembled and added to the work order; and if reimbursable work, the customer has been formally charged, through the accounts payable system, for the work.

3.1.17   Results recorded, analyzed, and measured

Once work has been closed out, it has to be analyzed and measured. This is performed to compare actual performance against predetermined standards. As a minimum, quantities of materials used, as well as the charges that may have been credited, should be spot-checked. Actual labor hour charges should also be compared to those estimated by the planner estimator in order to improve the value of the planner estimator's database. All of the above is critical in the continuous quality improvement loop. This step is important to determine if trends, lessons learned, or potential improvements can be synthesized from the work that took place. Tools that can be used to assess the effectiveness of the maintenance and PM programs can be incorporated into weekly and monthly management reports.

3.1.17.1   Weekly reports

- *PM inspection report*   This report is the result of follow-up inspections of work identified from PM work orders.

- *Open work order report*   This is a weekly status report of all open work orders. See Fig. 3.11.

- *Backlog report*   This is a weekly list of backlogged craft hours. See Fig. 3.12.

3.1.17.2   Monthly reports

- *Performance analysis report*   Measures the quality and quantity of work performed. Actual labor hours and materials used are compared to the estimated amount and variances are identified. See Fig. 3.13.

- *Closed work order report*   Summary of work orders closed out. See Fig. 3.14.

- *Maintenance budget report*   Includes expenses year-to-date compared to budgeted amount and depicts variances.

- *PM maintenance summary*   This is a 12-month trend report of expenditures by equipment. See Fig. 3.15.

- *Trend analysis report*   This report is used to project future maintenance trends by using historical data. Forecasts are generated by graphically displaying the historical information and analyzing the data to establish trends. Typical applications include projecting

Work Orders by Shop

Georgetown University
Washington, DC

Date range from 3/31/97 to 4/4/97

Shop Codes: 10
Shop Description: UTILITY PLANT

Work Order #	Requestor	Building	Room	Work Order Description	Employee Assigned	Start Date	Completion Date	Total WO Cost
16100	BROWN,MONICA	INTERCULTURAL CENTER ICC_L03	306	SEE NOTES		4/3/97		$0.00

Work Orders printed: 1

Shop Codes: 24
Shop Description: ZONE I MAINTENANCE

Work Order #	Requestor	Building	Room	Work Order Description	Employee Assigned	Start Date	Completion Date	Total WO Cost
15669	DIVEN ALICE	NEW RESEARCH BUILDING NRB_Z09	EP07 LABORATORY	AIR FLOW NOISY	BINSON J38	3/31/97		$0.00
15680	JOHNSON LINDA	MED-DENT MDL_Z01	4 SE402	INSTALL VACUUM LINES/SEE ATTAC UCKER	R98	3/31/97	4/2/97	$14.62
15984	DAVILA DR	MED-DENT MDL_Z01	4 NE420 LABORATORY	TUBE UNDER SINK BROKE/WATER F SEAL	M99	4/2/97		$0.00
16016	APPERSON, CHERYL	NEW RESEARCH BUILDING NRB_Z09	G00C1 CORRIDOR	ADJUST SWINGING DOORS/SEE NOT		4/2/97		$0.00
16024	KOZIKOWSKI ALAN	NEW RESEARCH BUILDING NRB_Z09	EP07 LABORATORY	HANG MARKER BOARD IN OFFICE		4/2/97		$0.00
16077	SINGLETON, MARY	LEAVEY CENTER LEA_P02	S-316 RECEPTION	DR BELL NOT WORKING	PRATT D17	4/3/97		$0.00
16096	JENNINGS JASON	LEAVEY CENTER LEA_P02	1319 VITAL VITTLES	REMOVE FREON FROM COOLER/SEE ILLIAMS R53		4/3/97		$0.00
16153	HENDRICKS CECILA	LEAVEY CENTER LEA_P02	S-530 MAIL ROOM	HANG PLAQUE ON WALL	PRATT D17	4/3/97		$0.00
16174	MICHELLE	NEW RESEARCH BUILDING NRB_Z09	E202A OFFICE	TOO HOT	COATS T58	4/4/97		$0.00
16199	MUNOZ, JOE M	NEW RESEARCH BUILDING NRB_Z09	B0M1 MECHANICAL ROOM	REPAIR BOOSTER PUMP	COATS T58	4/4/97		$0.00

Figure 3.11 Weekly status report, open work orders.

3.26

WO Backlog Report by Building

Georgetown University
Washington, DC

Building Code: COP_K01
Building COPLEY

Date range from 11/1/96 to 4/10/97

Days Late	Work Order #	Date WO Issued	Date WO Required	Room Code Description	Building System	Requestor/ Job Description	Employee Assigned	Vendor Assigned
	12168	2/25/97		516 LOUNGE	MECH	REPLACE FCU CONTROL SWITCH		
	12161	2/25/97		LOUNGE	CARP			
	15893	4/2/97		B-25 BOILER ROOM	15730	install window stops		
	15614	3/31/97		OUT OUTSIDE	16210			
	15432	3/31/97		B-24 LAUNDRY				
	15590	3/31/97		B-17	15510			
	13459	3/10/97		B-25 BOILER ROOM	15250			
	14776	3/27/97		BOILER ROOM	CARP	REPLACE MISSING & CRACKED LENSE		
	13854	3/17/97		530 BEDROOM	MECH	broken fcu speed switch		
	15604	3/31/97		B-17A MECHANICAL ROOM	15510			
	9657	2/4/97		B-25 BOILER ROOM	15730			
	15520	3/31/97		B-17	15510			

Figure 3.12 Backlog report.

3.27

Performance Analysis Report							
Week Ending	Work Order #	Estimated Hours	Estimated Material Cost	Actual Hours	Actual Material Cost	Labor Estimating Efficiency (1)	
13-Mar	72332	6.2	140.23	9.4	120.98	51.60%	
1) (Estimated Hours - Actual Hours) / Estimated hours X 100 = %							

Figure 3.13 Performance analysis report.

future costs for materials, utilities, hazardous waste disposal, labor, and equipment; budget and financial planning; and operational planning backlog.

3.1.18 Feedback

In order for a facilities department to improve its service and the way it conducts business it has to receive timely and accurate feedback. This feedback, in the form of review and analysis, provides the information required to ensure that actions are being performed as planned. It also allows the facilities manager to focus efforts for organizational improvement. For instance, the facilities manager and staff should meet periodically (quarterly may be opportune) to evaluate the efficiency of operations, review the quality of performance, measure the performance of management initiatives, and identify deficiencies and opportunities for improvement. Major accomplishments should be highlighted and disclosed to higher leadership. Remember: The "boss" doesn't necessarily know what you, as the facilities manager, are doing unless you tell him or her. The review and analysis process should be formalized and documented to ensure

Work Order Status Report

Work Order #	Current Date	Work Order Description	Status
15766	4/1/97	LIGHT OUT 3RD FL HALL NEAR MENSRM	Complete
15845	4/1/97	RAIL DOWN IN 91-96	Complete
15805	4/1/97	HANDLE BROKEN/KEY WONT TURN	Complete
15755	4/1/97	MIRROR BROKEN ON CABINET DR	Complete
15820	4/1/97	LATCH/BAR/ SCREWS OUT OF DOOR	Complete
15812	4/1/97	ORDER TOOLS	Complete
15741	4/1/97	REPLACE DOOR STOP-1ST FL MEN'S ROOM	Complete
15850	4/1/97	LECTURN HAS NO LIP	Complete
15753	4/1/97	CLEAN LAUNDRY ROOM FILTER	Complete
15830	4/1/97	SEE NOTES	Complete
15777	4/1/97	PLEASE CHECK DISHWASHER SEE IF WE	Complete
15765	4/1/97	10,LIGHTS OUT ON 3RD FLOOR	Complete
15780	4/1/97	REPAIR DOOR CLOSER	Complete
15762	4/1/97	POLE LIGHT OUT	Complete
15746	4/1/97	DRAIN AIR COMPRESSOR	Complete
15852	4/1/97	LIGHT OUT IN BATHROOM	Complete
15782	4/1/97	COIL ON SERVING LINE REACH IN FROZE	Complete
15758	4/1/97	OUTLETS NOT WORKING	Complete
15781	4/1/97	REPAIR DAMAGES DUE TO BREAK IN	Complete
15742	4/1/97	TOILET CLOGGED	Complete
15779	4/1/97	PLUMBER TO DISCONNECT/SEE BUZZ	Complete
15840	4/1/97	GARBAGE DISPOSAL CLOGGED	Complete
15817	4/1/97	TOILET CLOG	Complete
15822	4/1/97	COMPRESSOR LOCKED UP/W.COOLER/LADY	Complete
15745	4/1/97	DRAIN AIR COMPRESSOR	Complete
15833	4/1/97	5TH FL LADIES RM FAUCET BROKEN	Complete
15839	4/1/97	BRICK MISSING FROM OUTSIDE WALL	Complete
15841	4/1/97	COLD	Complete
15769	4/1/97	CHANGE LOCK AND CUT 6 KEYS	Complete
15844	4/1/97	TOILET CLOGGED	Complete
15752	4/1/97	LIGHTS OUT	Complete
15771	4/1/97	BATHRM LIGHT OUT	Complete
15835	4/1/97	CK COURTYARD DR/NOTLOCKING BEHIND	Complete
15829	4/1/97	CUT (2) KEYS #45 FOR SUPPLY CLOSET	Complete
15757	4/1/97	REPAIR HOLES ON WND FLOOR HALLWAY	Complete
15848	4/1/97	CALLED LOCKSMITH IN 4:00A.M.	Complete
15759	4/1/97	POWER OUT	Complete
15790	4/1/97	FRONT DOOR GLASS COMING OUT	Complete
15842	4/1/97	BATHROOM LIGHT OUT	Complete
15813	4/1/97	REMOVE WATER COOLERS/STORE IN VIC	Complete
15748	4/1/97	DRKNOB OFF DOOR CANT GET INTO RM	Complete
15749	4/1/97	CLOGGED SHOWERS-GROUND, 1ST,2ND FL	Complete
15828	4/1/97	LIGHT OUT	Complete
15778	4/1/97	7TH FL STORAGE CLOSET LOCK BROKEN	Complete

Figure 3.14 Closed work order report.

its success and continued emphasis. The facilities manager has to form a viable corporate group that will develop strategy and provide the necessary leadership to the rest of the organization. The results of this program should contribute information for each manager's annual performance evaluation.

Work Orders by Shop

Georgetown University
Washington, DC

Date range from 4/1/97 to 4/10/97

Shop Codes: 20
Shop Description: MAINTENANCE CONTROL

Work Order #	Requestor	Building	Room	Work Order Description	Employee Assigned	Start Date	Completion Date	Total WO Cost
15855		LXR LXR_A01	G00U3 MECHANICAL ROOM	BI-MONTHLY LESLIE READINGS	ANTER H55	4/2/97	4/2/97	$0.00
15856		LXR LXR_A01	G00U3 MECHANICAL ROOM	BI-MONTHLY LESLIE READINGS	ANTER H55	4/2/97	4/2/97	$0.00
15857		COPLEY COP_K01	B-25 BOILER ROOM	BI-MONTHLY LESLIE READINGS	ANTER H55	4/2/97	4/2/97	$0.00
15858		COPLEY COP_K01	B-25 BOILER ROOM	BI-MONTHLY LESLIE READINGS	ANTER H55	4/2/97	4/2/97	$0.00
15859		POULTON HALL PLH_D19	B07 BOILER ROOM	FILTER INSPECTION FOR CLOSED LO		4/2/97		$0.00
15860		LXR LXR_A01	G00U3 MECHANICAL ROOM	FILTER INSPECTION FOR CLOSED LO		4/2/97		$0.00
15861		LXR LXR_A01	G00U3 MECHANICAL ROOM	FILTER INSPECTION FOR CLOSED LO		4/2/97		$0.00
15862		WALSH WAL_A04	100M MECHANICAL ROOM	FILTER INSPECTION FOR CLOSED LO		4/2/97		$0.00
15863		NEVLS NEV_A05	B-028 MECHANICAL ROOM	FILTER INSPECTION FOR CLOSED LO		4/2/97		$0.00
15864		MED-DENT MDL_Z01	BSE B03A MECHANICAL ROOM	FILTER INSPECTION FOR CLOSED LO		4/2/97		$0.00
15865		LAUINGER LIBRARY LLI_E01	B-15E1 MECHANICAL ROOM	FILTER INSPECTION FOR CLOSED LO		4/2/97		$0.00
15866		LAUINGER LIBRARY LLI_E01	B-15E1 MECHANICAL ROOM	FILTER INSPECTION FOR CLOSED LO		4/2/97		$0.00
15867		PRECLINICAL SCIENCE PCS_Z05	B09 EQUIPMENT ROOM	FILTER INSPECTION FOR CLOSED LO		4/2/97		$0.00
15868		PRECLINICAL SCIENCE PCS_Z05	B09 EQUIPMENT ROOM	FILTER INSPECTION FOR CLOSED LO		4/2/97		$0.00

Figure 3.15 PM maintenance summary.

- Facility manager
- Division chiefs
- Members of the work control center

The following is a review and analysis suggested agenda:

- Update on open action items from the previous meeting (review of minutes from previous meeting).
- Discuss available resources versus required and make adjustments.
- Review goals and objectives and the obstacles to meeting them. Alternatives for corrective action are discussed.
- Assign responsibilities and suspense dates for solving identified problems.
- Review quality improvements and accomplishments.
- Outline topics and objectives for the next meeting.

The following should be included in a review and analysis meeting:

- Presented information and discussion should be captured in the minutes.
- Minutes should include date, time, attendees, discussion items, highlights of policy changes, customer feedback information (percentages and comments), open items, and taskers and accomplishments.
- Successful quality improvement stories and other accomplishments should be mentioned.

Customer feedback can be solicited in various ways. Mechanics can leave "while you were out" cards when they respond to a service order and the initiator is not available. See Fig. 3.16. By leaving this card, the initiator knows that the facilities department responded and either repaired the deficiency or defined a reason why it was not fixed. The key here is that customers are satisfied if they know what's happening. They only get disgruntled when they're "kept in the dark." A second means of feedback is for the work reception center to call back a sampling of customers daily, following up on completed work orders and soliciting feedback. In this case it would be worthwhile to standardize questions. These questions will provide immediate feedback and serve as a good measure of customer perceptions of the services they receive. These questions can be as simple as:

Georgetown University
Facilities Management

While you were out,
our staff had to enter your area to:

_____

Date: _____ Time: _____ Zone # _____

Employee Name: _____ Shop # _____

Work Completed: Yes/No   Work Order # _____

Comments: _____

Inquiries: (687-3432)

Figure 3.16 While you were out card. (These cards can be the size of 3×5 cards, which allows them to be carried in a shirt pocket, or be made in the form of 3×5 Post-it Notes® so they can be adhered to a refrigerator, door, desk, etc.)

- Was the work completed to your satisfaction?
- Were facilities personnel courteous?
- Was a "while you were out" card left?

Another, more formal, means of obtaining customer feedback is to use a formal survey (covered in detail later in this chapter).

3.2   Building Operational Plan

Facility managers must oversee the safe, efficient operation and maintenance of each facility they control. Their decisions impact every member of that organization every day. Within the numerous processes they are answerable for there are common issues: how to schedule and coordinate work efficiently, how to control startups and shutdowns of equipment and operating systems, how to handle emergency situations, how to coordinate diagnosis and repair of trouble, and how to use benchmarking and performance indicators to determine the operational status of equipment and systems.

3.2.1   Scheduling work

This work can be described before its actually being required. To be effective, scheduling must be founded on a detailed work plan. The work must be defined and identified in advance of accomplishment and must consider what is to be done and what coordination is needed.

3.2.1.1   Scheduling objectives

- Synchronize resources to meet requirements. Consideration must be given to:

  Workforce availability.
  Transportation requirements.
  Material availability (lead time).
  Conflicting work priorities.
  Job site availability.
  Anticipated weather conditions.

- Maximize productivity.

- Minimize cost.

- Minimize impacts on customers.

- Accomplish a higher portion of maintenance work on a scheduled basis.

- Reduce downtime of equipment.

3.2.1.2   Scheduling coordination.

Coordination is a positive action which implies cooperation to achieve common goals. It is a necessary step to ensure that the work of the facilities department meets the needs of the customer in a timely but considerate manner. Customers deserve to know what and when work is going to occur. Typical coordination issues include:

- Contact all customers and departments that will be involved in any planned outage. The proposed schedule of work should be coordinated with customers if there is a potential that they will be impacted. They need to know so they can plan their own work. For example, a facilities department may schedule a full-load emergency generator test in a medical center at 2 o'clock in the morning, thinking this is the best time not to disturb customers. However, a researcher could have a very important experiment ongoing and should be made aware of the test and its possible impact. This coordination could prevent a disastrous outcome.

- Contact the fire and police departments when standby fire protection is required and there is a possibility of traffic flow interruptions.

- Coordinate with the environmental program manager when environmental protection must be considered.

- Notify the company insurance carrier and the risk management office when fire protection systems are disconnected.

3.2.1.3   Coordination methods.   Coordination is essentially a communications process. The methods used to coordinate maintenance work with customers can be described as verbal and written.

- Verbal techniques

   Regular staff meetings
   Design review meetings
   General project meetings
   Construction meetings
   Conferences
   Technical meetings
   Committee meetings
   Telephone notifications
   Daily work relationships

- Written techniques

   Policy and procedure manual
   Operating handbook
   Engineering instructions
   Memos, letters, and reports
   Minutes of meetings
   Progress reports
   Engineering drawings
   Specifications (design, construction, and maintenance)
   Bulletins and newsletters
   Electronic mail

3.2.2   Shutdowns and startups

The facilities manager is continuously faced with many circumstances requiring equipment and utility shutdown or startup. Some of the reasons include emergency repairs, preventive and predictive maintenance, scheduled testing, bringing new facilities or equipment on-line, and holidays and vacation periods. It is essential that these operations be accomplished using documented procedures that will minimize the possibility of overlooking an important step in the planning process and also serve to disclose potential coordination problems.

3.2.2.1   Shutdowns.   Scheduled shutdowns are not considered emergencies, and advance notification to customers should be provided in writing at least 72 hours prior to the scheduled outage. Alternative support may be required if the shutdown will last more than several hours. This support could consist of lighting, heat, potable water, or cooling and can be planned for, and connected, in advance. Typical planned shutdowns include emergency generator testing; utility repairs; equipment repair, replacement or overhaul; sprinkler and fire protection system testing and repair; and building heating, ventilating, and air conditioning (HVAC) system repair or adjustment.

3.2.2.2 Startups. The objective is to get equipment operating in an efficient, safe, cost-effective manner as quickly as possible. The following points should be considered.

- Expect equipment problems. Depending how long the equipment has been down, fluids and lubricants will have drained from metal surfaces, systems will become airbound, and equipment will lose calibration.

- Be cautious of pressure systems. Because of no demand, systems that are normally pressurized may have higher than normal pressures.

- Ensure that all necessary materials and supplies are on hand. This will reduce workforce idle time and provide for a more balanced workload.

- Ensure that prior to startup all maintenance and utility crews are on hand early, available to minimize and respond to problems.

3.2.3 Emergency situations

Emergency situations require immediate action. When compared with scheduled maintenance, emergency maintenance is a very small percentage (usually less than 10). The detailed planning and scheduling that precede scheduled maintenance are impractical in emergency situations because of the immediacy of the situation. This is why emergency situations must be planned ahead of time and written procedures set in place. Section 3.3 will cover this in more detail. As a minimum, the facilities manager should plan for three areas of immediate concern.

3.2.3.1 Fire protection

- First step is to develop a fire protection protocol, unique for each building. It should define who the "authority having jurisdiction" is (prior to the fire department arrival: this could be the building security officer, the property manager, or the facilities manager); building evacuation procedures; description of the alarm system and location of annunciator panels, sprinkler valves, and fire pump; shutdown procedures; emergency generator location and description; and what to do if the systems fail.

- Training is extremely important. Occupants of the building should be familiar, through periodic fire drills, with how to evacuate and also know location and how to activate alarm pull stations. Maintenance and operations personnel should be trained, semiannually, on the location of specific monitoring panels and shutoff valves and disconnects and should have the protocol etched in their minds.

- Designate the spokesperson to coordinate public relations and news media issues and statements.

3.2.3.2 Power outage

- Control devices should always be set for "fail-safe."

- All electrical start-stop switches should be shut off. This will prevent damage to certain types of electrical equipment should partial electrical power be restored. Also, circuit breakers may switch off if they sense an overload when power is restored.

- Verify that all emergency systems are operational. If the outage lasts for several hours, check the fuel tanks of emergency generators.

- If in the winter season, check throughout the building for possible pipe freeze-ups. Crack open water faucets at various locations throughout the building.

3.2.3.3 Flooding

- If the building is located in a floodplain, contact the state flood control agency or the U.S. Army Corps of Engineers to obtain information on expected time of arrival and water levels.

- Provide property protection. Time permitting, all doors and windows should be sealed and sandbagged. Emergency generators should be obtained to operate pumps and provide emergency lighting. Electrical equipment should be deenergized. Where feasible, delicate equipment should be relocated to safe locations.

- Implement procedures that define who makes decisions to evaluate and shut down equipment, details evacuation procedures, and determines emergency crews.

3.2.4 Trouble diagnosis and coordination

As mentioned earlier in this chapter, coordination is extremely important. You can have the most competent mechanics and the best maintenance plan, but if there is no coordination, the facility department is looked on as incompetent.

3.2.4.1 Diagnosing a maintenance problem. Diagnosis is the identification of maintenance causes and effects. This eventually leads to correction and elimination of such causes as misapplication and operator abuse. The "diagnostician" plays an important role here. This individual must have the technical background and personality to be able to find a problem and teach mechanics and operators the proper maintenance techniques without coming across as a "know-it-all" or trying to find fault. It is not easy to find an individual with these characteristics. Before any diagnosis of a maintenance problem can begin, the technician must have the proper training and experience, understand the steps involved in diagnosing a problem, and be knowledgeable and skilled in the use of diagnostic methods and instruments. In some cases it may be more cost-effective to contract for this service.

3.2.4.1.1 Training

- Continuously attend training sessions to remain current in new technological advancements. Check with the Instrument Society of America and the IEEE Reliability Society for seminars and short courses involved with engineering and maintenance management.

- Because human beings normally learn by doing, technicians should be required to train workers. This also increases their technical credibility with those same workers.

- Bring in manufacturer's representatives to conduct training on site.

3.2.4.1.2 Steps in diagnosing

- Review previous inspection reports.

- Analyze equipment histories for unusual pattern development and trending.

- Analyze actual repair records for failure analysis and cause and effect correction.

- Identify where inadequate or improper operating procedures are having adverse impact on the reliability or maintainability of equipment.

- Review operation and maintenance manuals for frequency and procedure of required maintenance.

3.2.4.1.3 Diagnostic methods. Some of the more common diagnostic methods are outlined below. The purpose of their use is to decrease unscheduled downtime.

- Ultrasonic testing. Used to detect leaks in air and gas systems, steam valves, and tanks.

- Oil analysis. Used to determine the condition of the oil and the system it lubricates.

- Infrared imaging. Used to detect temperature differences.

- Vibration analysis. Used to monitor vibration of bearings on a piece of rotating equipment.

3.2.5 Coordination with third parties

If customers have been impacted by an equipment outage or if they are going to pay for the repair, they must be told what happened, why it happened, and the corrective action taken to prevent it from happening again. Unless this is done, the facilities department loses credibility. Even the best operating maintenance department, having an excellent PM program to minimize equipment downtime and maximize scheduled maintenance, will experience occasional emergencies wherein customers cannot be warned of unscheduled outages. This is where good customer relations comes into play.

3.2.6 Benchmarking

The purpose of benchmarking is to improve an organization's performance by comparing its practices to those of another. In order for facilities managers to conduct a benchmarking study, it is important that they know their own processes. This is necessary to get the maximum value from study comparisons with other organizations. Good preparation is important before a study is begun! Benchmarking equipment and systems in individual buildings are important in order to measure performance and make comparisons of that equipment and system with other organizations equipment and systems.

3.2.7 Statistical process control

Once thorough understanding of a procedure is attained, major advances in service quality, productivity, and cost can be achieved. To develop this knowledge a tool such as statistical process control (SPC) can be used. SPC is used in the industrial and manufacturing environment as a process to improve quality and minimize waste. It is a statistical technique suited for use whenever something can be measured or counted in order to show trends that will guide future actions. It is typically used in continuous improvement processes and can be creatively adapted to various facility management building operating systems. Facility managers can use SPC to help better understand their processes, measure and analyze quality, and ultimately improve maintenance.

3.2.7.1 Process tools. Various statistical tools can be used in processes and equipment performance. The information presented below is adapted, with permission, from GOAL/QPC, 13 Branch Street, Methuen, MA 01844-1953.

Pareto Chart:	▪ *Purpose:* To visually display collected information by type or category, in order or priority.
	▪ *Use:* Set priorities on what to work on first. To focus on a problem offering the greatest potential for improvement. Make comparisons.
	▪ *Example:* Track frequency of equipment breakdowns per (time frame); then determine a priority of work.
Histogram:	▪ *Purpose:* Summarize data collected over a period of time, and graphically display its frequency distribution, in bar form.
	▪ *Use:* Display data in graphic format, simplifying interpretation and analysis. To provide information for predicting future performance.
	▪ *Example:* Track frequency of unscheduled maintenance.
Run Chart:	▪ *Purpose:* Measure trends or patterns over time.
	▪ *Use:* Determine if a process has improved over time. Detect trends, shifts, or cycles. Compare performance measure before and after implementation to determine impact.
	▪ *Example:* Circuit breaker must be reset more frequently over time, indicating the possibility of a bad breaker or overload.
Control Chart:	▪ *Purpose:* Improve monitor and control process performance over time.
	▪ *Use:* Focuses attention on determining and monitoring process variation. Obtain immediate feedback on the performance of a process.
	▪ *Example:* Monitoring incoming electrical power to determine number of interruptions.
Cause and Effect Diagram:	▪ *Purpose:* Graphically display all factors relating to root cause.
	▪ *Use:* Focus on content of the problem. Diagram collective knowledge. Focus on causes, not symptoms.
	▪*Example:* Diagnosing equipment failure.

3.3 Comprehensive Facility Operational Plans

All facilities managers have the responsibility to effectively plan for and respond to an emergency situation and bring that situation to a suitable conclusion. To do this, preparation is vital! The establishment of a formal written plan which defines how to handle emergency procedures is essential. This plan should be disseminated to everyone who has responsibility to put emergency procedures into practice. Additionally, emergency drills and problem exercises should be periodically scheduled to ensure that facilities personnel are properly trained and conditioned to react quickly and skillfully when an emergency situation occurs.

3.3.1 Emergency response plan

Emergencies occur! A good emergency response plan defines duties and responsibilities, identifies various emergencies which may occur, explains what to do before, during, and after an emergency, and details the procedures for resolving most conceivable emergencies.

3.3.1.1 Organization. An emergency coordinator must be appointed by top management. This individual should work out of a central location (emergency operations center) and coordinate actions during both the emergency and the restoration period. The emergency operations center (EOC) is the primary control point for the coordination and handling of responses to emergencies. Staffing of the EOC is dependent on the type, magnitude, and location of the emergency.

3.3.1.2 Concept of operation. The concept of operation is a statement of the emergency coordinator's visualization of how an emergency is handled from start to finish. It is stated in sufficient detail to ensure appropriate action. Normally, after normal work hours, the security office will be the first to be notified of an emergency. This is due to several reasons: security officers patrol buildings and can physically detect a problem, alarm systems (fire and environmental) terminate at the security office, or individuals detecting an emergency situation call the security office to initiate the emergency response plan.

3.3.1.3 Command and control. During major emergencies, a positive chain of command must exist and be functional. There must be a clearly identified chain of succession.

- *Command* The responsibility for resolving any emergency situation rests with top management. However, the emergency coordinator should be the executor of the emergency response plan and assist top management in planning, training, responding to, and mitigating emergencies.

- *Responsibility succession* In the absence of the final top management responsible individual there exists a need to define a successive line of responsibility for decision making.

- *Control* Initial control responsibility of emergency scenes rests with the director of security. As a minimum this responsibility includes the following activities:

Immediate evacuation of affected buildings, necessary for the life safety of occupants.

Initial response to emergencies in the facility or building. This response determines type and size of the emergency.

Liaison coordination with external support agencies such as police and fire departments.

Exterior crowd control at all emergency scenes.

Assisting the emergency coordinator with initial activation of the EOC.

3.3.1.4 Communications. Adequate, effective, and redundant communications systems must be available to effectively respond and control emergency situations. These systems must be an integral part of the EOC. They must be reliable and capable of functioning during periods of power loss.

3.3.1.5 Types of emergencies. Depending on the type of facility or property the following emergency situations should be addressed. For each of the types of emergencies the plan should address preparations before an emergency occurs, training requirements, response procedures (which include description of personnel actions for each credible emergency requiring specialized response), and restoration procedures.

- Hazardous material emergencies

- Fire emergencies

- Natural disaster emergencies

- Bomb threat emergencies

- Utility outage emergencies

- Labor unrest emergencies

3.3.1.6 Support services. These are services that can be provided to measure, control, or mitigate emergency situations. Information which should be incorporated here includes:

- Resources available

- Temporary housing support

- Updated list of telephone numbers and points of contact

- Local building codes and regulatory requirements

- Insurance information

- Flowchart of the chain of command

- Reporting procedures and required documentation
- Federal, state, and city disaster assistance
- Information on how to track costs

3.3.1.7 Guidelines for facility specific plans. Specific emergency procedures for each individual facility should be prepared. The plans should contain, as a minimum, the following information.

- Brief description of the facility or type of operation conducted.
- Procedures for evacuating and accounting for any visitors and for all personnel normally working in the facility to include:
  Determination of when evacuation is necessary.
  Designation and use of assembly areas to account for personnel.
  Selection of evacuation routes out of the facility.
  Establishment of a personnel accountability system.
  Procedures to evacuate physically impaired personnel.
- Description of means of communicating the emergency situation to all personnel.
- Identification of specialized equipment needed by units responding to the emergency.
- Procedures for shutting down all utilities to the affected area and securing the facility from unauthorized entry.
- Floor plans and blueprints.
- Building systems information.

3.3.2 Hazardous materials plan

Hazardous and toxic materials are all materials (gaseous, liquid, or solid) that can cause physical or chemical changes in the environment, affect property, or affect the health or physical well-being of living organisms. A release or spill of hazardous material requires the controlling organization to immediately notify people in the surrounding vicinity that a release or spill has occurred. If deemed necessary, an evacuation of the area (room, floor, or building) will be accomplished. A specific hazardous material of concern to all facility managers is the use of refrigerants for cooling and refrigeration equipment. In accordance with international agreements (Montreal Protocol) and federal laws, chloride-bearing refrigerants (CFCs) were phased out of use on January 1, 1996. Phase-out of HCFCs will begin in 2003.

The purpose of a hazardous materials plan, then, is to:

- Define types of releases or spills.
- Identify procedures for personnel to follow at the scene.
- Define cleanup procedures (including contamination cleanup).
- Describe and identify spill management equipment and material needed.
- Describe transportation requirements.

3.3.2.1 Training. Facility managers are not only responsible for the facilities they maintain, but they also have an obligation to ensure their personnel are appropriately trained in hazardous materials in the workplace. The following training should be considered:

- Hazard communication (right-to-know)
- First responder awareness
- First responder operations

3.3.2.2 Assessment. According to the Code of Federal Regulations, 29 CFR 1910.132, employers are required to perform a workplace assessment to determine whether hazards are present or likely to be present that necessitate the use of personal protective equipment (PPE). If such hazards exist or are likely to, the employer must select appropriate PPE and certify that a workplace hazard assessment was performed.

3.3.3 Refrigerant management

3.3.3.1 Refrigerant accountability. According to the Environmental Protection Agency (EPA) Stratospheric Ozone Protection, Final Rule Summary, EPA-430-F-93-010, "Technicians servicing appliances that contain 50 or more pounds of refrigerant must provide the owner with an invoice that indicates the amount of refrigerant added to the appliance. Technicians must also keep a copy of their proof of certification at their place of business. Owners of appliances that contain 50 or more pounds of refrigerant must keep servicing records documenting the date and type of service, as well as the quantity of refrigerant added." Essentially, refrigerant must be accounted for from the day of purchase to the day of disposal.

Many computer software packages are available on the market designed to provide a means of tracking refrigerant usage in accordance with EPA requirements. These software packages provide the capability of matching usage in relation to an asset's basic refrigerant charge, which can help in determining leakage. An example of a

GEORGETOWN UNIVERSITY REFRIGERANT USAGE LOG

Technician Name:_____   Cylinder Number:_____

Issue Date:_____   Closed Date:_____

Refrigerant Type:_____   Starting Weight:_____

Date	SPAN Ticket #	Equipment ID	Refrigerant Cylinder #	New	Refrigerant Recovered	Contaminated	Total Refrigerant Added	Removed

Total New Refrigerant Used:_____   Cylinder #_____
Total Recovered Refrigerant:_____   Cylinder #_____
Total Recovered Refrigerant:_____   Cylinder #_____
Total Recovered Refrigerant:_____   Cylinder #_____
Amount Recovered from new Cylinder :_____
When Refrigerant was added was the reason Unintentional Venting?Yes / No (circle one)
Did you repair all leaks according to EPA 608? Yes /No (circle one)

Please Weigh all Cylinders before and after each use. Keep all Cylinders locked! Accurate reports are very important, please make sure that all new cylinders are emptied and add up to starting weight.

Technician Signature:_____   Date:_____

Figure 3.17   Refrigerant usage log.

refrigerant usage log can be seen in Fig. 3.17. To protect themselves and their organization, facility managers should establish a policy for purchasing, accounting for, and reclaiming refrigerant, then develop a system to implement that policy. This is essential in the event EPA decides to conduct an audit.

3.3.3.2 Technician certification. As of November 14, 1994, federal law requires that all mechanics working on air conditioning or refrigeration equipment be certified to correctly service that equipment, including recovery and recycling practices. These mechanics are required to pass an EPA-approved test. There are four types of certification:

- Type I: Servicing small appliances.
- Type II: Servicing or disposing of high-pressure appliances.
- Type III: Servicing or disposing of low-pressure appliances.
- Type IV: Servicing all types of equipment (universal).

Under Section 609 of the Clean Air Act, sales of CFC-12 refrigerant in containers smaller than 20 pounds are now restricted to technicians certified under EPA's Motor Vehicle Air Conditioning regulations.

3.3.4   Safety plans

All organizations want to reduce the loss of downtime, lost production, and Worker's Compensation related to accidents. In order to improve safety within an organization and, hopefully, decrease the number of accidents and injuries on the job, a formal written safety plan is needed. The safety plan provides a system to constantly monitor the work environment to minimize potential safety and health threats.

3.3.4.1   Purpose. To establish standard procedures for normally occurring activities. These procedures, when followed, save the organization money by keeping safety continuously in the minds of everyone.

3.3.4.2   Involvement. In order for employees to "buy into" the safety plan they have to see the policy put into practice. This requires that safety be included in the organization's mission statement; various safety topics, germane to the functions of the organization, are included in the plan; safety training; consistent enforcement of safety policies and procedures; and positive reinforcement of safe working behavior. Below are a variety of methods used by organizations to promote safe working areas and prevent accidents.

- Area involvement in the process. It is very important to have input from all areas affected by the safety topic. Include office areas, the loading dock, production areas, distribution areas, mail room, supply and storage areas, laboratory areas, mechanical rooms, and others.
- Safety and health committees. These committees are formed to serve several basic functions.

   Create and establish safety guidelines.
   Plan and implement facility safety programs.

Establish and review safety training programs.
Investigate accidents.
Conduct and evaluate inspections.
Initiate ideas and plans to correct safety problems.
Hold weekly or monthly safety meetings.
Have a safety suggestion program.
Design safety training sessions with employee input.
Stress long-term safety goals.
Institute a philosophy that accidents are not acceptable.
Use statistical techniques to categorize accidents into systematic shortcomings and human error.

3.3.4.3 Accident reporting. To facilitate a uniform response to accident investigation and provide for the safety and health of employees a standardized accident investigation report form should be implemented. This form should be completed by the appropriate supervisor, within a specific amount of time, for any accident or on-the-job injury. The facility manager should review each form for completeness, find out what happened, and take corrective action to prevent similar accidents in the future.

3.3.5 Fire protection plans

The facility manager normally has responsibility for fire protection and prevention within a facility. This responsibility can be divided into regular prevention and protection inspections and how to control fires that occur. A planned fire prevention and protection program must provide for the maximum protection against fire hazards to life and property consistent with the mission, sound engineering, and economic principles. Additionally, facility managers should have a plan to combat fire emergencies.

3.3.5.1 Prevention and protection. Inspections should be conducted monthly and are performed to reduce the potential of a fire hazard.

- *Prevention* Environmental situations known to be causes of fires are inspected first. These situations include storage and handling of flammable materials and liquids; operation of electrical equipment and associated wiring; high-temperature generating equipment, such as hot work involving welding and soldering equipment; smoking noncompliance; and finally, proper housekeeping procedures.

- *Protection* Inspection and testing of fire protection equipment is intended to ascertain the operability of that equipment. This inspection should identify and verify that each fire protection control valve functions correctly; verify the service availability of water supply; complete a fire pump checklist for each pump; examine the critical components of special extinguishing systems; test all sprinkler systems and alarms; check all fire extinguishers; identify each hydrant; determine that all fire doors work properly; verify that smoke and heat detectors operate properly; and finally, make sure that protective signaling devices function correctly.

3.3.5.2 Fire control. Upon discovery of a fire, the fire alarm should be sounded. If the fire is small and an extinguisher is nearby, the fire should be extinguished with the extinguisher. If the fire is larger and spreading, clear the building. If time permits:

- Close all doors and windows.
- Turn off all oxygen and gas outlets.
- Disconnect electrical equipment.
- Turn off all blowers and ventilators.

3.3.5.3 Plan. Fire emergencies can occur at any time. Having a fire plan developed is good insurance against total chaos. As a minimum the fire plan should cover the following:

- Authority having jurisdiction. Define the overall organization having ultimate jurisdiction until the fire department arrives.
- Define what measures should be taken before, during, and after the fire emergency.
- Training requirements.
- Formation of an organization fire brigade.
- Ensure plans and drawings for each building are available. These plans should identify electrical, chilled water, steam, telephone and data cable locations, and other utility distribution systems.
- Specific requirements should be spelled out for the operations and maintenance staff, contracts section, purchasing staff, and public relations staff.
- Requirements for fire watch.

3.3.6 Labor unrest

Facilities managers having a unionized workforce will have unique challenges facing them in the event of a labor strike or threat of any job action. A strike or walkout can leave a facility critically short staffed, vulnerable, and unable to deal with a crisis. Consequently, a detailed strike contingency plan should be prepared ahead of time. This plan should include sections on:

- Communications. Staff members, vendors, suppliers, contractors, utility suppliers, and police and fire departments need to be kept informed of the status of the strike. Customers must also be informed of the impact to them. Using voice and electronic mail broadcasts may be the best way to keep the staff and customers informed.

- Services that would:

  Continue to be provided.
  Be curtailed or suspended.
  Be contracted out.

- Staff reassignments to cover critical responsibilities and vital services.

- Protection of facility equipment and personnel.

- The in-house security department must be given specific instructions on how to respond to unrestful situations.

- A detail log including date, time, and description of the incident should be kept in the event later legal action is necessary.

3.4   Facility Occupant Support Plan

Today, one of the biggest challenges facing facility managers is to provide high-quality customer service. To do this, facility managers must ensure that not only routine scheduling of work is conducted but also that it is done in a caring manner. High-quality service exists only when the expectations of customers are exceeded.

Many facilities departments believe they provide excellent service because all the mechanical and electrical systems in a building operate and meet required life safety and building codes, burned-out lights are replaced immediately, maintenance and repair backlog work has been reduced, and service technicians respond quickly when a service request is submitted. All of this may be true and still customers may not be satisfied because their expectations are not exceeded. Possible cause for this may be the way the customers perceive the service that is being provided.

3.4.1   Improving perceptions

There must be caring and consistency in the way service is provided before customers will perceive facility support as efficient, effective, and customer-oriented. This change will come each time they have a positive experience as a result of the service being provided. Expectations must be exceeded every time. Every service opportunity must be seized to influence customer perceptions in a positive manner. Below are some ideas that will help.

- Minimize interference with customer activities.

  Curtail unplanned outages. Ensure customers have, as a minimum, 72 h notice.
  Coordinate with customers, just to provide them information. They will accept almost anything as long as they know the impacts and why it has to be.

- Provide proper and timely service for customer work requests.

  As discussed in Sec. 3.2, work requests are screened, evaluated, and then prioritized. Time frames established for each level of work priority must be adhered to.
  A system should be established to "close the loop" with customers if work cannot be accomplished in a timely manner. Electronic mail lends itself nicely to doing this.

- Don't let problems that occur go unanswered. If a facility manager learns of a problem wherein a customer is dissatisfied, investigate the reasons on both sides to "close the loop" with the customer. If the facilities department erred, acknowledge that, apologize, and let the customer know that you will implement corrective action... But respond to the customer! Again, customers will accept almost anything as long as they know the department is trying to improve and that it does care. Michael LeBoeuf, in his book *How to Win Customers and Keep Them for Life,* states, "A rapidly settled complaint can actually create more customer loyalty than would have been created if it had never occurred. Customers are much more likely to remember the 'extra touch,' fast action, and genuine concern that you exhibited when they felt dissatisfied."

3.4.2 Determining wants

To exceed customer expectations facility managers must determine what customers need and what they expect.

- *Need* One effective method of determining need is to go see the customer, face to face. Personal contact has a way of cementing relationships and developing rapport.

- *Expectations* To determine customer expectations a formal written survey is a good technique to use. Once you have it developed,

Customer Survey

Please indicate your level of satisfaction by circling 1, 2, 3, 4, or 5, 5 being the highest and 1 being the lowest, 3 being "satisfactory".

1.   Are your common areas maintained in a generally clean and attractive manner by our custodial staff?

     1   2   3   4   5

2.   Did we respond satisfactorily to your request for service?

     1   2   3   4   5

3.   Was a requested repair or correction completed to your specifications?

     1   2   3   4   5

4.   Was the work area left in a clean and proper manner?

     1   2   3   4   5

5.   Was the worker involved courteous in manner and professional in appearance?

Figure 3.18 Customer survey.

plan on distributing it annually. Start a mailing list of persons to mail the survey to and keep adding to it. Include people who have experienced problems with service in the past; budget and financial people; and key department, floor, or building coordinators.

1   2   3   4   5

Please give us any additional comments and suggestions.

_____

_____

_____

_____

_____

_____

_____

_____

_____   _____

Signed (optional)   Building/ Department (optional)

Your cooperation in this survey is appreciated as we strive to improve our

customer service.

Figure 3.18   _(Continued)_ Customer survey.

Remember the survey has to be simple in order for people to take the time and interest to complete. They don't want to spend a lot of time composing and writing responses, but they will check a box corresponding to a yes/no question. It should take them less than 5 minutes to complete. A sample survey is shown in Fig. 3.18. Returning the survey has also got to be simple. A card-type survey, preaddressed, with postage applied (if necessary) can be used. Once the person filling out the survey is finished, that individual can just drop it in the mail (campus mail, postal system, or building/department mail system). In addition to surveying external customers, try issuing an annual survey to the facilities workforce. This is a good way to take the pulse (measure morale) of the organization.

3.4.3   Provide feedback

Most customer complaints and dissatisfaction are generated because of a lack of communication. Implementing a simple customer communications program will eliminate many routine complaints before they mushroom into major customer relations problems. Below are some ideas that can be used.

- *Callbacks*   On a daily basis have the work receptionists make callbacks to customers where work they requested has been completed. A good percentage to use would be to call back 20 percent of work orders that have been completed. This technique lets 20 percent of the customers who called in a problem know that the facility department is measuring its performance, identifies follow-up problems which can then be proactively corrected, and finally provides a daily gauge to measure customer satisfaction.

- *Customer contact employees*   Train employees who have daily contact with customers to "close the loop" and provide information that customers should be made aware. These employees are the first line of contact with these customers and know when something is not right. They can serve as an early warning system to alleviate potential problems. Additionally, a periodic group meeting with all customer contact employees can be instructional for them, through the sharing of ideas, and can also alleviate customer concerns.

- *Reports*   Distribute an annual report to customers. This lets them know what, and how well, the facility department is doing in meeting their expectations.

- *Periodic visits*   On an occasional basis the facility manager as well as all supervisors and the customer service section should stop and visit customers in their work areas or send them electronic mail to determine how the customer is doing. This demonstrates genuine interest and care for the service being provided them. It helps to cement credibility and cultivate rapport.

3.5   Quality Control Plan

The facility department (FD) quality control plan (QCP) should conform in every respect to corporate policies and procedures. A priority and an important management tool used to ensure maintenance of high quality standards is a sustained quality control effort. The FD's QCP should ensure the customer's (tenant's, building occupant's etc.) satisfaction with services received, work that is properly done, adherence to work performance schedules, and disciplined cost control. The QCP should be based on procedures that include an inspection system, administrative procedures for recording and tracking deficiency corrections, and timely record keeping and reporting.

3.5.1 Specific QCP objectives

These objectives should ensure that the mechanical, electrical, plumbing, structural, energy management control system, custodial and related services, structural alterations and improvements, and preventive and predictive maintenance work, and materials utilized for the operation, repair, and modification of the facility will be of the best quality. This will include but not be limited to:

- Detecting and correcting of existing or potential deficiencies
- Ensuring that FD personnel correct deficiencies right the first time
- Ensuring that employees know the expected work quality standards and are qualified to achieve them
- Ensuring that work performed is within established corporate acceptance quality levels (AQL)
- Conducting ongoing corporate trend analyses to ensure uniform quality and reliability
- Providing accurate QCP records and timely feedback reporting

The facility manager has ultimate responsibility for the QCP program design, development, implementation, and execution monitoring. Usually the FD staff size does not justify the cost of a separately staffed quality control inspector. Rather, the onsite QC program should link the QC efforts of FD supervisors, mechanics, and contractors; corporate managers; and FD professional and technical personnel.

Specific objectives of the QCP should include clear definition of organizational and individual responsibility. The goal is to involve every employee in the pursuit of work performance excellence. The QCP will be based on three separate undertakings: an overlapping system of inspections by FD supervisors; procedures for identifying, tracking, and correcting deficiencies; and administrative procedures for tracking, recording, and analyzing inspection results and determining trends in the quality of work performed. The following paragraphs describe an overall QCP that will meet all criteria and provide the best result—the inspection system; the checklists, forms, and other tools used; and the techniques used to track and correct deficient work. The final focus is on the overall QCP objectives.

3.5.2   Inspection system

The FD staff should develop a detailed set of QC procedures, check-lists, and guidelines for each provided service. The facility manager should then institute a series of interlocking QC inspections to be conducted by personnel on several organizational levels. These will include the facility manager, chief engineer, operating engineers, maintenance mechanics, cognizant corporate personnel, and contractor account managers.

3.5.3   Inspections by FD personnel

The facility manager, chief engineer, and operations engineers should inspect completed work, including service calls and maintenance and repair work orders. The chief engineer, operating engineers, and maintenance mechanics should tour the facility each day, conducting QC inspections informally. When their daily QC inspections are completed, covering their QC activities for the day, their reports should summarize all deficiencies noted during the QC inspections and provide recommended corrective actions. The facility manager should review the QC reports and forward them to the FD administrative assistant for processing. All daily submitted QC reports should be maintained in a hard-copy file for review by corporate managers.

Discrepancies noted during QC inspections should be keyed into the FD's computerized maintenance management system (CMMS)—specifically into the QC module initially for tracking purposes and into the service calls and maintenance and repair work order modules ultimately for corrective actions based on the deficiency's dollar value. Using these modules, deficiencies recorded (during QC inspections) will be tracked by the facility manager, chief engineer, and operating engineers until corrective action is taken. Each deficiency will be assigned by the chief engineer to the appropriate mechanic (or subcontractor) for corrective actions.

Once corrected, the CMMS modules databases will be updated by the administrative assistant to include cognizant information and data; i.e., labor hours used, materials and parts used, equipment used, date completed, performing mechanic, etc. The facility manager will routinely assess pending workloads to plan for and assign the resources needed to correct the noted QC deficiencies. This action will be taken to prevent QC deficiencies corrections from being overlooked. Figure 3.19 illustrates the flow of QC inspection data in the CMMS from inception and discovery through to posting as completed jobs data bits ("cradle to grave" follow-up).

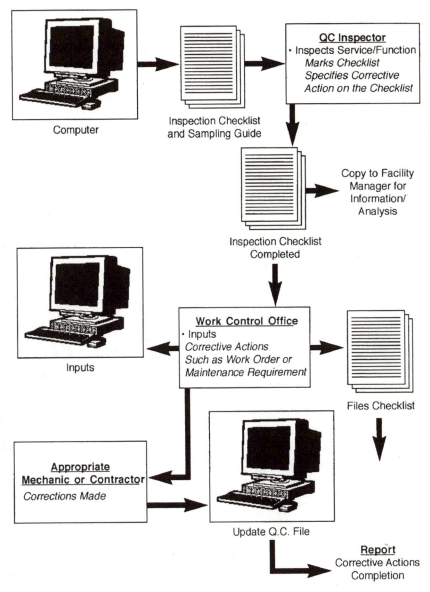

Figure 3.19 Flow of quality control inspections.

3.5.4 Identifying and tracking substandard performance

The QCP should provide a system of overlapping inspections by FD staff, corporate professional and technical personnel and managers, and contract managers to identify quality problems. Several techniques and tools should be used to assist these QC inspectors. The primary tool should be facilities and equipment specific inspection checklists to be developed by technical experts. Another QC performance evaluation tool is the completed preventive maintenance work ticket form. Still other tools include visual observations, service call reports, and work order data. In general, for QC purposes, items inspected should include PM work, service calls and emergency callbacks, work orders issued and completed, tenant comments, tours and watch reports and logs, energy management control system reports and data, and historical repair records.

If the QCP is to be truly successful, it cannot be a "top-down" initiative only. Nonsupervisory personnel must have a role in the process. Accordingly, mechanics, engineers, and other FD staff should be continually alert to note quality problems as they perform their duties. For example, mechanics often see poorly performed PM or repair work while engaged in preventive maintenance inspections on the same equipment item or one located nearby. When observed, deficiencies should be noted on the work order form. That information should then be annotated by the mechanic on the PM work ticket and keyed into the CMMS's QC module by the administrative assistant. The noted QC deficiencies should then be scheduled for corrective action by the facility manager.

Tour inspection reports are another practical means of identifying substandard maintenance work performance. Tour reports should be routinely reviewed by the facility manager, chief engineer, and operating engineer. QC problems should be processed as noted immediately above. Inspection reports and corrective actions taken should be maintained on file for further review, as required.

In addition to facility inspections, corporate managers and corporate professional and technical staff will review preventive maintenance work orders, service calls, and work order logs during their QC inspections. The priority given to a QC repair or deficiency observed should move higher the longer it remains on the service call log, either a manual or computerized log. The service call log is an important QC data source. The log should be reviewed daily by the chief engineer and operating engineers and reviewed periodically by the facility manager. Essentially, technical personnel should evaluate chronic calls, calls for the same problem, the number of calls received over time, etc.

3.5.5 Correcting deficiencies

Tracking deficiencies from detection to correction is an important QC process. Deficiencies reported or identified by QC inspections should ultimately result in corrective action work orders. Like any work order, those documenting QC deficiencies should be assigned a work performance priority based on the criticality of work to be performed and then tracked through the company's CMMS. Such work orders should be identified as a QC correction. A computerized suspense file should be maintained for all QC corrective actions. This suspense file should be updated as the work is completed.

In addition to correcting specific problems reported (or otherwise identified), it is essential to identify casual factors leading to the deficient work. If not identified and addressed, casual factors lead to a cycle of repeated deficiencies and corrections. This trend drains productivity, job resources, and staff morale. Several strategies can be used to avoid this problem. For example, copies of quality control reports can be cycled from the FD back to the corporate engineering office. Data from these reports should then be extracted, input to a computerized database, and analyzed. These reviews yield trends or "root cause" analyses to be used by corporate staff engineers to detect the source of trends in work quality. When quality trends in a specific phase of work, area of the building, equipment, or system can be identified, determining the source of that trend becomes easier. Once the source is known, long-term corrective measures can be taken.

While such analyses are used to identify specific sources for negative (or positive) trends in work quality, various standard measures can be employed to proactively ensure that tasks are performed properly. Examples include:

- Writing quality standards into employee job descriptions

- Group training or one-on-one employee and supervisor training

- Observing employees throughout tasks performance

- Round-table discussions with employees to elicit ideas that will improve work methods and resolve persistent quality problems

3.5.6 Total Quality Management (TQM) program specifically for O&M Procedures

An FD TQM program, in concert with the TQM program discussed in Chap. 8, should be instituted for the purpose of achieving effective, efficient, and cost-effective O&M support efforts. The program should be designed to accomplish the following three important objectives:

1. Improving service quality by focusing on *how* work gets done,

rather than *what* work gets done (i.e., methods as opposed to results)

2. Improving service without adding to staff or cost

3. Establishing TQM as a continuing process in the FD

To accomplish these objectives, several TQM methods should be used. These include but should not be limited to the following examples:

- Instructing supervisors in alternative or new methods of performing their jobs.

- Getting supervisors to listen to facility tenants effectively, ensuring that the provided services meet their needs and are efficiently delivered.

- Permitting nonsupervisory personnel to improve work performance by contributing their ideas and giving them a stake in the success of those ideas.

- Basing operational decisions on data, not guesswork.

- Grouping in sequence all the tasks directed at accomplishing one particular outcome.

- Use a TQM team as a means of pooling skills, talents, knowledge, and ideas—also to obtain differing perspectives on a task.

The approach to using TQM principles should be symbolized by the relationship between work quality, a "scientific" or disciplined approach to studying the process, and advantages gained by a team approach, which is a shared objective. Working in tandem, these elements can be very effective and sustainable. The first TQM objective cited above—focused on process, rather than results. It bears further discussion. Essentially, a process emerges by grouping in sequence all tasks directed at a particular outcome, for example, the service call process. The *service call process* entails every action from call reception through job completion and upload of job data to the CMMS service call module. In evaluating service calls processing, TQM methodology should consider factors such as the following:

- Overall lead time to perform each action in the process

- Quality of work for each task performed

- Methods of reducing interference with tenant activities

- Response time

- Tenant satisfaction with work performed

Another important characteristic of TQM is the use of teams.

Rarely does one person have enough knowledge to understand all phases of a job or process. Therefore, major gains in quality and productivity often result from the use of teams—a group of process-performing people pooling their skills, talents, and knowledge. With proper training, TQM project teams should be able to tackle complex and chronic problems and derive effective, permanent solutions.

Working as a team has another distinct advantage over solo efforts—the mutual support that arises between team members. The synergy that comes from people working together productively on an important project is usually enough to sustain the enthusiasm and support, even through a difficult time.

TQM project teams should be formed, over a period of time, to study all processes involved in FD service delivery—preventive and predictive maintenance, service calls, repairs, inventory control, warranty administration, energy management control system, etc. TQM teams should be organized by drawing personnel from the following sources:

- Corporate managerial, professional, and technical staff
- FD staff
- Tenant representatives

TQM teams should be formed, one at a time, in a manner least disruptive to organization operations and activities.

3.5.7 Tenant relations program

Facility managers should be the focal point of the tenant relations programs with the primary responsibility of managing the FD resources to accomplish good tenant relations.

The foundation of the program is frequent visits by the facility manager to each of the tenant contacts in the building(s). Tenants should be encouraged to raise concerns about the services they are receiving. Once a concern is raised, it is important to the program's credibility that the issue be resolved accurately, promptly, and to the tenant's satisfaction. The facility manager should follow up personally with the tenant to ensure satisfaction.

Each call to the service desk that requires a repair action is normally assigned to a mechanic for completion. When the call has been reported to the service desk as being completed, the administrative assistant should call the tenant that initiated the request, within 24 h, to ensure that the call has been attended to, that the problem has been corrected, and that the complainant is satisfied with the work. If the call cannot be completed in a timely manner because, for example, a part must be ordered, the tenant should be advised of this by the administrative assistant and given an estimated time of completion. FD employee training, with emphasis on tenant relations, ensures that mechanics performing work in a tenant suite are efficient and courteous and adhere to published strict standards of conduct and appearance. Craft personnel, for example, should be provided with and required to wear a fresh uniform each day and to be appropriately groomed when reporting to work.

3.5.8   Tenant visit report

Once each year, each tenant representative should be interviewed by the facility manager, who should have been trained in tenant relations and is thoroughly knowledgeable of all property management functions and building operations. Tenant representatives are encouraged to speak freely about their experience at the facility. An interview guide should be used to ensure that all topics are covered. The results of all interviews in a single building are then analyzed and compiled into a tenant visit report which is circulated among the company's top management. Problems are identified, and management assigns whatever building or corporate resources are needed to correct the problem. Tenants that expressed a concern are later recontacted by the interviewer to determine their satisfaction with the problem resolution.

Finally and most important, the tenant has a right to conduct business without distractions from the building. The building systems must perform properly, maintenance needs must be attended to promptly, and the property must be clean, safe, and comfortable. If these things are not achieved first, tenant relations programs will be a waste of time and money.

3.6   Summary

As can be seen throughout this chapter, the common thread rests with four words: *communicate, coordinate, care,* and *measure.* These words apply not only to external customers but also to internal customers (facility department workforce).

- *Communicate*   Remember: People will accept almost anything as long as they are told ahead of time.

- *Coordinate*   Coordination ensures that elements of planned action fit together as an integrated whole. It is essential for three reasons: to ensure consistent and thorough communication, to avoid conflict and duplication of effort, and to ensure all elements and circumstances are considered.

- *Care*   Customers (both internal and external) need to feel that the facility department cares about the type of service provided. This leads to building credibility, reliability, trust, and eventual dependence.

- *Measure*   Measuring what we do is basic to improving operations and maintenance services.

Maintenance Plans

William G. Suter, Jr.

Physical Plant Operations
American University
Washington, D.C.

Joel Levitt

Springfield Resources
Philadelphia, PA

4.1 Equipment, Systems, and Structure Building Tours

4.1.1 Introduction

Few facilities management activities have as much potential to influence facilities improvements as formal scheduled tours of inspection. Total building, equipment, and system tours are an essential part of an overall facilities management plan. Facility managers should think of formal tours not as a replacement for other types of maintenance and operations processes but as another process in the network of interdependent processes that make up the facilities management system. Throughout this section we refer to equipment and systems. We do this with the purpose of driving home the need for anyone involved in facilities tours to understand the systems perspective for the equipment involved. Data collection, deficiency reporting, and assessment activity for a piece of equipment is of very little value if the person performing the tour does not understand how a specific piece of equipment fits into the system of which it is a part. Every piece of mechanical equipment must be part of an interdependent system that results in the temperatures, volumes, control, and automation that support the occupant's activity. The same can be said for electrical, life safety, and structural components. They are all parts of systems that are compromised when something is done to a piece of the system absent knowledge about how all of the pieces fit in and how the system is designed to work. Well-intentioned employees working without appropriate knowledge and understanding of the system perspective create many problems.

4.1.2   Purpose

Formally planned and scheduled tours can serve a variety of purposes. Confirming the effectiveness of preventive, predictive, and corrective maintenance programs; verifying that equipment and systems are operating as intended; data collection; interaction with facility users; and formal equipment, systems, and structural deficiency assessments are examples of the various purposes these tours may have. Determining the goals of a tour well in advance of implementation of a tour system allows for various types of planning and preparation.

While there may be many purposes, there are three primary categories of tours that can assist facility managers in the development and monitoring of facility management operations and in determining facility conditions. A tour can be conducted to determine the existence of system and equipment deficiencies and confirm that equipment and systems are functioning as intended. Tours can be organized around data collection tasks. Tours can also be organized around the need to communicate regularly with customers and to experience equipment, structures, and mechanical and administrative systems from their perspective.

The challenges represented by the need to maximize the benefit of the time and expertise spent on tours requires high-quality planning, training, and documentation. Personnel looking for deficiencies and checking on equipment and system settings require a high degree of technical training in the technology they are inspecting. Inspectors conducting data collection tours or collecting data in conjunction with other tour activities need to understand how the data will be used and how the data relates to the functioning of the equipment and systems involved. Inspectors communicating with customers need to be trained in effective communication skills and need enough technical knowledge to explain equipment and systems to those who occupy the facility. For real estate the three most important things are location, location, and location. For tours of inspection the three most important things are planning, planning, and planning. Planning for successful tours must precede activity but must also be continuous to assure that the methods are constantly improving and that the maximum amount of benefit is derived now and in the future.

4.1.3 Implementation

The first step in implementing a successful tour system is to decide exactly what the tour is to accomplish. Tours of mechanical rooms require equipment inventories and training to assure that those conducting the tour know all of the equipment involved, understand the equipment, and understand the systems they are touring. Equipment inventories usually incorporate some sort of equipment tagging system so that equipment can be identified easily by tag number. Tagging systems often are organized around codes for different types of equipment so that some knowledge about the type of equipment and possibly the location can be derived from the tag number. Even the simplest daily walk-through of mechanical spaces requires a broad range of understanding of the equipment and systems being inspected. An annual inspection tour of a building envelope requires foreknowledge about the construction of the facility and the dates of additions, remodeling, roof replacements, etc. Data collection tours may require the installation of output monitors on equipment and outcome monitors in occupied space. It is very important to determine what to measure when designing a data collection tour. A significant amount of time can be spent collecting and analyzing data that cannot be developed into meaningful information. Measuring just the output of pieces of equipment does not provide information about whether this output measure is having the desired result in the space the equipment serves. Inventory and assessment tours require that inspectors know how to evaluate and identify equipment. If equipment will have to be taken out of service, downtime will have to be arranged with those served by the system.

Facility managers with computerized maintenance management systems will gain the most if they organize the instructions and scheduling of their tours within the same system used to organize other types of facilities work. These systems can be set up to provide specific instructions including tools and equipment lists to assure that inspectors have what they need when they get onsite. Nothing is more expensive than multiple mobilizations to perform one job.

Those without a computerized maintenance management system will have to prepare work assignment sheets consistent with whatever system they are using for the organization of other work. It is important to integrate inspection tours into the routine of the facility management organization. The tours have to be important to management for them to be important to others. If the work is not well organized and not integrated into systems used for other "important" work, the message is loud and clear.

Any task assignment sheets need to be organized so that information about the tasks to be performed is clearly laid out and logical. A building tour involving several mechanical rooms should be organized so that inspectors can work down a single sheet or through several sheets as they go through the tour and not require flipping around a page or back and forth between pages. The sheet should also include enough space for inspectors to record their findings. Boxes provided for the recording of gauge readings should be large enough to accept likely readings. Whenever possible, instructions should help the inspector identify readings or conditions that are abnormal and how to respond to abnormal conditions. Instructions may include a range of readings for a particular gauge that is considered normal. When an abnormal condition is experienced the inspector must know how to respond. Equipment may need to be deenergized and locked out and emergency calls may need to be made. Equipping inspectors with the right emergency information is extremely important. An example of a tour work assignment sheet is shown in Figs. 4.1 and 4.2.

Sheets such as the one in Figs. 4.1 and 4.2 can be completed well in advance of the tour. The variations on this generic form are extensive. Recording actual time spent may be preferable to time-in/time-out, or there may be a different way to capture measurement data. If the scale of the measurement is known (time, distance, temperature, volume) you may want to set up the data collection section on the assignment sheet to allow the recording of only the variable information.

Tours designed as purely assessment activity require significant research into the equipment and systems under review. For these tours, installation date and maintenance history are critical to successful assessments. Methods of expressing condition assessments also need to be determined prior to conducting this type of tour. Usually the purpose of assessment tours is to determine what work is required, including replacement, and when work is to be done. In the normal world of facility management budgets, there is usually a need to establish replacement and major repair priorities, and valuable information for this process is available during an assessment tour. Determining that a piece of equipment needs to be replaced is usually not enough. Even determining that it needs to be replaced within the next 6 months is not enough. A useful way to express more information about the condition of a piece of equipment is to express it as "percent of useful life left." This piece of information may assist in the difficult job of comparing and establishing priorities among many dissimilar pieces of equipment and/or building components. Inspectors

Daily Mechanical Room	
Inspection Tour	
Building:	
Room #s	
Equipment Inventory	1
2	3
4	5
Safety Precautions:	
Readings: *Normal range*	1
2	3
Measurements: Normal range	1
2	3
Logs:	1
2	3
Comments:	

Figure 4.1 Daily mechanical room inspection checklist, page 1.

making judgments like these need to be supported with information and training. If you are gathering this type of information through the work of a number of inspectors, the reliability of the assessments is a function of the quality of the inspector training. This is surely one type of training that should not be left to the last person to go

Daily Mechanical Room	
Inspection Tour Page 2	
Deficiencies discovered:	
Time in:	Time out:

Figure 4.2   Daily mechanical room inspection checklist, page 2.

through it but will benefit from a professional needs analysis and presentation. The validity of these assessments can also be called into question if the judgment is based solely on the experience or expertise of the inspector(s). Accuracy in these judgments is greatly facilitated through the use of manufacturers' maintenance and operations manuals and guidelines about useful life. If you know that a piece of equipment has been in place for 5 years, has benefited from a good PM program, has had few maintenance problems, and is expected by the manufacturer to last 10 years, then you can reasonably expect the equipment to last another 5 years or have 50 percent of its useful life left. It needs also to be mentioned that effective PM can often extend a piece of equipment's useful life well beyond that predicted by the manufacturer. Equipment that is monitored on a regular basis can provide data that may indicate that the frequency of preventive maintenance and/or data collection tours should be altered. This is especially true if equipment is being kept beyond its recommended life.

An often overlooked step in the implementation of an inspection tour is to ensure that the inspector has access to all the areas to be toured. It is necessary to equip inspectors with whatever keys, codes, and/or access cards they will need to access the areas required.

It is also necessary to establish logical tour routes so that the inspector can move logically from one area to another without needing to double back and travel past facilities that will be toured later. When tours need to be conducted in adjacent facilities, they should be scheduled at the same time. Establishing tour circuits is often a good strategy so that whole areas of a facility can be toured at the same time. This factor is especially important with complex facilities like a university campus or municipality where the distances between facilities can be significant.

The topic of inspector training has been mentioned in several contexts, and here it is mentioned because it is important to view it not only as an implementation step but also as part of a continuous improvement strategy to assure that inspectors always know what they need to know. Training should be a systematic process that is well integrated into the overall facilities management plan. Inspector training for facility tours involves the same steps as training in other arenas and briefly is composed of the following steps:

Job analysis

Establish all required competencies

Needs assessment

What competencies do trainees have and what do they need training for?

Basic training competencies

Given the variety in trainee competencies, what topics will be covered in training?

Scheduling of training

When and where, type of facilities, length?

Conducting training

Logistics, trainer readiness

Evaluation of training effectiveness

Evaluation of individual trainee

Evaluation of trainee groups

Evaluation over time/trainers/schedule/etc.

4.1.4   Scheduling

While the scheduling of inspection tours may seem a simple process, when done poorly it can have a dramatic impact on results. There must be different scheduling strategies for the different types of tours. Daily walk-throughs of mechanical areas are best done at the beginning of the day but benefit greatly by following a previous shift debriefing session. A tour to determine if HVAC equipment is meeting the needs of the building occupants is best made during the time when occupants are using the facility. Tours that require HVAC equipment to be taken off-line are best done when the occupants are not using the facility. Effective tour scheduling is dependent on a clear understanding of what the tour is intended to accomplish. Tours that involve the use of testing equipment or photography must be scheduled when this equipment can be used to best advantage. A thermal screening of a building exterior may be done best at night so that thermal readings of a facility are not affected by ambient heat energy. A thermal screening of an electrical panel may be done best during the day or at the time when the usage of electricity through the subject panel is high. Scheduling must be carefully thought through and viewed as an integral part of the planning and operational process.

4.1.5   Equipment and system logs

Collecting data on the output and outcomes achieved by equipment and systems, and data on the resources consumed in the collection process is the first step in documenting not only the performance of systems but the operation of the entire inspection tour program. Data collected and stuffed in a drawer or entered into a computer and forgotten is of no value and hardly worth the time it took to collect. Equipment and system logs provide the tools for technical, administrative, and management personnel to evaluate equipment, systems, and programs. Logs that lend themselves to statistical analysis are extremely valuable. Statistical analysis, such as statistical process control, can be used to establish ranges of normal that can guide preventive maintenance frequencies and analyses of operating parameter trends that could indicate problems before they affect a facility's occupants. Often, techniques such as these can identify problems whose detection prior to the implementation of a system of inspection tours was available only after a failure of the system causing a service interruption to facility occupants. Noncomputerized logs, while less valuable than computerized logs, can play an important role because some of the information that can be derived from log data can be seen at a glance once all of the data is logged in one place. Log sheets kept in equipment rooms are often used as quick references to trends in operating parameters over time. Noncomputerized equipment and system logs as permanent records must be maintained in locations were they can be preserved for the life of the equipment or system. Decisions that need to be made about replacement of equipment and systems are often well informed by a review of the logs kept on the equipment or system being replaced.

4.1.6 Technology

The advent of energy management systems and computerized maintenance management systems has resulted in a focus on the value of formal inspection tours. While one can take a "virtual" tour of a facility through an energy management system's screens, it does not replace the trained eye, ears, and touch of a trained and experienced inspector. An energy management system is also a system that requires preventive maintenance and monitoring to keep it operating at its best. Energy management systems have become integrated into routine facility management activity, and inspection tours can add value to the technological system. Many computerized maintenance management systems on the market have the capability to organize all manner of data collection through the use of bar codes. This technology has had a dramatic impact on the quality and quantity of data that can be collected and the ease with which this data can be integrated in decision support systems. Most of us experience bar code technology each time we go to the supermarket. This point-of-sale data gathering technology allows many retail businesses to keep daily track of inventory and analyze trends in product sales all while decreasing the amount of time it takes to go through the checkout and improving the accuracy of the data being entered. Inspection tours can be made much more productive through the use of bar code technology. Imagine the value of having every piece of equipment tagged with a bar code so that identifying the equipment is accomplished by a mere "swipe" of a pen-sized device over the bar code sticker attached to the equipment. Any routine activity, scaled gauge readings, times, and temperatures can have a bar code identity to be swiped as needed to collect the data. Upon the completion of a tour the data can be downloaded into a computer system and the analysis can begin automatically. Data collected at night can be ready for analysis first thing in the morning of the next day. Another type of technology that can add value to an inspection tour system is scanning equipment. Scanners can be set up to scan field data collection forms and inspection reports directly into a computer system. Careful training must be done to help inspectors understand what the scanner can and cannot read, but this can be powerful technology when applied well.

4.1.7   Staffing

Two fundamental approaches to staffing are possible when planning a system of inspection tours. Determine the overall need of your facilities, design a system to meet those needs, and then determine where the inspectors will be obtained. Determine what resources you have available and design a system to take advantage of what is available. If you can tolerate the emotional pressure of a system that generates work you know you cannot complete, the former approach to staffing generates a lot of data in support of requests for additional human resources. If you have to use only what you have, you must design a system that focuses on the most important equipment and systems. You are the only one who knows what is the most important for your facility, but cost of equipment, cost of downtime, life safety, and criticality to mission are a few items that have to be given serious consideration.

4.1.8   Documentation

No matter what type of inspection tour system you have, without documentation of the results you have gained less than you could. Systems of documentation range from filing cabinets full of paper to computerized databases of electronically stored data. Any type of documentation system must result in data that is accessible, secure, and organized in such a way that the logic and organization of the inspection system is reflected in the filing system.

4.1.9   Summary

There is little doubt that our facilities are going to become more and more difficult to maintain and operate. The amount of technology being built into modern facilities and the technology that promises to make us more productive is increasing at a rate that threatens our ability to adapt. Formal, scheduled, documented tours of inspection ranging from daily walk-throughs of mechanical spaces to biannual facility condition assessments can be an extremely valuable element in a comprehensive facility management plan. Those who are designing such plans, those who have succeeded, and those who have not done as well all have lessons for those yet to begin.

4.2   Engineered Performance Standards (EPS) Usage Plan

The development and use of labor time standards (EPS) in facilities has been an ongoing activity for over 30 years. The driving force was the need to contain, control, and cut costs of facility operations and maintenance. Most of the pioneering work was funded by the federal government.

Any study of EPS would be incomplete without a discussion of quality. Most labor time standards do not measure quality. You must independently measure quality in addition to work performance times. Without a quality measure your performance times might increase at the expense of job quality, which is a bad trade!

All EPS are based on knowledge of exactly what work is to be accomplished. The accuracy of this knowledge will dictate the accuracy of the EPS, regardless of the system you choose. Thus, one of the suppositions that underlie all of the standard techniques is that the job to be done has been defined, usually by a personal visit from a skilled supervisor, planner-estimator, or the tradesperson who will perform the job.

The job visit will generate a scope of work, specifications, and possibly a sketch. The scope of work or sketch has to be expressed in the terms and units of measure of the time standards system you plan to use. The planner-estimator then looks up the appropriate time standards based on the information shown in the scope of work and sketch. A single job might require several time standards to be added together to represent the total standard allowed time.

Four basic types of standards can be used together or separately in a facility department: direct observation, published standards, historical standards, and skilled estimate. We discuss the use of all four types of standards together with their advantages and disadvantages.

When seeking EPS, look first to the manufacturer of the equipment or materials for information. Other places to look include experienced tradespeople, your own written work order history (if available on work orders and CMMS), contractors involved in that type of work, and consultants. Published building maintenance EPS are also available and are described in detail in the following.

In this section two definitions will be useful:

Labor standard time   That is the labor time standard (EPS) itself. If a job has a time standard of 3.0 hours we say that the allowed standard time is 3 hours.

Labor standard hour   This is the amount of work that a mechanic or custodian can perform in 1 hour. If we allow someone 8 hours of work (by adding up the EPS times), a productivity report credits the worker with 8 hours for those jobs, regardless of how long or short the jobs actually were.

4.2.1   Basis for all types of labor time standards (EPS)

1. Similar types of equipment (like equipment in like service) should be grouped together for estimating purposes. Subtle differences can result in major changes in the EPS. A compressor replacement on one air conditioner make and model might take a very different amount of time than the same job on a different unit, even if both come from the same manufacturer, due to equipment location, age, level of maintenance, etc.

2. Concise description of repair or PM tasks to be performed (for example, simple Monroe toilet replacement with no complications).

3. Description of what work is to be accomplished (an adjustment to the toilet to stop a leak, a replacement of the valve and float, an inspection; in short the phrase "fix toilet" is not adequate for accurate labor time standards assignment).

4. Easy access to the room or area to work on the asset.

5. Number of people required (crew size).

6. Location of the item repaired or maintained (travel time from shop, distance to storeroom, etc.).

7. Complications and extra work to be recorded by the mechanic. A mechanic who finds other problems with replacing the toilet (such as having to replace a cracked flange on the ideal bend) might choose to complete the whole repair and write up the additional work. Written-up work is usually added on a standard labor hour per work hour basis.

8. EPS are usually based on a skilled and professional tradesperson,

with the right tools, materials, and equipment, in the right place, at the right time.

9. Factors like age, gender, and years of service of the mechanic, time of day, and parts cost are *not* considered in labor standards.

Reasonable returns should be expected from investments in EPS. A contractor who couldn't accurately estimate job duration wouldn't stay in business. On the other hand, organizations used to major planned estimating efforts cannot survive where the estimating is more informal and not planned. A facility/department should balance its investment with the potential returns available. Certainly non-repetitive jobs of short or moderate duration do not require elaborate estimates unless the timing of the return of the item to service is critical. Repetitive jobs such as PM tasks, custodial tasks, etc., respond well to advanced planning and estimating.

4.2.2 Computerization

Many facilities have computerized their work order system, inventory control, and PM task list generation. Most of the over 100 maintenance packages (called CMMS, computerized maintenance management systems) support some level of EPS usage.

The minimum level of support is a field on the work order called estimate. When the work order is opened, the dispatcher is expected to enter the description of the repair. They (or a planner-estimator and/or supervisor) are expected to manually add their best time estimate. The estimate could come from any of the four systems discussed in detail in Sec. 4.2. The field can then be used on reports. Most systems allow the user to create specialized reports with any field captured by data entry or "known" to the system (such as date, etc.). With these level systems there is no attempt to manage the EPS themselves or automate the assignment of EPS to jobs.

Some CMMSs support standards to a slightly greater extent. In addition to a field on the work order there is the ability to store repetitive jobs under a code or a mnemonic name. The code for an annual PM on an air conditioner might be X12345 and a mnemonic might be PMA-A/C-01. The system allows you to store the complete task list (by craft if necessary), parts list, tool list, and job time estimate. Anytime that job comes up again, the allowed time estimate comes with it. These named jobs do not have to be PM jobs only. Any job can be preplanned, entered in the database, and have a code assigned, and can be labor hour time estimated in this manner.

The third level of support to look for is the recording of performance against the job descriptions or job codes. Some systems have the ability to track and calculate historical standards and keep these numbers on file for reporting, job analysis, and employee comparisons. Some will take all of the actual work times that a particular job used for completion and develop a statistically valid mean time to repair.

Higher-level systems help manage the labor time standards as an independent database with time standards assignments, modifications, histories, and performance tracking. Several systems keep two different types of standards for each job. They are internal and external standards. The internal standards are historical and are updated by the system every time a repair is completed. The second is external and doesn't change with the jobs. This allows the supervisor, or manager, to compare the performance goals against the outside standard as well as the performance that can be expected based on historical data for actual job work times.

4.2.3 Why have labor time standards (EPS)?

There are six major reasons to develop, use, and track labor time standards (EPS). They include work order scheduling, long-term planning of repetitive tasks, evaluation of individual employees, budgeting, morale improvement, and improved customer service.

4.2.3.1 Work order scheduling.

The first reason to use EPS is to develop a work order scheduling system. Scheduling repairs and PM activity requires estimates of allowed labor time to be effective. The most effective ways to plan and schedule maintenance work require an estimate of the amount of time an individual job is going to take.

Facility managers can schedule their maintenance personnel or thus can let breakdowns schedule them. One way or another, maintenance personnel will be sent to jobs.

Work order scheduling is the execution step in the planning process. It looks at today's demand for service, calculates the amount of hours needed from the EPS, and matches that to today's available resources. The process of work order scheduling is an evaluation and a prioritization of the work backlog, pending jobs and emergent urgent requests. Managing that mass of (frequently) conflicting demands is where the work scheduling pays dividends. EPS are the lion trainer's whip to help get the conflicting demands into a timeline where efficient execution (the alternative can be mass bedlam) can take place.

Facility maintenance work scheduling means above all having the least possible disruption to the user-customer. Also the manager can decide to schedule equipment service when the equipment is not in demand. Work order scheduling in buildings, such as hospitals, is critical in that servicing certain systems has to wait until it is convenient for the patient (who might be in the ICU or might be having an operation) and staff.

The advantages of EPS in active work order scheduling are in the areas of efficiency (you can optimize travel time, storeroom trips, special tool usage, etc.), morale (assign constant level of work, smooth out the feasts and famines discussed in detail in the morale section), and customer service (discussed in the customer service section).

When the start time is known, the other elements of the job can be coordinated. Maximum efficiency requires bringing together in precise timing the seven elements of a successful maintenance job: the mechanical skills; the transportation; the tools; the materials, parts, and supplies; the drawings and information; the availability of the unit to be serviced; and the authorizations, permits, and statutory permissions.

Efficiency can also be improved by scheduling several jobs in one end of the facility together. By knowing how long jobs take, fill-in jobs can be added to a large job to fill out the day at that location. Another efficiency can be gained using the same technique of tagging jobs together by scheduling several jobs that use the same materials together. This will optimize trips to the storeroom.

Some jobs go bad. Hidden problems, inadequate skills, defective materials, or inadequate tools might make a job take twice or more longer than it should. EPS can help. If a large repair is falling behind schedule (the EPS shows that you should be done by now, and aren't), interventions can be instigated (such as bringing in the company expert, adding overtime, bigger and better tools, adding people, specialized contractor, etc.) to minimize losses. Interventions when the job is running can correct the problem and prevent a complete disaster. Waiting until a monthly meeting to find out that a job went bad is too late, and no action can be taken.

People work at a pace that is partially based on the amount of work they are given. Work is quantified in the work order schedule based on EPS hours. A reasonable amount of work is given to each person and is expected to be completed each day. This reasonable amount of work, with backup jobs, might be 9 to 10 hours. Schedules based on EPS hours free workers from a hurry-up atmosphere one day and a kill-time atmosphere the next. Backup jobs are given so employees can move to another job if they are impeded by a lack of parts or tools. This constant flow model helps improve output and quality.

4.2.3.2 Long-term planning of repetitive work. Preventive maintenance (PM) task list work, routine jobs (like policing the parking area of a sports stadium for broken glass after game nights), and scheduled equipment replacement are repetitive jobs that have to be completed on an ongoing basis. If these proactive PM-type jobs start to exceed a high percentage of your available hours, the natural tendency is to let them slip in favor of urgent jobs.

Here is an example of using EPS to smooth labor requirements (which will increase the probability that the system is used as designed and that the PMs get done in the month they are scheduled): Your organization searches the marketplace for a computerized PM and work order system. After a lengthy process you and management decide on a likely system. You start gathering nameplate data in June and are completed in September. From September to January you complete training and run the system parallel with your current manual system. On January 1 your facility department goes live on the new system.

An early step is to lay out PM task lists (both tasks and frequencies). You know that some of the equipment and system should be cleaned, lubricated, and looked at weekly or monthly. Other pieces should be inspected quarterly. Still other pieces need more extensive work yearly or biannually. Some items will have multiple task lists at all those frequencies.

The clock starts ticking on all PM lists on January 1 so that 30 days later, on February 1, all the monthly PMs are printed. Ninety days later, on April 1, all the quarterly and all the monthlies come out. Then beginning in January (when half the crew is off) the next year all the annuals, quarterlies, and monthly task lists print out. There is way too much work for the available hours. Short of calling a contractor you're confronted with missing PMs or excessive overtime!

Laying out the whole year's PM and routine work using EPS is a way out of this startup dilemma. The process is to set up a 52-week PM calendar. Go through all the task lists and assign standards to each one. Lay out the task lists by equipment ID numbers with the standard hours for each task list on a 52-week calendar. Stagger the start times so that different equipment has PM task lists come due in different weeks. Exclude holiday weeks. The goal is that the amount of PM hours in any week is no greater than 15 to 25 percent of the expected labor hours. Try to schedule late December and early July even more lightly to account for holidays and vacation.

All routine, and repetitive tasks, can be included on this work order schedule to ensure that there is adequate time for call-in work without excessive overtime or the need for contractors. This process is aided by many CMMS programs. In many programs you could enter the EPS times, tasks, and frequencies and the system will print out a labor requirements schedule for the next 52 weeks. You can modify the start dates and see the labor requirements change. These programs greatly simplify the process of leveling the annual PM schedule.

Yearly planning using EPS can reduce the need for overtime and contractors. It will also ensure that there is enough time for your PMs to get completed within the month they are scheduled.

4.2.3.3 Employee evaluation.

Another reason to have EPS is to be able to evaluate the proficiency and productivity of individual mechanics. Comparisons of an individual's work performance to EPS, over time, will uncover and, perhaps more importantly, prove who needs training, reassignment, or to look for another job.

The first measure is simply tracking your mechanic's work performance against EPS. One performance measure compares the work hours reported on the work orders against the EPS hours for the same repair. This measure is focused on the mechanic's ability to complete maintenance tasks. It does not measure the overall efficiency or the job quality (which would have to be looked at separately).

If your EPS calls for a faucet replacement of a certain type, in an hour, and the mechanic routinely takes 2 hours, an investigation may be necessary. The investigation would look into the mechanic's technique, ability to plan (a second trip to the storeroom would negate any attempt to achieve the EPS), aggressiveness to get the job started and completed, fatigue level, the tools used, and other factors. Problems with the overall maintenance management system tend not to show up when jobs are looked at one at a time.

A second measure adds up all of the EPS hours accumulated in a week or month and compares that number to payroll hours for the same period. Any of the EPS hour systems could be used. This measure is more a measure of the whole maintenance management delivery system because lost time throughout the day is included in the comparison. Your department's ability, or inability, to assign jobs, communicate critical information, get parts to the workers, and get rooms unlocked is really being measured with this type of metric.

The only EPS system that cannot be used for this measurement is historical standards. As explained below, historical standards include all the foibles, problems, and inconsistencies unique to your operation.

Example of employee evaluation on a job-by-job basis and on a monthly basis: Tom Duvane wanted a method to measure the productivity of his carpenters. Tom did all the work order estimates (he had spent 15 years with the tools himself). He wanted an outsider's opinion to ensure impartiality. He called around, and one of his friends got him a copy of job time estimates from the U.S. Navy's Engineered Performance Standard (EPS) series. His friend worked with the U.S. NAVFAC (the publisher of EPS) and knew a great deal about productivity issues. He warned Tom that 80 percent of the losses of time were due to problems with maintenance management systems and not with the employee.

At the time Tom ignored that advice. He chose to evaluate Joe Dillion, a responsible employee viewed as slow and sometimes difficult. Joe was from another country, didn't like the paperwork at all, and constantly complained about it. Tom figured if the EPS worked with Joe they would be a breeze with the others. Tom was not looking at his own operation, but focused his attention on the mechanic.

During the month of October, Joe Dillion had 84 hours and 45 minutes regular time standard hours* and was paid for 176 regular time hours. He completed 61 chargeable jobs. He logged almost 127 work order hours total. Twenty of the remaining hours were spent on non-standardized activities such as cleaning up and fixing the grounds department lawn mower, sweeping the shop, etc. If they are added in on an hour-per-hour basis to the standard time, he completed 104 hours and 45 minutes of standard timed work that month.

Engineered performance standards hours completed*
= 79.3% (std. hr/work order hr)   or   59.5% (std. hr/payroll hr)

Tom noticed two important items from the study. Based on his years of experience he couldn't see why anyone couldn't achieve EPS standards. They seemed very reasonable. He also was concerned about the overall amount of work being done. He never considered that his management style might have an impact on the amount of work completed.

As he observed Joe Dillion over the next few days, he saw that Joe worked at a pace which would easily make the EPS numbers on a pure work basis. Tom got very curious and looked deeper. On a closer look he saw that Joe spent an excessive amount of time completing the work order paperwork. The extra time to complete the paperwork negated his ability to make the EPS allowed time, particularly on shorter jobs. Since Joe did 61 chargeable jobs, his average job duration was less then an hour and a half.

Tom discovered that while Joe could speak English very well he could barely read and write. Remedial classes were offered, and accepted. In a few months Joe could sail through the work orders and improved his production against the standards by 20 percent. Problem number one was solved. EPS showed how a good employee was being held back by a solvable problem.

---

*To determine the standard hours, add the individual standards for each of the 61 jobs Joe Dillion completed plus the nonstandardized time on an hour-per-hour basis.

The second problem was stickier for Tom. Even if everyone on the crew attained 100 percent on every individual job, at the end of the month Joe would only be doing 72 percent of what he could be getting done. Tom realized that his style of management was the big culprit. Because he considered himself the only really responsible person in the department, he didn't delegate. Each person had to get jobs one at a time from Tom himself after an interrogation about the job that was completed. Tom also kept the keys to the tool crib, storeroom, and janitor's closet. Every time anyone needed anything they had to find Tom, and Tom was in constant motion. He realized big changes were needed to improve productivity.

4.2.3.4 Budgeting. EPS can positively impact the budget process. Accuracy and timeliness are the hallmarks of a budget system that supports the long-term goals of the organization. EPS allow the facility manager to answer the question, how much will it cost to maintain this new building? It can also help with the question on a special project, how much will it cost to set up a new laboratory for research into this new product line?

4.2.3.5 Morale builder. One of the continuing issues of maintenance work is that its workers do not get accurate feedback about the kind of job they are doing. The feedback they do get is more related to the customer's mood than to the speed or quality of the job. EPS will help by providing the mechanic with the expectations of management. When mechanics can keep up to the EPS, they know they are doing a good job by management's definition. This specific feedback improves morale.

4.2.3.6 Customer service. The last major reason for using EPS is to be able to accurately predict when an item will be returned to service for its users. Accurate estimates improve your customer service and increase the value of the service you perform for your users.

Good customer treatment requires the work control center or the supervisor to be able to answer the question, "When will the maintenance worker get here?" and the related question, "When will my repair be completed?" A work order schedule will lay out the EPS times for each mechanic and can estimate when he or she will be available to work on the customer's job.

Performing maintenance work, no matter how well, does not usually add value to your organization's product or service. Excellent customer service is the holy grail of all service organizations. Great industrial organizations such as IBM, AT&T, Honeywell, Trane, and Xerox built their empires on excellent customer service. In all cases the companies spent significant money predicting when the technician would show up and how long the repair would take.

The facility department staff should utilize the principles and techniques shown in the following, which provide excellent customer service support:

1. Attention is on the work outcomes for the customer. The biggest outcome after avoiding failure in the first place is when will you get here and when will the asset be fixed?

2. Thought has gone into the customer's needs, problems, pressures, and history. Truly uncomfortable (too hot or too cold) customers in the midst of a deadline have tremendous pressure on them. We can help them by showing up when we promised.

3. Mechanics are the key players, since they are the ones closest to the customer. Attending to the mechanic's needs means protecting them as much as possible from the shifting priorities of the organization and allowing them reasonable time to complete the jobs. This improves their morale and promotes pride in a job well done (a powerful motivator).

4. Customers are treated as individuals. Both the important ones and the less important ones are given consideration and time for their jobs.

5. Non-value-added coordination is kept to a minimum. EPS systems require a time investment. Be sure the investment is appropriate to the size and scope of your operation.

6. Information systems provide information that will help service the customer better. Constant review of the CMMS database will show you areas where your published EPS vary from the actual work performance times.

7. EPS times, and earned hour reports, are shared and understood by mechanics.

8. EPS-tabulated benchmarks reflect the service values of the individual customer. Two benchmarks you could track are MTBC and V (mean time between call and visit) and MTTR (mean time to repair) with variance to EPS.

4.2.4 Types of EPS

We have been through a discussion about why to use EPS. The second phase of the discussion is what systems of standards are available to the facility manager. There are four ways of developing standards.

These different ways can be used together to create a unique set of EPS for your facility. All of the techniques are related.

Estimate. Estimates are educated guesses. People in the maintenance field for any length of time get experience in judging how long jobs will take. The advantage of estimates is that judgment and experience require the least overhead and additional time. The disadvantage is that different people estimate differently. Some people are better at judgmental estimates than others. It does not work well if it is the first time you have ever seen that kind of job. It might also not be considered fair by the workers.

Historical standard. All of the elapsed times that it takes your personnel to perform individual jobs are recorded and averaged. When sorted by item identification and work accomplished, the accumulated workerhour result is the historical standard for that job. Usually the work order system is the date source. The advantage is that these averages are real numbers with your people, your tooling, your physical layout, etc. The main problems with historical standards are the same, your people, your tooling, your layout, etc. All of the lost time, built inefficiencies, and irrational conventions developed in your facilities are included in the averages.

Direct observation. This comes in two approaches called methods engineering (ME) and gross observation called reasonable expectancies (RE). Described later, these two strategies share the burden of very high expense to set up and continued expense to keep them up-to-date. They are the most accurate methods of assigning labor hour time standards because they reflect your people, layout, and tools. Unlike historical standards the observer can eliminate lost time, frequent breaks, and excessive trips to the storeroom.

Published standards. Published standards are usually derived from direct observations and published by both public organizations (such as U.S. NAVFAC for the U.S. DOD) and private companies for profit (such as R. S. Means). Published standards have the advantage of being inexpensive, particularly as a starting point. The disadvantage is that they have differing assumptions in skill level, tooling, and layout of the facility.

4.2.5 Estimate

By far the most common type of EPS is estimates made by supervisors, craftspeople, and planner-estimators. Anyone involved in maintenance has been asked how long it will take until this bathroom is fixed or this building is cool. The answer is an estimate.

4.2.5.1 Building an estimate. Review the job and be sure you include all the work that has to be done. Some additional investigation at this phase usually pays off. The single biggest problem in estimating is not pinning down the complete scope of work. The second biggest issue is dealing with what happens when the scope of work changes by the job as it is in process. This happens regularly in older buildings where a simple one-day job uncovers a mass of other problems and stretches the job to a week or more.

The scope of work also has a tendency to get larger on its own (called scope creep). The customer is the frequent source of scope creep (if we're replacing this, why don't we replace that at the same time?).

In most construction jobs there is a set of drawings to do a take off from. In maintenance there are sometimes a set of drawings, but not often. There are rarely drawings for smaller jobs. In any case reduce the drawings or your notes to a quantity of tasks with definable beginnings and endings. Also make a list of materials. Eighty percent of facility jobs do not require detailed analysis because they are one-step (or a few step) jobs.

The seven elements of planning come into play in the estimating process. A defect in any of the seven steps will cause the job to go over estimate. They include determining skill sets required and their sequence and listing the tools required with an emphasis on special tools (and where to get them); also, ensuring that transportation is available, if required.

The next step is to determine the materials, parts, and supplies needed and which ones will have to be special-ordered. Some jobs require specialized information such as old drawings (to locate utility runs), wiring diagrams, and programming links.

The last two areas have caused many jobs to run over estimate. The fifth area to be examined is access. Working out the issue of access to the room or area to be worked on, obtaining keys, or relocating workers is essential to starting the job. The last issue is permits and permissions to proceed. While these areas don't seem to be part of the planning and estimating process, they are essential for estimating time.

4.2.5.2 Slotting method of estimating. Slotting compares the job to be evaluated to a group of well-known and studied jobs. The supervisor determines where the job fits in (which slot). Is it bigger then slot 2 but smaller then slot 3? The time estimate is between the two (Fig. 4.3).

Slot	Job description	Duration
Slot 1	Replace electrical outlet or switch	15 minutes
Slot 2	Remove and replace toilet	60 minutes total crew of two
Slot 3	Install door with trim	240 minutes total crew of two
Slot 4	Remove and replace single family house hot air furnace where there is very little modification needed	600 minutes total crew of two

Figure 4.3 Chart showing the use of slotting for determining labor standards.

Usually it is easier to determine whether a job is bigger or smaller than another than how long it will take.

Your slotting chart could be broken up by craft and have 7 to 10 entries. The standard jobs could be picked by the tradespeople after a short discussion.

Sources of errors in estimation (partially adapted from Means Facilities Maintenance Standards):

Errors in thinking how fast you could do something when you were the fastest of the crew

Errors in defining the job to be done

Errors in limiting the scope of the work

Unanticipated problems with the job (frequently related to one of the seven planning items)

Errors in math

Errors in copying information from one sheet to another

Errors in estimating the costs and time of getting the material to the job

Errors in judgment about costs in relocating existing processes or workers

Errors in including outside factors like weather, traffic, and inability to get access

4.2.6 Historical work standards

Just about all computer systems track the time it takes to do every job. Some systems can evaluate all of the repairs of a particular type on a particular type of equipment. These standards can be useful, since (unlike the EPS rates) they factor in important variables such as the actual condition of your equipment, the skill level of your workforce, the layout of your shop, and your tools and equipment.

The disadvantage of historical standards is the accuracy of the data collection and of the definition of the tasks completed. For example, the work order might read "Fix pump." The mechanic might have to check valve positions, motor circuits, and source product before accepting that there is even a problem with the pump itself. All of these activities are lumped together in the historical standard.

Most organizations do not insist on accurate descriptions of work accomplished. Without detailed descriptions historical standards will be averages of very different jobs. If all toilet jobs were lumped together, the range of jobs might be from minor adjustment to removal and replacement with a rebuild to the floor!

Schedules based on historical standards include a full amount of lost and wasted time. When an employee is talking, eating, or not at his or her job for any reason, the time is still charged to the repair and will find its way into the historical standard. Published standards provide time for normal job delays, but not time for such items as eating lunch, abnormal restroom usage, non-job-related talk with colleagues, etc.

4.2.7 Direct observed standards

Direct observation has two different approaches. The micro approach is called methods engineering (ME). ME has two roots called time study (industrial engineers with stopwatches) and motion study. Time study breaks the task into units and precisely times each element.

Motion study studied motion in a laboratory and divided it into its elements, which are still used as a basis of methods engineering. These elemental motions are called therbligs.

The scientific methods spread widely in World War I, but there were not an adequate number of skilled practitioners. Mistakes were made and tremendous distrust was rampant. Methods engineering became popular once again in the 1940s after its acrimonious launch. By now there were adequate numbers of qualified practitioners. It was widely applied in war production in World War II and resulted in significant productivity gains.

The problem of ME is time. It might take a week or more to fully analyze a single repair. It is a very time-consuming process that results in accurate estimates when it is applied properly. The time study person is an industrial engineer. Their major tools are the stopwatch and clipboards. They break down the job into simple motions and time each section. They are also skilled in determining the best way to perform certain activities.

The second method of direct observation is a macro approach. It was developed as a reaction against the methods engineering orthodoxy. It is based in part on the theory that most of the benefit can come from getting to 80 or 90 percent accuracy. The last 10 to 20 percent of accuracy was the most expensive. In the second method, a person would observe the whole job at once and record the time while noting the nonproductive time. Stopwatches are not used. The standard developed is called a reasonable expectancy. The reasonable expectancy (RE) is a work standard based on several macro observations. The RE is a reasonable, observable amount of work performed in a given period.

The concept of reasonableness is very important. By observing several craftspeople performing the repair you have a good idea of how long it should reasonably take. No speed-up is needed to improve productivity. REs will improve productivity by *recapturing time lost in nonproductive and marginally productive activities.*

It is far more informal (and less accurate) than methods engineering study and easier to develop and maintain. With REs we are looking for the conditions of having employees at work. We aren't necessarily concerned with the individual's technique or skill. Time study examines the individual worker's technique to evaluate its effectiveness. Hundreds of times more information is required to properly time study a repair compared to observation and assignment of an RE for the same repair job.

Because of the inherent inclusion of lost time in the historical standard and not taking into account your buildings, environmental conditions, and skill levels in the published standards, either of the ways of direct observation is the most accurate method of developing EPS. They also by far are the most expensive.

REs can be developed by anyone. It does not take extensive training to directly observe a repair with enough accuracy to set a standard.

4.2.7.1 Preparation for establishing REs

1. Determine what activities you want to observe. Start with repetitive repairs and PMs. Review your repair history and list the most frequent repairs.

2. Be sure to capture the whole job.

3. Determine if the job is performed by a single person or crew. If a crew is required, ask yourself the question: Will you be able to observe all of the members at the same time? If not, add an observer and compile your observations afterward.

4. Discuss the study with all affected employees. Sell the concepts of improved customer satisfaction (for your own user group), improved quality of life for shop people, and easier recognition of

excellent performance. Include a discussion that it will be easier to isolate bad actors who make more work for everyone.

4.2.7.2 Instructions for observations

1. Position yourself so you can see the entire repair.
2. Use discretion in recording observations. No stopwatches.
3. Be on the scene before the activity starts and stay until after it is completed.
4. Do not carry on a conversation with employees. Answer any questions very briefly.
5. When recoverable lost time is observed, document in detail what occurred. If you, for example, observe a number talking, note with whom and the content (job-related?).
6. Make an entry with a start and finish time every time a function changes. Function changes include travel, setting up, acquiring materials, parts, and tools, actually working, and both recoverable and nonrecoverable lost time.

To assist in making observations, observe the following tips:

1. Try to limit observation periods to 20 to 30 minutes. Divide large jobs if convenient or assign another observer.
2. Three or four observations for each repair is usually enough if there is not a great unexplainable variance in the times.
3. Try to observe different people doing the same job.
4. Record and document all data relevant to the observation. Decide and record if the circumstances of the observation were "normal."
5. Make new observations for substantial changes in tools, shop layout, or procedure.
6. Don't act immediately on lost time. Gather information to ensure you have all the relevant facts. This study is not to be used as a whip but rather to determine a reasonable expectation of a day's work. Later when your schedule is installed is the time to act on lost time issues.
7. Recap all observations by the end of the day while they are still fresh is your mind.

4.2.8 Published engineered
performance standards

Engineered performance standards are published by organizations for their own use (many contractors have developed standards for their own use in preparing estimates and for evaluating mechanics), by equipment manufacturers, or by a third party for profit publishers. They are developed by a variety of methods. The most widespread and oldest collection of standards is the Engineered Performance Standards published by the U.S. NAVFAC. It was developed by industrial engineers using time and motion study, work sampling, and methods time measurement techniques to set the standards. Others are developed by comparison to other related jobs (similar to slotting) and by timing skilled mechanics.

In all cases, skilled mechanics with proper tools should be able to meet or exceed the published EPS. Activities such as diagnosis do not lend themselves to standards. The EPS for a repair is sometimes too fast if the item is old, hard to get to, or has had very rough service and the components are rusted in place or the mechanic is not fully trained. If rusted bolts or studs break during the repair, even a highly skilled mechanic is likely to miss the standard rate with good reason.

Published EPS are a good place to start your standards system as long as they are compared to the other standard techniques. If you start using published standards to estimate repair times, 100 percent would be the average productivity rate for skilled people. Work order schedules based on published standards should be derated by a percentage based on the relationship between the standard and the experience of your worker (you can use reasonable expectancy to check the published standard).

The reason that published EPS are essential in the mix is the clarity brought to the situation by an outside, impartial source of productivity data. The published standard was developed somewhere else by, presumably, expert mechanics observed by expert observers. This outside opinion is a good check on the skill and veracity of your own crews. Historical standards and observed standards can be mitigated by mechanics that are trying deliberately to undermine your efforts at developing standards or by employees who are not knowledgeable about a job or a particular kind of work. The published standard will immediately highlight that deficiency.

4.2.9 Engineered performance standards (EPS)

The U.S. NAVFAC's EPS is the oldest set of maintenance labor time standards in use. They were developed starting in the 1950s. The Navy's Public Works Departments and Public Workers Center were responsible for maintenance and operations of all U.S. Navy bases and installations worldwide. At the time they had little capabilities in maintenance management, work planning, or more scheduling.

NAVFAC decided to develop a series of engineered performance standards to assist the Public Works Departments and Public Works Center plan and schedule work. They started with an industrial engineering contractor on painting and carpentry skills and then expanded into other skills in the Departments of Center.

The EPS contractor used time and motion studies, work sampling, and methods time measurement to assign standard times. Their observations and times were based on studies of experienced mechanics. Each element of motion in the repair was divided into sections and analyzed. These lists of elements were cataloged and added together to make up the standard time of a repair act. This would be a massive job anytime, but before computerization it was a mammoth undertaking.

It took approximately 10 years before the first version was ready for use. During the long development phase of the entire system, the U.S. NAVFAC brought the project in-house. NAVFAC assigned an industrial engineering group to accomplish this task. Each of the 14 engineering field divisions, at the time, was assigned a craft to analyze and develop copyright EPS data. At its peak, 200 industrial engineers were working on the project nationwide. The next problem was keeping up with the changes. Every time materials, tools, and techniques changed, the standards had to be revised.

In the early 1970s NAVFAC was assigned the job of keeping the EPS up-to-date for use by all three branches of the military (Navy, Army, and Air Force). The last major change in concept was the computerization of the whole effort in the early 1980s.

EPS was based on the actual pure work content of the job. Factors for travel, preparation, and delays were added based on the trade and the location of the job. The standards were aimed at maintenance and repair functions, not major alterations/improvements.

The EPS has five parts:

1. The first part is the task area. This is a standardized description of the work to be done. The system is cataloged in the task area. The first word is the word you would be likely to be looking up. For example, if you wanted to look up sweeping the floor you would look under "Floors and stairs—Dust mop and sweep." If you wanted to install some insulation fittings on pipes you would look up "Insulation—fittings—install."

2. The second part is a description of what is included and what is not included in the job. The comments in this section relate to all the individual tasks mentioned later. This section includes job content, tools to be used, short description of technique (if appropriate), whether travel is or isn't included, example of calculating standard times, and references to other sections.

3. The detailed job description and trade reference is in the third section. A job description could be as simple as "Fiberglass, elbows, 90° and 45° cover with steel jacket." Other job descriptions could be several lines of descriptive text. There is also a reference (in this case QT-100) for more information.

4. EPS assigns standards based on groups. The groups are letters A, B, C, D, E, F, or higher. Each letter is assigned a craft time such as on the "Insulation—fittings—install" sheet. The groups range from A to E. Other sheets might start with E and go up to H or higher.

Group A allows 0.13 h
Group B allows 0.3 h
Group C allows 0.6 h
Group D allows 1.1 h
Group E allows 1.8 h

On different sheets the groups have the same hourly values. So group C is always 0.6 h.

5. This is the detailed standard section in spreadsheet. This section is a matrix. One direction, the x axis, is the group listing and the other direction, the y axis, is the detailed description. For our example in elbows EPS assigns one fitting to group C (or 0.6 h). Two to three fittings is group D (1.1 h) and four to five fittings is assigned to group E (1.8 h).

One advantage of EPS to facility managers is their availability for use. The United States government has made the standards available through the National Technical Information Service. To order, call the NTIS sales desk at (703) 487-4650. This is an agency devoted to disseminating government technology transfer. All of the documents start with Engineered Performance Standards Public Works Maintenance. The general handbook is called *General Handbook*, 5 Oct 72, 61 pp, $21.50, Order number TB 420-2ING, Report no. TB4202.

4.2.10 Management manuals

Real Property Maintenance Activities Unit Price Standards Handbook, 1 Aug 83, 572 pp, $67.00, Order number TB 420-33ING, Report no. TB42033.

Engineer's Manual, 5 Oct 72, 205 pp, $41.00, Order number TB 420-1ING, Report no. TB4201.

Job Elements and Performance Standards—Samples for Supervisors, 18 Sep 96, 110 pp, $28.00, Order number PAM 690-27ING, Report no. PAM69027.

Maintenance Activities—Planner and Estimator's Workbook (Instructor's Manual), 1 Dec 73, 25 pp, $19.50, Order number TB 420-31ING, Report no. TB42031.

Maintenance Activities—Planner and Estimator's Workbook, Student's Manual, Jan 87, 304 pp, $49.00, Order number TB 420-32ING, Report no. TB42032.

Maintenance Activities—Preventive/Recurring Maintenance Handbook, 1 Mar 84, 228 pp, $44.00, Order number TB 420-34ING, Report no. TB42034.

4.2.11   Trades manuals

Carpentry Formulas, 25 Jul 96, 259 pp, $47.00, Order number TB 420-5ING, Report no. TB4205.

Maintenance Activities Emergency/Service Handbook, 25 Jul 96, 92 pp, $25.00, Order number AFM 85-55ING, Report no. AFM8555.

Electric, Electronic Formulas, 5 Oct 72, 284 pp, $49.00, Order number TB 420-7ING, Report no. TB4207.

Heating, Cooling, Ventilating Formulas, 5 Oct 72, 32 pp, $21.50, Order number TB 420-9ING, Report no. TB4209.

Janitorial Formulas, 5 Oct 72, 89 pp, $25.00, Order number TB 420-11ING, Report no. TB42011.

Activities—Janitorial Handbook, 1 Apr 81, 74 pp, $21.50, Order number TB 420-10ING, Report no. TB42010.

Machine Shop and Repairs Formulas, 5 Oct 72, 220 pp, $41.00, Order number TB 420-13ING, Report no. TB42013.

Maintenance Activities—Machine Shop, Machine Repairs Handbook, 1 Apr 83, 310 pp, $49.00, Order number TB 420-12ING, Report no. TB42012.

Masonry Formulas, 5 Oct 72, 122 pp, $28.00, Order number TB 420-15ING, Report no. TB42015.

Maintenance Activities—Pipefitting, Plumbing Handbook, 1 Aug 83, 224 pp, $41.00, Order number TB 420-20ING, Report no. TB42020.

Moving, Rigging Formulas, 5 Oct 72, 40 pp, $21.50, Order number TB 420-17ING, Report no. TB42017.

Maintenance Activities—Moving, Rigging Handbook, 1 Apr 81, 65 pp, $21.50, Order number TB 420-16ING, Report no. TB42016.

Paint Formulas, 5 Oct 72, 136 pp, $31.00, Order number TB 420-19ING, Report no. TB42019.

Facilities Engineering Management of Maintenance Painting of Facilities, 30 Oct 73, 27 pp, $19.50, Order number TB 420-51ING, Report no. TB42051.

Pipefitting, Plumbing Formulas, 5 Oct 72, 170 pp, $35.00, Order number TB 420-21ING, Report no. TB42021.

Roads, Grounds, Pest Control, Refuse Collection Formulas, 5 Oct 72, 108 pp, $28.00, Order number TB 420-23ING, Report no. TB42023.

Roads, Grounds, Pest Control, Refuse Collection Handbook, 1 Mar 84, 193 pp, $38.00, Order number TB 420-22ING, Report no. TB42022.

Sheet Metal, Structural Iron, and Welding Formulas, 5 Oct 72, 149 pp, $31.00, Order number TB 420-25ING, Report no. TB42025.

Maintenance Activities—Sheet Metal, Structural Iron and Welding Handbook, 1 Mar 84, 209 pp, $41.00, Order number TB 420-24ING, Report no. TB42024.

Trackage Formulas, 5 Oct 72, 45 pp, $21.50, Order number TB 420-27ING, Report no. TB42027.

Maintenance Activities—Trackage Handbook, 1 Nov 79, 148 pp, $31.00, Order number TB 420-26ING, Report no. TB42026.

Wharf Building Handbook Maintenance Activities, 1 Nov 79, 88 pp, $25.00, Order number TB 420-28ING, Report no. TB42028.

Figure 4.4 is an example of an engineered performance standard spreadsheet for installing fixtures, fluorescent or incandescent, using a ladder.

Figure 4.5 is an example of an engineered performance standard's tabular listing spreadsheet for brush painting flat concrete surfaces.

TASK AREA Install fixtures, fluorescent or incandescent, using ladder.

All tasks include the following work content: fixtures unpacked and assembled and installed as required; stem components assembled and installed as required; conductors pulled through stems or fixture troughs and wire splices or connections made as required; and bulbs or tubes installed.

		DESCRIPTION	Group C Allow .6	Group D Allow 1.1	Group E Allow 1.8	Group F Allow 2.9	Large Quantity Multiple Factors
Fluorescent Lights	Mounted to Junction Box — Surface Mounted	Individual fixtures, 2 or 4 tubes, open reflector or diffuser/louver type GT-240	2 fixtures (For 1 fixture allow .3 hours)	3 or 4 fixtures	5 - 7 fixtures	8 - 11 fixtures	over 11 fixtures, allow .31 hrs. for each
		Interconnected fixtures, 2 or 4 tubes, open reflector or diffusor/louver type GT-241		2 fixtures	3 fixtures	4 or 5 fixtures	over 5 fixtures, allow .65 hrs. for each
	Stem Mounted	Individual fixtures, 2 or 4 tubes, open reflector or diffuser/louver type GT-242	1 fixture	2 fixtures	3 fixtures	4 or 5 fixtures	over 5 fixtures, allow .69 hrs. for each
		Interconnected fixtures, 2 or 4 tubes, open reflector or diffuser/louver type GT-243		2 fixtures	3 fixtures	4 or 5 fixtures	over 5 fixtures, allow .70 hrs. for each
	Mounted Adjacent to Junction Box — Surface Mounted	Individual fixtures, 2 or 4 tubes, open reflector or diffuser/louver type GT-244	1 fixture	2 fixtures	3 or 4 fixtures	5 - 7 fixtures	over 7 fixtures, allow .48 hrs. for each
		Interconnected fixtures, 2 or 4 tubes, open reflector or 4 diffuser/louver type GT-245		2 fixtures	3 fixtures	4 or 5 fixtures	over 5 fixtures, allow .70 hrs. for each
	Stem Mounted	Individual fixtures, 2 or 4 tubes, open reflector or diffuser/louver type GT-246	1 fixture		2 or 3 fixtures	4 fixtures	over 4 fixtures, allow .73 hrs. for each
		Interconnected fixtures, 2 or 4 tubes, open reflector or diffuser/louver type GT-247			2 or 3 fixtures	4 or 5 fixtures	over 5 fixtures, allow .71 hrs. for each
Incandescent Lights		SURFACE MOUNTED GT-248	3 or 4 fixtures (for 1 or 2 fixtures allow .3 hours)	5 - 7 fixtures	8 - 11 fixtures	12 - 18 fixtures	over 18 fixtures, allow .2 hrs. for each
		STEM MOUNTED GT-249	1 fixture	2 fixtures	3 or 4 fixtures	5 - 7 fixtures	over 7 fixtures, allow .5 hrs. for each

TASK AREA: Install fixtures, fluorescent or incandescent, using ladder.

Labels: Information Section, Task Group, Group Time, Large Quantity Multiple Factors, Work Categories, Task Time Standard.

Figure 4.4 Example of engineered performance standards spreadsheet for installing fixtures, fluorescent or incandescent, using ladder.

TASK AREA: Brush paint, flat concrete surface.

1. The times are the same for prime, finish, or semigloss coat.
2. Surface preparation not included - refer to applicable surface preparation spread sheets as may be required.

Operation / Surface Area in Square Feet / Task Time Study Number	No Ladder Required		Ladder Required		Wall	
	Any One Coat	Any Two Coats	Any One Coat	Any Two Coats	Any One Coat	Any Two Coats
	PT-11	PT-12	PT-13	PT-14	PT-15	PT-16
10	.053	.106	.055	.110	.054	.107
20	.106	.212	.110	.220	.108	.214
30	.159	.318	.165	.330	.162	.321
50	.265	.530	.275	.550	.270	.535
70	.371	.742	.385	.770	.378	.749
100	.53	1.06	.55	1.1	.540	1.07
200	1.06	2.12	1.10	2.2	1.08	2.14
300	1.59	3.18	1.65	3.3	1.62	3.21
500	2.65	5.30	2.75	5.5	2.70	5.35
700	3.71	7.42	3.85	7.7	3.78	7.49
1,000	5.3	10.6	5.5	11.	5.40	10.7
2,000	10.6	21.2	11.0	22.	10.8	21.4
3,000	15.9	31.8	16.5	33.	16.2	32.1
5,000	26.5	53.0	27.5	55.	27.0	53.5
7,000	37.1	74.2	38.5	77.	37.8	74.9
10,000	53.	106.	55.	110.	54.0	107.
20,000	106.	212.	110.	220.	108.	214.
30,000	159.	318.	165.	330.	162.	321.
50,000	265.	530.	275.	550.	270.	535.
70,000	371.	742.	385.	770.	378.	749.
100,000	530.	1,060.	550.	1,100.	540.	1,070.
Per square foot	.0053	.0106	.0055	.0110	.0054	.0107

TASK AREA: Brush paint, flat concrete surface.

Figure 4.5 Example of engineered performance standards tabular listing spreadsheet for brush painting flat concrete surfaces.

4.2.12 General Services Administration: public buildings maintenance guides and time standards

The General Services Administration (GSA) is one of the largest landlords in the world. They manage over 7500 buildings for federal agencies. The GSA uses contractors extensively and uses its standards system as a contractor management tool. The time standards themselves were originally based on the engineered performance standards discussed earlier.

The GSA guides are very useful because in addition to the time standards for most PM jobs there are detailed procedures for preventive maintenance for most kinds of equipment found in buildings. These detailed procedures are excellent starting points for a PM task list.

Detailed specifications aid the contracting function. In the GSA's case the descriptions and times in the guide are used to both bid and monitor performance of the contractors.

The 1995 guide is divided into three parts. The first part is general information on safety, good maintenance practices, standard tools, and a description of GSA forms to be used.

The second section is the labor time standards. In this manual the labor time standards are listed with a guide number which refers to the third section. They are organized alphabetically so that boilers are under B and electrical items under E.

Pumps are under P. The designation P4 is the labor standard for an annual PM for a 25-hp centrifugal pump. The time given is 5.5 h. Looking up the guide P4, we find a 3½-page description of all the tasks that make up that PM.

4.2.13 R. S. Means maintenance standards

R. S. Means publishes several sets of books of labor standards and estimating guides. They are the largest publisher of information of this type for the facility manager. Their specialty has been building construction.

In addition to their books, R. S. Means maintains a database of standards. Any facility can purchase a subscription. Many contractors (their primary market) and any department that does a good deal of estimating could use the Means database to good effect. The data can be integrated into many of the popular estimating software packages.

A recent catalog lists six estimating guides applicable to facility maintenance. They also list two specific books of maintenance standards. One of these books, *Means Facilities Maintenance Standards* by Rodger Liska, lists over 180 standards for common maintenance tasks. His 575-page book goes well beyond time standards to include

PM task lists, listings of tools and techniques, and information on managing a facility maintenance department.

The roots of this book stretch back into the late 1970s. Rodger Liska, now Associate Dean of the School of Architecture Arts and Humanities of Clemson University in South Carolina, was asked by AT&T to look into its building maintenance practices. AT&T management felt that it was spending too much money on building maintenance and wanted a scientific expert's view on the issue. His investigation uncovered a lack of consistent standards and practices. He created a course of study over a several-year period that became the predecessor to the *Means Facilities Maintenance Standards* book.

His book is designed as a reference work to provide a starting point for facilities leadership's journey toward work standards, PM, agreed-upon methods of doing jobs, and maintenance improvement. The labor section was distilled from governmental sources, including the EPS mentioned earlier. He then took those standards and sent them to experts throughout the country for their input. Some times were modified.

Liska feels that all standards systems have to be adjusted for the particular facility. Some of the risk factors that could modify the published standards are:

1. Size of asset (impact travel time, communications)

2. Type of construction, materials, and workmanship (many standards do not fully take into account the original materials and the original quality of construction)

3. Quality of design, which can have a major impact on maintainability

4. Age of building (different ages had different techniques; some techniques are difficult to work with)

5. History

6. How asset is used (standards might change in the same building if it is used as an elder day care or juvenile detention center or research lab)

7. Location in the country (there might be an impact from the local culture)

8. Knowledge and dedication of staff

9. Availability of spare parts

10. Type, knowledge, expectations of users (really a measure of the accuracy of the work request)

11. Laws, building codes, statutes that affect buildings

12. Competition (are there excellent local outsourcing contractors?)

13. Amount of deferred maintenance

14. Hours of use a day

15. Time of day that the maintenance is carried out

His advice to facility managers is to start with published standards and gradually create your own standards for your unique situation. The standards in his manual are designed for planning. Expect to modify them to reflect your environment.

If we return to manual dust mopping, the Means standard is a minimum of 5 to a maximum of 20 minutes per 1000 square feet. The facility manager would have to modify that based on the degree of obstruction as well as the 15 factors above. A complete description of technique, tools, and materials is included for each repair or maintenance activity .

For the same sweeping activity the EPS has a half page of descriptions and different cases. One category is corridors, lobbies, and landings. EPS time is 0.3 h (Group B) for 1901 to 3800 square feet or a range of about 5 to 10 minutes per 1000 square feet. (See Fig. 4.6.)

4.2.14 How to use standards, a step-by-step guide

1. It is imperative that your work order system is accurately reporting detailed job descriptions and actual time. Look into the accuracy of the overall job reporting system. CMMSs (computerized maintenance management systems) are a great help but are not absolutely necessary.

2. Enlist the aid of the mechanics in the overall design of the program. The goal is to accurately predict when a job will be complete. The secondary goal is to smooth out the work flow to the workers, not to increase the amount of work. A discussion of productivity (would the worker rather spend time working or rooting around looking for parts, walking to the basement, or using an efficiently set up cart, etc.) might be in order at this point.

3. Standards usage can be started on a pilot basis. The best pilot would be a remote area or a part of your main operation that can be separated from the rest of the activity. The other key is to start with an area, equipment, or activity that consumes a good deal of time. This could include frequent PMs, routine jobs (such as filter change routes), or common repairs.

4. By far the easiest place to start is with published standards. For a few hundred dollars you can obtain copies of all of the R.S.

Facilities Maintenance Worker Hours		
	Unit	
	Minimum	Maximum
Blast cleaning		
White-metal	100 sq ft/h	
Near-white	175 sq ft/h	
Commercial	370 sq ft/h	
Brush-off	870 sq ft/h	
Paint application		
Brushing	125 sq ft/h	
Rolling	125 sq ft/h	
Spraying	500 sq ft/h	
Plaster cleaning		
Wall dusting	2 s/sq ft	3 s/sq ft
Vacuuming	4 s/sq ft	5 s/sq ft
Spot washing	125 sq ft/h	175 sq ft/h
Thorough cleaning	275 sq ft/h	
Plaster repair		
Gypsum and lime repair	5 sq yd/h	10 sq yd/h
Ceramic tile repair		
General	7 sq ft/h	10 sq ft/h
Adhesive tile setting	9 sq ft/h	12 sq ft/h
Pointing tile joints	10 sq ft/h	15 sq ft/h
Floor cleaning		
Manual sweeping	10 min/1000 sq ft	25 min/1000 sq ft
Dust mopping	5 min/1000 sq ft	20 min/1000 sq ft
Buffing	15 min/1000 sq ft	40 min/1000 sq ft
Spray buffing	20 min/1000 sq ft	50 min/1000 sq ft
Damp mopping	15 min/1000 sq ft	30 min/1000 sq ft
Wet mopping	30 min/1000 sq ft	50 min/1000 sq ft
Scrubbing	50 min/1000 sq ft	140 min/1000 sq ft
Scrubbing using electric floor machine		
General	15 min/1000 sq ft	30 min/1000 sq ft
Stripping	100 min/1000 sq ft	200 min/1000 sq ft

Figure 4.6   Example of labor standards for building maintenance activities. (*From Means Facilities Maintenance Standards by Rodger W. Liska, P.E., AIC. Copyright R. S. Means Co., Inc., Kingston, MA, 617-585-7880, all rights reserved.*)

Means books. You could also purchase the EPS through the National Technological Information Service mentioned in that section.

5. Start off assigning standards to all PM task lists and all repetitive work. Also as work comes in against your pilot area, assign standards from the books.

6. When completed work orders come back, note performance against the standard. The actual times (as reported on work orders) are the historical standards. In the beginning you will start with three columns, the published time, the historical standard, and the observed time for each described repair or PM.

7. Perform observations of some of the more common jobs after getting historical standards back from the field. Follow the rules for taking an RE (reasonable expectancy).

8. Compare the published standards, REs, and historical standards on jobs where all are available. Usually the RE would be the fastest, the published standards the second fastest, and the historical standards the slowest of the standards. Seek any patterns between the standards. A pattern of use might be that the RE time is 55 percent of the historical standard.

9. Determine the relationship between the different standards and determine factors for assignment for your facility, crews, and tooling. Work order scheduling and job assignment should be based on the most realistic of the standards with or without extra factors. You might decide that 85 percent of the historical time is a good starting point.

10. The job time as determined by a properly observed RE is the goal. Reengineering the maintenance task or machine, retooling, or redeployment of parts or people might be necessary to achieve the goal of beating the RE time. The best mechanics can train the others in their technique for common jobs. Overall discussions about how to do jobs are useful. Testing new ideas is essential to productivity improvement.

Bibliography

Delmar Karger and Franklin H. Bayha, *Engineered Work Measurement,* Industrial Press, New York, 1987.
Joel Levitt, *Handbook of Maintenance Management,* Industrial Press, New York, 1997.
Rodger Liska, *Means Facilities Maintenance Standards,* R. S. Means, Boston, MA, 1987.

5

Preventive Maintenance (PM) Plan

J. R. Ruta and Tony Graff

Client Services Division
MicroWest Software Systems, Inc.
San Diego, CA

5.1 PM Procedures

Preventive maintenance procedures (PMs for short) comprise the very backbone of maintenance and contribute to life as we know it in the western world. Simply put, preventive maintenance (sometimes called "preventative") means to effect repairs *before* a failure or breakdown is likely to occur. The goal is to eliminate or significantly reduce actual failures or breakdowns, ensuring higher reliability and reducing costs through managing work and downtime and offering greater utilization of any facility.

Where would we be today if all our wonderful and now essential services and conveniences were prone to unexpected breakdowns? We couldn't count on heat or electricity, medical services, transportation, or communications. Could we trust our lives to airplanes that had no regularly scheduled maintenance? Ride in elevators that were never inspected? Work in buildings plagued by surprise power failures, heating and cooling problems, plumbing and phone problems? The point is clear: If it wasn't for some kind of PM program, every motor, pump, valve, seal, compressor, boiler, toilet, light, ad infinitum, in a facility might break down at any time. All things have the need for inspection and repair. "Murphy's law" especially applies to maintenance: If it can break, it will break, and at the worst possible time.

Preventive maintenance also means *scheduled* maintenance vs. *emergency* maintenance. Unexpected breakdowns can be dangerous, can cause delays, and are costly. Preventive maintenance affords us a true asset: the ability to choose a date and time for repair.

Scheduling downtime for preventive maintenance allows for better utilization of a facility, as repairs can be scheduled in consideration of peak demand and available maintenance labor and spare parts resources.

Practicing preventive maintenance sounds like common sense but actually represents a recent shift in policy away from the "run it into the ground" philosophy. That policy, more than anything else, was intended to save money by not spending any: Less maintenance must certainly mean less labor and fewer parts. This isn't how it works out. Downtime costs money, big money, and reputations are at stake. The market demands reliability and accessibility of products and services. How long will your facility last if it gains a reputation for falling apart or is even *perceived* as being unhealthy or unsafe?

5.1.1 Dollars and cents

Maintenance is now being recognized as a cost center instead of just some "necessary evil" in an organization. A good preventive maintenance system contributes enormously to efficiency, reliability, and life expectancy of equipment and structures. With a solid preventive maintenance program, we can expect:

- Increased utilization and better overall appearance
- Less equipment failure
- Increased safety and fewer resulting accident insurance claims
- Increases in productivity
- Reduction in overtime or call-in hours to respond to emergency
- To have the right spare parts, and enough of them on hand
- Money saved paying premium prices for parts you have to have— *NOW!*

Preventive maintenance tasks can be applied to anything which benefits from routine inspection, repair, or renewing. We have learned over time which things are likely to cause problems in every trade. Critical components dealing with safety or maintaining production are given priority. Every facility, for example, must keep its exits unobscured and well marked in case of emergency. Fire extinguishers must be inspected and recharged so that they'll work when needed, filters must be changed for proper air and water filtration, mechanical equipment must be inspected or lubricated to ensure good working order, and in general, all wear parts must be replaced before failure.

PM tasks are cyclical and are repeated at regular intervals. The length and kind of interval will vary depending upon the object and tasks. Many maintenance managers pattern PMs after calendar standards, weekly, monthly, every 90 days, and so on. However, some PMs are best done based on an amount of usage, sometimes known as *run time*. This helps ensure that PMs are not getting done too often or too late. In general, PMs are set up by either:

- *Calendar Days*   A fire system test every 30 days

- *Usage (by Meter Reading)*   Changing lamps every 10,000 hours of use

- *Both Calendar and Meter Readings*   Changing the oil on a forklift every 90 days or 500 hours, whichever comes first

PMs will also have a *reschedule type,* also know as either fixed-or floating-type PMs. This refers to how the PMs are rescheduled:

- *Fixed*   Every X days or X meter units—no matter what

- *Floating*   Every X days or X meter units—from *whenever* the PM was *completed* by date or meter

Nowadays we can estimate with some accuracy when a failure is likely to occur, based on testing and historical performance. Newer or larger pieces of equipment will almost certainly come with some manufacturer-supplied PM schedule with procedures for routine maintenance. Many vendors now supply maintenance routines for their products. These routines may include recommended frequency of the procedures, labor and spare parts requirements, special tools or permits required, and other information. Obtaining and gathering this information together is the first step. Having it centralized and making it accessible is the next. Prioritizing and planning is another.

Some PM procedures may be already in place in most facilities, and those procedures can be the genesis of PM "templates." A PM template is the entire description of a specific PM procedure or procedures, including the equipment asset, interval or *frequency,* reschedule type, required resources such as parts and labor, and other detail. Every time a PM is due, this template is the start of a PM work order. The PM work order will hopefully be updated with actual labor and parts used, as well as comments on what was found or what was done. This work order will then become a part of work history.

There may well be an excellent manual "card" system existing with key equipment and assets identified including procedures and parts lists. If, on the other hand, a facility has no or few PM procedures in place, then the "opportunity" exists to create a PM program based on the experience of what maintenance work has proved the most useful in that facility. How often it should be done should be determined by that "seat of the pants" feel that many tradespeople have. Experienced tradespeople typically have an excellent grasp on priorities and frequencies, as they are the ones performing the repairs. As much information as possible should be gleaned from the tradespeople and broken down into written procedures.

Many begin using a box and filling it with this manual "card" system, with typical PM procedures for critical equipment and assets. Try for as much accuracy as possible, quality over quantity. This manual system may ultimately find its way into an automated PM software program. Table 5.1 lists some of the basic information you may wish to collect for your new PM system. It can be copied and put to use immediately. Completed cards should be kept in a separate box and filed either by date or by equipment and asset. This is work history.

5.2   Automated Program—Software

PM programs can begin with a humble or quite elaborate manual system, and work fairly well...for a while. Facility maintenance today is more than just keeping a few boilers or HVACs running or doing a few safety checks on fire extinguishers or smoke detectors. Facility maintenance is about everything from rebuilding generators to fixing toilets, from refrigeration to groundskeeping. Everything from floor to ceilings, inside and outside, needs some kind of maintenance sometime. There are so many things, it's next to impossible to keep track of them all, let alone prioritize and schedule them.

Good recordkeeping is a problem. What if some auditor asks for proof of the last time a stack was cleaned? It most certainly was done, but where is that work order? What about precise cost tracking? It's not quite "rocket science," but like NASA, the answer is to use a computer. PM programs can be automated to use the power of a computer to do the real grunt work. Have the computer do what it does best, remember things, retrieve things, and crunch numbers. Have the computer manage PMs and other work, so you don't have to.

Fair warning: A computer itself has no arms or legs and if left alone isn't going to do anything. There are several critical components to consider before implementing a formalized maintenance management software system:

TABLE 5.1   Basic PM Template Info

Equipment / Asset Number	
Name	
Location	
Job #	
Dept.	
Account #	
Remark	
Frequency in Days	
Next Due Date	
Frequency by Meter	
Next Due by Meter	
Meter Units	
PM Reschedule Type	Fixed_____ Floating_____
Special Tools:	
Drawings: Folder Name:	
Labor: Craft and Est Hours	
Parts:	
Job Description:	

1. Commit to and budget the time and money to purchase and implement a program. *A program not implemented is no more than good intentions.*

2. Adopt or create a formalized numbering system to identify each equipment and asset, and inventory spare parts.

3. Gather information and create meaningful PM templates and procedural methods.

4. Ensure that all PM tasks and frequencies are entered into a computer program.

5. Have a dedicated person or people willing to use a computer in place.

6. Choose a flexible enough software program especially designed for preventive and non-PM maintenance that can change with you as your procedures evolve.

Such a software program is known as a CMMS (computerized maintenance management software). A few respected maintenance managers like to use the acronym CAMMS for computer *assisted* maintenance management system, stressing the point that a computer is just a tool. The computer allows a CMMS to store vast amounts of information and search and sort on any information in the system with amazing speed.

A CMMS is designed to store complete information on each equipment or asset in your facility. This information includes PM procedures and must include non-PM work such as breakdowns and inventory as it relates to spare parts for maintenance. The program in itself acts as a well-organized and cross-linked filing system. Separate files or *databases* store specialized information. Examples of the databases typical in a CMMS are an equipment file, a preventive mainte-nance file, a work order file, a work history file, a labor file, a schedule file, an inventory file, an inventory activity (transaction) file, a vendor file, an order parts (requisition) file, a purchase order file, and a purchase history file. An example of a CMMS system is the PC/windows-based AMMS advanced maintenance management system from MicroWest Software shown in Fig. 5.1.

This menu graphically presents the files that are used on an everyday basis. Today's software is easier to use than ever, immediately available, and cost-effective. It is important to find a generic CMMS that is *flexible* enough to adapt to your environment and your way of doing things. It is much easier to change software setup than to reengineer your way of doing things. A CMMS can quickly total expenditures by cost center, unit and machine, floor, building, etc., and account for all labor, materials, and outside costs. A CMMS can help you identify and *quantify* problems in dollars and cents, and help bring costs down.

What benefits can you realistically expect from a CMMS? The following are actual comments from users and a list of some of the achievable results you might expect:

"It forced us to make a list of all our equipment and number them."

"A CMMS makes it easy to access equipment history data."

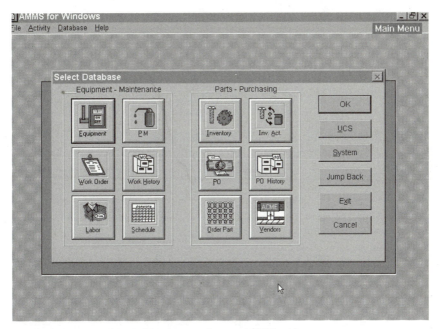

Figure 5.1   Example of a CMMS program layout: AMMS main menu.

"*Much* better control of PMs."

"We are finally able to control our work order schedule."

"Formalized our work order system."

"Developed a work history."

"Better PMs helped us lower labor costs."

"Eliminated several clerical tasks and duplicated efforts."

"We are able to relay information more readily."

"Other departments now realize how effective we are, and our efficiency is up."

"Getting it going was clockwork with good software vendor support...and simplified user-friendly software."

These comments support the following average savings or benefits realized by CMMS users.* Additionally shown in Table 5.2 are achievable results that may be obtained from using a CMMS.

Industry Week, "Move the Wrench Over and Pass Me the Computer," February 1990, Therese R. Welter.

TABLE 5.2   Achievable Results with a CMMS

RESULTS:	DUE TO:
*Overall Better Organization	Identifying and Numbering Assets Gathering, PM, WO, Parts information together
*Instant and Accurate information	Computerized Reporting: no lost pages, instant searches, sorts, calculations
*Reliable PM scheduling	Less Failures
*Improved Equipment / Asset Availability	Less Failures Scheduling Preventive/Corrective Maint
*Reduced Repair Cost	Less overtime Parts availability, price shopping
*Increased Labor productivity	Having Parts and Procedures available
*Reduction in Paperwork	Easily accessible electronic records No more lost paperwork
*Better interdepartmental Relations	Have dates and costs on anything readily available
*Reduced inventory stock costs	Buy from the best Vendors,
*Reduced purchasing costs	**Reduced number of stock- outs**
*Fewer expediting costs	**Ability to Plan Purchases Ahead of Time**

Planning / Scheduling

*Shorter repair times thus reduction in downtime
*Reduction in delays through systematic planning and scheduling

28.3 percent increase in maintenance productivity

20.1 percent reduction in equipment downtime

19.4 percent decrease in material costs

17.8 percent reduction in MRO inventory

"The way maintenance people work with those in other departments can become more productive as well....When you have a formal work-order tracking system, there's a joint effort between maintenance and operations in the planning of what should be done next."

These are the overall results:

- Better service to other departments
- Access to maintenance information, dramatically expanded
- Longer useful life for equipment
- Better capital expenditure decisions based on history
- Improved quality

- Increase in overall plant productivity
- Higher profitability

"Purchasing agents too can be positively affected when maintenance uses a CMMS.... Instead of rushing out and paying whatever it costs to buy the material...purchasing has the time to go out and find it at the right price."

5.2.1 Getting started with a CMMS

A few things to consider when selecting a CMMS:

1. Get the support of management.
2. Create a list of your most pressing current maintenance needs.
3. Try to guess at what the most pressing needs will be 5 years from now.
4. Budget enough for quality software *and training and implementation services.*
5. Have someone ready, willing, and dedicated to use a computer program.
6. Hardware: Take stock of your tools. Do you own any computers currently? Can you buy new ones or "appropriate" some from other departments?
7. Test the CMMS to make sure it will run well on the computers you have.
8. Do you need the CMMS to be *networked?*
9. Try to establish *current* "baseline" numbers on: What are your labor and parts costing now? How often are breakdowns occurring and where?
10. Have a plan: set goals, guidelines, or timetables. A set of realistic, achievable goals is vital.
11. Include the right people to participate in the project early.

Identify resistance and try to make them part of the solution.

5.2.1.1 Management support is vital. Plan on management support of some kind to allot time, approve funds, and provide enforcement of policy to implement a CMMS once you're started. Selecting and implementing a CMMS is the first step. The next is to ensure that PMs actually get done. What if your tradespeople refuse to do these wonderful new PMs (too busy, someone else's job, etc.)? Completed

PM information such as who did the job, how long it took, what parts were used, date completed, and comments on what was done or was found *must* be entered into the system for proper work history. In some cases, like with a *floating PM,* a new PM job will never occur if the *computer* doesn't know that the last one has been done.

5.2.1.2 Will everyone use a computer? The answer is: It depends on management commitment. Some tradespeople are receptive, some not. There can be resentment toward any "new" ways of doing things. They may feel things are going fine and don't need a computer to track their time, tell them how to do their jobs, or make their hard-earned experience available to any new employee. Prepare for this possibility. These people will need extra attention to make them comfortable with the idea.

When setting the system up, it is important to make *realistic* goals for yourself and not set yourself up for disappointment. Two classic examples of *unrealistic* goal setting are:

- Expecting universal acceptance (even in your department)
- Expecting immediate cost savings

It will take a while for everyone to get "on board" with the program. Start working on this before bringing the software in. As far as payback, it takes a full year on the average for most CMMS projects to hit full stride. Payback will usually occur sometime in the second year. The first months will be spent offsetting costs related to:

- Initial investment of the CMMS and CMMS support services
- Time set aside for CMMS training for shop personnel
- Worker hours spent loading data into the system

Software selection is one of the most important decisions you will make. The key to a good CMMS is flexibility and a logical enough layout so that supervisors and tradespeople alike can become comfortable with it. The *Number One reason cited for failure* of a CMMS (reported by users to an *Industry Week*/AT Kearney Survey) is a program that is not flexible or "user-friendly." The *Number One thing users would do differently* is provide more vendor-supplied software training. It has been said that every dollar spent in training will pay you back threefold. Tables 5.3a and 5.3b are an example of a checklist which may be followed to select your software.

5.2.1.3 The software. General software and software vendor recommendations:

TABLE 5.3a   Sample Vendor Software Feature Analysis

Product__AMMS
Vendor__MICROWEST SOFTWARE, INC.

Max. Users _____
Inc. Maint and Stores

Item	Yes	No	Remarks
General Information			
System is IBM-PC Windows Compatible			
System is Windows '95 / NT Compatible			
Software is Network Compatible			
System is Available in "Modules"			
Easy Customization- on-Screen Contents and Fields			
Ability to Remove Excess Fields on Screens			
Minimal Number of Screens to Generate Action			
Software Operation Between Screens -Max. 0.6 Seconds			
Minimal Keystrokes per Action			
Unlimited Security Passwords w/ Different Access Levels			
Able to Move Between Security profiles w/o having to exit system			
Vendor has 500+ Installations / In Business 10+ Years			
Vendor Offers Training / Technical Support			
Vendor Offers Implementation Assistance			
Work Order			
Complete PM and Work Order Scheduling and Printing			
Track Work Order Status, Cause of Failure, Corrective Action			
Crafts, Employee Information, Labor Rates			
Extensive Reporting Capability			
Able to View Reports On-Screen w/o Printing			
Automatically Remove Parts from Inventory			
Easy to Modify Work Orders			
Generates Automatic Work Orders			
On-screen W.O. approval			
Graphic File Interface for Equipment / Parts Illustration			

1. Software flexible enough to change as your needs change. Don't spend all of your time trying to fit a square peg into a round hole.

2. "User-friendly," logical software: different screens for different users.

3. Software that does what you want. Some programs are geared toward vertical solutions.

TABLE 5.3b   Vendor Software Feature Analysis (*Continued*)

Item	Yes	No	Remarks
Preventive Maintenance			
Unlimited Number of Different PM Jobs for each Equipment / Unit			
Choice of Floating / Fixed PM Type for each Equipment / Unit			
Complete Description of PM Tasks including Labor, Parts			
"Seasonal" Range PM's			
Breakdown of Equipment by Part / Stock Number			
Equipment Cost, Location, Usage, Repair Costs, Downtime			
Historical Data by Discreet Equipment Number			
Budgeting -Track Maint. Cost (Actual vs. Budgeted)			
Equipment warranty/service info included			
Inventory Management			
No Double - Entry of Information to Handle Stock Items			
Build Database of Equipment Spares from Parts Usage			
Inventory Reporting -Qty.. on Hand, Re-Order, Parts Usage			
Cross-Reference Stock Numbers and Vendor Part Numbers			
Separate Files for Open / Closed Purchase Orders			
Barcode Capability			
Purchasing & Vendors			
Paperless Purchase Requisition process including Approvals			
No item ID Code number needed for purchasing			
Allows Easy Transfer of ASCII files			
Provides Separate Tax Rates by Vendor			
Tie non-code items to maintenance through charge acct. #'s			
Real time desired versus batch time (stock item PO's)			
No restrictions on Units of Measure			
Add or Update an existing vendor record			

4. Software cost: Be careful! Be prepared to spend a little more up front. You usually get what you pay for. Look closely at how you add it up. Some vendors literally give away the software and charge a fortune for support, upgrades, and services.

5.2.1.4   The vendor

1. Vendor offers responsive, quality technical support. (The human factor is one of the most important. Make *sure* you try it.)

2. Vendor offers quality training and implementation services.

3. Vendor has a solid track record. (Rule of thumb: In business for 10+ years, 500+ software installations.)

5.2.2 Flexibility and user friendliness

Look out, this term has been so overworked it means almost nothing! By definition, all software is "user-friendly," since no software vendor will want to claim their product is "inherently challenging" and expect to sell many programs. However, software must be geared toward providing an average person (*not* a programmer type) a modicum of comfort. One example of a true user-friendly CMMS program is AMMS. This program is "generic" in that it can be used for many applications, from hospitals to steel mills. The maintenance function is generic in that it contains common elements, doing some kind of work on something that will require labor and sometimes parts and / or outside contractors. Although maintenance itself is generic, people will like to see programs that appear to be aimed at their particular niche. These "vertical" niche programs tend to be few, inflexible, and expensive.

AMMS allows a person with little or no computer experience to customize the generic screens, forms, and reports to reflect the individual needs of each customer. It is easy to make changes in terminology, field length, and layout on-the-fly from inside the program using menus and mouse. Forms and reports are also created and modified on-line from inside the program. Seeing what they want and having this kind of control to change as they change helps users feel comfortable with the system. It helps them feel that they are in control of the software instead of having the software control them.

Besides customizable screens, forms, and reports, a logical layout and streamlined procedures for everyday activities, i.e., generating PM work orders, opening and closing work orders, ordering and receiving parts, are very important. Performance and minimum hardware requirements are other very desirable attributes of any system. Make sure you obtain a working demo of a CMMS system before you invest. A product may look great when showcased by a slick salesperson on a warp-speed million-megahertz laptop. It should. How well will it work for you on your computer? Will everyone on the network be able to run it just as well?

5.2.2.1 Customized screens. The ability to customize screens can be very helpful in making the computer comfortable for all kinds of users. Different user-defined screen layouts let users see just what they need and allow for quick data entry. Figure 5.2 shows an exam-

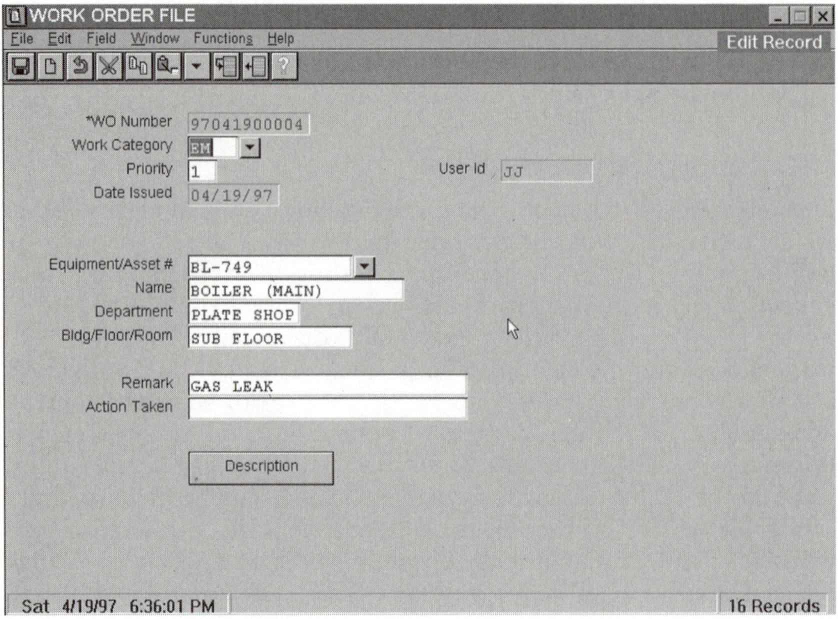

Figure 5.2   Example work order screen as seen by a tradesperson.

ple of a work order screen streamlined for the quick work order entry of a tradesperson.

5.2.3 Generic CMMS function

In general, all CMMS systems will have the following areas in common.

1. *Equipment and Asset Information*   Nameplate information, specifications, spare parts lists, PM procedures, repair history, and cost tracking.

2. *Complete Inventory Management*   Part descriptions, families, quantities on hand, on order, min/max quantities, reorder points, issuing and receiving, transaction history, ordering and purchasing.

3. *Employee Labor Information*   Who did the job, how long did it take, four different pay rates per persons.

4. *"Text Codes" Standard Operating Procedures for Jobs, Tasks, Safety.*

Figure 5.3 shows a work order screen for a PM job as seen by a supervisor. The work order integrates the equipment and asset being

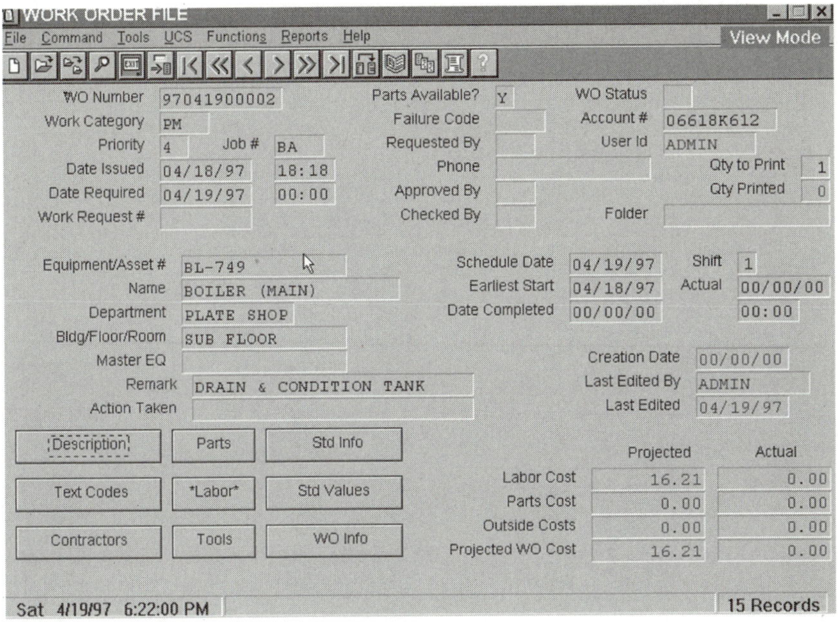

Figure 5.3 Example work order screen as seen by a supervisor: integrating equipment and asset with inventory, labor, standard procedures.

worked on, a cost center (account, linked to the budget file), with on-hand quantities of parts and their costs (from the inventory file), and labor and labor costs. An employee ID or clock number links a tradesperson's record in the labor file to the work order (Fig. 5.4). Multiple tradespeople can do work on the same work order over many days. Each tradesperson is defined with four separate pay rates, regular hours, overtime, double time, and call-in rates.

When a PM or non-PM work order is opened, the job is initiated and progress or status can be tracked over time, ending with a completed work order sent to work history. The completed information for that job now becomes an important part of work history for that asset and is used in reports and graphs. Completed work orders remain in work history forever, or until you decide to "archive" or delete them. Deleting information is not recommended, as fate seems to dictate that you will be sure to need it just as soon as it becomes permanently unavailable. Archiving involves selectively extracting information from a file and saving it as a separate file where it can be stored for later retrieval. Having all closed work orders in a central history file allows for easy searches based on any information entered. The work history file, inventory activity file, and purchase history file are the primary files for reporting and analysis.

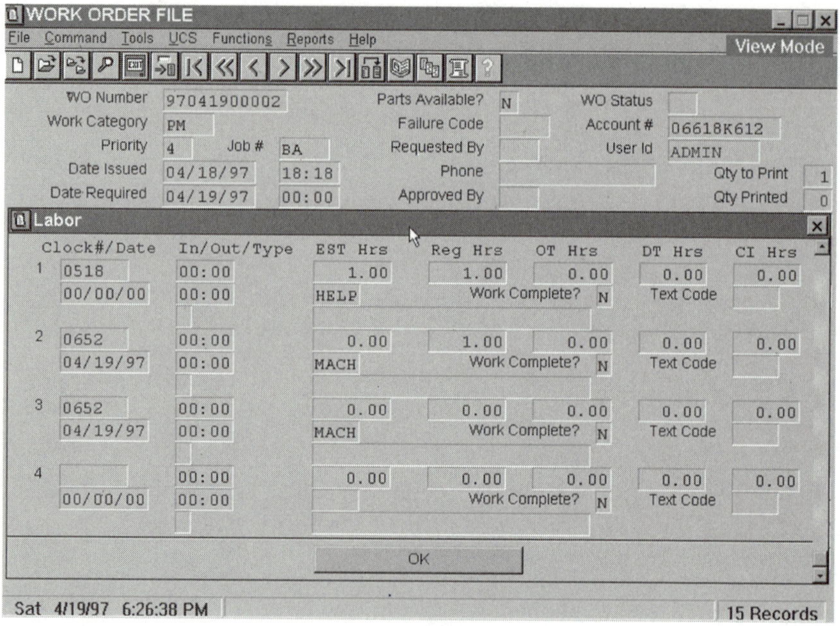

Figure 5.4 Work order screen with labor tracking window. Buttons or windows are subscreens that expand to show related information on a record.

5.2.4   Implementing a CMMS

No CMMS system you buy, whether you spend $100 or $100,000, comes with a magic wand that mystically loads the software with your very own information on each asset, part, PM procedure, tradesperson, pay rate, and other specific information. Most software vendors can, however, probably provide services to do this for a price. This is called software implementation, and you will find the cost of the software is just the tip of the iceberg when it comes to bringing a system up to speed.

In addition to program setup, the true investment and the key to successful implementation is proper training and motivation for your people to use the system. Training your people on a system's capabilities and making them comfortable with it will help ensure that they will be more willing to use it and enter *useful* information. Far too many times "problem" descriptions have been listed as "broken" and the action taken listed as "fixed." This provides no help to the next person.

Initial system setup is by far the hardest part of the job. This can be made significantly easier if you have some data already stored in a computerized format such as Lotus, Excel, dBASE, etc. This data can usually be imported into a CMMS, eliminating manual entry. If you have no "electronic" data but do have hard copy, it is simply a typing job to get it all in. To expedite this process and utilize your labor resources more efficiently, you might try hiring temporary help, preferably with some computer experience and typing skill. Be aware, though, that your hard copy data must be legible and organized, as temporary help will not intuitively know your equipment, procedures, and parts.

If you don't have any kind of organized information or have *obsolete or incomplete* data, this is your opportunity to clean up and correct the data as you input it. "Cleaning" up means to get rid of all obsolete parts, equipment, labor personnel, task procedures, etc., that are no longer in service or applicable.

Alternatively, you have the option of putting in whatever information you have and then cleaning it up as you use the system. Obviously, the "no time better than the present" concept is appropriate, and it would be best to have nice complete, correct data on all of your assets including PM procedures, spare parts, vendors, labor, etc.

5.2.4.1 Worst case scenario. You have nothing: *Solution:* No spreadsheets, no data cards, no handwritten procedures are required if you select a CMMS that lets you build your databases on-the-fly. This way you can start using the system immediately and backfill information as you go or as time permits. Here's how.

5.2.4.2 Create PM tasks on the most critical components. Create a PM template to facilitate entry of information into the computer, such as equipment and unit number, name, location, department, account and cost center, etc. The system should tell you that the equipment and asset is not in your equipment and asset file. It should then ask you if you wish to add it to the file. Answering "yes" will tell the system to add that information into the equipment and asset file, where its cost life to date will now keep a running total on the cost of all PM and non-PM work orders. This new information in the equipment file will be automatically added to a lookup table and will transfer over to the PM and work order files when called up. Entering spare parts information will likewise prompt you to add those parts to the inventory file.

If you can dedicate a set amount of time each day to the chore of rounding out this information, you'll be surprised to see how fast things happen! To wait to have thorough, complete, up-to-date data installed in a CMMS before putting it on-line to start using it only wastes valuable time during which the system could be working for you.

It's also not necessary to have the whole system fully operational before you start receiving benefit from it. Although the system may be completely integrated, where each database file relates to each other (i.e., the work order looks into the inventory file for spare part information, into the asset file for asset information, into the labor file for pay rates, etc.), it is broken up into separate modules.

- The equipment management system (EMS)
- The inventory and purchasing system (IPS)

When setting up a CMMS, some people choose to implement part of the system first, then the other part later. If you are plagued with inventory problems like:

- *Running out of parts* (usually at the worst times!)
- *Having trouble locating parts* (wasted time looking for parts, ordering parts you already have, but don't know it!)
- Having too many of a part in stock (excessive inventory overhead)

...then inventory is your main concern and your efforts should be focused on setting up and implementing the inventory side of the system. If, on the other hand, you find yourself constantly "putting out fires," then you really need to concentrate on the PM side of the system. With the PM side implemented, you'll see that the number of emergency work orders drops dramatically as the number of PMs increases.

This implementation process can be broken down one step further. You could choose to implement the equipment, PM, and work order part first. Again, considering that time is of the essence, start by inputting the 20 percent of the equipments and assets that cause you 80 percent of the trouble. You have the most to gain by setting up PMs on this equipment as soon as possible. The same logic holds true if you choose to set up the inventory part first. Start with the parts you use all the time or the ones you have been having trouble with (i.e., the ones you have been running out of or the ones that have long lead times or are hard to get).

5.3   Maintenance File Cards

As discussed earlier, a good place to start to build up a PM program is with a manual maintenance card system. These 3×5 or larger hard-stock cards are reused over and over, perhaps being updated with just a date done and a set of initials of the tradespeople who did the job. These cards are set up for each equipment and asset indicating the

PM frequency, the work to be performed, parts needed, special tools needed, last serviced, and so forth.

The manual card system is the precursor to the CMMS and is little different in theory, except that a computer enables record keeping, searching, and summations to occur at light speed. PMs set up from the file cards are the procedures that are triggered automatically by a computer system according to calendar date or run time (Fig. 5.5).

SUPERIOR BUILDING AUTHORITY   **#974-108**

Maintenance Schedule

EQUIP # : _____

NAME   : _____

JOB #   : _____

BLD.. / FL / RM : _____

DEPT/TENANT : _____

COST CENTER : _____

WORK DESCRIPTION:

_____

_____

_____

PARTS NEEDED	DATE DONE	INITIALS

SAFETY IS EVERYONE'S JOB!

Figure 5.5 Sample maintenance file card with all the primary elements needed to build a PM into a CMMS system.

5.4   Equipment History Files

CMMS systems are designed to store complete nameplate, manufacturer, spare parts lists, vendor, capacities, and other specifications in the equipment and asset files. All work performed becomes a permanent record in work history.

Importance of keeping equipment history:

- *Centralized list of all equipment and assets, with information readily available* (not confined to a set of torn manuals locked away in an office).

- *Satisfy regulatory audits* (health, safety, ISO 9000, etc.) that assure equipment is regularly checked and inspected to be in good working order.

- *Breakdown analysis:* number and severity of occurrences, finding common failure conditions.

- *Cost tracking:* total investment in equipment maintenance in dollars, by department, repair category, etc.

- *Eliminate the paper chase:* the speed of computerized filing vs. the inefficiency and inconvenience of traditional paper filing.

- *Troubleshooting:* "What did we do the last time that motor went bad?"

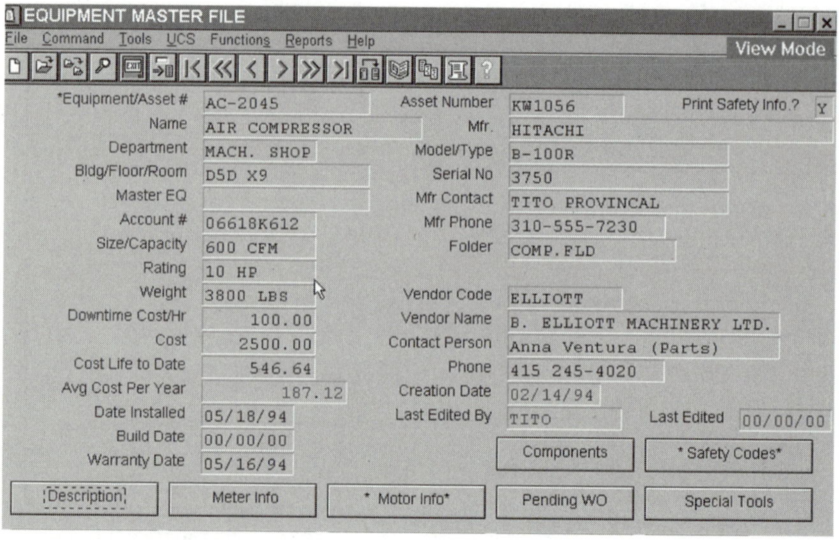

Figure 5.6 Equipment master file.

The equipment file is intended to store detailed information in a central record. This record has buttons, which open to show subscreens, for spare parts, meter information, standard procedures (test codes), open (pending) and closed work orders for that equipment, and more. One sign of a good CMMS is to consolidate related information in a single screen, making it easier for the user (see Fig. 5.6). The equipment file in AMMS also links graphic images such as scanned photos or cutaways of the equipment and asset in a folder for viewing or printing. There are also "safety codes," or safety procedures linked to this equipment and asset, defined in the text code database. The "print safety info" is a "set and forget" mechanism that automatically enters all the linked safety information to both PM and non-PM work orders for this equipment and asset, which is then printed out on the work order form.

5.5   Verifying Facility Inventory

5.5.1   Computerizing inventory

One major benefit of a successful CMMS implementation is the taming of what some like to call the "black hole," or their *inventory*. Wouldn't it be nice to be able to look up a spare part and instantly find out anything you wanted to know? Things like: "Do we have the part in stock? How many? Where are they? What do they cost? If we don't have it, is it on order? When will we get it? Did we order it locally or from out of state? What's the P.O. number?"

A CMMS can provide you with immediate answers to these questions. A CMMS will provide the structure and centralized storage capability for a formalized system. Implementing a program will require you to standardize part numbers and descriptions and keep detailed information on parts including physical dimensions, the material it's made of, storage locations, special handling or disposal instructions, and so forth. You can even store pictures or technical drawings of the parts, to look up for reference or to print out on PMs, work orders, requisitions, and purchase orders.

After some decision making on parts numbering and description standardization, and a little initial setup in a computer, regular use of the system will keep the information up to date and immediately accessible to everyone who needs it. This work in progress or *setup* is called the task of *computerizing inventory*.

The key to computerizing your inventory lies in having a good part numbering system. If you don't have a good numbering scheme, now is the time to develop one, because computerization will show its flaws quickly. The part number is an extremely important and perva-

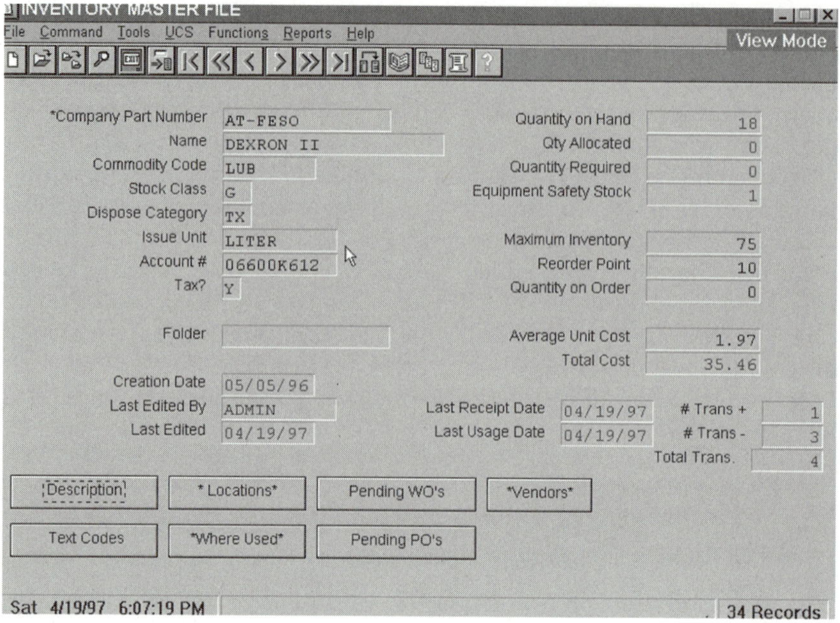

Figure 5.7 Inventory master screen.

sive bit of information in a CMMS, is difficult to change once in place and linked to equipment and assets, PMs and vendors, and has a transaction history (see Fig. 5.7). It is definitely worth taking the time up front to ensure the numbering scheme makes sense and is logical in the way a computer thinks about things. More on numbering later.

Once you've decided on a good part numbering scheme, you can begin entering inventory master information such as minimum/maximum quantities, reorder points, average unit cost, location or locations where parts will be kept, quantity of parts at each location, and linking vendor and ordering information to each part.

In the vendor part of the inventory file shown in Fig. 5.8, separate ordering and purchasing information for each vendor, including the purchase unit, the number of issue units in a purchase unit, a definition of the issue unit, purchase unit, and issue unit prices, as well as economic reorder quantities and a manufacturer's and vendor's part number cross-reference is maintained. Descriptive information on each part, and the locations where the parts are kept are fundamental for tracking specific parts.

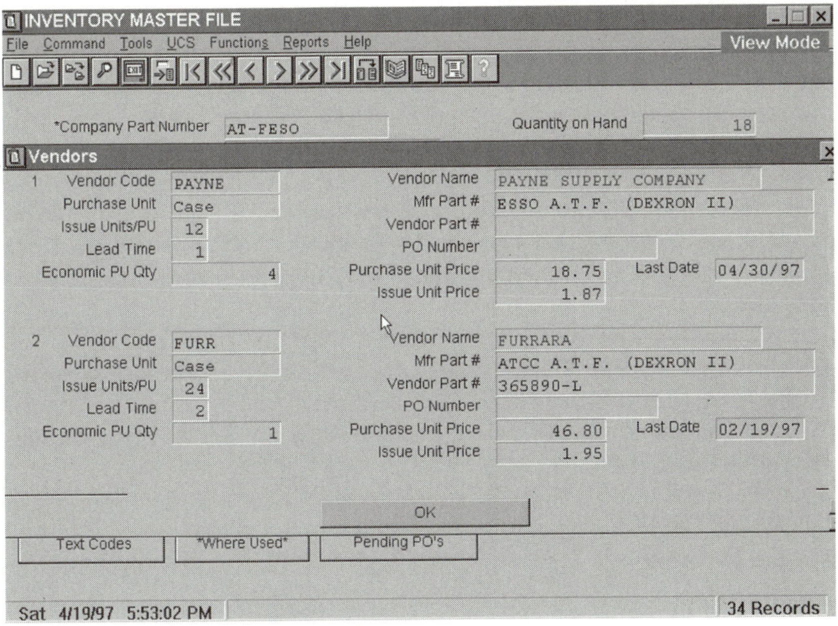

Figure 5.8   Inventory vendor associations.

5.5.2   Inventory control (tracking)

Once you have *computerized* your inventory, you will have the capability to know what you have, where it is, and how many there are. This is no small feat. Let's also assume for the sake of argument that tradespeople report every part taken out of stock, and every part purchased is received into the system. (A veteran parts man once described the difficulties of true inventory control. He said that there are really only two kinds of parts: the ones that have "legs" and the ones that have "babies," *and that a lot more parts have "legs" than "babies."*...

How does a CMMS enable parts *tracking?* Most programs will have a special file that will keep a record for each issue or receipt (transaction) of every inventory part. The file records the part number, the date, the time, quantity, location drawn from or added to, and transaction type (work order, purchase order, physical count, correction, bar code download, or nonspecific issue). This is called the inventory activity file in AMMS, and a record is noneditable and nondeleteable once saved. If a mistake is made, a correction needs to be made into the activity file to offset the mistake. This is also sometimes referred to as an *inventory audit* trail.

5.5.2.1   Example: PM work order transaction.   Once a PM is triggered by calendar or usage, the PM template information is then copied from the PM file to the active work order file and is given a work order number. If spare parts are listed for that PM work order, they will update the inventory quantity required and quantity allocated fields for each part listed. (Required refers to the quantity requested by the PM work order. Allocated is the quantity "reserved," not to exceed the quantity on hand.) At this point the parts are not yet removed from stock, as the actual job requirements may deviate from these projected quantities.

Next, when a work order is updated with actual parts quantities, or the work order is closed, the system will automatically create a record in the inventory activity file for each different part used to document the transaction. Some of the items updated in this kind of transaction are the following:

- Company part X
- Date and time
- Quantity Y
- To and from location A
- Used on equipment and asset B
- Work order C for a total cost of Z
- User ID

System remark: "work order activity"

Figure 5.9 shows an inventory activity screen.

Similar transactions are also created for each *receipt* of a part on a purchase order, updates due to physical counts, and even *manual* inventory issues and receipts. A manual entry need not refer to a work order or purchase order. A manual entry of this kind might involve issuing batteries or gloves, where you want to track the issues but don't necessarily need to tie them to an asset. With this level of documented parts tracking, a wide variety of reporting opportunities exist. Database "filters" can be set on selected fields to search for a particular time period, family of parts, transaction type, user ID, etc. Some reporting examples include:

- Work order activity report
- Purchase order activity report
- Part utilization breakdown

    Part usage by asset, department, location, account, tenant, etc.

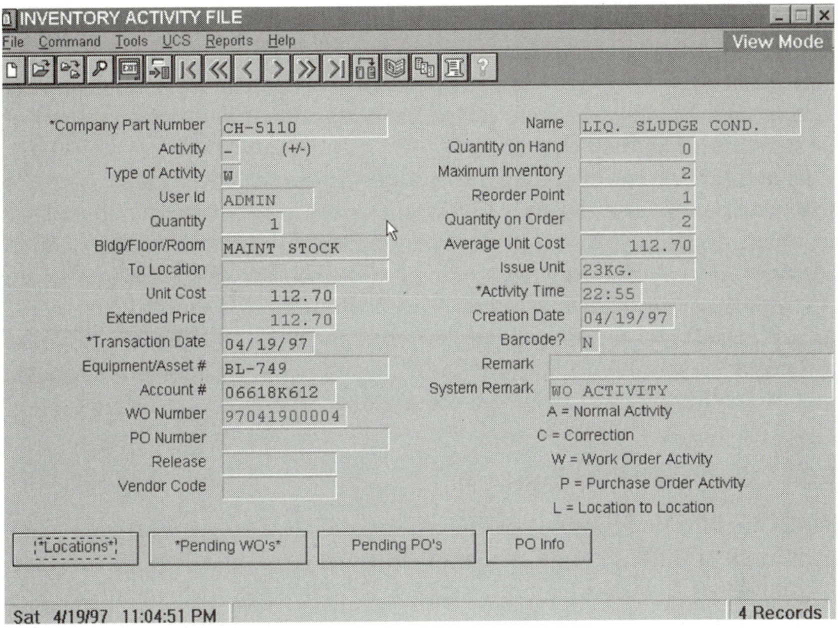

Figure 5.9 Inventory activity record.

- Stock justification:

  Which parts move the most?
  Which parts are not active enough to keep in inventory?

- How many times did I issue a certain part...to what equipments?

- Are my min/max quantities on target?

5.5.2.2 Locating parts. Knowing where parts are can save countless hours of searching. One customer reported saving up to 40 person-hours per week, just looking for parts!

5.5.2.3 Reduced overhead. Many customers report drastic cuts in overhead. The reason is simple: Before they had a CMMS system set up to provide them with an accurate location and count for parts, they simply ordered more parts for every job they did just to make sure they had them.

5.5.2.4 Critical work scheduling. Realistic planning and scheduling demands accurate knowledge of available resources: parts and labor. The ability to quickly check inventory status (are the parts in stock? on order? allocated to someone else? etc.) is essential.

5.5.2.5 Automatic reordering.   Once inventory has been entered and set up with accurate quantities and reorder points, a CMMS can be set to automatically reorder parts when they reach certain minimum quantities. This saves time and helps ensure that you don't run out of parts when you need them.

Spare part inventory can be a huge financial investment and an integral part of maintenance. Not only does computerizing inventory in a CMMS help you save costs by reduction of excess inventory but it saves time in part searching and provides better organization. More importantly, downtime is reduced because automatic reordering helps ensure that you won't run out of critical parts when you can least afford to. Lastly, complete part information is easily accessible to anyone at any time, right from the computer.

5.6   Labeling Facility Equipment

When labeling equipment or inventory, the first thing which must be considered is to develop a logical, informative numbering scheme. Utilizing a meaningful numbering methodology is not specific to any industry or department within a facility. Numbering schemes are used with spare parts, equipments and assets, accounts,... even people! Just consider how many account numbers you may have...Visa cards, bank accounts, mortgage or loan accounts, passports, driver's licenses...the list goes on and on.

An efficient numbering scheme helps find the item or piece of information sought and is essential to optimize the logical sort and search capabilities available through computerization. Before embarking on a new numbering scheme, consider these questions:

1. How does our current organizational procedure work?

2. What key information is most important to know to identify an item?

3. How do we want our organizational procedure to work?

The work flow, procedures, and paperwork used to coordinate activity between maintenance, storeroom, purchasing, and vendors must be examined to determine what works well and where the system fails. Understanding how a manual system functions helps provide a framework for how the new, computerized system should work. This analysis is important in helping you determine how you would *want* the procedures to work.

Once you understand the procedural flow, you can decide what information you want from a number. One important rule is to never tie an equipment or part number to something that is likely to change. What happens when you move a pump if its equipment number is tied to another piece of equipment? Or a part is moved from one location to another and its part number is tied to a location?

There are many ways equipment and inventory can be numbered, but the key is to have the number contain meaningful categorization. The few times that a haphazard parts numbering scheme may work is when tradespeople use the *manufacturer's* part numbers, already known to them all by heart, and never use other manufacturer's parts. Manufacturers, however, can change their part numbers at any time!

5.7   Annual PM Schedule

Many facilities tend to have cycles with peak and slow periods, and sometimes even planned shutdowns. Traditionally, these periods have revolved around seasonal changes. Even in a facility such as a house, "spring cleaning" is both a familiar and applicable annual PM. Historically, maintenance managers have taken the opportunity of slow times or shutdowns to perform yearly maintenance. This makes best use of available resources.

Of all the PM tasks being set up for your facility, annual PMs are usually the most intensive, owing to the relatively long interval. It is likely that many annual PMs will incorporate one or more smaller PM tasks which have been previously defined. We will use a boiler for an example. You could have an annual PM including:

- Visually inspect for tube or other leakage.

- Inspect and clean the burner assembly.

- Drain and chemically clean and treat the tank.

- Test all water-level controls, floats and switches; replace O-rings and gaskets.

- Brush and clean tubes.

- Run through a system startup-shutdown cycle.

- Perform stack test.

Some of the tasks in a detailed, annual PM may normally be done in smaller, more frequent PMs. When the annual PM comes due, it is redundant for the smaller "subset" PMs to be done in addition to the annual PM. Some CMMS systems can group or link PMs together with an ordering sequence. When an annual PM comes due, the smaller "subset" PMs (i.e., 30-, 90-, 180-day PMs), linked to the annual PM will be skipped but have their next due date advanced so they will be due at their next regular interval following the annual PM.

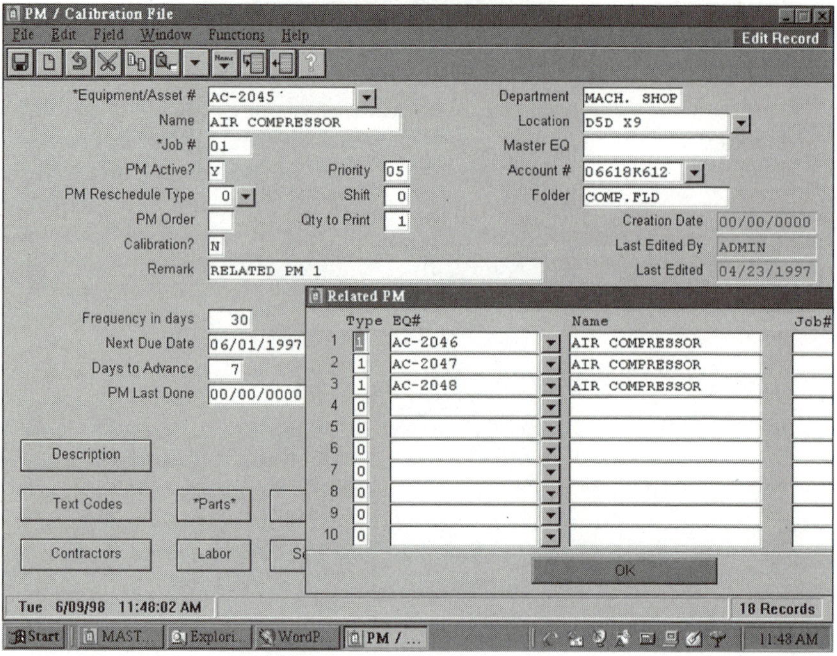

Figure 5.10 Linked or "related" PM.

In AMMS, the PM linking need not only be for annual PMs; any PM can be linked to another. A weekly PM might be a subset of a monthly PM, itself a subset of a semiannual PM. AMMS refers to these smaller PMs as related PMs to the larger, longer-interval PM (see Fig. 5.10).

5.8 Setting Special Schedules

A good PM program must be flexible enough to account for fluctuations in environment and utilization. Most facilities will have peak and off times, and seasonal PMs represent the most obvious use for specialized schedules. For example, refrigeration equipment is utilized in summer and heating equipment in winter.

In the AMMS system, these seasonal periods are treated as ranges for the PM template to actively cycle. The PM is "turned off" when it falls outside the seasonal range. Figure 5.11 shows a PM template with a seasonal frequency.

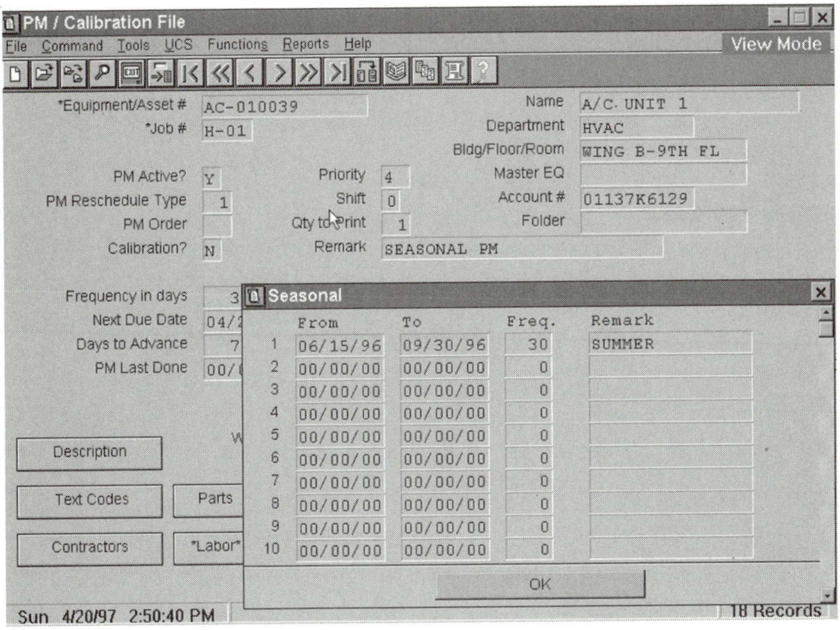

PM / Calibration File

File Command Tools UCS Functions Reports Help   View Mode

*Equipment/Asset # AC-010039   Name A/C UNIT 1
*Job # H-01   Department HVAC
                                          Bldg/Floor/Room WING B-9TH FL
PM Active? Y   Priority 4   Master EQ
PM Reschedule Type 1   Shift 0   Account # 01137K6129
PM Order   Qty to Print 1   Folder
Calibration? N   Remark SEASONAL PM

Frequency in days 3
Next Due Date 04/   Seasonal   x
Days to Advance 7   From   To   Freq.   Remark
PM Last Done 00/   1 06/15/96 09/30/96   30   SUMMER
                     2 00/00/00 00/00/00   0
                     3 00/00/00 00/00/00   0
                     4 00/00/00 00/00/00   0
   Description   W 5 00/00/00 00/00/00   0
                     6 00/00/00 00/00/00   0
   Text Codes   7 00/00/00 00/00/00   0
             Parts   8 00/00/00 00/00/00   0
                     9 00/00/00 00/00/00   0
   Contractors *Labor* 10 00/00/00 00/00/00   0

                                  OK

Sun 4/20/97 2:50:40 PM   18 Records

Figure 5.11 Seasonal window ranges.

5.9 Opening or Dismantling Equipment

When opening or dismantling anything, descriptive and visual guides are more than useful. Where would we be without those cryptic yet vital instructions on toy assembly before Christmas?

A CMMS system can greatly enhance your ability to operate on both everyday and exotic equipment. PM procedures can include detailed descriptions on precisely how you want work to be performed or on how work has been performed in the past by expert mechanics with a lifetime of experience.

Procedures in a CMMS can also include graphic images like cutaway diagrams, parts lists scanned out of manuals, and CAD drawings. Checklists can be created, and nameplate and other reference information can be printed with work orders to virtually provide a small manual on specific repairs. When you have progressed to a point where your CMMS possesses this wealth of knowledge, it can be utilized by less-experienced tradespeople to work on unusual, new, or outdated equipment.

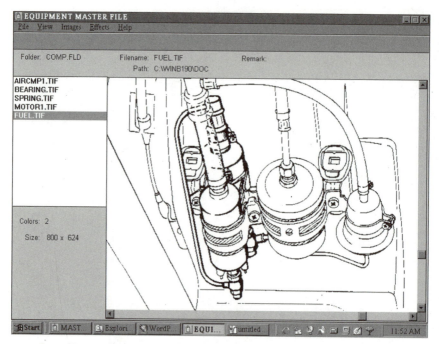

Figure 5.12 Viewing and printing equipment images from a CMMS.

5.10 Computerized Records and Reporting

Once a CMMS system has been set up, and PM and non-PM work orders are being closed, it becomes the permanent repository for all work orders on every asset in your facility. The real beauty of harnessing a computer utilizing a CMMS is its ability to search and sort thousands of work orders in seconds and pull out only the information you need.

Computerized records not only make it quick and convenient to look up information on essentially anything in the system (work orders, assets, parts, vendors, personnel, etc.) past or present but allows you to search and sort this data quickly and easily for reports. The AMMS system comes out of the box with over 100 useful standard reports and allows you in addition to easily create your own custom reports. Table 5.4 shows an example of some of the standard reports that come with AMMS.

Since you will want to run many reports out of the work history file (these are all the work orders that have been closed) let's look at one from the AMMS called the work order cost variance report (Fig. 5.13). In the history files, you also have the option to print or view graphical reports. Figures 5.14 and 5.15 are two popular graphical reports from the work history file.

TABLE 5.4   Examples of Standard Equipment and Inventory Reports in AMMS

Equipment/Work Order Reports	Inventory Related Reports
Equipment Cost Summary	Inventory On Hand Report
Equipment Location Summary	Reorder Report
Equipment Usage Report	Parts Activity Report
Work Order Material Report	Parts Usage Report
Work Order Labor Report	Charge Account History
Work Order Cost Variance Report	Open/Closed Purchase Orders Report
Repair Cost Summary	Vendor Bid Report
Downtime Summary	

WO COST VARIANCE REPORT

12:51 04/20/97 Page 1

ALL RECORDS SELECTED

WO Number	Equipment #\Name		Parts Cost	Labor Cost	Total Cost
95010100001	LT-010	Actual:	0.00	55.73	55.73
	LATHE	Projected:	1.50	64.02	65.52
		Variance:	-1.50	-8.29	-9.79
95010300001	LT-001	Actual:	166.60	99.33	275.93
	LATHE ENGINE	Projected:	83.30	89.76	183.06
		Variance:	83.30	9.57	92.87
95011000001	LT-010	Actual:	119.60	56.74	196.34
	LATHE	Projected:	149.50	81.05	250.55
		Variance:	-29.90	-24.31	-54.21
95011500002	DR-054	Actual:	300.00	78.24	478.24
	DRILL RADIAL	Projected:	300.00	94.86	496.86

1 / 10 x 0.75

Sun 4/20/97 12:57:38 PM   94 Records

Figure 5.13 Work order cost variance report: work history.

Another benefit of computerizing your records is that you can provide legitimate computerized documentation if ever needed for certification or regulatory inspections.

Inspections and certificates are used with the PM "calibration" work orders. These "cal" work orders require different forms than

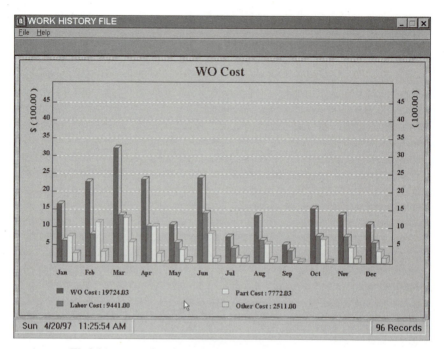

Figure 5.14 Work history work order cost graph.

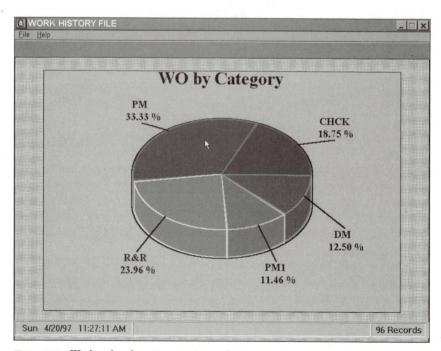

Figure 5.15 Work orders by category pie graph.

CALIBRATION   CERTIFICATE

MICROWEST
SOFTWARE SYSTEMS, INC.

98060900006	14:08	06/10/98	Page 1

Instrument Description:
HVAC UNIT 1
Equipment: HV-0021

Manufacturer: TRANE
Model #:   HC-2001
Serial #:   791H71
Room: WEST END
Test Freq: 30
Next Due Date: 07/09/1998

Reference Standard   1:

Standard Name:	CAL-C		Range:	Low	High	Units
Std./Tag #:	001			10	20	DEG

Calibration Type: Routine__X__   Event____
Event:_____

Last Approved By: Calibration Last Done: 06/09/1998

Std. Used	Tol.	Ref. Values	As Found	As Left	Dev. or P/F	Comment
CAL-C	+-5	15	18	18	3	PASS

Comments on Calibration: _____

Figure 5.16   Calibration certificate.

standard work orders as they need to record columns of numerical data as well as signatures. Other than that, they store and generate calibration work orders much the same way PMs are handled. They can be generated separately, and produce a calibration certificate.

In many facilities (such as hospitals or pharmaceutical companies), many instruments will need to be calibrated regularly with resulting certificates of completion for these jobs mandated. Figure 5.16 shows an example calibration certificate from AMMS.

In summary, a good CMMS system is comprehensive, flexible, and easy to use. It must be able to accommodate data entry in a quick, systematic, and logical way so that the end user will be motivated to use the system. The desire to maintain such a system from a company standpoint comes from the increased productivity, efficiency, and cost savings which will be achieved with a successful installation.

6

Predictive Maintenance Plan

Joseph E. Humphrey

Potomac Electric Power Company
Upper Marlboro, MD

Predictive maintenance can be defined as a maintenance philosophy in which equipment condition is monitored at appropriate intervals to enable accurate evaluation to use as input when determining whether maintenance action or "no action" is required without sacrificing equipment reliability.

Effective, reliable operation of a facility dictates that the facility's availability and intended functionality be optimized. One aspect of the optimization process is designing, implementing, and following through with an effective predictive maintenance program. Facilities maintenance has evolved from a "breakdown" (if it's not broken, don't fix it) philosophy in the fifties and early sixties to a preventive maintenance (time-based) philosophy in the seventies, progressing to a predictive philosophy at present. This chapter is a summary of some of the most effective predictive maintenance technologies and processes which can help to optimize the availability, reliability, and profitability of both large and small facilities.

6.1 Benefits of Predictive Maintenance

Knowing the condition of your equipment at any point in time provides valuable information which can then be used as input to:

1. Anticipate the need for repairs by identifying faults in the early stages of development.

2. Order required parts only when needed to avoid stocking unnecessary inventory.

3. Schedule appropriate resources for repairs well in advance to minimize the impact on the process.

4. Eliminate unnecessary preventive maintenance tasks.

5. Avoid catastrophic failures and minimize additional damage which may occur.

With these benefits in mind, it is easy to understand how a formal, comprehensive, and effective predictive maintenance program will provide cost savings which will improve the profitability of any facility. The actual savings realized by a particular facility will vary based on the size of the facility, the quantity and type of equipment involved, and the extent and effectiveness of the program.

6.2   Predictive Maintenance Technologies

Many technologies or "tools" of predictive maintenance are available which aid in predicting the need for maintenance action or no action. Perhaps the simplest tool is that of the operating and maintenance personnel. They know the equipment and are in close proximity to it on a daily basis. Therefore, their eyes and ears are capable of detecting very subtle changes in its operating condition. As such, they should be directed and encouraged to be alert for these changes and, when changes are detected, to act to have the equipment's condition analyzed further for abnormalities. When looking past the personnel for predictive maintenance tools, four specific technologies stand out as the most effective for application in most facilities. These are:

- Vibration monitoring
- Infrared (IR) thermography
- Oil analysis
- Ultrasound

6.2.1   Vibration Monitoring

The overall levels of vibration which are considered "acceptable" can be obtained from a number of sources including:

- Equipment manufacturer's recommendations
- Industry guidelines
- Historical levels of similar machines

Figure 6.1 is one of many charts published to provide a guide to operating and maintenance personnel to help determine the overall condition of their machinery. However, it is just a snapshot look at vibration severity and must be used only as a guide.

Once it is determined that the vibration on a machine exceeds the acceptable limits, additional analysis must be performed to determine

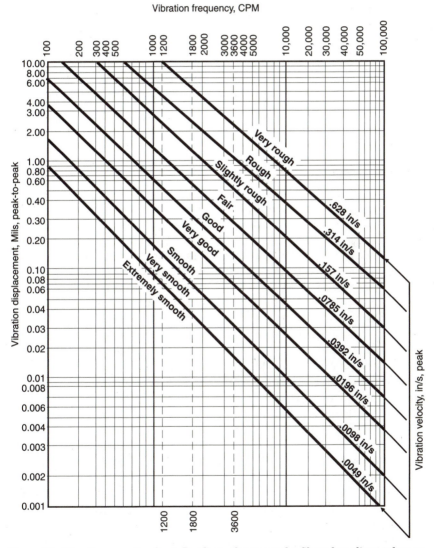

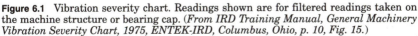

Figure 6.1 Vibration severity chart. Readings shown are for filtered readings taken on the machine structure or bearing cap. (*From IRD Training Manual, General Machinery Vibration Severity Chart, 1975, ENTEK-IRD, Columbus, Ohio, p. 10, Fig. 15.*)

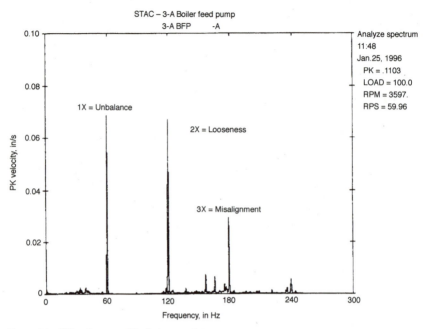

Figure 6.2   Vibration amplitude versus frequency.

the source. This is accomplished by identifying the frequencies of the vibration generated by the machine. The specific frequencies can be related to specific faults and components of the machine. For instance, for a machine that operates at 3600 rpm (60 Hz), any unbalance of the machine's rotor will be represented by the existence of vibration at a frequency of 60 Hz. The unbalance may be the result of wear, loss of material, or material buildup. Once identified as "unbalance," steps can be taken to reduce the unacceptable levels of vibration by balancing the rotor, repairing worn parts, or cleaning the buildup from the rotor. Similarly, a looseness in the same machine or its mounting may exhibit a frequency of 120 Hz, or two times its rotating speed, as a result of the rocking motion resulting from looseness. Figures 6.2 and 6.3 show typical plots of vibration amplitude versus frequency, one of the more effective displays of a machine's characteristics used in the diagnostic process.

Additionally, many machine faults can be identified by the frequency, direction, phase relationship, and other characteristics of the vibration exhibited by the machine. Table 6.1 shows the relationships of various characteristics of vibration to specific machinery faults. It should be noted that the relationships are based on probabilities and are not intended to be hard and fast rules. Machinery analysts are

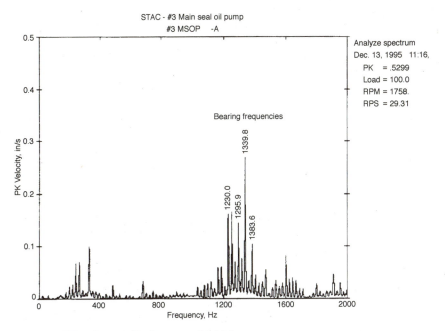

Figure 6.3 Vibration amplitude versus (high) frequency.

continually finding machines which exhibit unusual vibration characteristics which challenge their analytical skills when they fail to fit the generally accepted rules of diagnostics.

In order to have an effective program the vibration exhibited by a machine must be trended over time. To accomplish this there is a considerable range of vibration data loggers on the market. The basic instrumentation will display, in analog form, overall vibration levels which can then be recorded and trended manually. The latest microprocessor-based data collectors are very powerful and function as fully functional fast Fourier transform (FFT) analyzers capable of recording frequency spectra, time waveform, and other relevant vibration data in various forms. This data is then stored, trended, viewed, and manipulated through standalone or networked PCs. The current software packages available for this purpose can be very comprehensive, with the more advanced versions offering expert systems to aid in the diagnosis of the vibration data.

Regardless of the particular hardware and software package chosen, vibration monitoring is an imperative predictive tool for any facility which relies on rotating machinery. Many vibration monitoring programs in a wide range of industries have recovered their costs in less than 1 year.

TABLE 6.1   Vibration Frequency and the Likely Causes

Frequency in terms of rpm	Most likely causes	Other possible causes and remarks
1 × rpm	Unbalance	1. Eccentric journals, gears, or pulleys 2. Misalignment or bent shaft—if high axial vibration 3. Bad belts if rpm of belt 4. Resonance 5. Reciprocating forces 6. Electrical problems
2 × rpm	Mechanical looseness	1. Misalignment if high axial vibration 2. Reciprocating forces 3. Resonance 4. Bad belts if 2 × rpm of belt
3 × rpm	Misalignment	Usually a combination of misalignment and axial excessive clearances (looseness)
Less than 1 × rpm	Oil whirl (less than ½ rpm)	1. Bad drive belts 2. Background vibration 3. Subharmonic resonance 4. "Beat" vibration
Synchronous (ac line frequency)	Electrical problems	Common electrical problems include broken rotor bars, eccentric rotor, un-balanced phases in polyphase systems, unequal air gap
2 × synchronous frequency	Torque pulses	Rare as a problem unless resonance is excited
Many times rpm (harmonically related frequency)	Bad gears Aerodynamic forces Hydraulic forces Mechanical looseness Reciprocating forces	Gear teeth times rpm of bad gear Number of fan blades times rpm Number of impeller vanes times rpm May occur at 2, 3, 4, and sometimes higher harmonics if severe looseness
High frequency (not harmonically related)	Bad antifriction bearings	1. Bearing vibration may be unsteady—amplitude and frequency 2. Cavitation, recirculation, and flow turbulence cause random high-frequency vibration 3. Improper lubrication of journal bearings (friction-excited vibration) 4. Rubbing

6.2.2   Infrared (IR) Thermography

In the last several years monitoring equipment temperatures using infrared measurement technology has become a very effective predictive maintenance tool. One reason for this is that the cost of the measurement instrumentation has in some cases become more reasonable while the performance has improved considerably. Also, we have

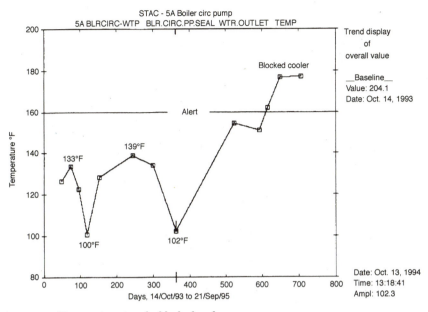

Figure 6.4 Temperature trend—blocked cooler.

learned that most equipment experiences an increase in temperature as faults develop. Identifying those faults early in their inception provides valuable time for planning repairs, procuring necessary parts, and allocating resources with minimum disruption of the facility's process.

The measurement instrumentation may be a simple infrared measurement device which can be interfaced with the same data logger used for vibration measurement. Measurements such as machine bearing temperature, cooler inlet and outlet temperatures, motor air inlet and outlet temperatures, etc., can be stored, trended, and automatically alarmed using the same computer hardware and software used with the vibration monitoring portion of the predictive program. Figure 6.4 is an example of a trend of temperature using this method.

Figure 6.5 indicates the resulting reduction in temperature after the cooler was found to be partially plugged and subsequently cleaned. This action prevented serious damage to the machine.

More sophisticated thermal imagining cameras are used as comprehensive analysis tools and are most effective for quick scanning of a large number of components for fault identification, as well as for applications where heating or cooling is a process variable. Infrared imaging has proved particularly effective on:

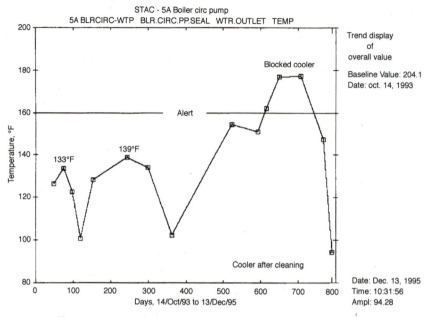

Figure 6.5 Temperature trend-cleaned cooler.

- Motors
- Motor control centers
- Load centers
- Breakers
- Circuit board components
- Transformers
- High-voltage disconnects
- Transmission and distribution systems

Figure 6.6 is an example of the type of electrical fault which is easily identifiable with a quick scan on an energized circuit. It is important to realize that not only are the faults identified, but during a typical thermal survey better than 90 percent of the equipment can have routine, preventive maintenance deferred. No longer will the periodic, time-consuming task of checking each connection for corrosion and looseness be necessary. As such, only the maintenance that is required will be identified and accomplished making effective use of resources.

In order to classify the severity of electrical heating anomalies, guidelines have been established by various institutions such as the

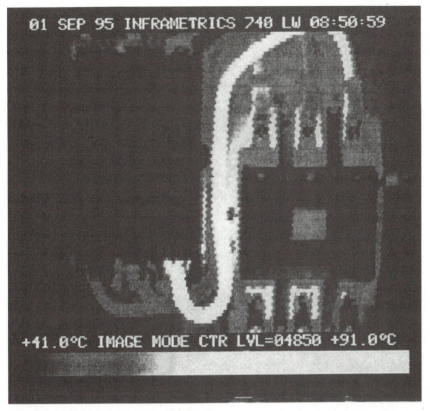

Figure 6.6 Thermal image of electrical fault.

TABLE 6.2 Temperature Rise Classification Guide

Minor problem	1–10°C rise	Increase monitoring frequency
Intermediate problem	10–20°C rise	Repair within 1 year
Serious problem	20–30°C rise	Repair in near future
Critical problem	30°C or greater rise	Repair immediately

Electric Power Research Institute (EPRI) to aid in the planning and scheduling of repairs. Table 6.2 is one example of these guidelines.

In addition to the electrical applications already mentioned, infrared (IR) thermography can be effectively applied to the following:

1. Insulation degradation

   ▪ Building siding and roofing
   ▪ Steam piping
   ▪ Heated and refrigerated spaces, tanks, etc.

- Boiler and furnace walls
- Electric motor hot spots

2. Steam traps

3. Leaking valves

4. Oil, water, and other liquid tank levels

5. Below-grade piping leaks and ruptures

6. Rotating machinery faults

   - Seal rubs
   - Faulty bearings
   - Coupling misalignment
   - Misaligned belts
   - Uneven heating and cooling due to blockages

7. Restricted flow due to blockages

   - Water or fluid lines
   - Coal supply lines
   - Boiler tubes

8. Process applications

   - Mass production utilizing a heating and cooling process
   - Bonding and lamination processes

9. HVAC

   - Leaks, blockages, restrictions
   - Proper heating and cooling distribution
   - Compressors, fans, and motors

Routine temperature monitoring at regular intervals is a necessary aspect of any facility's predictive maintenance program. However, the extent and frequency of monitoring will vary with the equipment monitored and its importance to the process. For instance, monitoring and trending the temperature of the bearings of the critical rotating machines weekly or monthly is reasonable. Using a thermal imaging camera to scan all motor control centers and breakers may be required only on an annual basis. Again, these intervals will vary and should be defined during the initial setup of the predictive program and reviewed and revised as needed.

6.2.3   Oil Analysis

Effective monitoring of equipment condition includes development and implementation of a formal lubricant analysis program. The program serves two major functions:

- Identifying lubricant condition
- Identifying equipment wear

Knowing the condition of the machinery lubricant provides valuable knowledge which can be used as an integral part of a facility's overall maintenance program. Many facilities have machinery which requires large quantities of lubricants. These lubricants are periodically replaced even if, based on their condition, they do not require replacement. When the oil is removed and discarded, valuable information about the wear of the components may be discarded with the oil. Each time the lubricant is replaced, environmental concerns must also be considered.

6.2.3.1 Lubricant condition. Monitoring oil to determine its condition includes a look at contamination from water, fuel, dirt, and other substances. Knowing the type and quantity of contaminants will often lead to identification of other machinery problems which can then be addressed. The lubricants have been chosen by the equipment manufacturer for their particular properties for each application. Some of these properties are achieved by additives. If a facility is interested in deferring unnecessary oil changes, knowing the oil has the desired properties is imperative. Therefore, appropriate tests are performed to monitor these properties.

Many oil monitoring programs, particularly during the initial sampling, have identified improper oil grade or type for a particular machine. This condition could explain premature failures and the need for unnecessary repairs. This discovery alone may be justification for initiating oil monitoring.

6.2.3.2 Equipment wear. The predictive maintenance process is aimed at predicting when maintenance is needed and when it is not. While vibration monitoring is certainly the most widely used tool for determining rotating machinery condition, oil analysis will, in many situations, provide an earlier indication that abnormal or premature wear is in progress. Oil monitoring and analysis is especially appropriate for slow-speed machines, reciprocating machines, and gearboxes, as they usually show developing faults earlier using oil versus vibration analysis. As internal machine components wear they leave the wear particles in the lubricating oil. Identifying the existence, size, shape, and elements of the wear particles leads to identifying the particular component experiencing the wear. This valuable information can then be used to aid in determining the ability of the machine to continue operating, planning for repairs, ordering neces-

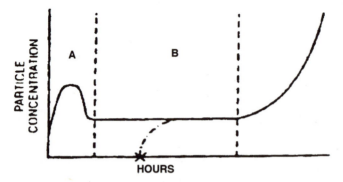

Figure 6.7   Idealized wear curve (Predict/DLI). [*From Wear Particle Atlas, Predict/DLI, Idealized Wear Curve (Ferrographic Data), p. 131, Fig. 3.1.1.4.*]

sary parts, and preventing unnecessary, unplanned downtime. The wear may be the result of many mechanisms including improper maintenance practices, misalignment, overheating, improper lubrication, improper operation, or poor design application. By monitoring and trending the wear particles, initial levels can be established and changes can be recognized. Figure 6.7 illustrates the typical wear pattern of machine components over time.

The initial break-in period will indicate an increase in wear particles as the parts wear in and the particle concentration stabilizes. With continued monitoring at appropriate intervals, excessive wear will become evident by an increase in the wear particle concentration. At that time a comprehensive analysis may be initiated to determine the particle type, wear mechanism, and the source.

6.2.3.3   Oil testing.   Numerous tests and instruments have been developed to provide accurate information about the condition of machinery lubricants. Initial testing should be a screening process to identify those cases which require additional assessment and to eliminate the need to further evaluate the vast majority of machines which are operating properly. As with any predictive technique, each machine should be evaluated to determine which tests are appropriate and to assign the appropriate monitoring frequency.

Sampling methods have a great impact on the results of the oil analysis program. The sampling is where the analysis process begins, and if not done properly it will affect the results, including the possibility of performing unnecessary maintenance. Formal sampling guidelines should be established and adhered to. Many oil analysis laboratories will provide sampling guidelines for the program.

The specific tests which are determined to be appropriate and cost-effective can vary considerably depending on the particular machine, its service and environment, the type of lubrication used, and the laboratory performing the testing. Some large facilities may have the resources to perform some or all of the tests in-house. However, most facilities find that contracting the testing from a credible laboratory is effective, accurate, and reasonably priced. Many modern laboratories have mechanized testing devices which allow for large quantities of samples to be processed each day while maintaining very competitive costs. Also, some laboratories provide the sample results in electronic form which can be downloaded to a facilities predictive maintenance database for trending, analysis, correlation with other predictive technologies, and reporting. Table 6.3 shows some of the tests which may be used to determine the condition of the lubricant. Table 6.4 indicates some of the more common tests performed to identify wear particles used for determining machine condition.

TABLE 6.3 Lubricant Condition Tests

Test parameters	Test options	Comments
Viscosity	ASTM D445	Measures oil's resistance to flow
Oxidation	Total acid number (TAN) ASTM D974/D664	Determines acidity level
	Total base number (ASTM D2896)	Determines alkalinity level. Used to detect fuel or coolant contamination
	Fourier transform infrared analysis (FTIR)	Screening test only. Exceptions require TAN test
Water	Appearance	Visual check
	Karl Fischer reagent (ASTM D1744)	Gives water content in ppm. Accurate to 10%
	Fourier transform infrared analysis	Exceptions require Karl Fischer reagent test
	Crackle test	Oil on a hot plate. Subjective test
Solids	Light extinction	Oil passed through photo-detectors. Particles are counted and classified by size
	Mesh obscuration	Oil forced through different-sized meshes and counted under microscope

TABLE 6.4   Machine Wear Tests

Wear particle test	Test description
Spectroscopic analysis	Determines concentration (ppm) levels of key elements in oil
Direct ferrography	Provides large and small ferrous particle count
Analytical ferrography	Quantifies extent, type, and distinguishing features of particles

These are only some of the many tests available for analyzing oil condition. There are also a variety of opinions as to which tests are the best. Discussions with suppliers of the services and instrumentation will provide insight into which combination of tests is appropriate for a particular facility.

Although most suppliers of lubricating oils are reputable, a facility's oil monitoring program can also be used to provide a means of acceptance testing on deliveries of new lubricating oils. This also provides a baseline of values for comparing future samples drawn from the machinery.

Knowing the condition of the lubricating oils provides valuable information which can be used as one of the many facets of a facility's predictive maintenance program. When properly developed, implemented, and periodically updated, it will be an effective tool for controlling maintenance costs.

6.2.4   Ultrasound

High-frequency sound is generated by many mechanical and electrical systems. This sound is not audible to the human ear, which hears in a frequency range up to 20,000 Hz. Instead, the high-frequency, low-level sound generated by the early stages of improperly lubricated bearings, leaking valves, and other faults appears in the frequency range above 20,000 Hz. To detect these faults early in their development, monitoring instrumentation has been developed which detects and converts this high-frequency sound into the audible range for detection by the analyst with headphones and for recording on data loggers and printers if desired.

The ultrasound frequency range used for fault detection is typically in the range of 20,000 to 100,000 Hz. This high frequency has a relatively short wavelength and therefore tends to travel in straight lines and over short distances. This characteristic provides an opportunity to use ultrasonic detection to zero in on particular faults, such as an improperly lubricated bearing, a leaking valve, or leaks in pressurized systems. Identifying the exact component early in its deterioration will provide opportunity to plan for repairs, order necessary parts, and schedule resources with minimum impact on a facility's process.

Ultrasonic detection can be useful for airborne noise as well as contact measurements. In either application the filtering of sound below approximately 20 kHz effectively blocks out the typical machinery background noise, allowing concentration on fault frequencies. Although different faults generate typical frequencies, the specific frequencies may vary with varying situations. Table 6.5 can be used as a guide to help identify some of the typical faults and the frequencies they generate.

6.2.4.1 Ultrasound applications. As seen in Table 6.5, there are a number of applications for this technology. One very effective and widely used application is identification of leaks in pressurized systems. A system under pressure that develops a leak to atmosphere will generate high-frequency noise owing to the expansion of the gas or liquid moving through an orifice from a high-pressure to a low-pressure environment. Similarly, vacuum leaks may also be detected, although this is usually more challenging.

Detecting leaking valves is also an effective application of ultrasonic technology. If the system containing the valve is pressurized, a closed, leaking valve will also generate a high-frequency noise. With valve leak detection, using a contact probe is more appropriate than using airborne detection because it eliminates the background noise from the surrounding area.

Antifriction (ball and roller) bearings in rotating machinery are also candidates for ultrasonic detection of developing faults. These bearings, even when new, generate high-frequency noise. However, when faults develop, the amplitude of the high frequency will increase. Based on NASA research, the following gains in decibel levels above baseline indicate various stages of bearing failure:

8 dB	Prefailure or lack of lubrication
12 dB	Beginning of failure
16 dB	Advanced failure
35–50 dB	Catastrophic failure

---

*Adapted from UE Systems, Inc., Ultrasonic Applications, Bearing and Mechanical Inspection.

TABLE 6.5 Ultrasound Frequency Chart

| | Suggested module | | | | | | | | | | |
	Fixed band	20 kHz	25 kHz	28 kHz	32 kHz	40 kHz	50 kHz	60 kHz	100 kHz	Meter mode	Selection
Steam traps	X		X			X				Log	Stethoscope
Valves		X				X				Log	Stethoscope
Compressors (valves)	X		X					X	X	Log	Stethoscope
Bearings	X			X						Lin	Stethoscope
Pressure and vacuum leaks	X				X	X				Log	Scanner
Electrical arcs (and corona)	X					X				Log	Scanner
Gears		X	X							Log	Stethoscope
Pumps (cavitation)	X	X	X							Log	Stethoscope
Piping systems (underground)	X	X				X				Log	Stethoscope
Condenser tubes	X					X				Log	Scanner
Heat exchangers (tone method)	X									Log	Scanner

Source: *Ultraprobe 2000 Instruction Manual*, Frequency Selection Chart, UE Systems, Elmsford, NY.

When electrical equipment begins to fail, it may produce arching or corona. If heat is generated and the equipment is accessible, infrared thermography may be used to detect the problem. However, if no heat is generated or the equipment is isolated, ultrasound can be an effective tool for fault identification. Both of these faults generate ionization and produce high-frequency airborne noise which is detectable with the ultrasound instrumentation.

Reciprocating compressors used for compressed-air systems are often critical to the operation of a facility. Their unexpected failure is expensive in terms of both resources and lost production. Therefore, it is important to be able to identify potential problems with these machines. They have valves which must operate properly. It has been found that when they are leaking or sticking they exhibit a change in their sound pattern. This change can be detected using ultrasonic instrumentation. Figure 6.8 shows a comparison of a good valve and a failing valve using an ultrasound detector output providing input to a FFT analyzer.

Ultrasound is a relatively easy to use and inexpensive predictive tool which, for most facilities, will provide a very rapid return on investment.

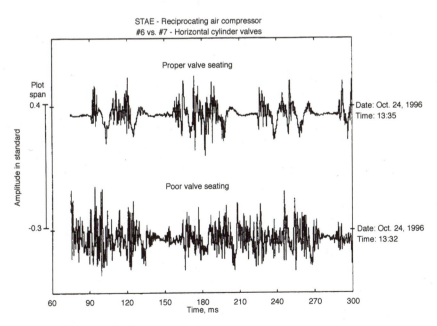

Figure 6.8 Ultrasound valve comparison.

6.2.5 Integration

Although not actually a tool, integration of all the predictive maintenance technologies is a powerful diagnostic function. Also, integrating the predictive program with all the maintenance functions should be the goal of every facility.

Preventive maintenance (PM) has largely been a function where certain inspections and replacements have been performed at certain time intervals such as monthly, quarterly, or yearly. Countless worker-hours have been spent inspecting equipment that has no problems. Many machines have experienced failure following preventive maintenance action and may be referred to as "maintenance-induced failure." If a facility's maintenance program were integrated, it would use the results of the predictive program to influence the need for preventive inspections and repairs. There are certainly justifiable reasons for performing time-interval maintenance. However, integrating the condition of the machine, previous history, and manufacturers' recommendations would make the program more effective by eliminating unnecessary PMs, reducing maintenance-induced failures, and approaching "just in time" maintenance.

To illustrate the potential benefits of integrating a predictive maintenance program's technologies, consider a scenario where a large machine, critical to the process, is identified as experiencing an increase in the wear particle concentration in one of the bearing oil sumps. The recommendation from the lab is to increase the frequency of oil sampling. A subsequent oil sample shows another increase in wear particle count and, through analytical ferrography, identifies components of the bearing cage as the major contributor. A recommendation is made to shut down the machine for inspection of the bearing. With several days left in the production cycle, shutting down the machine now would result in a considerable loss of output. However, continuing to operate the machine in this condition may result in a catastrophic failure which would be even more costly in terms of repairs and downtime. In this situation the use of vibration analysis and perhaps temperature monitoring may provide additional information which could be used to make a more informed decision regarding the course of action for this machine. For example, if the high-frequency vibration on this bearing showed a somewhat elevated level and, through a review of previous similar cases, it is known that at those levels these bearings have operated for several weeks without failure, the course of action becomes more clear. At that time the frequency of monitoring vibration and temperature should be increased to several times a day until the production cycle is complete and repairs can be accomplished.

Knowing the condition of the facility's equipment by using and integrating all the predictive tools enables managers to make more informed maintenance decisions.

6.2.6 Expert Systems

Expert Systems pertaining to predictive maintenance programs are currently under development, with some fairly comprehensive packages available for vibration analysis. These systems still rely on interaction with a knowledgeable analyst; however, they do provide a means of speeding the process of reviewing large amounts of data. The rule-based Expert System is useful for providing suggestions for probable fault based on the information provided by the recorded data and by the machinery analyst. Vibration analysis is still considered to be an art as well as a science. This challenges the current Expert Systems and dictates the need for trained and experienced machinery analysts.

As the Expert Systems become more advanced and utilize information learned through the use of neural networks, they should prove to be a very effective tool for integrating and speeding the predictive maintenance process.

6.3 Predictive Process

Throughout the development of a predictive maintenance program it is important to remember the goal of the program. That goal is to provide the information needed to make decisions on the need for maintenance action or no action without sacrificing reliability. In order to achieve this goal, the program must be set up so that the proper parameters are monitored on the appropriate equipment at the appropriate frequency. There may be any number of approaches to the predictive process. What follows is one approach which will highlight some of the important considerations necessary in the development of the predictive maintenance program.

6.3.1 Program Startup

Once the decision is made to develop and implement a formal, integrated predictive maintenance program, the manager must ensure there is the needed support in the way of adequate resources to ensure the success of the program. Many predictive programs have failed because they lacked the needed support from management. This is true for any program at any facility. Without the support, it will fail. With that understood, the following steps need to be accomplished, though not necessarily in the same order. The extent to which each step is carried out will be somewhat different for each facility. It will depend on the size of the facility, the facility's process, and the level of commitment desired.

1. Select personnel.
2. Identify equipment to be monitored.
3. Select predictive technologies.
4. Establish monitoring frequency.
5. Select instrumentation (in-house applications).
6. Implement program and establish baseline data.
7. Set alert levels.

6.3.1.1 Select personnel. For the predictive maintenance program to succeed it must have dedicated, motivated personnel with the authority to perform the required functions. They must also be held accountable for the results of the program. These requirements are no different from those of any other program if the expectation is success. Many predictive programs have failed because the personnel perform the predictive maintenance function only part time when "things are slow" or when there is an emergency which requires immediate attention. Consider the manager of a facility who has other duties and only manages the facility part time! The results are in proportion to the effort. The same goes for the predictive maintenance personnel. This is not to say that the individual(s) cannot perform other functions. However, the program must be taken seriously in order to have a chance of success.

Another important factor in the success of the program is the training of the personnel. The entire field of predictive maintenance including its tools and techniques continues to evolve along with the development of other technological areas that depend on microprocessor-based instrumentation and software. To keep informed and continually more effective, continued formal training is imperative. Fortunately many vendors are available to provide the required training. Many of the suppliers of the predictive instrumentation will provide, free of charge or for reasonable fees, valuable training specific to their tools.

6.3.1.2 Identify equipment to be monitored. When selecting the equipment to be included in the monitoring program, a number of considerations need to be addressed. These include:

- Criticality of the equipment
- Cost to repair or replace
- Cost of downtime
- Past maintenance history
- Manufacturers' recommendations
- Program resources and limitations

The more critical the equipment to the continued operation of the process, the more seriously it should be considered for monitoring. If the failure of a piece of equipment reduces the facility's output, it is imperative to know the condition of the equipment and to know that it can be relied upon to provide continued service. Also, knowing the existence of a developing fault and the extent of the fault will provide information for scheduling repairs at the earliest opportunity without adversely affecting facility output—the goal of the predictive program.

The cost to repair or replace a piece of equipment should be one of the factors to consider when selecting candidates for the program. Equipment which has a long lead time for obtaining parts or replacement should be considered for the program. Generally, the greater the cost of the equipment, the greater the need to monitor. A comparison must be made between the cost to monitor and the cost of repair or replacement. For example, spending resources to monitor an inexpensive ventilation fan for bearing failure is not reasonable when the cost of the bearing repair or fan replacement is insignificant and it does not adversely affect the facility's process significantly. Let common sense prevail. Predictive maintenance is, after all, a commonsense maintenance philosophy.

As mentioned above, if the equipment's failure causes downtime for the process, the condition of the equipment should be known. This avoids surprises and reactionary maintenance, which is almost always the most expensive.

A machine's past maintenance history may dictate whether it is a candidate for the predictive program. With experience and accurate machinery history records, personnel can effectively direct the program resources where they will be best utilized. A machine which historically has had very few maintenance problems, requires little attention, and just continues to operate as designed should be considered for only basic monitoring of critical parameters at infrequent intervals. In contrast, a machine which has a history of failure and adversely affects the process should be included in the predictive process to provide the early warning of possible failure and can be repaired during a scheduled shutdown.

Some equipment is supplied with recommendations from the manufacturer describing the parameters to monitor, the frequency of monitoring, and the allowable limits. These manufacturers' recommendations should be incorporated into the predictive program for that piece of equipment. In some cases a phone call or written request to the manufacturer will provide recommendations of parameters to monitor.

Many facilities will find that as they develop their predictive program they do not have the resources available to include all of the equipment, measurement points, parameters, or optimum monitoring frequency to achieve a complete monitoring program. When first establishing a program it is advantageous to start small and grow gradually. This approach provides time to establish the process and make adjustments while the program is more manageable at the smaller size and complexity. Trying to take a program too far, too fast can lead to failure. As every facility is different, every predictive program is unique and must be developed, implemented, and periodically modified as it grows. It is much easier to make these initial modifications while the program is still small and more easily managed.

6.3.1.3   Select predictive technologies. Once the equipment to be included in the program is selected, the predictive technologies which are appropriate for each piece of equipment must be determined. This process involves a number of considerations including the following:

- Equipment's likely failure mode
- Manufacturer's recommendations
- Maintenance and failure history
- Benefit versus cost of monitoring
- Preventive maintenance schedule and activity

After these areas are evaluated, the applicable predictive technologies should become apparent. Again, common sense should prevail. If, for example, the temperature of a machine bearing increases and it has been predetermined that no actions will be taken until the vibration indicates action is necessary, perhaps temperature measurements are unjustified.

6.3.1.4   Establish monitoring frequency. With the appropriate monitoring technologies identified for each piece of machinery, the next step is to select the proper monitoring frequency. Throughout the initial setup of the program it is important to remember that all aspects of the program can and should be modified periodically. Therefore, it is not critical that the optimum mix of technologies or monitoring frequencies be established initially. When selecting the frequency for monitoring equipment the following must be considered:

- Criticality of the equipment
- Maintenance and failure history
- Preventive maintenance schedule
- Lead time to failure after fault initiation
- Benefit versus cost of monitoring

If resources permit, it is better to monitor more frequently at the start of data collection to establish a reasonable baseline value for each of the parameters monitored on each of the measurement points. Along with selecting the frequency of monitoring, the actual measurement points need to be identified. Of critical importance is the recording of data from precisely the same point each time. This will help to ensure accurate and reliable information from which maintenance decisions can be rendered.

6.3.1.5 Select instrumentation. At some point in the setup process the determination must be made as to whether the program will be administered in-house or contracted from a supplier of the service. Factors such as cost, program ownership, and the size and complexity of the facility need to be considered. Some technologies are best applied through the use of a vendor. Oil analysis, for example, is often performed by an established oil lab and only the sampling is performed in-house. For those technologies which will be applied in-house by the facility's personnel, the necessary instrumentation hardware and software must be evaluated and purchased. Many vendors are available to supply both monitoring and diagnostic hardware and software. Some of the same vendors, as well as others, will provide the service under contract.

6.3.1.6 Implement program. Once the program completes the initial setup, implementation begins. The implementation may be accomplished in phases to limit initial costs, provide training time, and allow for modifications during the learning process. As mentioned earlier, it is often advantageous to start small to give the program the best chance of success.

The initial recorded data, whether it is vibration-, temperature-, or oil-related, will be used to establish baseline levels to which subsequent data can be compared. Ideally, baseline data should be recorded when the equipment is installed and acceptance testing complete.

However, reality doesn't always provide that opportunity. Experience has shown that, during initial data collection, there will be what seems like a large number of parameters which are above alert levels or appear out of the expected range. This is typical, and as the program matures and initial problems are addressed, incidents of expectations will decrease and stabilize. As significant maintenance is performed or equipment replaced new baseline data must be recorded.

6.3.1.7   Set alert levels.   Alert levels must be established for each parameter measured to provide a flag or trigger for additional action. Initially these levels may be established based on recommendations from the equipment manufacturer, industry standards or guidelines, or the instrument suppliers. Once a representative number of data are recorded and trends of typical operating levels are established, the alert levels can and should be evaluated and adjusted. These modifications will prevent the repeated flagging of equipment which is operating satisfactorily when alert levels are set too low. It will also minimize, for levels set too high, not being alerted to potential problems.

6.3.2   Measure Program Effectiveness

As with any program, the predictive program must be monitored for effectiveness. This will provide the opportunity to identify the value of the program, compare it to the other aspects of the integrated maintenance functions, and provide a basis for justifying and measuring expansion of the predictive effort. Programs that add no value to the process will eventually fail. Therefore, it is necessary to document and even publicize the value of the predictive program if it is expected to survive. Knowing how the program compares in value to the other maintenance functions provides insight into how resources should be directed.

Measurement of the program's effectiveness can be accomplished in a number of ways including:

- Overall maintenance expenses
- Unscheduled repairs
- Process downtime
- Savings from deferred maintenance
- Avoided catastrophic failure

The measurement process may seem unnecessary at first, but it will prove valuable when modifications to the maintenance program are considered and when justification of the predictive program becomes necessary.

6.3.3 Review and Revise Program

The predictive maintenance program of any facility should be viewed as a living program. It should be routinely evaluated for its impact and effectiveness. The entire process should be periodically scrutinized to determine whether the appropriate equipment is monitored, the proper parameters and measurement points are included, and the proper frequency of monitoring is applied. This will help to ensure the resources are being utilized fully and not being expended on unnecessary monitoring or analysis. After all, the predictive maintenance philosophy embraces performing maintenance only when required without sacrificing reliability. To achieve this, the program must be reviewed and revised periodically.

Suggestions for Further Reading

Crawford, Arthur R., *The Simplified Handbook of Vibration Analysis,* vols. 1, 2, Computational Systems, Inc., Knoxville, TN, 1992.

Harris, Cyril M., *Shock and Vibration Handbook,* 3d ed., McGraw-Hill, New York, 1988.

Jackson, Charles, *The Practical Vibration Primer,* Gulf Publishing Co., Houston, 1979.

Mann, Lawrence, Jr., *Maintenance Management,* D.C. Heath and Company, Lexington, MA, 1989.

Mitchell, John S., *Introduction to Machinery Analysis and Monitoring,* 2d ed., Penwell Publishing Co., Tulsa, OK, 1993.

Toms, Larry A., *Machinery Oil Analysis, Methods, Automation and Benefits, A Guide for Maintenance Managers and Supervisors,* Pensacola, FL, 1995.

7

Indoor Air Quality (IAQ) Plan*

Richard M. Silberman

Healthy Buildings International, Inc.
Fairfax, VA

7.1 IAQ and the Contemporary Facility Manager

When the public considers pollution they mostly think of an industrial scenario. And thus when they consider their exposure to pollution, they as a logical matter of course focus on the outdoor environment. This was, after all, the prevailing view during many facility managers' formative years, and subsequently most discussion on pollution always centers on the outside environment. But there is now a school of thought which suggests that if pollution exposure is the criterion by which we formulate environmental policy and allocate scarce resources, then the *indoor* environment is where our country should be putting its energies. And as a result, as we approach the new millennium facility managers will be in a position of increasing importance as the significance of their responsibilities becomes more and more evident to society.

Note: Nearly all the information contained in this chapter is the by-product of unparalleled field study by Healthy Buildings International, Inc. (HBI), of commercial buildings for indoor pollution and other indoor environmental issues. Since 1981, HBI has conducted indoor air quality investigations in over 2,000 major buildings for more than 550 clients in 42 countries. HBI has also pioneered the concept of proactive IAQ monitoring programs and healthy building design studies.

The U.S. Environmental Protection Agency (EPA) estimates that our exposure to pollution is consistently two to five times greater indoors than outdoors. This is of concern because we now spend upward of 90 percent of our workday indoors. Few of us today are farmers. Nor do many of us engage in other vocations in which we work to a great extent outdoors. We now live in urban America in a high-tech information-based society. For the vast majority of us this means spending our days in buildings.

7.1.1   Indoor Pollutants

Literally thousands of pollutants are found in today's buildings. New building interiors, common office supplies and equipment, furnishings, detergents, pesticides, paints and lacquers, and even humans themselves contribute to a mix of gases, fibers, dusts, compounds, dander, and microbial contamination in a building. Human exposure to this pollutant mix can take place in numerous ways, but the method which receives the most scrutiny is inhalation.

After the oil crises of the 1970s energy costs skyrocketed. Economic pressure bore on architects and engineers to design buildings which were more efficient, i.e., "tighter." This translated into structures with sealed windows and mechanically induced ventilation only. Similar pressure bore on facility managers to adopt building operation strategies which minimized the introduction of outside air to the occupied spaces. The environment within these buildings was now hospitable for these omnipresent pollutants to quickly and steadily accumulate.

When this mix of indoor pollutants builds to critical levels, facility managers will notice that occupants start to suffer from certain physical symptoms. They mirror what are thought of as classic cold, flu, and allergy symptoms: headaches, lethargy, weakness, and upper respiratory ailments such as runny nose, dry throat, coughing, and sinus congestion. Often these symptoms worsen in the afternoon after prolonged exposure and then mysteriously disappear upon exit from the property. When this situation occurs to more than just a few building occupants, it is called sick building syndrome (SBS).

7.1.2   Sick Building Syndrome

Some people mistakenly think that SBS describes the phenomenon that occurred in Philadelphia in 1976 which came to be known as Legionnaires' disease. Though the building contributed fundamentally to the problem (a deadly bacteria was spread from the cooling tower through the building's air via the air-conditioning system), Legionnaires' disease is technically an example of a different condition known as building-related illness (BRI). BRI results in a clinically defined illness or disease caused by a specific agent. The victim remains ill until recovery (or death). SBS, on the other hand, is multifactorial in origin and has no specific cause. Also, it is not a pathological illness but rather a compilation of many vague symptoms. These symptoms lessen or disappear when occupants leave the "sick" building, only to return the following day when the building is reentered. Though BRI can be fatal and hence more apt to be a headline maker, SBS is far more common and overall disruptive to American business and society.

Though it is unknown exactly how many buildings today are "sick," the problem is commonly regarded as widespread and nondiscriminatory by geographical location or climate. The World Health Organization estimates that 30 percent of all new and remodeled buildings in the industrialized world suffer from SBS.

What *is* known is that the upper respiratory tract irritations which characterize SBS are highly prevalent among the U.S. workforce. Medical researchers report that approximately half of the health problems plaguing American laborers are in fact these cold, flu, and allergy types of ailments. These problems prove to be quite costly to American business. The EPA estimates that sick buildings could cost the economy as much as $12 billion annually in lost productivity alone—not to mention absenteeism costs. Certainly people get sick and miss work for a myriad of reasons other than sick buildings. But there is now a growing body of data which indicates that the state of the indoor environments in which we spend so much time can have a profound impact on the probability that we will develop these upper respiratory infirmities.

What makes the discipline of indoor air quality (IAQ) so challenging to facility managers is the overlapping symptomatology of these various indoor pollutants. Virtually any indoor pollutant can cause any of the aforementioned symptoms. Consequently it becomes nearly impossible for a facility manager to isolate a culpable contaminant from the mix and eradicate it. Often what is more practical is to analyze the overall building and identify those baseline conditions which may be allowing any or all pollutants to build up to unacceptable levels. Altering those conditions can now result in a more fundamentally effective and long-lasting change for the better in the indoor environment. This holistic treatment of the building as a living, breathing, interdependent entity is often referred to as a "building systems approach" in the field of IAQ.

7.1.3 Causes and Considerations

Healthy Buildings International, Inc. (HBI), a Virginia-based environmental consulting firm specializing in the diagnosis of sick buildings, has conducted detailed indoor environmental studies in over

2000 major buildings around the world since 1981. The results of these studies form what may be the most comprehensive database in existence pinpointing the causes of indoor environmental problems. HBI has identified three main common denominators facilitating the buildup of indoor pollutants to critical levels: inadequate ventilation, poor filtration, and unacceptable mechanical system hygiene.

7.1.3.1   Inadequate ventilation.   Outside air ventilation is one of the keys to a healthy indoor environment. When facility managers purposely limit the amount of outside air entering their building in an effort to save energy, they miss a valuable opportunity to dilute pollutants to benign levels. These pollutants are now free to amass if other conditions are right. (Recent studies have shown that the idea that reduced ventilation will result in tremendous energy dollar savings is a myth.)

7.1.3.2   Poor filtration.   Using cheap, inefficient filters to save money is unwise as well. Traditionally, filters were judged by their ability to remove the larger particles in the air as these were the ones which could cause the most damage to mechanical equipment downstream. Today we are also concerned with human health implications. This means we must focus on the ability to remove the *smaller,* respirable-sized particulates, as these are the ones which are able to pass through our pulmonary system's natural filtration. Facility managers should understand that good high-quality filters are well worth the money. Also, care should always be given that they are well fitted within the air-handling unit. Gaps in the filter bank will allow air to take the path of least resistance and rush through the breach unfiltered. This will also result in higher air velocities which could hinder heat-transfer efficiency with the coils and facilitate moisture carryover.

7.1.3.3   Unacceptable mechanical system hygiene.   Since the mechanical systems in modern sealed buildings are often the only source of air to the occupied spaces, their cleanliness is paramount. Far too often dirt, grime, rust, and refuse are left in air-handling systems in a position of direct contact with the passing supply airstream. Furthermore, there is often water present in air-handling systems from humidification and/or condensation. This key ingredient, if allowed to stagnate, along with dirt, darkness, and even temperatures combine to produce an ideal environment for microbial growth. This growth, as well as its metabolic by-products, can now be released into the air and wreak havoc with sensitive and nonsensitive building occupants alike.

There are other important indoor environmental issues in a building that can impact the occupants. These include air pathways, air pressurization differentials, thermal comfort parameters, control strategies, air balance, lighting, and ergonomics.

7.1.4   Facility Managers Go Proactive

For many facility managers, a detailed indoor environmental study of their building is an important weapon in their professional arsenal. During the 1980s such a study, whether performed in-house or through an independent professional third party, was almost always done on a reactive basis. Often the driving force behind a reactive study is employees who have been experiencing discomfort and/or health problems linked to the building for some time. Without fail, these people (and hence the organization) suffer from decreased productivity and increased absenteeism. But in addition they may be influencing fellow workers who up until this point were reporting no problems, or they may be threatening to resign or may be considering legal action. Whatever the reason, the situation has boiled over and management sees no alternative but to take action.

The last few years, however, have seen a shift in thinking on the part of facility managers regarding the indoor environments for which they are responsible. This shift has been to change their approach from one of "reactive" to one of "proactive." The facility manager who embraces a proactive strategy is one who understands the advantages of prevention over cure.

Consider the facility manager who must react to an indoor environmental problem after a situation has built to a crescendo. The organization will pay a steep price in terms of shattered public relations and damaged reputation. Once a building has been deemed "sick," the stigma is virtually impossible to shake.

On a more direct financial note, it will cost more to rush in a qualified professional to "put out the fire." The environmental investigator will command a premium for time under these conditions. Also, some very specific and atypical tests will possibly have to be conducted, the costs of which will ultimately be borne by the client. Finally, the results, ensuing recommendations, and their documentation will have to be an expensive rush job as well.

By contrast, the facility manager who takes a proactive approach to the indoor environment is in a far more advantageous position. Proactive analysis allows the facility manager to direct the building's engineering staff to potential hot spots before they worsen into conflagrations. Voluntarily compiling a proactive database on the state of the indoor environment also serves as a potent insurance policy against not just possible future accusations concerning a particular area but also the rising threat of litigation.

On the economic theme, a professional indoor environmental investigator can schedule a proactive inspection in concert with the usual schedule, thus offering more attractive fees. Under this scenario the investigation is usually no less comprehensive, but materials, methodologies, travel, and reporting can be carefully planned under more economical conditions. In addition, if an immediate problem arises between proactive investigations, ensuing reactive work can often laser in on the specific problem area, as a deep understanding of the entire building already exists.

But perhaps the biggest payoff that proactive indoor environmental studies provide to facility managers is their use as a communication tool. The indoor environment is heavy on the minds of virtually all segments of the workforce these days as sick building horror stories in the media multiply. What a bonus for a facility manager entrusted with people's safety and comfort to truthfully declare to them that an extra effort is being made to ensure a healthy indoor environment! This manager will command great respect and admiration within an organization. The declaration will infinitely boost morale and ultimately will make that organization a more successful one. When that happens, everyone benefits.

7.2   The Proactive Facility Inspection and Sampling Techniques

7.2.1   Objectives

The objective of the proactive facility inspection is to confidentially test, inspect, and diagnose the IAQ in a building, in a reasonable period of time, during normal conditions and occupancy. More specifically, it is to inspect and analyze systems, building construction, building operation, test data, and samples in order to produce an objective, comprehensive, and confidential report which will aid management in identifying and correcting existing problems, as well as help identify potential problems and offer corrective suggestions regarding ways to prevent them from manifesting themselves.

7.2.2   Scope of Proactive Facility Inspection

The proactive IAQ facility inspection should consist of a thorough review of each building's mechanical systems where they relate to indoor air quality. These inspections are integrated with measurements of selected indoor contaminants which help verify the performance of the building systems in providing good-quality indoor air. An initial inspection should then be followed with regular 6-monthly reinspections to form a proactive preventive monitoring program designed to monitor trends, identify potential for future problems before they manifest themselves by complaints, and document and communicate building maintenance efforts.

The following protocol should be used at the initial inspection, with specific reviews of any particular areas where there is concern.

7.2.3 Analysis of Building Data

Building technical data should be reviewed and extracted from mechanical drawings of air-handling systems, building specification documents, test and balance reports, maintenance data, and interviews with building engineer(s). Data to be collected where available include:

1. Building configuration, size, and use.
2. Renovation details.
3. Special-use areas such as smoking lounges, copying, kitchens.
4. Air-handling unit type, design, and operation.
5. Building exhaust and return fan operation and fan tracking system.
6. Heating and cooling system type and operation.
7. Economizer system type, operational parameters, and other energy management device details.
8. Filtration type and maintenance program.
9. Air-supply system operation and controls, including variable air volume (VAV) zoning, regulation, and sequencing.
10. Humidification type and control details, if present.

The collected data should be analyzed and used where appropriate to interpret and evaluate the operation of the building and its mechanical systems as they relate to indoor environmental quality.

7.2.4 HVAC Inspection

1. A visual inspection should be made of the internal chambers of the main air-handling units, including coils, condensate drainage, fan chambers, humidifiers, controlling hardware, etc. A representative selection of any supplementary, perimeter, or other auxiliary air-handling units should also be visually examined.

2. An assessment should be made of the type, installation, and condition of filters present in the main and any supplemental air-handling system with regard to their impact on indoor air quality.

3. Fiber-optic examination should be made of representative sections of the internals of the main and secondary ducts leaving each air-handling unit with a borescope. When necessary, access can be gained to these ducts by the insertion of access points into the ductwork. Access points should be adequately sealed after examination has been completed. These inspections should embrace representative sections of ducts throughout the building.

4. Microbial surface samples should be taken from voids and relevant surfaces of the air-handling equipment and ductwork, especially in areas where water leaks are evident. Samples are analyzed for bacterial and fungal types present with species identification.

5. Static air pressure measurements should be taken inside ductwork at representative locations in main and secondary air supply systems.

7.2.5   Air Quality Tests

The following samples for laboratory analysis should be collected and analyzed as part of the proactive facility inspection:

1. Airborne microbial samples should be taken in selected areas followed by analysis and counting of bacterial and fungal species present with species identification.

2. Formaldehyde levels should be taken in selected areas using impinger sampling and analytical techniques.

The following instrument tests are performed at representative test locations throughout the building. Each test location is recorded and the methodology used for analysis, and the standards used to compare the results are defined.

1. Measurement of airborne particles (respirable suspended particulates) should be made using electronic aerosol mass monitors.

2. Measurement of carbon dioxide levels should be made using nondispersion infrared absorption techniques. Levels are compared against carbon dioxide levels in outdoor air and used as an indicator of ventilation rates.

3. Measurement of carbon monoxide levels should be made using electrochemical sensors.

4. Tests should be made for miscellaneous gases in selected areas, including screening for low- and high-molecular-weight hydrocarbons using a precision pump with direct-reading calibrated detector tubes.

5. Temperature and relative humidity levels should be recorded using an electronic thermohygrometer.

6. Measurements should be made of airflow direction and/or air pressure differentials between building shell and the outside, as well as between key locations within the building. This data is then used to identify existing and potential pollutant pathways.

7.2.6   Outdoor Air

On the same days of indoor sampling, samples of outdoor air should be collected for analysis of dusts, gases, bacteria, fungi, temperature, and relative humidity. Outdoor samples are taken for control purposes and are used to help evaluate the indoor readings.

7.2.7   Water Quality Testing (Associated with HVAC Systems)

1. Microbial analysis of selected water sources such as cooling towers, condensate pans, and humidifier reservoirs in or on the building should be conducted to evaluate likely sources of waterborne microbial contamination of the building.

2. *Legionella* bacteria identification and counting should be conducted in the cooling tower water or other potential source of this organism. A risk evaluation for exposure to *Legionella* from the source should also be conducted.

3. A total microbe count should be made on the selected water source.

7.2.8   Drinking Water Sampling and Analysis

1. A sample should be taken from each domestic main water supply in the building. This sample is then analyzed for a wide range of contaminants, including microbiologicals, metals, other inorganic chemicals and physical factors, organic chemicals, and additionally in agricultural areas for pesticides, herbicides, and PCBs. The concentrations of these pollutants should then be compared with the relevant EPA maximum contaminant level for drinking water for 91 parameters.

2. Lead in water analysis: A sample should be gathered from one representative drinking source on each floor of the building and analyzed for lead content using atomic absorption spectrometry. The results of this analysis are then compared with current EPA standards for drinking water in commercial buildings.

7.2.9   Other Tests

There are a host of other relevant tests and analyses which the facility manager can add to the facility inspection. These tests and analyses can be augmented within the framework of a proactive screening of the building or can be used as valuable diagnostic tools in the consideration of a reactive situation.

1. Air volume measurements of actual outside and exhaust airflows at representative points of the air system to assess if proper building ventilation standards are being met.

2. Analysis of design airflow specifications (if available) at representative points of the air system for comparative analysis with actual airflows.

3. Asbestos-containing material bulk samples can be obtained, with subsequent laboratory analysis using either the polarized light microscopy (PLM) method or the transmission electron microscopy (TEM) method.

4. Asbestos-containing material surface samples can be obtained using surface wipes with subsequent laboratory analysis by the TEM method.

5. Airborne asbestos fiber counts can be obtained with subsequent laboratory analysis by either phase contrast microscopy (PCM) or TEM method.

6. Analysis for lead in paint in the building can be performed using an x-ray fluorescence lead paint analyzer in general accordance with applicable Department of Housing and Urban Development (HUD) guidelines.

7. Airborne lead concentrations can be sampled by passing a known volume of air drawn through a filter cassette equipped with a cellulose ester membrane per NIOSH Method 7082. Subsequent laboratory analysis is performed by flame or graphite furnace atomic absorption spectrophotometry.

8. Screening for radon can be performed via the installation, collection, and analysis of Alpha Track long-term radon detectors in occupied areas in contact with the ground floor. The detectors should preferably remain in place for 90 days.

9. Testing for a range of other volatile organic compounds (VOCs) can be accomplished using absorbent tubes followed by analysis with gas chromatography/mass spectroscopy. VOCs to be tested for should include benzene, carbon tetrachloride, chlorobenzene, 1,4-dichlorobenzene, 1,2-dichloroethane, ethylbenzene, limonene, 4-phenylcyclohexene, perchloroethylene, tetrachloroethene, toluene, 1,1,1-trichloroethane, trichloroethene, trichlorofluoromethane, vinyl chloride, xylene, and total VOCs.

10. Measurements regarding the power frequency environment can be taken using an ELF (extremely low frequency)/power frequency EMF (electromagnetic frequency) survey meter.

11. Noise level testing can be done using an audio dosimeter.

12. Measurement of components of environmental tobacco smoke (ETS), including airborne concentrations of carbon dioxide, carbon monoxide, respirable-sized particulates (RSP), and airborne nicotine concentrations, can be taken to determine ventilation efficiency in smoking areas and extent of ETS migration to nonsmoking areas. The nicotine components of environmental tobacco smoke are measured using adsorbent sampling and analysis with gas chromatography.

13. Measurement of lighting and glare levels can be taken in selected or random locations, with subsequent analyses with consideration to specific activities in those locations.

14. Testing of, and analysis for, ozone levels in various locations. The purpose of this test is to quantify the levels of this lung irritant and determine its contribution from electronic equipment in these areas.

7.2.10 Facility Inspection Reports

A comprehensively written report should be written following each proactive facility inspection. This report serves as professional documentation of effective ongoing management and maintenance efforts to create a healthy indoor work environment for building occupants. The following information should be contained with the facility inspection report:

1. A description of the air-supply, air-exhaust, and air-conditioning system highlighting aspects that relate to indoor air quality.

2. The tests conducted, including a description of the test procedure and the instruments used.

3. The test results.

4. A discussion and comparison of test results to all relevant environmental standards and to suggested standards for the most sensitive members of the general population.

5. Where practical and helpful, the report should include photographs which help to explain any unusual conditions or areas needing attention.

6. Practical recommendations should be made for any corrective actions for improvement of the indoor environment.

7. A one-page executive summary of each report can also be prepared, designed as a handout to occupants who inquire about the air quality within the building.

7.2.11   Facility Reinspections

The key to a good facility indoor air quality plan is the establishment of a database from the initial facility inspection. This database is built via the execution of proactive IAQ facility reinspections on a regular basis. (Usually biannual inspections are recommended so that building conditions during both the heating and cooling seasons may be evaluated.) The scope of work followed during a facility reinspection should in most ways mirror that of the initial facility inspection.

The initial facility inspection is used as a reference point against which the subsequent reinspections are judged. The reinspections then comment on recent trends (positive and negative), compare the building with similar structures which have been inspected, and verify the effectiveness of any plant, maintenance, and/or operation changes which have been made in the building since the last visit. A new report should be issued following each reinspection summarizing the status of the air-handling systems with respect to their internal structure, corrosion conditions, filtration systems, contamination status, and general air quality. When relevant, attention should be drawn to changes or trends, and the rationale for such changes should be discussed.

7.3 Avoidance Procedures

7.3.1 Preventive Operations, Maintenance, and Policy

There are a plethora of preventive measures which facility managers can initiate within buildings in order to prevent future IAQ problems. These measures, often simple commonsense issues, fall under the categories of operations, maintenance, and policy.

7.3.1.1 Operations

1. Always operate your building with sufficient outside air. The amount of outside air should follow ASHRAE Standard 62-1989. By introducing sufficient volumes of outside air, two benefits will be achieved. First, indoor contaminant concentrations will be sufficiently diluted so as not to irritate building occupants. Second, by mechanically introducing more outside air into the building than is exhausted the facility manager will ensure that the building is maintained at a positive pressure with respect to the outside. This, in turn, will prevent a natural inflow of unfiltered, unconditioned air through the building shell, especially at ground level where pollutants are plentiful. Some buildings' mechanical systems can no longer deliver sufficient outside air owing to changes in building design, layout, occu-

pant characteristics, etc., over the years. Facility managers of those buildings consider retrofitting their systems with some of the fine auxiliary equipment which exists today for boosting ventilation, such as heat pipes, desiccant units, heat exchange unit ventilators, etc.

2. If the building utilizes a VAV system, ensure that the VAV boxes are never able to fully close. This will ensure that a minimum of outside air is always introduced into the occupied spaces regardless of the occupant density. This can be accomplished through creative control sequences or the retrofitting of minimum setpoints into the VAV boxes.

3. Try to maintain a lower temperature within the occupied spaces. This will automatically increase the relative humidity and lessen occupants' sensitivity to indoor contaminants via the drying out of their mucous membranes (especially in the winter). In addition, studies have shown that people perceive cooler air to be "cleaner" and "fresher."

4. Care should be given that supply air diffusers serving the occupied spaces are not covered over. Often in an effort to stop conditioned supply air from blowing directly on them, building occupants will cover supply air diffusers. Ceiling diffusers are often covered with cardboard and/or tape. Sometimes plants, books, and other objects are put over auxiliary perimeter unit diffusers. While these actions are well intended, the result is that the occupied spaces are denied the proper amount of outside air originally designed for the occupancy level. Consequently indoor pollutants build up and complaints ensue.

In addition, covering supply air diffusers can ruin proper air balance in the occupied spaces. Some areas receive too much supply air, while other areas receive too little. Also, improper air pressure relationships between the building and the outdoors, as well as between key areas within the building, can also develop. Pollutant transfer is then facilitated.

Supply air diffusers should never be covered over. If occupants are complaining of drafts, several remedial options exist: occupant locations can be changed, diffuser locations can be moved, or a different diffuser design can be utilized. Some of these options may require that the air system be rebalanced.

5. The ceiling voids, the area between a floor's slab and suspended ceiling, should be kept clean and free of obstructions.

In most commercial buildings the ceiling voids on each floor serve as the primary return air pathway back to the building's air-handling system. Return air from the occupied spaces travels up through grilles or slots in the drop ceiling and enters the ceiling void. The force creating this movement is usually back pressure from the air-handling system's supply fans or dedicated exhaust fans located elsewhere in the building. This return air then travels back to an air-handling unit where some is exhausted from the building, and the majority mixes with new outside air to repeat the conditioning processes and return to the occupied spaces through ductwork as new, clean supply air.

If the ceiling void is contaminated with dirt, microbes, damaged insulation fibers, fumes from stored chemicals, etc., these contaminants can easily become introduced into the return air. The contamination then stands a good chance of at least partly being reintroduced into the occupied spaces via supply air.

If there are obstructions in the ceiling void which prohibit proper return airflow, the resultant improper pressure differential in the occupied space will not allow new, fresh supply air to enter. The result will be improper airflow, improper air balance, and pockets of poor ventilation and contaminant buildup in the occupied space.

7.3.1.2 Maintenance

1. A formal, written preventive maintenance plan should be codified by the facility manager for the building and its HVAC system. Thereafter, a system of checks and balances must be instituted to ensure that preventive maintenance tasks and procedures are indeed carried out on a timely basis. Many preventive maintenance programs fail because well-designed schedules are simply filed away and no action was taken. Also within this plan a chain of responsibility should be designed to establish who will make decisions regarding appropriate remedial action if during a maintenance task a problem is uncovered.

2. On a daily basis, confirm that all garage, kitchen, printing, graphic, etc., area fans are on and functioning properly. Verify that the building and areas within are properly pressured by use of quick smoke tube tests. Where necessary validate that thermostats are properly set and in the "fan on" mode. Use the smoke tube in several random locations on each floor of the building to ensure that air is flowing from supply air diffusers.

3. On a weekly basis, inspect floor drains and add water where necessary. Either check all outside air dampers or randomly inspect as many as possible to make sure they are operating properly. Use a smoke tube to ensure air is entering, not exiting, outside air dampers. Inspect all plenum rooms for storage of hazardous or potentially irritating chemicals, such as paints, herbicides, cleaning agents, etc. Perform a complete building walkthrough to look for any signs of water leaks or other items that may cause IAQ problems.

4. On a monthly basis, inspect all air-handling units for cleanliness

and confirm that all internal components are in good repair. Also within air-handling units, inspect filter condition and fit. Confirm that condensate trays are draining properly and that chemical packets have not expired. Inspect all induction fan VAV box filters for condition and fit.

5. On a semiannual basis, perform a comprehensive operational inspection and review of all HVAC system controls. Be sure that the operation of all related mechanical devices is physically confirmed. Randomly measure the maximum and minimum air volumes of as many supply air diffusers as possible to confirm the systems are still properly balanced. Randomly check the air volumes of exhaust fans to confirm proper air volumes.

7.3.1.3 Policy

1. Even if a facility manager believe there is no credibility to an IAQ complaint, the complaint should be taken seriously. To do otherwise is to invite worse trouble. Commonly, there is a real, but temporary, technical incident which triggers an IAQ complaint(s), e.g., an office renovation. However, what all too often occurs is that even after the incident concludes or is rectified, if the original complaint was not handled with professionalism it takes on a life of its own and snowballs throughout the organization.

2. Appoint an IAQ manager. Whether it is the facility manager or a designated representative, it is good policy to have one focal point to field and screen IAQ complaints to see that they are kept confidential and handled in an appropriate manner.

3. Facility managers who are proactive about IAQ should hang it out on a lamppost for all to see. Utilize in-house newsletters, memorandums, working lunches, and other forums to let your constituents know that management is "going the extra yard" to attempt to provide them with as healthy an indoor work environment as possible. Many IAQ complaints are partly rooted in the fear of the unknown. By giving people on a regular basis proactive IAQ information resulting from a solid proactive IAQ monitoring program the facility manager significantly reduces the chance that this topic will come across his or her already crowded docket.

4. It is imperative that employees understand that they themselves play a pivotal role in the achievement of a healthy indoor environment. Occupant activities which negatively impact IAQ such as covering of air diffusers, ill-planned renovations, removal of ceiling tiles, and poorly maintained office plants need to be identified. Good communication between the facility manager and all building occupants is a key toward avoiding IAQ problems.

7.3.2 Designing Healthy Buildings

In the quest to avoid IAQ problems, there is now a budding move-ment in the facility management community to pay attention to the issue of sick building syndrome (SBS) at the *design* and *construction* stages of a building's life. This approach recognizes the benefits of prevention over cure in treating building system ailments. Often such a strategy can head off SBS problems before the building is occupied and an operating history is established. Such foresight also sets the foundation for a permanently healthy building and lessens the chances of SBS problems later during its occupied life. Striving to attain a healthy building when the property is merely being conceived is the ultimate proactive approach to the commercial workplace envi-ronment.

There are many indoor environmental issues which must be consid-ered when designing a healthy building. Many converge on the future optimization of the three main common denominators which facilitate the buildup of indoor pollution to critical levels (ventilation, filtration, hygiene). Other issues attempt to manage the introduction of pollu-tant resources into the building in the first place.

These environmental design issues can be divided into five main areas: site plan and building configuration, HVAC system, maintain-ability and durability, product selection, and building commissioning.

7.3.2.1 Site plan and building configuration.   The first step in designing a healthy building is to understand how the building's configuration will bring the outdoor environment in contact with the indoor envi-ronment. Prevailing weather and wind patterns, ambient air quality, and major outdoor sources of pollution are analyzed.

Also conducted is a review of design documentation procedures and assured compliance with any local indoor air standards (if they exist). Ventilation rates and air distribution systems are critiqued under dynamic circumstances, including the full range of operational and outdoor weather conditions. Conspicuous sources of pollution such as photocopiers, printing operations, and dedicated smoking lounges must have adequate exhausts adapted to them. The locations of *gener-al* air intakes and exhausts are studied to ensure that they are not poorly positioned near sources of outdoor pollution, such as garages, loading docks, busy street corners, cooling towers, and garbage dumps. Additional pollutant pathways of importance are within the building vertically from one floor to another, and horizontally across floors.

The building's configuration will also impact indoor environmental issues such as noise, thermal comfort, microbial pollution, and light-ing. The strategic locating of corridors, choice of wall material, and selection of glass and glazing based on acoustical ratings will provide buffers from noise pollution, especially near industrial and/or high-traffic sites. These same issues can also help control the building's thermal environment by retarding heat transfer into the building in the summer and out of the building in the winter. By eliminating thermal bridges in the building envelope one also helps to avoid moisture buildup via condensation, which is key for subsequent microbial infestation. This can be accomplished by insulating connecting members to assure that no break in the insulation occurs and by using nonconducting materials (such as plastic) in building components connecting exterior and interior surfaces. By properly locating and orienting windows natural daylight can be bountifully harvested. Daylight will improve the well-being of occupants in the building, can improve productivity, and may save a considerable amount of electric energy with the appropriate design, glazings, and electric lighting controls. Windows must be designed properly to prevent disability glare, a situation where an overabundance of light within the eye causes irritation, reduced productivity levels, and sometimes momentary "blindness" to building occupants.

7.3.2.2 HVAC system. Occupant density, activity, and location must be outlined and documented to ensure proper interface to the HVAC system. This includes a judgment on the configuration of office partitions to maintain good air distribution, mixing, and exchange rates in the occupied spaces. Only then can ventilation requirements as documented in the American Society of Heating, Refrigerating, and Air-Conditioning Engineers (ASHRAE) Standard 62-1989 and local presiding building codes be adhered to.

A thorough review of the building's filtration system is imperative. This includes type and design, materials employed, and location within the ventilation system. Filters' future impact on human health and comfort should always be judged by the ASHRAE Atmospheric Dust Spot Test efficiency rating. This rating is based on the discoloration potential of airborne particles rather than on particle weight, which has an inverse relationship to indoor air quality. A new ASHRAE filter efficiency rating which focuses specifically on a filter's ability to remove respirable-sized particulates (RSP) is expected soon. Good filtration will not only create a marked improvement in the quality of the indoor air as compared to the outside air, it will protect expensive mechanical room equipment from accelerated degradation. In special-use and/or high-pollutant areas supplemental filtration may be in order such as electrostatic precipitators, adsorbent beds, and HEPA (high efficiency particulate resistance) filtration.

HVAC design precautions should also be taken to prevent the pooling of water within the system, a key ingredient in subsequent microbial growth. HVAC system components of particular concern are humidity control equipment, internal fibrous glass insulation, filters, and especially condensate trays. Regarding this last piece of equipment, condensate trays should be designed with sufficient slope, drainage, size, and depth. For enhanced surface preservation against microbial contamination, as well as corrosion, all the internal surfaces of the various chambers of the HVAC systems, whether lined or unlined, should be treated with a coating containing an approved antimicrobial compound.

7.3.2.3 Maintainability and durability. Another way facility managers and planners can strive to design a healthy building is to recommend the use of materials and physical plant within the building that will be durable and maintainable over the test of time. Materials should be selected which will resist corrosion, erosion, and contamination by weather and the many indoor pollutants which can accumulate. Discussions with material manufacturers give insight into the appropriateness of different materials for a given application.

Nothing can compromise the durability of building materials and physical plant more than standing water. (Standing water can also have profound impacts on indoor air quality. When mixed with dirt it becomes an ideal breeding ground for microbial growth. Airborne spores from microbial growth are a major pollutant connected with SB Syndrome.) Therefore, the consultant checks to see that condensate drains, water baffles, mist eliminators, humidifiers, cooling towers, and other pieces of equipment that utilize water in their operation will be used correctly to avoid water stagnation (see Sec. 7.2.4).

For proper air system hygiene and maintenance, checks are made to assure available access via doors and ports to chambers of air-handling systems, plenums, and ductwork. The same access is important for reheaters, turning vanes, smoke detectors, filters, coils, VAV boxes, perimeter fan coils, and induction units, etc.

Good design also mandates that attention be paid to the integrity, material type, and location of insulating materials associated with HVAC equipment, ductwork, and ceiling plenums. Poorly fitted and unsealed insulation used in an air system, for example, can easily fray over time and release irritating fibers into the airstream, which are then delivered to the occupied spaces.

7.3.2.4 Product selection. The materials for the interior spaces should be selected carefully with regard to their future impact on the indoor pollutant loads. The general principles of making sound environmental material selection decisions are customized to the specific indoor environment planned. Material selection criteria to be judged include offgassing, fiber release, microbial support, sink effect, and durability. Products of concern include constituents of carpets, flooring, linen, adhesives, wall coverings, partitions, ceilings, insulating and fireproofing materials, sealants on windows, walls, and floors, preservatives, paints, varnishes, and other architectural coatings.

The first step in product selection is a review of manufacturer's safety data sheets (MSDS). MSDSs are government-mandated records of various fire, physical, health, reactivity, and other hazards associated with new materials and chemicals. Unfortunately, the information contained on these resource sheets is often sketchy and incomplete from an IAQ perspective. This often necessitates consultation with an organization which maintains a well-maintained database on environmentally friendly interior materials. If little information is available publicly, one must call material manufacturers and suppliers directly for information, which usually exists but is rarely openly disseminated.

Sometimes the chemicals used to install a certain product and the installation procedure are more important than the product itself in terms of impact on the indoor environment. In particular, contaminants emitted from various adhesives can be very irritating to certain individuals, and there is evidence of adhesives interacting with materials with which they come in contact to form a third, even more irritating chemical. If these risks can be predicted when interior materials are being specified, the benefits are obvious.

The future maintenance requirements of a product are extremely important for long-term contaminant concentrations in a building. Products that require little maintenance or can be maintained with non-solvent-based cleaning products are preferred whenever available.

7.3.2.5 Building commissioning. The period of time shortly before a building's completion and subsequent occupation can be the most problematic. New interior materials, an untested HVAC system, and unfamiliar building management personnel combine with the usual stresses of a building move to create a sensitive situation. Often a sick building reputation can be acquired during this phase, and future efforts to shake the reputation are difficult. A formal plan encompassing final HVAC installation and startup, initial ventilation strategy, design documentation, operating and maintenance training, and remaining pockets of construction goes a long way toward avoiding unnecessary complications.

Some initially high indoor pollutant levels (such as solvent fumes and various volatile organic compounds) are the direct result of unique conditions present during the earliest months of a new building's life. This involves the offgassing of paints, lacquers, adhesives, carpets, and other new interior materials. To mitigate this condition, special operating practices should be enacted during the commissioning period. This involves temporarily intensifying ventilation and filtration to the maximum levels possible. This policy should remain in effect 24 hours per day and on weekends during the first few months of the building's life to "flush out" these pollutants during their off-gassing periods.

Proactive attention during the building commissioning process should also involve periodic visits to the construction site. These construction site visits are essentially visual inspections to ensure that all recommendations made by the environmental consultant at the design stage are being complied with. Reports are also made concerning any ongoing construction practices which may negatively affect indoor air quality during the subsequent occupied life of the building.

As construction of the embryonic healthy building proceeds to more advanced stages, protocol is established for the formal transfer of information and hand-over procedures from the design team to the building maintenance personnel.

7.3.3 Managing IAQ Risk within Leases

Most organizations lease workspace, and for them the contemporary facility manager is upper management's technical confidant in the landlord-tenant relationship.

Few businesses were affected as heavily by the recession that struck the U.S. economy in the late 1980s as that of commercial real estate. Suddenly new construction ceased and a glut of vacant office space abounded in commercial centers from Seattle to Miami. Rates for office space plummeted, resulting in an attractive rental market for those businesses which exhibited the efficiency and adaptability to weather the turbulent times. In addition, incentives arose for building owners and managers to make every extra effort to retain existing tenants, as well as secure new ones. Though the more recent economic recovery enjoyed by the nation has altered this situation somewhat, still significant office vacancy rates in many large urban centers allow these new facts of life in the commercial real estate rental market to endure today.

This basic example of the law of supply and demand has allowed both landlords and tenants to add new and creative twists to leases which would have been unthinkable only 10 years ago. One such area which is being formally addressed within lease language at an increasing rate is that of the indoor environment in general and IAQ in particular. This phenomenon is an outgrowth of the overall attention IAQ has received in American society over the last few years.

The fact is that the achievement of good IAQ in a building is a team effort between tenants *and* landlords. While data over the years has shown that general building maintenance and operation issues (i.e., typical landlord responsibilities) are responsible for many IAQ problems, quite often tenants themselves can be at fault. As our society has become more and more litigious, both tenants and landlords have found that the cost of poor IAQ can grow beyond the marketplace as outlined above. Both sides of these business agreements are beginning to seek ways to shelter their risk and protect themselves from the other side, and the opportunistic environmental lawyers are all too eager to represent them.

Incorporating IAQ language into leases allows both landlords and tenants to safeguard themselves from future ill-advised actions of the other. IAQ lease language minimizes the likelihood that either organization will unfairly suffer the consequences of future IAQ problems.

One of the most important items to be added to a lease is language requiring that an indoor air quality inspection be conducted in the space prior to the official consummation of the lease agreement. Regular proactive inspections throughout the term of the lease can be required too. These inspections should be performed by a credible, experienced independent third party with a proven track record of conducting IAQ investigations. How the costs of these inspections are shared between tenant and landlord is negotiated. Some tenants are lucky enough to already rent from landlords who understand the market advantages of paying proactive attention to IAQ and are presently conducting IAQ investigations. The following language, or similar, can be used within a lease agreement:

> Lessor shall cause to be conducted and delivered to lessee prior to lessee's execution of this lease, and over the term of the lease, environmental assessments of the air and water in the deemed premises and the building, in a form reasonably satisfactory to both the lessor and lessee. Lessor and lessee represent and warrant to each other that as of the date of lessee's execution of this lease and through to its termination, the air and water quality of the demised premises and buildings meet or exceed current legal requirements and/or professional standards. Should these legal requirements and/or professional standards not be met, the executor of the aforementioned environmental assessments shall determine the cause, and the appropriate lessor and/or lessee actions necessary to make satisfactory.

A legal responsibility on the part of the landlord to pay rigorous ongoing attention to some of the more specific landlord issues contributing to future IAQ problems can be incorporated into the lease by the tenant as well. Examples of these include:

1. Requirement of minimum ventilation rates, carbon dioxide levels, and other indicators of proper outside air ventilation.

2. Proof of supply air testing, adjusting, and balancing to achieve proper air distribution within occupied spaces.

3. Assurance of proper drainage from air-handling unit condensate trays, changing of air filters, and other appropriate HVAC maintenance activities.

4. Achievement of proper air pressure flows and differentials between outside and inside, and between key locations within the building to prevent the migration of pollutants.

5. Designation of acceptable contaminant levels above which common indoor pollutants may not rise in the occupied spaces.

6. Establishment of an appropriate response to deal with future IAQ complaints.

Presently one of the more formal sets of tenant IAQ lease requirements along the above lines is put forth by the U.S. General Services Administration (GSA). The GSA in their solicitation for lease offers demands provisions for IAQ complaint response and IAQ investigations, and sets action levels for carbon dioxide, carbon monoxide, formaldehyde, lead, radon, and asbestos.

Likewise, landlords can require within the lease that tenants be liable for IAQ problems resulting from careless actions which are traditionally theirs. Examples of these actions include the covering over of air diffusers, unacceptably high staffing densities, introduction of atypical pollutant sources (such as printing machines) without proper supplemental ventilation, and poorly designed and planned office renovations.

The smart landlord-tenant team addresses IAQ within the structure of a lease agreement before it is signed, so there can be no future finger pointing and unfair apportionment of risk. A good lease agreement is like a good marriage: both parties derive a lot of satisfaction.

8

Technical
Improvements Plan

Clive Shearer

Management Consultant
Bellevue, WA

Harlen E. Smith

Value Engineering Services Division
Parsons Brinckerhoff Construction Services, Inc.
Dallas, TX

Alan Pearlman, P. E.

Electronic Systems Associates
New York, NY

8.1 Total Quality Management (TQM) Program

A management tool that is misused will sooner or later fail. Place the same tool in competent and willing hands and success awaits. This section shows how to use and reap the rewards of TQM. But first a review of what TQM is and is not.

8.1.1 TQM—What it is

TQM can lead to an organization consistently doing everything right, not something an organization does just to solve problems. It does this through the participation of the people who are involved with the facility. Ideally these long-term improvements stem from infusing TQM into the overall business plan, not just the facility management plan. It develops improved teamwork as owners, occupants, tenants, and users become partners of the facility staff, thereby fostering a greater understanding of the potential for their facility. TQM's multiple benefits include:

- Refining communication procedures
- Strengthening outsourcing methods
- Enhancing environmental risk management
- Optimizing safety, accident, and fire protection
- Fine-tuning real estate acquisition and leasing guidelines
- Building better equipment and operational procedures
- Improving maintenance and repair guidelines

By fostering a continuous improvement mind-set, TQM becomes the heart of excellence for facility management.

8.1.1.1   The three phases of problem solving

1. Problems can be solved before they start by anticipating them and thus avoiding them.

2. Problems can be solved as they surface through early intervention and course correction.

3. Problems can be solved after people start complaining, when crises arise, when disaster looms, or worse, when disaster has already occurred. TQM is useful for all three phases, but it excels in the first phase.

8.1.1.2   The two kinds of improvements

8.1.1.2.1   Refining processes and activities that currently add value. Value-added processes and activities are the actions that provide perceptible or tangible benefits to owners, occupants, tenants, guests, and those who work in the facility. Value-added refinements commonly include:

- Upgrading computer hardware and software
- Refining a well-organized inventory control system
- Publishing a new version of a cyclical maintenance manual
- Simplifying outsourcing procurement paperwork
- Developing an improved churn tracking system
- Simplifying equipment inventory procedures
- Limiting outsourcing by getting users involved in daily continuous improvement

- Redesigning inadequate performance standards
- Implementing alternative officing, including telecommuting
- Updating contracts used for specialty suppliers
- Implementing computer-aided facility management software (CAFM) to track "what it is, where it is, what it is doing, and what it costs"
- Going beyond the tracking capabilities to use CAFM as a planning tool and a cost-saving tool

These improvements are worthwhile, but because they are improving what is already value added, the gains may be small compared to first considering the far greater return on investment available through the elimination of wasted effort.

8.1.1.2.2 Eliminating wasted resources and effort. Perhaps one can define the goal of facility management as managing a facility where people delight in interacting with the building, its setting, and its contents, with optimal cost-effectiveness. The facility manager is mostly "invisible" because from the occupants' point of view, the facility seems to run itself. Any tasks, processes, methods, and operations that do not support this goal are called "waste" if they add nothing of value to the facility or people's environment. Some of these non-value-added activities include tasks tackled out of habit, perhaps appearing on an old checklist, used for decades. Other non-value-added tasks may be performed out of fear, ignorance, or stubbornness, or simply because of a management directive. When such wasteful activities are eliminated, morale grows, money is saved, and facility management becomes valued for quiet efficiency and effectiveness. This will be difficult for those facility managers who thrive on the daily drama of putting out fires and heroics.

Waste-elimination tasks that must have priority over value-added refinements include:

- Implementing a centralized, managed preventive maintenance program
- Developing and implementing a comprehensive plan for "indefinite quantity" contracting
- Finding ways to reduce the churn rate
- Scrapping and recreating an outdated safety plan
- Rectifying dangerous equipment shutdown procedures
- Working toward a reduction in workplace violence

8.1.1.3   Contractors. Working effectively with contractors brings several benefits:

- Maintaining in-house control of all needs, activities, and results while partnering with experts in each field
- Having those experts available in the right dose at the right time
- Competitive pressure by contractors to perform

Developing a partnership mind-set with contractors strengthens the delivery of service.

8.1.1.4   Global thinking. Thinking globally means looking at management structures and systems in terms of relationships. TQM provides the tools to improve interactions of services, departments, and regions, track activities, and evaluate trends.

8.1.2   TQM—What it is not

8.1.2.1   TQM does not fix people. TQM is a means to improve management systems and not a means to make people perform better, improve their attitude or morale, or support a personal agenda.

8.1.2.2   TQM is not a training program. TQM is not about training, team spirit, or empowerment. It is about improving facilities, procedures, and systems and it is about satisfying owners, users, and occupants. It involves getting better at everyday activities, not getting better at training, team spirit, and empowerment. Yet focused, well-timed training does have a role.

8.1.2.2.1   Senior management training. Provide an overview of TQM and the value of continuous improvement as a core value of the organization. Clarify management's role as visible and active champions of continuous improvement.

8.1.2.2.2   Facilitator's training. Provide knowledge of TQM, measurement processes, and team facilitation skills, such as how to orchestrate and guide a team without "leading" it. Negotiation and conflict resolution skills will also prove to be valuable.

8.1.2.2.3   Training is rarely the solution. TQM teams often study an issue and conclude that the solution is simply to institute more training. This represents a failure of the team to:

- Search for deeper causes

- Get past politics
- Overcome the tendency to fix people rather than fix systems

8.1.2.3 TQM does not mean empowering people or teams.

While TQM can be empowering, this is merely a by-product of the process—and it is not a necessary by-product. If empowerment happens, that is good, but it is not a goal of TQM. Too much emphasis on empowerment can weaken an organization; one cannot have the tail wagging the dog.

Yet empowerment within a framework, with boundaries established and with set guidelines and ground rules, can be very valuable. In fact, improvement teams that have clearly established operational parameters generally work far more efficiently and produce far more usable and practical results than those that have total empowerment to wander along the paths of their own choosing, exploring avenues that seem appealing. These may not be the paths and avenues that urgently require revitalization, and when senior management refuses to implement the suggestions of these totally empowered, self-directed teams, everyone loses. Senior management loses respect in the eyes of the team members, the organization loses the team's drive and energy, and the team members lose enthusiasm for improvements and sometimes even for their organization.

8.1.2.4 TQM is not a quick fix.

Short-term thinking means wearing blinkers and focusing on "emergency room" repairs rather than on wide-ranging vision and long-term healing. It takes time to identify what is really wrong, correct it, and make sure that the improvement works. The quick fix often leads to future visits to the emergency room for more quick fixes, and this soon becomes the accepted way to conduct business. One must balance short-term fixes with long-term global thinking.

8.1.2.5 TQM is not a life jacket.

TQM is not an emergency repair kit. When the *Titanic* starts to sink, TQM will not come to the rescue. However, if TQM was the builder and navigator of the ship, disaster might have been: (1) avoided altogether through navigation adjustments, (2) limited through damage control procedures, (3) mitigated by means of more efficient communications and practiced lifeboat drills. The planning exercise of asking "what if?" is always valuable before the event, and demoralizing when asked too late.

8.1.2.6 ISO 9000 is not as comprehensive as TQM.

ISO 9000, the international quality standard first published in 1987, provides a way to document procedures and apply them consistently. However, it provides no direct verification that those procedures will satisfy owners, users, and occupants, or improve the financial bottom line. TQM, on the other hand, embraces all of the ISO 9000 benefits and furthermore develops a continuous improvement methodology without the expense of obtaining and maintaining ISO 9000 registration.

8.1.2.7   Senior management cannot delegate quality.   Because today's facility manager typically reports to the CFO or controller, short-term cost-benefit thinking may loom larger because many financial managers operate under profit production pressure. This makes TQM not just helpful but essential, and senior management must play an active role. Senior managers that encourage and support a culture of continuous improvement enjoy the benefit of a happy and loyal staff, improving productivity, and an increasing bottom line. To ensure buy-in, involvement, and results, everyone has a role to play in quality improvement.

8.1.2.8   TQM, quality assurance, quality control, and reengineering are not the same.   Consider the requirement for pilots to perform a preflight check before every takeoff. *TQM* utilizes continuous improvement to design the instruments to function reliably, to make sure that there is sufficient redundancy in systems in case one system malfunctions, and even to make sure that airport operations and communications are efficient, trouble-free, and safe.

Quality assurance would be the creation of the comprehensive, clear, and concise preflight checklist. *Quality control* would be the pilot going through the checklist to ensure that all the conditions exist to allow for a safe takeoff and flight.

TQM deals with getting right the things that matter the most and then improving them over time. Quality assurance is a facet of TQM that establishes the "best practice" methods for use, and quality control implements those best practices. One can no more rely on quality control to improve a facility than one can rely on a vehicle's oil change to improve the performance of the vehicle.

In the facility management context, quality assurance includes:

- Developing thorough procedures
- Creating or improving manuals to be comprehensive
- Developing operations checklists to ensure consistency

Quality control includes:

- Making sure that procedures are correctly followed

- Ensuring that the manuals are used as designed
- Using checklists at the right place and at the right time

An important aspect of quality control is its use both as a verification tool and as a data-gathering tool to spot the need for further improvements.

Reengineering is the process of "wiping the slate clean" and recreating all or part of the organization. It tends to lead to a focus on the bottom line, and many people have lost their jobs as reengineered organizations slashed costs and trimmed their workforces. Not to say that reengineering is bad. The concept has value, but many reengineered organizations turned out to be mirror images of the original organization, without much actual change except an increase in workload for those who survived the layoffs.

8.1.3   Rewards to be reaped

8.1.3.1 Long-term solutions.   Going into a garden and cutting off weeds at ground level will usually result in a temporary "cosmetic" improvement, until the weed grows back—perhaps even stronger than before. In the same way, cosmetic changes are often used in the workplace, or band-aids are taped over gaping wounds while placating words are uttered. Yet the underlying causes, lurking below the surface, are left untouched. Then a "new" crisis arises which is simply the same old problem coming to visit in a new place in a new guise. Wouldn't it be nice to get rid of the problems once and for all? This can only be done by removing the roots, and TQM can provide this reward through "root cause analysis."

8.1.3.2 Developing consistency in a variable world.   Every process, changed or unchanged, will have some variation. Ask someone to sign a sheet of paper ten times and it will be highly unlikely that any two signatures will be precisely the same. Yet each signature would be acceptable at their bank because variation within certain bounds is normal.

With regard to facility management there will always be variations in the way the work is implemented. The trick is to identify the distinction between what is an acceptable variation and what is a danger signal. While there is no way to do this scientifically without a statistical chart, financial managers have successfully run giant corporations without resorting to statisticians, because they know what makes or breaks a business. While the facility manager can always become more educated in statistical methods, it is reassuring to know that by implementing sound TQM unproductive variation is usually reduced and consistency and positive trends are established.

8.1.4   How does TQM work?

Reading about TQM without implementing it would be like reading a menu but not getting the meal.

8.1.4.1   Step 1: Develop a list of problems.

Identify the current needs of the facility users and what they might need in the future to be more productive and happier in their environments. This is best done by surveying owners, colleagues, junior staff, clients, and users.

The survey must be designed to steer answers away from issues such as dealing with employees who don't pull their weight, are discourteous, or are not reliable. This is because, as previously described, TQM is not meant to be a way to "fix" people; it is a means to improve the systems within which people work. If some people are a problem, one needs intervention by management, not TQM. However, improvement ideas may include avoidance of missed deadlines, and dealing with frequent frustrations, important tasks that never seem to get tackled, and projects that never get completed.

8.1.4.2   Step 2: Develop a detailed cost-benefit plan.

What is the cost to the organization to continue doing things the way they are currently being done? This can range from portfolio asset management to the cost of moving a partition. How much could be saved by doing things more effectively? The easy and shortsighted way is to cut staff. The smart way is to be more efficient and keep morale at a peak. An improvement plan without a cost-benefit component starts in the clouds and ends in the clouds. It is important to commence on the solid ground of prioritized needs and end each improvement with demonstrated cost-benefit success. However, the cost-benefit analysis must recognize the huge expense involved in a drop in morale and productivity resulting if cost-thinking overrides benefit-thinking. There must be a balance.

8.1.4.3   Step 3: Get management involved.

The driver to successful implementation will usually be money. This is why each recommended improvement must have an estimate of savings expressed in time or money. This will always get the attention of financial managers, so present them with the list of needs and cost-benefit estimates for their approval and prioritization. This also forces TQM into the current of everyday operations where it rightly belongs.

8.1.4.3.1 The steering committee. The steering committee oversees and approves the implementation of the improvements and so must include senior managers; otherwise all the hard work of the group developing improvement ideas will fall into the black hole of rejection or neglect. The diversified steering committee might consist of the chief financial officer, controller, a division manager, a safety officer, a maintenance engineer, the facility manager, a human resources specialist, and two or three front-line staff who could rotate in and out of the committee every 6 months. This scheme gives wide exposure to TQM and allows the staff to get involved in continuous improvement decision making.

The first task of the steering committee is to develop a budget for the improvement effort. When the improvement list is linked to potential cost savings, it is easy for senior management to invest in it. The budget will be used for both the implementation of improvements and tracking and monitoring to demonstrate the savings incurred.

8.1.4.4 Step 4: Assemble a team

8.1.4.4.1 The process improvement team (PIT). A PIT is the group formed to investigate problems, find root causes, and improve processes and systems that exhibit wasted effort, wasted money, and wasted time. They also explore resource utilization improvements or examine the causes of and solutions to errors and omissions. The team's goal is to effect positive controlled change and avoid random, uncontrolled change.

Ideally this team would consist of a mix of managers and staff. They are not randomly chosen or chosen because they happen to be friends or supporters. They are chosen because they either have knowledge of the process under review or their everyday work is directly or peripherally impacted by the issue being examined. They may be from different departments, divisions, or branches. The PIT is not a permanent body. Once their recommendations have been made and implemented, they disband.

PITs typically meet once or twice a week for 1 to 3 months, depending on the complexity of the issue that they are investigating. If the matter needs urgent resolution, they might meet full-time over several days to get the solution implemented as soon as possible. This is especially useful when the information the team will need for their work is readily available.

The facilitator. The PIT facilitator need not be popular but must have the respect of the team. Nor should this person have dictatorial leanings because then team members may get cut out of the process and lose interest. The facilitator should be more orchestrater than leader.

Facilitation skills. An effective facilitator will exhibit (1) logical thinking, with left brain processing and analysis, linked to prior experience and (2) emotional thinking, with right brain empathy and creativity, linked to intuition.

This logical thinker with people skills will also need to:

- Have the respect of senior management and be comfortable interacting with them

- Be a good time manager to start and end meetings on time

- Have no personal agenda or power game

- Encourage all team members to participate

- Keep the team focused, for example, by using a "time out" signal when conversations wander away from the topic at hand

- Be reliable and organized, including keeping accurate records of discussions

Define the problem. If one is going to solve a problem, it is essential that the problem be well defined. For example, if the problem is expressed as "poor communications," it will be very difficult to satisfactorily resolve it because:

- This is open to differing interpretations.

- The mode could include in-person, telephone, written, or nonverbal communications.

- The term "poor" is vague and relative.

One way to adequately define a problem is to go from the general to the specific. In the above example, the facilitator might follow up with "give me an example of poor communications" and keep probing until the problem is exactly defined, perhaps as "management doesn't listen to the facility manager regarding maintenance costs." Only when the exact problem has been expressed will one be ready to resolve it. Only when a surgeon knows exactly what is malfunctioning can the operation proceed with precision.

Root cause analysis. There are sophisticated techniques to establish root causes, such as the cause and effect diagram (Shearer, pp. 185–189). While this is a very useful procedure, yielding a full range of causes, it does take time and may not be required for most of the investigations that facility management PITs will perform.

A quicker yet still effective method is to ask "why?" not just once, but three or more times. It is analogous to peeling off the skin of an onion, one layer at a time. For example, if the occupants of a room frequently complain that they are cold, ask "why?" The answer may be because the heating system is not functioning properly. Again, ask "why?" The answer may refer to the fact that it was not maintained properly. A third time, ask "why?" This may reveal that there was a misunderstanding as to who was responsible for maintaining the system. A fourth "why?" is asked. Now it may be learned that there are no properly developed maintenance procedures with detailed assignments. Ask a fifth "why?" and one may learn that no budget was set aside to develop those procedures. "Why?" once more may reveal that management did not think it was urgent because they expected people to put up with a little inconvenience from time to time. Now the layers have been peeled back and the root cause has been exposed. The PIT ask themselves, "How can we change management's perspectives?" If the team leader goes to management with this issue there will be a rebuff, frustration, and demoralization. Here is where measurement comes to the rescue.

Measuring results.   TQM without measurements is like driving across the United States blindfolded. You never know where you are, how close you are to your destination, or even if you are headed in the right direction. A huge amount of improvement effort can be quickly smothered by senior management saying "I don't think this has made any difference," or "I believe that the current system is working and doesn't need to be changed." The only way to destroy these arguments is to use measurements.

The goals of measurement are twofold: (1) know where one is and where one is going, and (2) reduce the magnitude of random variation.

One could measure the cost:

- To produce rarely reviewed monthly reports

- Of inadequate methods to predict maintenance needs

- Of not outsourcing inefficient operational tasks

- Each time old performance standards are used

- Per service call for frequently malfunctioning air-conditioning equipment

- Of the time taken to interrupt value-added tasks to troubleshoot departmental complaints

- Of having staff become sick owing to poor indoor air quality

- Of emergency diesel generator shutdown when the power fails

- Of legal assistance to defend a claim of unsafe conditions

There are so many ways to measure estimates, rework, performance, marketing, and financial operations (Shearer, pp. 117–119). If measurements are not taken, one will never be able to quantify the cost of stagnation: the cost of taking no action to improve.

Measurements must involve a benchmark. A map is useless unless one knows one's current location. In the same way, the continuous improvement journey demands a benchmark so one can measure progress and savings as improvements are implemented. The benchmark can be the starting point of one's own improvements, or one can benchmark similar processes performed by similar operations in one's own facility or even in competing facilities. However, what works in one body may be rejected in another. It is not practical to take an entire viable TQM system from one organization or division and transplant it whole into another organization or division. Placing an Indy 500 engine on a wheelbarrow would be very limiting until the frame was redesigned to realize the full potential of the engine.

Variables include the current processes of the support frame, the abilities of the people, and the links to other systems. As these will always differ from entity to entity, one may transplant parts of another's system, but it will be necessary to adapt, fine-tune, and recreate some elements to suit the organization or division being improved.

Measurements do not have to involve complex statistics. There are many learned publications and books with intricate statistical techniques, discussions on statistical process control, and other complex mathematical methods. These can be very valuable, yet for everyday facility management much of this is unnecessary.

Measurements can be just as helpful and meaningful if limited to:

- Totaling up the time taken to complete a task
- Counting and categorizing the number of times that an incident occurred
- Recording the breakout of dollars spent on maintenance
- Tracking time wasted by using inadequate manuals

Indeed these numbers could form the basis of statistical analyses, but it is suggested to the reader that this would be a later refinement of TQM. Most important is to get continuous improvement under way by collecting measurements before the improvement to (1) have a baseline to compare to future improvements, and (2) demonstrate to senior management the impact of the improvement.

A measurement plan. Think about what needs to be done, demonstrated, and tracked. How will the measurements be used? Who will use them? Then develop a plan:

- What data must be collected?

- How much data will be needed to form realistic conclusions, and how long will it take to collect it?
- How will the data be sorted?

Once the data is collected, first look for patterns and then spot trends that emerge from the patterns.

One good technique is to identify the cost of non-value-added activities per incident, including salary, benefits, and materials. This can then be multiplied out per month or year.

In the heating system example the team may obtain data on how many people have been off sick in that department, versus departments where there are no heating breakdowns. This has a cost impact on the organization that is easy to calculate. They can also gather data on the efficiency of those occupants, season by season, to see if the discomfort has an impact, and this too will have a cost impact. It is the cost impacts, presented to senior management, that will get the improvement program on track and allow the team to then start to solve the problems by peeling back the causes, one layer at a time. Once the cost impact facts are demonstrated to management, their attitude will be resolved. Because the cost of the problem over several years has been identified, it might now be clear that the costs to fix the problem are small in comparison, and the budget layer is resolved. This budget leads to the establishment of proper maintenance procedures with detailed assignments. The next layer clarifies who will now be responsible for maintaining the equipment, and this leads to the final layer of increasing productivity by people working comfortably. Without the root cause analysis and cost measurement, the PIT may not have been able to solve this problem other than cosmetically.

The idea bank. Teams often get sidetracked by premature solutions or useful ideas that apply to other situations. While the facilitator must keep discussions focused on the issue at hand, it is also important not to lose these random and serendipitous inspirations.

The way to accomplish this is through the use of the author's idea bank (Shearer, pp. 170–172). Concepts that do not apply to the matters at hand are captured on paper and become a bank of ideas to draw upon later:

- When the team searches for solutions
- When future teams require solution ideas

The key benefit of the idea bank is that the team keeps focus without diversions, detours, debate, or disgressions while the nonrelevant ideas are recorded without anyone feeling discounted.

The task team. It is not necessary to always form a PIT when resolving issues. The task team is a simplified version of the PIT. A task team is comprised of a small ad hoc group who usually tackle the improvement of value-added processes and systems. This means that they do not have to delve into root causes of wasted effort because they do not deal with problems; they are merely improving what is already working. Typically they involve themselves in developing or improving checklists or guidelines. They might develop a new program or research new regulations or create new standards or forms. While these tasks may take some time to research and develop, they are usually simpler and more direct than those tackled by the PIT. Some task teams may complete their work in as little as 30 minutes.

Task teams may also be formed to help a PIT with data gathering or tracking the implementation of an improvement. In these cases the task team may be composed of a few members of the PIT who volunteer for the task team.

The concept of trial improvements. It is strongly recommended that all improvements are implemented on a trial basis. If ships need to be put through commissioning procedures, if aircraft need test flights, if stage plays have rehearsals, and if software has beta versions, why would one not implement solutions on a trial basis? This affords an opportunity to see how it works in practice, fine-tune it, strengthen it, and get people used to it. In fact, the latter factor could be critical, even if the improvement works very well. Getting the staff, users, or occupants involved in testing and fine-tuning leads to buy-in and acceptance. No one likes to have change foisted upon them. People don't mind change, as long as they have a say in what is being changed.

8.1.5   Summary: What to do to keep TQM on track

1. Inspire management to participate as partners in TQM.

2. Embrace global thinking and ignore the trivial.

3. Identify and eliminate waste before improving value-added activities.

4. Remember that with continuous improvement, not everything can be accomplished overnight.

5. Develop goals based on cost-benefit analyses.

6. Be specific in defining problems.

7. Benchmark existing conditions before implementing solutions.

8. Avoid simplistic, cosmetic fixes—find the root causes.

9. Put improvements on probation—allow trial solutions to gather buy-in while you measure their value.

8.2   Value Improvement Program

The nineties is a decade of developing and maintaining value and optimizing return on investments. We have experienced TQM, reengineering, outsourcing, and cosourcing. Our customers are constantly demanding more for less. While we have paid close attention to optimizing repetitive activities, we may not have evaluated the one-of-a-kind activities as closely. We have considered alternative and innovative methods of project delivery ranging from the traditional design/bid/build/commission; design/build; turnkey; fast-track; and hyper-track. Are facility managers evaluating new or renovation construction projects for the best returns as a function of the investments?

Have we considered long-term maintenance aspects of the design? The physical facility is not an income-producing asset for most organizations. It is therefore imperative for the facility manager to effectively manage this support function, at least overall cost of ownership. When, during the project development, should we undertake this value evaluation?

We will discuss value engineering (VE), which is also known as value analysis (VA), and value management (VM) as an opportunity for an organized life-cycle oriented evaluation of a project by a multidisciplinary team. Value engineering is conducted to achieve the defined functions at the lowest total cost of ownership consistent with established requirements for performance, reliability, safety, quality, and maintainability. It is an effective tool to identify and implement change. VE can be viewed as an adversary to the status quo.

VE works because it is a simple and effective approach with a predetermined action plan. Most designs include some costs from designer preferences. The VE team members are motivated to look beyond the normal procedures and patterns of habit.

VE is a synergistic process that results in creative problem solving and value improvement.

8.2.1   Validation of value

Individuals have differing perceptions of value because of our different experiences, culture, and education. A minimal investment of resources in a value study can provide assurance that value is being delivered by the design and construction approach shown in the design. A value study can help minimize capital costs and ensure that the owner and user requirements will be met by using a function-oriented analysis. It also helps in integrating all systems for the project, enhancing quality, performance, and/or reliability. Such an evaluation can also retain the focus on issues like high churn rate and high space usage.

Prior to any attempt to assure value, we must define and agree to a common basis of value. Value can be interpreted as the usefulness of the product, its exchange or economic value, esteem, aesthetics, political statement, moral, cultural, and religious value. Clearly, not all of these components will be rated high in every situation. The relative importance of each of these elements constitutes the basis for value assurance. It is important that the VE team operate under the same set of values as the design and construction team. A systematic process, as enumerated here, can help the facility manager validate that all aspects of value, as interpreted within the corporate context, are being met with optimum usage of resources.

One commonly used technique is to return to the fundamentals of why a project is being undertaken, what function it is expected to fulfill, and what is its perceived value. It hinges on the current axiom of "question everything." Yet the most common response to this question is "this is the way we have always done it." The validation process seeks win-win solutions, with the primary focus being the project under consideration.

8.2.2   Value engineering

Value engineering is an organized effort at analyzing the functions of systems, equipment, facilities, procedures, and supplies by an independent multidisciplinary team for the purpose of achieving the required functions at the lowest total cost of effective ownership, consistent with requirements for performance, reliability, quality, maintainability, and safety.

VE stems from a systems-oriented approach with a formal work plan that systematically eliminates frivolous functions from the project. To ensure that all the desired functions are met, it is approached by a multidisciplinary team comprised of experienced design, construction, and facility professionals in all aspects of the work, led by a value practitioner. VE is conducted with a life-cycle orientation, examining and comparing total ownership and operational costs for the project being considered. It is a proven management technique which uses a predetermined work plan, a process that has proved itself in many types of industries and diverse applications. It relates the required functions with the value received.

VE is not a design review, nor is it intended to rectify design omissions or to review design calculations. VE is certainly not a cheapening process with the aim of cutting costs by sacrificing expected performance and reliability. VE is not a requirement for all projects. It is a tool that allows an assessment of value received for the committed resources.

8.2.3 Synergistic opportunities

VE brings together a talented team of specialists that spend several days together, analyzing alternatives to achieve the desired functions using innovative tools and techniques. These professionals draw from their experiences and build on ideas. VE isolates team members from the day-to-day interruptions, putting aside immediate solutions generated by force of habit, removing roadblocks to solutions, and changing attitudes about achieving the desired functions. It fosters review of the project divorced from the prejudices of the design process and encourages creativity.

8.2.4 Successful value engineering

Success in VE is attributed to the SAVE International job plan that utilizes a carefully selected team of experienced persons with expertise that addresses disciplines required in the project under consideration. The team must take the time to understand the project goals and current processes in depth before attempting to develop alternatives. An important responsibility of the facility manager is to ensure that during this education process, the team understands the true needs of the users and the organization. Cost comparisons must be evaluated in relation to the defined functions, with life-cycle costing approach. Simply using initial costs as comparative factors can be very detrimental to the overall project and create ill-will toward the extended team, besides creating a nightmare for the facility professional.

8.2.5 Increase value

Value can be defined as a relationship between performance and cost. Performance can be further simplified as success in attaining the desired function(s). We may increase value in two ways; either by increasing the functionality of the elements or by reducing the cost to provide the function(s). An inherent goal of VE is to increase value at no additional cost, when feasible. Contrary to popular usage of the terms *value engineering, value analysis,* and *value management,* it is not necessary to gain one at the expense of the others. Ideally, value engineering does not emanate from a desire to reduce cost but from a desire to achieve the best return from the capital investment. Inspection and corrective action, after the effort has been expended, detracts from rather than adds value. Quality should be built in by the process, an effort supported by VE.

8.2.6   Value perceptions

There are several types of value. The use value of an object is perceived differently in relation to the function to be achieved. While a shovel has a high value for digging a hole in the ground, it has a low value for turning a nut. The shovel has not changed, yet its perceived use value has been altered dramatically depending on the function that we wish to perform.

The exchange value of the object or service is modified by its demand in the marketplace, by various individuals, and by the government. A hundred dollar bill has an exchange value for goods and services because the government has a reserve of precious metal, gold, to support that value, and we as individuals accept it without question. If the quantity or relevance of this precious metal were to be drastically reduced, the one hundred dollar bill could become a worthless piece of paper.

We assign esteem value to an object or service which is determined by society based on its scarcity, or level of difficulty to obtain. Is the service provided by a certified facilities manager, architect, or registered professional engineer any different between the day the person obtained the designation and the previous day? We assign designations an esteem value, owing to the time and effort expended in obtaining them. Some objects with esteem value include the Hope diamond and a Stradivarius violin.

Beauty is in the eye of the beholder. Aesthetic value is a result of the pleasure and fulfillment evoked by an object's beauty, grace, or style. Perceptions change with what is considered pleasant. There are multiple examples of facilities that were considered aesthetically pleasing by some while evoking the opposite response from others. The same is true of art and music.

Items which embody or define the rules and agreements by which society functions are deemed to have judicial or political value, like laws, ordinances, and statutes. Other types of value include moral and religious value.

8.2.7   Why does poor value occur?

Since all of these manifestations of value are understandable, why does poor value occur? A series of issues face the facilities manager which, either individually or in various permutations and combinations, result in poor value. Primary among these are:

- Inappropriate definition of goals and functions from the customer
- Insufficient management direction
- Lack of time and/or budget
- Lack of data
- Lack of creativity
- Misconceptions
- Temporary conditions that inadvertently become permanent
- Habits
- Attitudes
- Politics
- Poor problem-solving approach

Addressing these constraints in an organized manner can overcome these hindrances and result in enhancing value. Planning the VE process with a detailed set of actions and taking action on that plan are key ingredients to obtaining high value. Nonperformance on either of the two will guarantee less than optimum return on a project. VE is one of the most misunderstood processes in industry and is routinely mistaken for cost reduction, cost containment, and other means of cheapening the end product.

8.2.8 Value engineering process

The traditional VE process comprises five phases—information, speculation, evaluation, development, and presentation—to be followed in succession. These steps must be carefully planned prior to assembling the VE team to conduct the exercise. The need for an experienced value specialist to lead the team cannot be overemphasized. The process of the VE team depends on the VE team leader.

8.2.8.1 Information phase. During the information phase, the VE team gets educated about the project, its functions, program, costs, the value of its functions, and other relevant facts. During this phase, the leadership of the management, design, and construction team should present pertinent details, to include the rationale, design criteria, site conditions, regulatory requirements, history of the project, existing utilities, and community requirements. The facility manager should be an active participant, providing guidance about the organizational culture, value perceptions, and special constraints on the project. It is important to identify specific issues that may not be VE candidates, like seeking another site, when you have obtained an environmental impact agreement after several years of community and municipal process. This allows the VE team to be firmly cognizant of what the realm of the possibilities are, so they are less likely to present alternatives that are not acceptable.

The VE team should define the functions of the project being studied to clearly understand the performance criteria. Among the questions that will help identify the functions are, what is it, what does it do, how does it work, how is it used, who uses it, and when is it used. Other aspects to be addressed are the evaluation of functions by asking such questions as what must it do, when and where is it used, and who uses it.

One of the basic rules of VE is to define a function using only two-word definitions: an action verb followed by a measurable noun. The action verb answers the question what does it do, and the measurable noun answers the question what does it do it to. This two-word definition of a function results in clarity of thought, avoids combination of functions, assists in separating basic functions from specific actions, and helps in disassociating from specifics. Listed below are a few examples of VE function definitions.

Item	Function
Electric wire	Conduct electricity
Screw driver	Transmit torque
Column	Support load
Beam	Transfer load
Fence	Enclose area
Roof	Enclose space
Door	Control access
Carpet	Cover floor

After functions are defined, it helps to also understand the worth of each function. Worth is the lowest cost of providing the function and is synonymous with maximum value for VE purposes. Worth is based on the value of the necessary function and not on the cost of the specific design. The worth is equal to the lowest cost for which the basic, or essential, function(s) can be obtained. Secondary functions are functions that are not basic functions. They are not evaluated for worth in VE. Value is determined by the lowest cost of attaining the basic function(s) without sacrificing essential criteria.

Function analysis is the primary difference between value and cost studies. VE asks what the item does before seeking alternative solutions and then new costs, while cost reduction asks how the item can be modified and determines its new cost.

Once all the functions are believed to have been identified, they need to be arranged in some logical order to see and understand how they relate one to another in accomplishing the basic function. We accomplish this by building a function analysis system technique (FAST) diagram. The VE team asks a series of how, why, and when questions to determine the sequence by which functions are logically related one to another as well as to help determine functions that may be needed that have not yet been identified.

FAST diagraming is one of the most challenging and critical concepts of VE. It brings logic to the function-determination process, visually displays relationships of functions to each other, validates the functions under study, and enhances the comprehension of the project by the VE study team. It integrates the relationships of the various functions and assures correct determination of the basic function(s). An example of a FAST diagram is shown in Fig. 8.1.

8.2.8.2 Speculation phase. The speculation, or creative, phase of VE uses brainstorming and other techniques to help the VE study team to identify as many alternatives as they possibly can for the project under study. When looking for VE team members, leaders intentionally seek individuals with many years of experience in their fields. They do this because they want team members who have had many challenges in solving problems in their discipline. This is what they bring to the study, this background and knowledge of having solved problems in different ways successfully.

Human beings are the most creative until approximately age five, when we begin to undergo formal education. The regiment of our formal education system seems to stifle creativity. There are several organizational barriers to creativity, including preoccupation with status, job insecurity, excessive bureaucracy, time constraints, lack of information, and intolerance of ambiguity and uncertainty.

Creative problem solving can take several forms. Brainstorming offers a free flow of ideas in a nonjudgmental framework. Checklists retained from earlier VE studies can be used to spur creativity. Morphological analysis ranks alternatives with all potential combinations of elements in deriving solutions. Some common roadblocks to new ideas include lengthy regulatory review, blaming inaction for the sake of safety, and pushing off responsibility to maintenance or operations.

The VE process offers each team member the opportunity to contribute to the dynamic process. It generates results because of the premise that "all of us are smarter than any one of us."

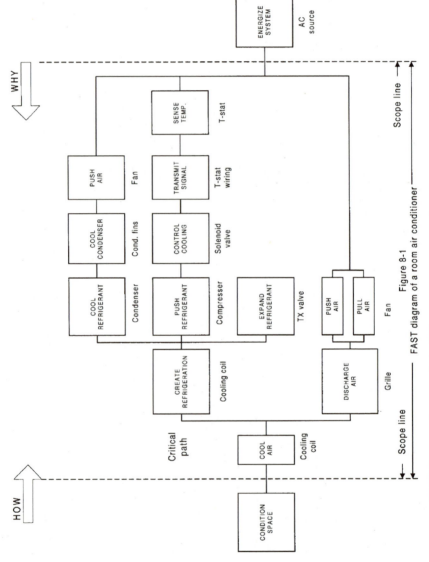

Figure 8.1 FAST diagram of a room air conditioner.

8.2.8.3 Evaluation phase. Creative ideas developed during the speculation phase need to be organized and systematically evaluated. It is helpful to establish a weighting and scoring system with a set of common definitions that are shared by all members of the VE team. Consensus is an important ingredient of establishing the evaluation criteria.

The VE team should work together to determine those criteria that are important to the project being studied and use those criteria to evaluate each of the alternatives suggested in the speculation phase. Care should be taken not to try to generate a long list of criteria, as it would become cumbersome to evaluate a long list of alternatives against a long list of criteria. After defining the criteria, weight, and scoring systems, the team can determine the maximum score any alternative could receive. The team should then determine the minimum score that any alternative must receive in order to move on to the development phase. After this has been agreed to, the team can proceed to evaluate the list of alternatives. Care should be taken to evaluate each alternative by a given criterion before moving to another criterion. An evaluation matrix is shown in Fig. 8.2. This type of matrix can be used to list the alternatives, the criteria for evaluating the alternatives, along with the assigned weights for each criterion and the scores. For each alternative, the raw score for each criterion may be shown in the top of the score entry and the weighted score (the score multiplied by the weight of the criterion) may be shown on the bottom of the score entry. The total scores and ranking are also shown in this matrix.

8.2.8.4 Development phase. Those alternatives that meet the criteria from the evaluation phase are now considered for further development. They may be reviewed and a determination made if any number of them should be combined because they may complement one another. At any rate, team members are assigned to develop alternatives to determine if there is an opportunity to increase the value to the project by these alternatives. This is done by analyzing the data available on the alternatives to determine if they really will work for this project. The development includes preparing sketches and/or drawings to show the current design and the proposed alternate, as well as a life-cycle cost comparison of the two methods and text to explain the alternative. Care must be taken in developing this information so that the pros and cons are addressed, as well as any impacts on the rest of the project if the alternative is accepted. The VE team should strive to include as much factual information as possible in order for the client and the designer to understand the alternative and then be able to make an informed decision as to whether

	VALUE ENGINEERING
	EVALUATION MATRIX WORKSHEET

OBJECTIVES OR CRITERIA

LIST THE ALTERNATTIVE SOLUTIONS. C1:
DETERMINE WHICH ONE C2:
RANKS BEST AGAINST C3:
DESIRED CRITERIA. C4:
WORK DOWN, NOT ACROSS. C5:

ALTERNATIVES	WEIGHT	C1	C2	C3	C4	C5	C6	C7	C8	C9	C10	TOTAL	RANK	COMMENT

Figure 8.2 Worksheet for evaluation matrix.

Present (show sketch)			Proposed (show sketch)				
Nonrecurring costs (estimate)				Direct Expense	Direct Labor	Variable portions of burden	Total Variable Cost
	Hrs.	$					
Design			Present Cost				
			Proposed Cost				
Drafting			Net savings:			Proposed by:	
			1. Difference in variable costs, proposed vs present _____				
Admin.			2. Less nonrecurring costs _____				
Maint.							
Permits			3. Net savings _____				
Contract Admin.							
Misc.							
Total							

Figure 8.3 Function and cost analysis worksheet.

to accept or reject the alternative. A function and cost analysis worksheet is shown in Fig. 8.3.

Life-cycle cost comparisons provide visibility for all cost factors included in the analysis and offer a clearly defined methodical framework for logically and fairly appraising alternative solutions. Most of all, they provide interdisciplinary cooperation embodied in an analytical tool that kinks planning, design, and long-term costs.

In addition to initial costs, life-cycle costs must include administrative costs, facility operating and maintenance costs, finance charges, energy consumption, cyclical replacement, and modifications of use, as appropriate. It is conceivable that some decisions could impact operating costs for the users. These costs must be included in the comparison along with the time value of money, salvage, and residual value of the decisions.

8.2.8.5   Presentation phase. Improvement requires change from the status quo, and change requires salesmanship. The presentation phase is the VE team's opportunity to sell their alternatives to the project decision makers, the owner/user, and the designer. This is a formal presentation where the team members present the alternatives and explain exactly what is envisioned if accepted. Afterward, the owner/user and designer may ask questions to clarify any items that they may want to understand better. At this point, everything may be in place for a decision to accept or reject the alternative. If a decision is not possible at this time, a clarification should be made as to who has the action item and the schedule required to perform the action(s) so that a decision can be made. The objective is to be able to come to closure on the VE study as soon as it is feasible and prudent.

8.2.8.6   Facilitator role. One of the most important roles required in a successful VE study is the VE team leader. This individual facilitates the process by coordinating between management, designer, and the VE team members. This individual should be equipped with all the facilitation tools and techniques and should be an experienced VE professional. The facilitator must focus all involved on the content of the study workshop and maintain a balanced participation with all persons involved. The facilitator must manage the process in such a way that all information acquired and the details of what was done in each phase of the study are recorded accurately and included in the report.

8.2.8.7   Implementation. VE does not end with presenting alternatives. The important action required is implementation of accepted alternatives. The VE team leader has the responsibility to maintain communications with the project management to determine the status of those proposals not actioned at the presentation meeting for acceptance or rejection. After all alternatives have been actioned, the VE team leader has the responsibility to produce a final report to the project management to document the entire process.

8.2.8.8   Ideal time for value engineering. Value engineering can prove to be effective at any time during the life of a project. The best return on investment from the VE process is obtained at the earliest stages of a project, from conceptual to approximately the 35 percent design milestone. This is true mainly because VE after this time most likely would have effects on the project schedule and budget because of possible redesign considerations. Prior to the 35 percent milestone, there is seldom detail design work; however, sufficient information exists to understand the intent of the design and how the project is expected to progress. There is a requirement for cost information on the project that coincides with the level of development of the project. This allows the VE team to use information generated by the project to analyze and use in conducting the study.

The use of the VE process should result in a win-win situation for all. The project owner and user win because the required functions are affirmed and met at the lowest life-cycle cost. The designer wins because the design is affirmed and those areas where alternatives make sense have been accepted. The VE team wins because they have provided a valuable service to conduct an independent review of the project to determine that the project needs are being met at the lowest life-cycle costs.

8.3   Energy Management Program

8.3.1   Overview

In today's facilities the goal of a facility manager is to maintain comfort conditions within the space while at the same time managing energy costs. In order to attain this goal facility managers must be concerned with both the mechanical systems and the control systems that regulate the mechanical systems. In order to maintain comfort within the space the facility manager must recognize that the goal of the heating, ventilating, and air-conditioning systems (HVAC) is to supply required ventilation to nullify the effects of contaminant buildup within the space and to maintain required space temperature conditions conducive to worker productivity. Some facilities also require precise humidity conditions which are also controlled via the HVAC systems.

The basis for an energy management program will require a baseline energy audit of the facility to have a reference point for comparison to determine actual savings. This energy audit can be obtained by tracking utility bills, energy costs, and maintenance costs over a period of time, usually 1 to 2 years. These costs must be correlated against weather conditions that have been experienced during those time periods. This comparison will give you an energy usage per degree day for both the heating and the cooling mode of the building. It is important to remember that internal spaces within buildings will require cooling in both the heating and the cooling season. It is only the perimeter spaces that will require heating in the heating season and cooling in the cooling season.

When a baseline energy usage has been determined, the facility manager can then implement plans to decrease this total energy usage. These plans can include modifications to the mechanical systems and/or energy management programs using the building control system to optimize the energy usage. To determine how the building energy usage baseline compares to similar buildings in similar climates it is possible to refer to the Building Owners and Managers Association (BOMA) bench marking guidelines. This is a compilation of energy usage from buildings of various styles and locations, and allows a facility manager to determine where on the scale of energy usage his or her building is located. If, in comparison to the benchmark, the energy usage costs are relatively low, it will be difficult for the facility manager to decrease the cost further. If the energy costs are high in comparison to the bench marking number the facility manager must determine why the energy costs are not within typical parameters.

If energy costs are higher than the benchmarking standards the total facility must be reviewed. This review should start from the outer skin of the building to determine the thermal conductivity of the glass area, the tightness of building construction, and the orientation of the building with regard to solar and wind load conditions. Once the physical aspects of the building has been reviewed the next step would be to review the mechanical systems.

This mechanical system review would include the quantities of fresh air inserted into the building for ventilation, the amount of exhaust air being removed from the building, the type of fan system and the controls system strategy, and the mechanical efficiency of the various components used.

8.3.2   Mechanical systems

HVAC systems can be grouped by common mechanical characteristics. These groupings would include constant-volume systems, variable air volume (VAV) systems, 100 percent outside air systems, minimum outside air systems, and heating and ventilating units.

8.3.2.1   Constant-volume air-handling units (CAV).

A constant-volume air-handling unit consists of a supply fan, a cooling coil, a preheat coil, outside air and return air dampers, a return air fan, and exhaust dampers (see Fig. 8.4). The supply and the return fan volumes in cubic feet per minute (cfm) are constant and are set by the air balancer during the initial unit setup. The cfm figures for these fans should be checked periodically to maintain an accurate differential

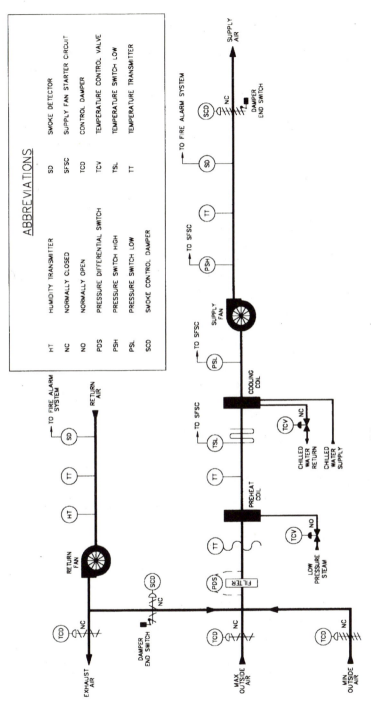

Figure 8.4 Typical constant air volume system control layout.

between the two fans. This differential should be equal to the minimum outside air requirements for ventilation. The minimum outside air requirements for ventilation are set by the design engineer using the ASHRAE standard for ventilation of 20 cfm per person or 1 cfm per square foot of space (ASHRAE 62) or as required by local codes or the particular area that is being served by the unit. To this standard configuration can be added reheat coils and humidification devices. Controls for this type of unit consist of motorized actuators for the dampers and control valves for the heating and cooling coil. Temperature sensors are located in the return airstream, mixed air plenum, and on the discharge of each coil. Additional safety devices such as smoke detectors and low-temperature alarm sensors may also be located on the air-handling unit. Since this is a constant-volume unit the discharge airstream temperature is controlled to maintain this temperature setpoint. The typical control setup will have the dampers open to minimum position and the heating coil modulate to elevate the discharge temperature. On a call for cooling, the outside air dampers will open first to use free outside air cooling (outside air economizer) when the outside air temperature is suitable for cooling. When the outside air cannot be used for cooling the dampers will remain at the minimum outside air ventilation position and the cooling coil will be modulated to maintain the discharge temperature setpoint. The discharge temperature setpoint for the air-handling unit will usually be reset from a return air temperature sensor or from a space temperature sensor.

To improve energy usage at the air-handling unit you must determine that all controls are first working properly. The outside air economizer section of the dampers must be reviewed to ensure that the outside air economizer cycle is being utilized as much as possible. Early control systems based outside air economizer on dry-bulb air temperature for lockout of the outside air economizer cycle. It was found that using outside air enthalpy vs. return air enthalpy will provide a savings of approximately 8 percent over a dry-bulb economizer lockout. To determine the outside air and return air enthalpy it is necessary to mount temperature and humidity sensors in the return airstream and outside air temperature and humidity sensors in the outside air plenum or as a master outside air weather station. In climates where outside air humidity sensors are not practical it would be necessary to measure outside air dewpoint. The outside air humidity or dewpoint and temperature can then be used to calculate outside air enthalpy. The enthalpy of the return airstream and outside airstream are compared for total heat content and the airstream with the lowest total heat content will be used as the primary airstream for the air-handling unit.

The control valves on the heating and cooling coil should also be reviewed to determine if proper flow is obtained within the coils and that the heat transfer is taking place as designed. This review should include a review of the control valve and its actuator to determine that the actuator closes the control valve fully in the closed position with no leakage and opens the control valve fully for the full open position. In the full open position a GPM reading should be taken to determine if the balancing valve within the water system is still set to the balancer's original mark and is allowing the full design flow to the coil. The control valve sizing should also be reviewed. Control valves are sized to have a pressure drop across the valve at full flow conditions of 5 lb or less for a chilled water valve or 80 percent of the inlet pressure for a steam valve. For water valves the size of the valve is given in both pipe size and the valve characteristic called C_V. C_V is a measure of the valve flow coefficient and is based on the flow and the pressure drop across the valve per the following formula:

$$C_V = \frac{Q}{\sqrt{\Delta P}}$$

where C_V = valve coefficient of flow
Q = flow in gallons per minute
ΔP = pressure difference between inlet and outlet ports

Normal design parameters are that the pressure drop ΔP at maximum flow should be less than 5 psi between the inlet and outlet. If the valve is sized too small, the pressure drop across the valve will be high and full flow may not occur. If the valve is sized too large, it will be difficult to control the flow properly, as full flow will occur when the valve is only partially open.

Dampers must be checked for proper sequencing between the outside, exhaust, and return air mixing dampers. Under normal operation with the outside air damper closed (minimum damper will be open or damper will be open to minimum position) the return air damper should be fully open and the exhaust air damper should be fully closed or opened to minimum position. With the outside air damper fully open the return air damper should be closed and the exhaust air damper should be open fully. If the dampers are not sequencing in this manner, a flow constriction could occur which will imbalance the amount of outside air and return air being utilized by the system.

Control sequences should be formulated so that there is no overlap between heating and cooling sequences. The maximum outside air damper and cooling coil must be fully closed (outside air damper to minimum position) before the heating coil begins to open. The heating coil must be fully closed before the dampers begin to open. The dampers should be fully opened to take maximum advantage of outside air cooling before the cooling coil is allowed to open.

Any overlap between heating and cooling sources will result in a waste of energy as you will be cooling heated air or heating cooled air. The only exception to this would be that if the outside air source is extremely cold a low limit on the heating coil valve will make sure that mixed air temperature will not drop below 50° to prevent freezing of the cooling coil and extreme discharge low temperature.

8.3.2.2 Variable-air-volume air-handling unit (VAV). A variable air volume unit (see Fig. 8.5) is configured in a similar manner to a constant air volume unit with the addition of supply and return volume varying devices and airflow measuring stations. Early systems used dampers in the return and discharge ductwork to vary the air volume by creating a pressure restriction. This was not efficient and was a waste of energy. Manufacturers then began placing dampers on the inlet section of the fan casing. These dampers restricted the amount of air going into the fan scroll, thereby reducing volume and decreasing motor horsepower required to pump the air. These devices did save some energy but the placement and resulting restriction resulted in high-velocity airflows. These high velocities caused the destruction of the damper blades and linkages due to vibration. The location also caused mechanical problems in the operation and high maintenance. With the advent of microprocessor technology low-cost variable-frequency drives are available to vary the frequency and speed of the fan. These drives accurately vary the air quantity delivered and reduce electrical consumptions of the fan motors, thereby saving energy costs.

Airflow measuring stations come in a variety of styles and types. Pitot tube arrays are placed across a duct section with straightening vanes resembling an egg crate. The total and static pressure are sent to a differential pressure transmitter which determines the difference in pressure. This difference in pressure is the velocity pressure, which is proportional to the velocity. To obtain an accurate reading it is best to have the velocity as high as possible in the duct section. The velocity varies with the square root of the velocity pressure, so as the velocity decreases by half, the velocity pressure decreases to one-quarter the original pressure. At low velocities the velocity pressure is so low that accuracy is a problem. To try to increase accuracy and reduce the quality of straight duct run required for airflow stations manufacturers have experimented with different methods of determining the velocity and different placements for the airflow stations. Airflow stations have been placed in the inlet bell of the fan to keep them in the

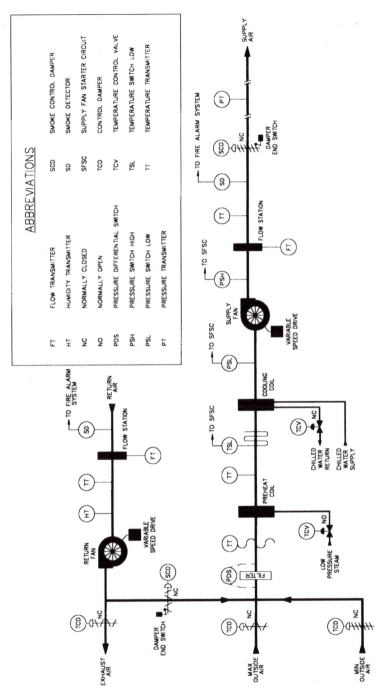

Figure 8.5 Typical variable air volume system control layout.

ABBREVIATIONS

FT	FLOW TRANSMITTER	SCD	SMOKE CONTROL DAMPER
HT	HUMIDITY TRANSMITTER	SD	SMOKE DETECTOR
NC	NORMALLY CLOSED	SFSC	SUPPLY FAN STARTER CIRCUIT
NO	NORMALLY OPEN	TCD	CONTROL DAMPER
PDS	PRESSURE DIFFERENTIAL SWITCH	TCV	TEMPERATURE CONTROL VALVE
PSH	PRESSURE SWITCH HIGH	TSL	TEMPERATURE SWITCH LOW
PSL	PRESSURE SWITCH LOW	TT	TEMPERATURE TRANSMITTER
PT	PRESSURE TRANSMITTER		

highest possible laminar flow condition. This placement has been very successful. Other methods of velocity measurement have been used including hot-wire anemometers which use a heated resistance wire to measure the airflow by the cooling effect across the wire.

8.3.2.3   One hundred percent outside air systems. One hundred percent outside air systems (Fig. 8.6) use complete outside air with no return air mixing. In a 100 percent outside air system the preheat coil must be sized correctly to prevent freeze-ups and to maintain discharge temperature within comfort limits. Such systems are usually constant-speed fans, and the discharge temperature setpoint is changed based upon space or exhaust air conditions.

8.3.2.4   Minimum outside air systems. Minimum outside air systems (Fig. 8.7) are configured similar to a VAV or constant-volume system except that there is no maximum outdoor air damper. Because of this lack of a maximum outdoor air damper the amount of outdoor air taken in and the amount of air exhausted are constant. No outdoor air economizer using outdoor air for free cooling is available in this unit configuration, so a waterside economizer is utilized. Minimum air units are designed into a project so that outside air ductwork is minimized. This reduces the shaft space required for the outside air ductwork and increases rentable space.

8.3.2.5   Heating and ventilating units. Heating and ventilating units (Fig. 8.8) are heating and outside air only; they do not contain a cooling coil. These units serve unconditioned spaces and because of their low energy consumption are rarely considered in energy management programs.

8.3.2.6   Chiller plant. The chiller plant generates chilled water for usage by the air-handling units and for other needs within the building. Chiller plants consist of chiller machinery, associated pumps, and cooling powers. Energy usage to generate chilled water is a major consideration in building energy consumption. Energy savings can be realized by using variable-speed drives on the cooling towers and pumps and by using energy management programs described later to optimize the electrical usage in the generation of chilled water.

8.3.3   Control systems

Energy management techniques have evolved as the sophistication of the control systems on the mechanical equipment has evolved. Early control systems maintained control within large deviations from set-

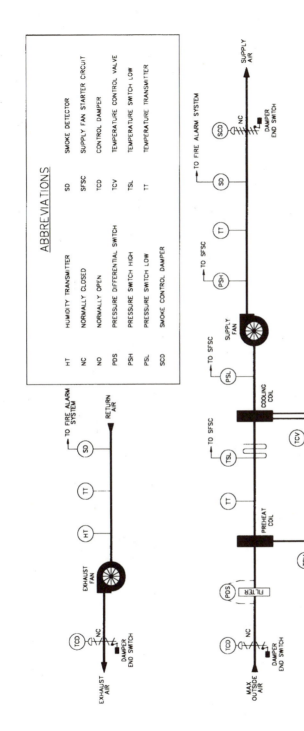

Figure 8.6 Typical 100 percent outside air system control layout.

ABBREVIATIONS

HT	HUMIDITY TRANSMITTER	SD	SMOKE DETECTOR
NC	NORMALLY CLOSED	SFSC	SUPPLY FAN STARTER CIRCUIT
NO	NORMALLY OPEN	TCD	CONTROL DAMPER
PDS	PRESSURE DIFFERENTIAL SWITCH	TCV	TEMPERATURE CONTROL VALVE
PSH	PRESSURE SWITCH HIGH	TSL	TEMPERATURE SWITCH LOW
PSL	PRESSURE SWITCH LOW	TT	TEMPERATURE TRANSMITTER
SCD	SMOKE CONTROL DAMPER		

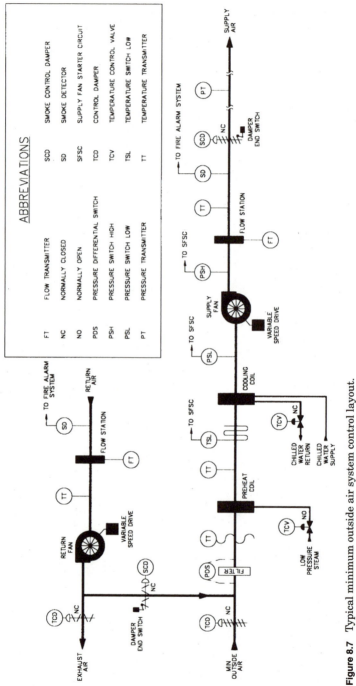

Figure 8.7 Typical minimum outside air system control layout.

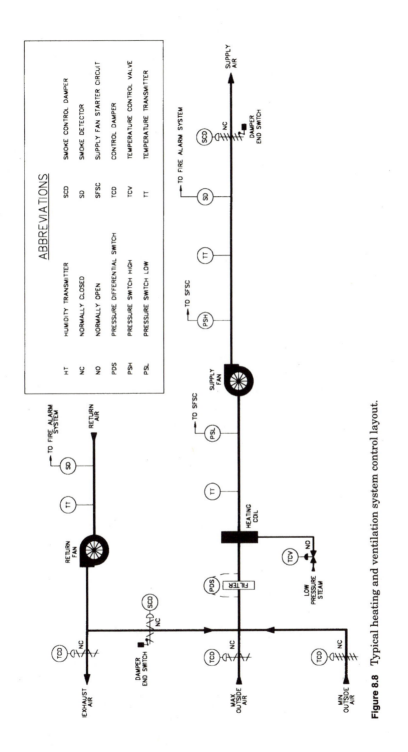

Figure 8.8 Typical heating and ventilation system control layout.

point. As these controls were refined, deviations of temperatures from setpoints became smaller and smaller. Control systems have progressed from early pneumatic and bimetal electric systems to electronic controls utilizing sophisticated circuitry.

8.3.3.1 Pneumatic control systems. Pneumatic control systems are utilized in larger buildings where long-term stable control systems and distribution of power via pneumatic air tubing is an economic alternative. Pneumatic controls use rolling rubber diaphragm-type actuators with spring return on loss of air pressure to move valves and dampers. Pneumatic control systems have existed for a long period of time and have proved themselves to be reliable and easily maintained. Pneumatic controls are, however, relatively unsophisticated and energy management possibilities are limited. Pneumatic control systems use a proportional force balance lever system to maintain relative air pressures between sensors and controlled devices. Temperature sensors utilize a bimetallic strip that curves as temperature changes. This strip is positioned above a calibrated orifice, and the distance from the strip to the orifice determines the back pressure on the circuit. Pneumatic controls are usually proportional controls only and therefore will only maintain a temperature close to the setpoint. The difference between the setpoint temperature and the actual temperature that is being maintained is called the offset. In order to minimize the offset, the proportional band, the range that a difference in temperature will make on the pneumatic signal output, should be as small as possible without becoming so small as to induce a hunting mode of operation (see Figs. 8.9, 8.10, and 8.11). Pneumatic actuators are frequently combined with electric, electronic, or direct digital controls to obtain the power benefits and spring return operation of pneumatic end devices such as valves and dampers while obtaining the sophistication of electric, electronic, or direct digital control systems.

8.3.3.2 Electric control systems. Electric control systems using bimetallic elements were the first control systems applied commercially. These control systems were the precursor of the simple thermostat currently in most homes across the county. When the bimetallic element is subjected to a temperature change the difference in expansion between the two metals creates a curvature. This curvature causes a contact on the end of the bimetallic strip to make contact with a fixed contact block, allowing the completion of an electric circuit to energize a home boiler or a commercial HVAC system. Later the first true energy management techniques evolved when a clock was added to a fan starter circuit to allow the fans to automatically

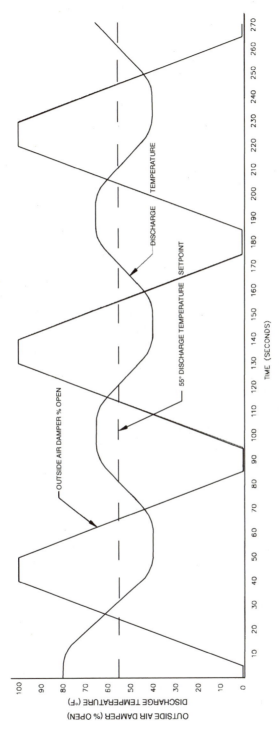

Figure 8.9 Control system with a severe hunting problem—as the outside air damper closes completely, discharge temperature fluctuates between return air and outside air temperatures.

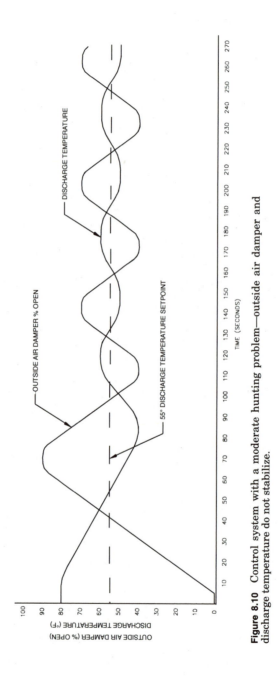

Figure 8.10 Control system with a moderate hunting problem—outside air damper and discharge temperature do not stabilize.

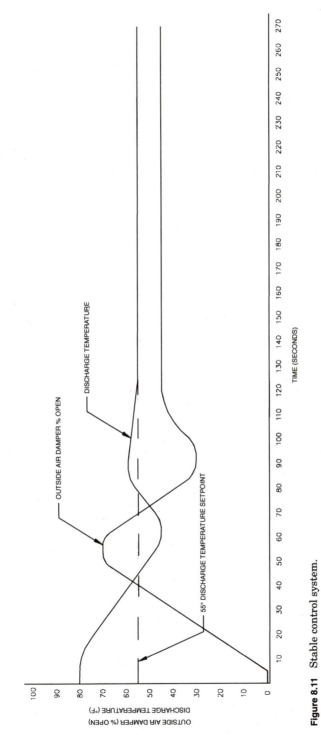

Figure 8.11 Stable control system.

come on during occupied hours and to turn off for unoccupied periods. Electric controls have increased in sophistication to allow modulating control systems via a resistance balanced Wheatstone bridge. Electric actuators are usually slower-acting and less powerful than pneumatic actuators. Electronic controllers have proportional circuitry that can be adjusted more accurately and can have built-in circuity to eliminate offset so that exact temperature setpoint is maintained.

8.3.3.3 Direct digital control systems. With the advent of microprocessing technology direct digital control systems with sophisticated programming techniques have become the current standard for the industry. These controls with interface devices can use pneumatic or electric actuators but have electronic discrimination circuitry that can utilize proportional, integral, and derivative (PID) calculations. These calculations enable direct digital controls to quickly and accurately stabilize the system and achieve precise setpoint temperatures.

8.3.3.4 Industrial control systems. Industrial control systems in the early days of pneumatic and electric control systems were more sophisticated control systems used for precise industrial controls. These controllers had early proportional and integral controls circuitry built into the controller. This enabled industrial controls to more precisely control process control variables. Industrial controls were also built with redundant circuitry and hardened components to survive process floor temperature and environmental conditions. As direct digital control becomes available in a commercial market the accuracy of industrial controls can be duplicated by commercial control systems. Industrial controls are still available with built-in redundancy, environmental hardening, and communications protocols that enabled process control to be accurately maintained.

8.3.3.5 Control panels. Control components are usually configured and assembled in system panels. Early system panels contained mixtures of components required to control systems. These components could be pneumatic controllers, electropneumatic solenoid valves, relays, timeclocks, and pneumatic gauges on the face of the panel for indication. Electric control panels had a similar assemblage of components. They were usually placed near the system to be controlled and did not communicate with other panels in the nearby area or within the building.

Current control systems use a modular or dedicated microprocessor components assembly along with an accessory panel to control systems. The modular panel consists of a systems communication card, a processor and memory card, and assorted analog and digital input and output cards as required for system operation. This type of panel has no preloaded software programs to control systems but can utilize library programs that have been derived from other projects.

Dedicated microprocessors with preprogrammed chips are also available for system operation. These panels have predefined input and output functions and are preprogrammed at the factory for predefined systems operations. Manufacturers have various names for this type of dedicated panel such as terminal equipment controller (TEC). This type of controller is used extensively on variable air volume boxes, small air-handling units, and heating and ventilating units. The advantages of factory preprogramming must be balanced against the lack of custom program capability that may be required in certain applications.

8.3.3.6 Communications protocol. With direct digital technology and electronics has come the requirement for systems communications. Communication is dealt with on a number of different levels. There is intrapanel communications, sublevel communications, peer-to-peer level communications, and systemwide (global) communications. Each of these communication levels may have different protocols and may operate at different speeds.

8.3.3.6.1 Panel communications. Communications inside a panel are usually handled on a dedicated internal communications channel. This allows all input and output modules to communicate with the controller directly. These communications are across very short distances and are handling on a high-speed internal bus. Faculty managers do not generally have to worry about this level of communication, as it is built into the panel at the factory and has little effect on overall systems operations.

8.3.3.6.2 Sublevel communications. Many building management systems today have sublevel communication pathways that extend from main panels to terminal equipment controllers. This allows the terminal controllers to communicate with the main panels, which then encode and send data over the network to the command station. Some local-level decisions are made at system panels and then communicated downward to the terminal equipment controllers. These lower-level buses that support the terminal equipment controllers are usually of a slower speed than the main communication bus because data communication is less frequent on this bus. Terminal equipment controllers by their nature are self-contained units that require few overall commands from the next level panel. Some commands that are frequently sent to terminal equipment controllers include day-night, occupied-unoccupied, and heating-cooling.

8.3.3.6.3 Systemwide communications. Main system panels are on an interconnecting bus that utilizes standard communications protocols such as Arcnet, Ethernet, Bacnet, or LonWorks, or they may use a proprietary protocol. This allows panels to talk to each other, share information as required, and communicate to a global head end. On early control system manufacturers encoded their communications to proprietary protocols. This did not allow different systems to communicate to each other. Recently opened standardized protocols such as Bacnet and LonWorks are utilized by some system manufacturers to allow ease of system interoperability. While the industry is heading toward this open protocol of the future, many major manufacturers are still not adhering to the open standards but are trying to establish their own protocol as the standard.

8.3.3.6.4 Gateways. To allow interoperation with other systems such as power monitoring, boilers, chillers, elevators, and lighting it is frequently necessary to use a gateway. A gateway is a communications protocol device that inputs one type of communications protocol, reassembles the information, and outputs it in a different communications protocol. Most major manufacturers of control systems manufacture these devices to allow other systems information to be displayed on their system and their system to have limited control functions. This allows an intermarriage of the lighting, chillers, power monitoring systems, and the HVAC system to give a facility manager unprecedented control over a facility.

8.3.4 Energy management programs

8.3.4.1 Time-of-day programming. One of the earliest energy management programs was the timeclock directly connected to a fan starter. This timeclock turned the fan off at the end of the business day and restarted the fan at the beginning of the next business day. This saved the operational energy cost of the fan system over the entire unoccupied period. The timeclock supplanted the operating engineer who had to go to the control board and manually turn fan systems on and off. The timeclock, being a low-cost device, was easily justified.

Today's control systems with direct digital controls and built-in microprocessors can do much more sophisticated time-of-day programming. Direct time-of-day programming for occupied and unoccupied cycles is still a useful feature that is built into most systems. It is not utilized on 24-hour occupancies but to put to beneficial use for energy savings in most other types of facilities. With microprocessors we may now expand time-of-day programming to include other features that can save additional energy. These features can include optimized start-stop, warm-up and cool-down features, and occupied-unoccupied modes.

8.3.4.2 Optimized start and stop. Optimized start is a program to set back the start time of the unit closer to actual building occupied time depending upon the space temperature and the capacity of the unit to bring the space temperature to comfort levels before occupancy. As an illustrative case you may think of a building in the middle of summer. The space temperature at 6 in the morning is 80° and the building is occupied at 8 A.M. You may want the air-handling unit to start at 6:30 to allow sufficient time to bring the building base temperature to 72°. The other case would be in the winter in that same building at 6 A.M. The space temperature may be 65°. Under this condition the building systems may not have to be started until ½ hour before occupancy to allow for some circulating and heating. With a straight timeclock scenario, and owing to the conservative nature of building engineers, units would be started for the worst case scenario and would therefore be running additional hours that are not required when the building is unoccupied.

For optimized start-stop to work properly representative space temperatures for each air-handling unit must be imported, and a system history must be developed to determine the least time required for system operation to bring space temperature within comfort limits. Control system manufacturers have developed a program that is self-learning and adapts from previous experience to determine the recovery time of the system. Optimized start should be enabled in all facilities to bring the greatest advantages and energy savings while still allowing a building's comfort levels to be maintained at occupancy. The energy savings from using optimized start can be calculated by determining the number of hours that the start has been deferred and multiplying that by the energy cost of the unit.

Optimized stop is a similar program that reviews space temperature before the end of occupancy to determine if the unit can be shut down and the temperatures allowed to drift without causing occupant discomfort. Optimized stop should be considered by facility managers but used sparingly. Problems connected with optimized stop can include the following: loss of space conditions in some areas, violating of ASHRAE ventilation guidelines, occupant dissatisfaction with loss of airflow, and minimal energy savings.

8.3.4.3 Warm-up and cool-down. In certain seasons it is possible to take advantage of cooler nighttime temperatures to allow for a nighttime building precool to save energy. This program involves looking at the outside air temperature during nighttime conditions and comparing it with the space temperature. If the building will be in the cooling mode during the occupied period, it is advisable to pump the cooler nighttime air into the building to precool the building before occupancy. This precooling will decrease building energy consumption required to bring building temperature down before the occupied mode. Building systems should be programmed to compare the outside temperature and the building space temperature. If the outside air temperature is below 72° and the building space temperature is above 76° it is advisable to open the outdoor air dampers fully and start the air-handling systems with mechanical cooling locked out. This would push the cooler outdoor air into the building in preparation for later building occupancy. Fan systems should be stopped when building space temperature reaches outdoor air conditions.

Warm-up mode is a method of heating the building as quickly as possible before occupancy. During winter conditions when heating is required, the space temperature can decrease below comfort levels in the building during the unoccupied hours. To warm up the building as quickly as possible it is desirable to reset the discharge temperature of the air-handing unit to a higher setpoint than normal or open the heating valve fully. Programming can be done to allow the heating coil valve to open fully to obtain the maximum possible discharge temperature to heat the building as quickly as possible. To help decrease energy loads during the warm-up and cool-down modes it is advisable to operate the units with no outdoor air into the system. Outdoor air is required for ventilation in the occupied period by ASHRAE standards and codes. Since warm-up and cool-down are implemented before building occupancy outdoor air is not required. Under normal circumstances (except for precooling) outdoor air is an additional load source for the building systems. By not utilizing outdoor air during the unoccupied period you decrease warm-up and cool-down time and save energy. To implement the outdoor air shutoff it is necessary that the outdoor air minimum damper be a controlled point from the building control system to allow the damper to close during occupied modes.

8.3.4.4 Economizer operation. Economizer operation must be implemented in facilities to save energy. It has been a standard program from the early pneumatic days where it was manually implemented via a pneumatic switch to the current days of direct digital control systems that obtain the last few percentage points of outdoor air economizer by utilizing temperature and humidity considerations. An economizer can be implemented in many functions based on building design. It can be either air side or water side.

8.3.4.4.1 Outside air economizer. In an air side economizer the maximum outdoor air damper is modulated to maintain the mixed air temperatures to give required cooling. An outdoor air economizer can be used whenever the outdoor air temperature is below the discharge air temperature requirement or the outdoor air total heat content is less than the return air total heat content. The energy content of air is determined by adding the sensible heat content (temperature) of the air in BTUs per pound to the latent heat content of the air. The latent heat content of the air is proportional to the moisture in the air. In the cooling mode it will be necessary to condense some of this moisture, which will require additional cooling. When this total energy content, which is called the enthalpy, is determined for the outdoor airstream, it is compared to the total energy content of the return airstream. The airstream with the lowest total energy content will be the airstream that is selected for use and mechanical cooling. Early pneumatic and electric systems and some early direct digital control systems utilize the dry-bulb economizer changeover. In dry-bulb economizer changeover the enthalpy content of the airstreams is not determined. The determination as to using the outside or the return airstream was made only on the outside air dry-bulb temperature. Above the predetermined value the return airstream was utilized, and below the predetermined value the outdoor airstream was utilized. It was found that under certain conditions outdoor airstreams with high humidity content actually had more total heat than the return airstream. It was therefore disadvantageous under those conditions to use the return airstream.

Many facility managers have found that the additional savings from enthalpy calculations are outweighed by the cost and the maintenance requirements of the humidity or dewpoint transmitters necessary for determining the moisture content of the outdoor and return airstreams. Enthalpy calculations were also hindered by the inaccuracies in some early humidity and dewpoint transmitters. The current generation of solid-state humidity transmitters are maintenance-free and accurate. Enthalpy should be utilized in most facilities to obtain the greatest energy savings in the outside air economizing mode.

8.3.4.4.2 Water side economizer operation. Waterside economizers are utilized in facilities where air handlers are not capable of air side economizers. These facilities have minimum air outside systems that introduce only the minimum outside air required for ventilation. In order to utilize free cooling in these systems it is necessary to obtain chilled water without using mechanical cooling methods. To obtain this free chilled water, condenser water from the cooling tower is used as a source of cooling. In this mode the cooling tower setpoint is reduced to a lower level (typically 42°) and the condenser water is run through a heat exchanger or transferred to the cooling water loop through a strainer. This allows chilled water generation without the use of mechanical chillers. Water side economizers can be effective only when the outdoor air temperature is low enough to generate condenser water in the 42 to 50° temperature range. Some systems do not use a heat exchanger to isolate the condenser water from the chiller water system but use an effective filtration device before intermingling the condenser and chilled water systems. This intermingling of the two systems is usually detrimental to both systems. On the chilled water side it allows contaminants from the condenser water to enter the chilled water system and also allows condenser water chemical treatments to infiltrate the chilled water system. These contaminates and chemical treatments can have an injurious effect upon chilled water coils. During the changeover mode between mechanical cooling and free cooling, if a heat exchanger is used, the changeover can be made seamlessly in a very short period of time. If a filter and intermingling system is used, the changeover is labor-intense, as it requires prefiltration of the condenser water system and cleaning of the chilled water system when the change is made back to mechanical cooling.

8.3.4.5 Load shedding and duty cycling. An early favorite of controls systems manufacturers for energy savings was the idea of load shedding and duty cycling systems. We will review the duty cycling concept first.

8.3.4.5.1 Duty cycling. Duty cycling was based on the assumption that most systems are oversized in all but extreme conditions. Because of the spare capacity built into the system it is possible to turn the system off for a period of time and then turn it back on again. The excess capacity within the system will bring space temperature back into design conditions in a short period of time. Energy savings were calculated based on the off times of the system, and temperature control manufacturers promised savings of 8 to 25 percent. While there is some savings during the off mode of system operations, these savings are usually offset by the initial inrush current required to restart the motor, wear and tear of the motor belts and pulleys during the frequent on-off starting mode, additional demand capacity during the rapid cool-down period if chillers are used, and occupant discomfort on loss of airflow during the off period. Additionally during the off load ASHRAE standards for ventilation are not being complied with. Duty cycling must also not be used during high cooling load periods, as units are at their maximum capacity during these times, their excess capacity is minimal, and recovery time would be far too great. To anticipate this problem temperature control manufacturers promoted temperature compensated duty cycling. In this program duty cycling was allowed but the off time was determined by comparing space temperature to space temperature setpoint. If the space temperature was at or under setpoint maximum duty cycle time was allowed. If the space temperature was above setpoint, duty cycle time was decreased based on the distance between the base temperature and setpoint; this naturally also decreased savings. Duty cycling times are determined based upon unit type and motor horsepower. Per the National Electric Code motors are allowed to be cycled on and off only a certain number of times per hour based on the horsepower. A larger-horsepower motor must be cycled less frequently than a smaller-horsepower motor.

8.3.4.5.2 Load shedding. Load shedding is a technique to reduce the peak electrical demand charges from the utility company. Load shedding involves monitoring the electrical consumption in the building and the building demand. A target is established on the building demand and operations is given notification when demand reaches this target number. If programmed correctly the building management system will automatically start to shed noncritical loads to reduce the electrical consumption below the demand target. Most facilities pay demand charges which are based upon the highest demand over a 15-minute period. These demand charges are carried for the next year unless a higher demand peak is reached. It is therefore in the building manager's best interest to reduce these demand charges as much as possible owing to their long-lasting impact. By shedding load, building demand is reduced for the period that the load is shed. The disadvantage of load shedding is that it typically occurs during a period of high ambient temperature conditions when building demand is highest owing to high mechanical cooling loads. Another disadvantage is that there is a demand surge when equipment that has been shed is restarted. When building space temperatures are outside of setpoint range load shedding may be inhibited to maintain occupant comfort. A facility manager must look at the load profile of the facility to determine if a load shed scenario can be effectively used. If the facility has demand peaks that occur unrelated to temperature conditions, load shedding may be a viable option.

8.3.4.6 Reset programs. Owing to the flexibility of direct digital control systems it is possible to reset air-handling unit temperatures to reduce energy consumption while maintaining space conditions. In the simplest scenario a constant-volume unit serving a space can use a space temperature to reset the discharge setpoint of the unit. This will reduce the heating and cooling loads to match the requirements of the space more accurately. This type of space temperature reset of discharge temperature has been standard since the early pneumatic and electric systems.

With direct digital controls it is possible to do additional reset algorithms. Variable-volume systems are designed to utilize constant discharge temperatures from the air-handling unit. Airflow is varied based upon the demands of the terminal units. With a decreased air volume the fan power required to move air is reduced. If the terminal equipment controllers on the variable volume boxes are in communication with the panel that is controlling the variable air volume system, it is possible to do a reset program on the discharge temperature of the variable air volume system to maintain comfort conditions and still minimize energy usage. One of the major problems of the variable air volume unit is that at low airflows not enough air is passing across the diffusers to properly mix within the space. This results in occupant discomfort due to air stratification. To increase the airflow it is necessary to increase the discharge air temperature from the variable volume unit, resulting in a greater air delivery. To protect the facilities manager the program must maintain a discharge air temperature that is low enough to satisfy the worst space conditions. The worst space condition is indicated by the variable air volume box that is opened a maximum amount. It is therefore possible to write an algorithm that increases the discharge temperature by a finite amount (usually 0.25°F every 5 minutes) when no VAV box is open greater than 80 percent. If a VAV box is open more than 95 percent, the algorithm is written to decrease the fan discharge setpoint in a similar manner.

8.3.5 Chiller plant

The chiller plant is one of the major energy users within a building. During the cooling season it is necessary to generate chilled water for use in the building's HVAC systems to maintain space comfort conditions. Many chiller plants are run independently by the building management system and have interaction with the building only via the operating engineers. Operating engineers have learned that chilled water temperature can be varied based upon the weather projection and occupancy load for that day. A building manager has many energy-saving possibilities within the chiller plant. These possibilities include variable-frequency drives on the cooling tower fans, condenser water setpoint controls, chilled water discharge setpoint reset, and optimizing chiller selection.

8.3.5.1 Cooling tower fans. Many buildings still have single- or multiple-speed cooling tower fans for use in controlling condenser water temperature. Condenser water temperature setpoint is maintained by starting and stopping of these fans or cycling the speed of the fans. These cooling towers should be retrofitted with variable-frequency drives to more exactly control the fan speed of the condenser water setpoint. If variable-frequency drive fans are implemented the control scheme used to operate these fans has not been optimized for building energy savings. Controls should be implemented as follows:

- A PID loop should be implemented on the condenser water temperature.

- On a call for cooling, the first cooling tower fan should start at minimum speed (usually 15 Hz).

- If temperature does not maintain the setpoint, fan speed should be increased up until 80 percent of fan speed (48 Hz).

- At this time a second fan will start and slowly ramp up to the same speed as the first fan. This slow ramping is to allow the speed of the first fan to decrease so that both fans are now being controlled at the same speed. The two-fan combination should then be modulated in unison to maintain the condenser water setpoint.

- If required, a third, fourth, and additional fans as available should be added in stages to maintain the setpoint.

- On condenser water temperature going below the setpoint fans should be slowed in unison until a desired minimum fan speed (40 percent) is reached. At this time one fan should be shed and the balance of the fans ramped up to control the condenser water at setpoint.

- Control systems must have built-in delays to prevent short cycling of fans on and off.

With sensors available to measure the outside air temperature and humidity it is possible to optimize the condenser water setpoint to gain energy savings. Most mechanical chillers work efficiently when the condenser water temperature is as low as possible. Knowing the outside air temperature and humidity allows the calculation of the outside air dewpoint. The condenser water setpoint can then be reset to be the outside air dewpoint plus 8°. Care must be taken to put a low limit on this value, as many chillers will not operate if the condenser water is too low a temperature.

If multiple towers are available and these towers have different capacities and characteristics it would be advisable to have the control system determine which tower would be most efficient under the anticipated load and select that tower as the primary tower. If additional towers are required, the most efficient of the smaller towers would then be utilized next.

8.3.5.2 Chiller reset. If all air-handling units are networked to a central command station it is possible to determine the valve position at each unit. If this information is available, the control system can make an intelligent decision as to when the chilled water temperature can be raised or lowered.

If an air-handling unit system valve is 100 percent open and discharge temperature is not at setpoint, a colder chilled water temperature is required and a command can be sent to the chiller to reduce its setpoint temperature. Before this command is sent, it must be determined that this valve is operational and that the chiller is capable of producing a reduced chilled water temperature. It must also be determined if this is a temporary aberration or if this chilled water requirement could be "ridden over" before action is taken.

If all air-handling unit valves are only partially open it is possible to increase the chilled water discharge temperature. Care must be taken to determine that there are no other cooling usages that require a minimum chilled water temperature, i.e., computer room units. If conditions are met it will be possible to increase the chiller water temperature setpoint and thereby reduce the load on the chiller.

8.3.5.3 Chiller selection. Chiller selection has been done by building engineers based upon their knowledge of chiller and energy efficiency and building engineers based upon their knowledge of chiller capabilities, energy efficiency, and building load requirements during the upcoming day, or it can be done automatically by a building management system to optimize the selection of the chiller. If building management system optimization is used, chiller energy profiles must be inputted along with the chiller startup lead times and requirements. My experience has been that on large chiller plants or on steam turbine plants, an advisory message should be sent to the operator as to optimal chiller selection. The operator can then acknowledge this advisory message and make appropriate selections based upon the knowledge. If, for example, a peak period requiring additional chiller capacity is reached at 4 P.M., but the operator knows that building vacancy is at 5 P.M., instead of starting an additional chiller the operator can decide to ride this out until vacancy time. Since enough chilled water will not be produced, the air-handling unit discharge temperature will not be maintained, allowing some variations from the optimal space temperature conditions.

8.3.6 Hot water reset

Many facilities use steam piped through a heat exchanger to generate hot water used for heating in perimeter units and in the heating coils of the air-handling units. This hot water supply temperature should be reset based upon the requirements of the load being served.

On a perimeter hot water system the hot water supply temperature should be reset based upon the outdoor air temperature, the solar load on the building, and the building exposure or in a manner similar to the chilled water reset. If the valve position of the perimeter units is known the water temperature can be regulated so that no valve is fully open and at least one valve is open more than 80 percent.

The hot water for the air-handling units should be operating only when one of the units is calling for heating or the outside air temperature drops below a predefined setpoint. The optimum temperature can then be determined in a manner similar to the perimeter system above.

8.3.7 Lighting controls

Lighting controls as either a standalone package or integrated into a building management system can be utilized to turn lights on and off during occupied and unoccupied time periods. Lighting is a major energy user in a building. Significant savings can be obtained through reductions in lighting. Care must be taken to allow overtime operations and to make sure that the cleanup crews do not override the lighting controls.

8.3.8 Conclusions

With the advent of direct digital controls the facility manager has more options open to save energy than ever before. The facility manager must know the operation in order to determine what programs are available to optimize energy usage. These requirements must then be communicated to a consulting engineer or control company to allow the fine tuning of the hardware and software components in the building control system.

Reference

Shearer, Clive, *Practical Continuous Improvement for Professional Services,* Quality Press, Milwaukee, 1994.

Bibliography

Cotts, Davis G., and Michael Lee, *The Facility Management Guidebook,* Amacom, New York, 1992.

Binder, Stephen, *Corporate Facility Planning,* McGraw-Hill, New York, 1989.

Equipment and Systems Operations and Maintenance Procedures

Equipment and Systems Operations Procedures

Dennis E. Mulgrew

Consolidated Engineering Services, Inc.
Arlington, VA

9.1 Equipment Inventory

Easily the most important aspect of developing a comprehensive operations and maintenance plan, the initial inventory of building systems, utilities, and controls provides the basis for all work plans, maintenance schedules, specific tasking, and operating procedures. As all preventive and predictive maintenance, corrective maintenance, and operating plans emanate from an understanding of the facility equipment, systems, and needs, this is the first step in the development of comprehensive operating procedures.

While extremely important, this aspect of facility management is often neglected or given resources that are not consistent with the level of importance. Many firms and facilities maintenance staffs, whose responsibilities include the development of operating plans and procedures, apply less than adequate professional resources to this effort, resulting in operating plans and procedures that are deficient with respect to the building systems. Comprehensive management plans require all of the operating parameters to be considered when establishing work plans, priorities, and strategies to affect the long-term utilization of the facility. The equipment inventory should then be considered as part of an overall intake assessment to develop these strategies.

Collection, identification, and planning aspects for facilities management begin with a careful review of all operating systems and the needs they are expected to satisfy. The inventory is therefore only one small piece of the intake process. All of the following aspects of facility management need to be considered when performing this function:

1. What are the needs of the facility? More specifically, what are the individual requirements of each functional area within the facility?

2. Do the mechanical, electrical, and control systems meet the needs of the functional areas of operation? This needs to be viewed from both aggregate and location-specific needs. It is insufficient to identify that the ventilation provided by an air handler meets the aggregate cfm per square foot requirement of the space being served if specific zone requirements cannot be met with the existing distribution.

3. Does the configuration of mechanical, electrical, and control systems meet the needs of the facility in a cost-effective manner? Often, to accommodate specific zone-related issues, large generation systems are operated when compartmentalized systems would offer a more effective solution.

4. What is the overall quality of the maintenance practices? What are the current maintenance standards for the building systems? Have the standards been applied uniformly? Are the standards representative of the needs of the equipment? Should the maintenance tasking or frequencies be modified to address specific equipment needs?

5. Do the building systems require specialized training in order to provide the needed operational oversight or maintenance practices?

The issues and questions raised above are not limited to the development of an operating strategy for newly constructed or acquired facilities, rather they serve as functional reminders in any facility operating program. Buildings are collections of dynamic processes that require regular and recurring reviews to ensure the adequacy and appropriateness of intended operating plans. Much as preventive maintenance addresses the recurring needs of the systems and equipment in place, a regular assessment of the ability of the installed systems to meet the functional needs of the facility is just as important.

Therefore, the inventory process can be defined as a two-part process. The first aspect deals with the ability of the equipment and systems to function as designed, while the second part deals with the appropriateness of the design itself.

9.1.1   Conditions assessment

The conditions assessment serves to define the needs of the installed equipment. By visiting each piece of equipment, the facility manager establishes a baseline of operating condition, remaining useful life of the device, and general understanding of the level of maintenance performed to date and whether the device is operating in accordance with the design. This visit is equally important for newly commissioned installations, where the systems and equipment may not have been subjected to the true loads of the facility during the commissioning process. A recommended frequency of this type of assessment is annual.

The focus of the assessment should include:

1. Document the working order of the equipment and systems.

2. Is the frequency of maintenance appropriate? Should the frequency of maintenance visits increase or decrease?

3. Is the level of maintenance being performed thorough enough to maintain the equipment in sound operating condition? Should additional tasks or tests be added to the current maintenance checklist?

4. What is the quality of the maintenance performed? Is it lacking or acceptable?

5. Are there any deficiencies that need to be addressed?

6. How critical are the deficiencies? Should they be performed immediately or within a reasonable time frame?

The following checklist serves as a guideline for the performance of the conditions assessment. Items to be visited should include mechanical, electrical, lighting, plumbing, utility, controls and building automation, fire and life safety, doors, roofing, and structural components.

Create a baseline of the current conditions by:

- Documenting the working order and condition of all equipment and systems

- Documenting items missing, such as belt guards, fan shrouds, or electrical box covers

- Documenting any and all deficiencies noted and generating a deficiency report

- Prioritizing the deficiency list

- Identifying which deficiencies can be repaired during the course of routine preventive maintenance tasks versus those that require additional labor or specific expertise

- Developing a cost estimate of the identified deficiencies

- Documenting the current quality level of maintenance
- Reconciling the equipment inventory to the maintenance work order system
- Identifying any devices which require increased or decreased frequencies of maintenance
- Updating the maintenance work order system to reflect any changes needed in the maintenance practices
- Documenting the estimated remaining useful life of the equipment in the maintenance work order system
- Ensuring that the maintenance work order system database matches the nomenclature of the systems and equipment installed
- Identifying suspect operating conditions or systems which may warrant professional inspection

9.1.2   Design assessment

This aspect of the conditions assessment is important, as many facilities were designed and constructed in an effort to accommodate multiple-tenant usage patterns. Particularly true of facilities constructed as speculative office facilities, the original engineering design was developed without any specific information on the tenant needs. Beginning with a "cold dark shell" (no HVAC or reflective ceiling plan) many assumptions regarding the aggregate electrical and air-conditioning loads were incorporated during the design process. Facilities constructed in this manner up to the late 1970s and early 1980s often were based on anticipated loads that are no longer valid assumptions in today's electronic offices. Conventional approaches used in estimating the occupant and appliance loads were 1 person for every 200 square feet and 0.5 watt of receptacle power per square foot, respectively. The advent of compartmentalized office furniture systems in open floor plans with personal computers at each workstation has easily rendered those assumptions invalid.

Even newer facilities may suffer from an inability to meet the current intent of ventilation codes and standards adopted to achieve enhanced indoor air quality. Where only a few years ago energy conservation was given more attention than the concern for occupant health, ventilation rates in most facilities constructed throughout the 1980s are deficient with respect to the current trend in the development of ASHRAE standards.

The concern over occupant health is not likely to wane in the foreseeable future, so the challenge facing facility managers and designers today is to meet these increased ventilation requirements without incurring substantial operating cost increases. Systems to be evaluated include those identified in Sec. 9.1.1; only the focus of the evaluation is related to code compliance, occupant safety, and energy performance.

In performing these inspections, several key questions should be addressed:

1. Have any areas of the facility undergone substantial changes in the operating hours, deployment of personnel, or electrical loading?

2. Have the building codes been modified since the construction of the facility?

3. What changes are necessary to bring the existing systems into compliance with the updated standards and codes?

4. Is the existing control system capable of operating within the intent of the updated codes and standards?

The result of this evaluation should form the basis of a capital improvement plan, designed to modernize the facility, improve indoor air quality, and promote occupant safety and health while mitigating unnecessary operating or energy expense. These objectives are not mutually exclusive where the modernization includes the necessary controls and automation to ensure compliance with the program objectives. Some modernization measures may be attainable without expensive modifications to the HVAC systems and can be implemented through changes in the base operating plan of the facility.

9.2   Operational Requirements

Forming the basis of the operating procedures, the building operations plan (BOP) outlines standards and guidelines for operating building equipment such that all temperature, lighting, and other environmental conditions required by individual zones are provided during building operating hours. The guidelines establish the desired conditions based on time of day, weekdays, weekends, and holidays and reflect the desires of the occupants as stated in leases, departmental objectives, or occupant agency directives. The BOP provides guidance for developing maintenance and operational staffing levels to facilitate the needs of the tenants or occupants within the framework of the mechanical systems installed at the facility. In office buildings, it also serves to define the separation between building standard operation and those additional requirements that specific occupants may require.

Key to the development of the BOP is achieving understanding of the needs of the occupants, the resultant requirements of the mechanical and electrical systems, and the level of staff needed to ensure the safe and effective operation of the building systems. As there are a wide variety of building designs and operating conditions, the following plan is just a sample of the elements that are routinely addressed in the BOP. Conditions affecting the actual plan include weather, equipment condition (which is strongly influenced by prior maintenance activities, equipment installations, operations, etc.), tenant-use patterns, and special requirements, in addition to a host of other variables. Therefore, the BOP is a dynamic plan that must be able to respond to many varying conditions. Facility personnel constantly adjust and modify the plan as needed to respond to severe weather patterns, extended operational needs by occupants, and changes in the performance characteristics of the building systems.

Items regularly addressed in the operational plan include:

Equipment startup and shutdown procedures

Building temperatures (both general and special areas)

Operational checks

Energy conservation requirements and objectives

Use of fresh air and economizers

Air filtration

Potable hot and cold water

Lighting levels

Relamping

Controls and automation systems

In general, the BOP addresses the intent of the facility manager's policies and procedures to ensure tenant comfort and safety while minimizing operating expense. Further, it serves to define acceptable practice for the scheduling of maintenance activities including operational tests of large energy-consuming devices. Examples of policy and procedural objectives include:

- Systems and equipment will be maintained in the automatic mode to the maximum extent practical.

- Routine procedures will guide startups and shutdowns.

- Equipment will not be cycled at frequencies that lead to premature failure.

- When duplicate systems exist, equipment will be rotated to equalize usage. Another school of thought on this subject is to run one of the two devices much more than the other, so that both will not reach the end of their useful lives at the same time.

- Utility meters will be read on the same date as they are read by the power company.

- Inventories will be maintained for all fuels received, stored, and used.

- Inventories will be maintained for all refrigerants received, stored, and used.

- Preventive maintenance and testing of large electrical devices will be scheduled at a time when the operation of the device will not impact the electrical demand peak.

9.2.1 Equipment startup and shutdown

Daily facility startup procedures will vary according to external environmental conditions and average building space temperatures. In general, the air-conditioning and heating equipment will be started (and shutdown) at the same time each day. Startup times should be coordinated with the arrival of the startup operating engineer, whether the equipment is started automatically via timeclock or energy management system. Prudent operating practice includes the presence of the operating engineer during primary machinery operations.

The following tasks should be undertaken at startup:

- Upon arrival, the startup operating engineer should assess external environmental conditions (weather, light, etc.) and any other building-use factors to ensure that the controls (energy management system or timeclocks) have started or will start the mechanical equipment in order to have the building within prescribed environmental limits by the start of normal building operating hours.

- Observe the mechanical system startup. Inspect the systems as they become operational, noting excessive vibration and equipment operating temperatures and pressures. Verify chiller startup purge operations and transition.

- Perform a watch tour of the systems after startup to ensure the proper operation of the equipment.

- Any major equipment or systems, including elevators, security, and fire alarm systems found to be inoperable, need to be addressed immediately by site personnel or subcontractor labor. The intent of the startup process should be to provide a reasonable amount of

	Outside Air Temperature	Perimeter Temperature	Start-Up Time
1.	60° or below	below 76°	fans as needed after 7:00 a.m. for free cooling; chiller as needed after 7:00 a.m.
2.	60° or below	above 76°	fans at 6:30 a.m. for free cooling; chiller as needed after 6:30 a.m.
3.	60° to 70°	below 76°	fans at 6:30 a.m. for free cooling; chiller as needed after 6:30 a.m.
4.	60° to 70°	above 76°	fans at 6:30 a.m. for free cooling; chiller as needed after 6:30 a.m.
5.	70° to 80°	below 76°	fans 6:00 a.m., chiller 6:00 a.m.
6.	70° to 80°	above 76°	fans 6:00 a.m., chiller 6:00 a.m.
7.	above 80°	---	fans 6:00 a.m., chiller 6:00 a.m.

NOTES
1. During prolonged shut down (weekends, holidays) it may be necessary to advance the start-up schedule by several hours.
2. During extreme heat wave conditions it may be necessary to operate for prolonged periods.
3. These are only typical guidelines, and need to be adjusted to the need of the facility.

Figure 9.1 Typical summer startup guidelines.

time to assess inoperable equipment and take corrective action before occupant comfort or safety is compromised.

Figures 9.1, summer startup, and 9.2, winter startup, provide a brief description of startup measures taken during each season of the year to ensure that interior building temperatures and environmental conditions adhere to occupant requirements. Equipment start times and settings will be governed by the following conditions:

- Outside air temperature
- Perimeter and interior building space temperatures
- Weather forecast for the day
- Current weather conditions (i.e., precipitation, sun load, humidity, wind conditions, etc.)
- Time elapsed since last operation (and/or weather change)

Shutdown procedures are designed to ensure that the systems and equipment are properly secured and that the desired mode of operation for the ensuing startup functions can be achieved (Figs. 9.3, summer shutdown, and 9.4, winter shutdown).

- Before leaving for the day, the shutdown operating engineer

	Outside Air Temperature	Perimeter Temperature	Start-Up Time
1.	50° and above	below 72°	Fans and heating equipment at 7:00 a.m.
2.	32° to 50°	68° to 72°	Fans and heating equipment at 6:30 a.m.
3.	32° to 50°	below 68°	Fans and heating equipment at 5:00 a.m.
4.	27° to 32°	below 68°	Fans and heating equipment at 4:30 a.m.
5.	27° to 32°	68° to 72°	Fans and heating equipment at 5:00 a.m.
6.	27° and below	below 68°	Fans and heating equipment at 4:30 a.m.
7.	27° and below	68° to 72°	Fans and heating equipment at 5:30 a.m.

NOTES

1.	During prolonged shut down (weekends, holidays) it may be necessary to advance the start-up schedule by several hours.
2.	During extreme heat weather conditions (extreme cold, high winds, sever wind chill, heavy snow, etc.), it may be necessary to operate for prolonged periods or around the clock.
3.	These are only typical guidelines, and need to be adjusted to the needs of the facility.

Figure 9.2 Typical winter startup guidelines.

	Outside Air Temperature	Perimeter Temperature	Shut-Down Time
1.	60° or below	76°	All equipment 5:05 p.m.
2.	60° to 70°	76°	All equipment 5:05 p.m.
3.	70° to 80°	76°	All equipment 5:05 p.m.
4.	above 80°	76°	All equipment 5:05 p.m.

Figure 9.3 Typical summer shutdown guidelines.

reviews the weather forecast for the evening and following day, indexing the equipment to the mode of operation that will best meet the needs of the facility the following operating day.

■ Equipment should be indexed for automatic operation through the EMS or timeclocks as to when to initiate shutdown procedures and when to schedule the subsequent startup.

Note: These procedures ignore the energy management program

	Outside Air Temperature	Perimeter Temperature	Shut-Down Time
1.	50° or above	70°	All equipment 5:05 p.m.
2.	32° to 50°	70°	All equipment 5:05 p.m.
3.	27° to 32°	70°	All equipment 5:05 p.m.
4.	27° and below	70°	All equipment except freeze protection equipment and heating pumps which run all night--5:05 p.m.

Figure 9.4 Typical winter shutdown guidelines.

function of optimal start and optimal stop. These functions can be automatically initiated through most energy management systems. Care should be taken with respect to this operation, as energy management programs can only make decisions based on information they possess at any point in time. They cannot obtain forecasts and adjust program objectives on data that has not been received. This can result in increased costs of operation by failing to enact precooling modes that serve to mitigate anticipated demand peaks. As most utility structures penalize the demand aspect of electric service more than the consumption element, it is routinely more cost-effective to start equipment early and avoid demand peaks than to delay the startup when demand will become an issue later in the day.

9.2.2 Operational checks

Operating watch tours should be constructed so as to confirm that operations of the mechanical and electrical systems are in accordance with the building operating plan. Not limited to conformance with plan objectives, the operating watch tour is a proactive assessment of the ability of the systems to meet the needs of the facility and ensure the useful life of the equipment. These tours should:

1. Identify the operating parameters of the major building systems.
2. Provide feedback as to the ability of the systems to meet the needs of the facility.
3. Identify areas where the systems and/or equipment are not being operated or maintained in accordance with the established objectives.
4. Identify areas where conditions warrant modifications to the operating parameters.

5. Identify excessive vibration or motor operating temperatures that will inhibit the ability of the systems to maintain reliable operations.

6. Identify opportunities to enhance the energy conservation of the facility.

7. Identify equipment and system deficiencies that will result in unscheduled repairs and/or failures that jeopardize the life of the equipment.

8. Identify lighting systems and equipment that fail to meet the needs of the usage areas.

9.2.3 Lighting

Lighting systems are designed to meet the aggregate requirements of the tenant spaces. In general, the standards for illumination vary with the intended usage of the space. The following illumination levels represent "normal" light levels required to facilitate office usage and provide occupant safety.

Illumination levels during tenant normal work hours:

Public areas within buildings	10 foot-candles
Normal work stations	50 foot-candles
General work areas	50 foot-candles
Storage areas	10 foot-candles

These lighting systems standards should not be increased or decreased without consideration given to all requirements within the areas being modified. In case of insufficient light levels, it is far more productive to combat the specific need with task lighting than to increase the aggregate lighting levels of an entire area to accommodate a specific function. The prescribed lighting level foot-candle readings should be randomly spot-checked during quality control inspection tours and/or operating watch tours. Degradation in the illumination levels can usually be linked with a need for cleaning of the fixtures or the aging of the lamps. Periodic checks of the illumination levels will serve to identify the rate at which the lighting levels are degrading.

9.2.4 Relamping

When areas of the facility indicate lumen degradation to a level that hinders the safe and secure usage of the space, group cleaning or relamping should be scheduled. As the major expense associated with either activity is labor, relamping prior to the calculated end of the useful life of the lamp should not be a significant deterrent.

9.2.5 Energy conservation

Temperature controls should be set to maintain an indoor environment that is consistent with general industry standards and practice during normal facility work hours. With the exception of specific areas that may require special environmental conditions, temperature controls should be maintained between 72°F (22.2°C) and 76°F (24.4°C). This temperature control range is understood to be the generally acceptable temperatures established for human comfort. Operating outside of this range under the auspices of energy conservation usually results in excessive temperature complaints by the occupants. Conservation should be an attempt to maximize the effectiveness with which desired conditions are met—not an effort to force environmental conditions on the occupants that are undesirable.

In maximizing this effectiveness, special consideration should be given to the use of economizers or "free coolers" to meet the needs of the facility. In general, the operating plan should emphasize the use of air-and water-side economizers to the maximum extent practical. Mechanical cooling and heating operations should be limited to the occasions when these economizer systems cannot maintain the desired occupant conditions.

Building operating policy should include the following objectives:

1. During moderate weather, use outside air, mechanical economizers, and/or other energy-saving equipment installed, to the maximum extent possible.

2. The use of outside air, equipment, or similar strategies should be based on outside temperature and humidity levels.

3. Building ventilation should be provided to the maximum extent allowed by the configuration and design of the installed mechanical equipment.

4. Fresh air should be adequately filtered at all times by using only air filters capable of at least 50 percent particulate removal. Filters should be changed frequently, in accordance with best industry standards and practices.

5. In the case of conflict between what is installed in the building and what the building design specified, the more stringent air requirements should apply.

6. Ventilation standards are defined and specified in the EPA's Clean Air Act and OSHA's Indoor Air Quality Standards, *Quality Standards for Design and Construction Handbook* PBS P 3430.1, Appendix 5-R; *The Energy and Water Handbook* PBS P 5800, Chapter 8-7e; and the American Society of Heating, Refrigerating, and Air Conditioning Engineers Standard 62-1981, Ventilation for Acceptable Indoor Air Quality; and CFR Title 40, Part 141 PCB procedures.

Note: ASHRAE Standard 62 is currently being revised to create a more stringent requirement for minimum fresh air levels in facilities. This is in response to the growing number of indoor air quality complaints across the country.

9.2.6 Use of demand control ventilation

With the focus on increased ventilation rates, many facilities managers are turning to demand control ventilation (DCV) as a means of identifying and controlling the level of fresh air provided to a facility. DCV is a method of modulating fresh air intake to a facility in response to the level of carbon dioxide (CO_2) present in the building system return air. In theory, the DCV application serves to increase fresh air intake volumes when CO_2 levels rise above a setpoint [nominally 500 to 600 parts per million (ppm)]. Concentrations of CO_2 above these values is construed as insufficient oxygen for human comfort and safety. Clearly, sampling CO_2 can provide a measure of the amount of fresh air that is being utilized for breathing but does not indicate other needs of the fresh air. Expulsion of outgasses commonly found in facilities is one of the most important elements of indoor air quality. Using only CO_2 as the barometer for controlling ventilation rates could result in failing to address one of the prime candidates for IAQ complaints.

While growing in popularity, this application is not a sure cure to the issue of ventilation standard compliance. ASHRAE Standard 62 relates to specific ventilation rates required for specific functions within a facility. These rates are defined as cubic feet per minute (cfm) per person or per square foot of area, varying with the type of occupancy and usage. Using aggregate sampling at common building exhaust or return air systems could indicate an aggregate compliance with the provisions of the standard, while specific areas within the facility could be underserved. Another major challenge to the utilization of this technology rests with the accuracy of the sensors themselves. Commercial-grade apparatus is commonly available with accuracies of 5 to 10 percent of the range of the device, with specified minimum values of potential offset (e.g., ±5 percent or 75 ppm, whichever is greater). When considering the values being measured, a 5 percent inaccuracy for a reading of 600 ppm (with the stipulated minimum error of 75 ppm) is actually an error of 12.5 percent. While this may represent a reasonable degree of measurement, using this value as a control point could lead to inappropriate control actions or the creation of a false sense of security. If the application of this technology was contested as the cause of IAQ-related illnesses, it would be difficult to defend this range of accuracy as a means of addressing such a prominent tenant health issue.

9.3   Facility Management Tools

Energy management systems (EMS) can be utilized as a valuable resource in performing facilities management. While the systems cannot hear unusual noises or see excessive vibration, they can be utilized to augment and enhance the operational effort. With facility staffing spread thin, the EMS can be used as a strategic tool to assess the operations and maintenance requirements of facilities. It is invaluable in providing the necessary information to identify problematic areas before they become critical failures.

Conventional thinking applied to EMS is focused on the ability of the connected points to be controlled, thereby controlling expenses. This strategy addresses only the utility cost portion of the operating expenses. While important, the most disruptive element in facilities management is the unplanned expense.

Utility expenses can be predicted with reasonable accuracy, accounting for variances in weather, occupancy, production, rate schedules, and usage changes. Using historical data, these variances can be projected to establish a reasonable expectation of utility consumption and costs. Unplanned repairs, however, are not so easily predicted. They are dependent on a number of issues:

1. The quality of maintenance being performed

2. Quality of equipment being used

3. Effectiveness of the system design

4. Level of staffing (adequate to maintain the systems)

5. Complexity of the systems and equipment

6. Ability of the operating personnel to diagnose the causes of the problems

Facility management systems (FMSs) can be utilized to provide the needed information to assess the conditions noted above. By extending the telemetry to include non-energy-conservation monitoring and control points, the FMS can become a management information system for system reliability and operations oversight.

Specifically, poor maintenance practices on mechanical systems can be determined by monitoring vibration, motor temperatures, oil temperatures, or the effectiveness with which the equipment is performing its required functions. Btu meters measuring the delivered chilled water and hot water can be compared with the electricity, oil, or natural gas utilized to create the commodity, resulting in the development of an overall effectiveness of generation. As heat-transfer surfaces begin to foul, the effectiveness of the system will diminish. This is indicative of poor maintenance to the heat-transfer surfaces. Corrective measures needed to overcome the poor transfer rates would include coil cleaning, eddy current tests and tube cleaning, and heat exchanger tube cleaning. Performing maintenance based on the degradation in the effectiveness would prevent emergency repairs needed when the quality of heat transfer degrades to a critical point, where systems or equipment could fail.

Similar evaluations of the effectiveness of HVAC systems to meet the needs of the facility may indicate design issues that are not in accord with the needs of the facility. The design condition assessment can be augmented dramatically using the FMS as the collection point for system operating profiles. A survey by consulting engineering personnel typically involves a walkthrough of various systems with small samples of data taken from the operating systems during the brief visits to the machinery. The FMS clearly can provide historical measured values for the same piece of machinery on an extended basis. Trend logs and point histories then become the basis of a perpetual survey of operating issues.

Information unto itself is useless unless the operating personnel are trained in the analysis of the data and the impact it has on the facility requirements. Careful consideration should then be given to the amount of data to be collected, the frequency of the collection, and the results or level of analysis desired. As with any computer system, there is a finite amount of processing power resident in any FMS. Extending trend logs and point histories to every connected point may diminish the ability of the FMS to perform other functions, such as operating programs, control loops, and alarm notification functions.

In using FMS technologies to mitigate the unplanned maintenance and repair expenses, it is therefore critical that the FMS itself become the most reliable operating system within the facility. This is easily accomplished by applying the same level of maintenance planning, scheduling, and tracking as is applied to the systems the FMS controls and monitors. Most facility maintenance management programs can easily accommodate the FMS monitoring and control points as part of the equipment inventory. The level of maintenance required to assure the proper operation of the FMS is fairly consistent across the multitude of vendors and manufacturers. In the same fashion that mechanical systems are defined as generation, distribution, or terminal use components within the overall HVAC system, the FMS architecture can be defined as:

1. Workstation or head-end devices comprising the computer, operating systems, worker-machine interface, display terminal, remote

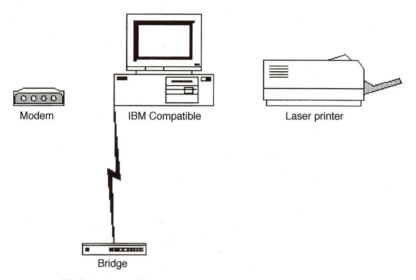

Modem   IBM Compatible   Laser printer

Bridge

Figure 9.5 Workstation architecture.

interrogation devices (modems), printers, and communication devices to poll the data collection–gathering panels (see Fig. 9.5).

2. Data-gathering panels. Whether for data collection or command processing, this device represents the primary component in the EMS architecture. Typically a standalone-capable device, it can measure and record values, execute program instructions, and command field devices without the presence of the operator workstation (see Fig. 9.6).

3. Field devices. These are the relays (start-stop or two-position status), variable-condition sensors (temperature, humidity, pressure, etc.), and variable-output devices for positioning valves, dampers, and other control devices.

9.3.1 Energy management control system (EMCS) operations

EMCS is a computer-based energy and temperature control system, featuring microprocessor-based remote field panels capable of stand-alone operation, that is used to administer building operations. These systems typically have modular architecture permitting expansion by adding computer memory, application software, peripheral equipment, and field hardware. Whether identified as an EMCS, EMS (energy management system), BAS (building automation system), or

Comm. Bus

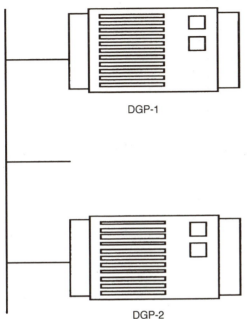

Figure 9.6 Data-gathering panels.

FMS (facility management system), these systems consist of the following components:

- Building environmental monitoring and control system, subsystems, and associated equipment
- Central control center console system, subsystems, and associated equipment
- Software programs (internal)
- Energy management control system, subsystem, and associated equipment

While operation of the FMS generally falls to the engineering staff, maintenance and repair services for these critical building systems are subcontracted—most often to the manufacturer of the system. As these systems perform a vital service in the operations and maintenance of facilities, they are deserving of a level of attentiveness consistent with the mission.

Understanding the pieces of the system and how they relate to the successful implementation of the building operating plan is essential to managing this valuable resource. In developing this aspect of the building operations plan, it is important to identify the tasks requisite to properly maintaining the FMS and the craft and trades best suited to accomplish these functions. Far too often, facility operators rely exclusively on the installing manufacturer for all aspects of the maintenance of the system. This practice translates into higher maintenance costs and fosters an extremely dependent relationship between the owner and operator with the system manufacturer.

FMS manufacturers are well versed in the requirements of their individual systems, products, and programs. They should not be construed as being experts in the application of HVAC control technologies, energy strategies, or control engineering. They are product experts and should be relied upon to maintain the portions of the product that engineering personnel cannot.

As noted above, the FMS is a combination of computer technology and electronic controls. The head-end computers are generally off-the-shelf components with custom software applications resident to poll the remote panels, follow operating sequence programs, and effect monitoring and control commands. The remote panels are essentially data collectors and microprocessor-based receiver controllers that emulate the same basic logic of conventional temperature controls. The primary benefits between these systems and the conventional control products are:

1. The ability to embed many receiver controller functions within one microprocessor

2. Higher degree of accuracy without constant calibration

3. The ability to alter or modify the operational response of many control loops from one location

4. The ability to monitor the operating parameters of an entire facility from one location

5. The ability to coordinate the operations of many devices in a manner that promotes energy conservation, while maintaining required comfort conditions

These benefits are mostly related to operational enhancements in the effort to mitigate energy usage and/or demand. When integrated as part of the building operations plan, they are utilized to perform the following routine functions:

■ Start and stop the building HVAC equipment according to predetermined schedules established in the building operations plan (see Sec. 9.2.1 including Figs. 9.1–9.4).

- Monitor and log any abnormal operational readings of the building equipment.

- Assess the operating efficiencies of the mechanical systems by developing operating profiles of the systems over varying load conditions.

While these systems are vital to the effective operation of the facility, an emergency plan of operation must be developed for those instances when the FMS is out of service or compromised.

9.3.2   Emergency and extended FMS operations

As the FMS is a vital tool in the performance of facility management, its reliable operation is critical to the successful facility manager. As such, the ability of the FMS to perform its functions cannot be compromised. The startup operating engineer arriving on site in the morning must know that the FMS is available to perform the requisite operating functions in order to allow the engineer to perform watch tours and other engineering functions. Finding the FMS in an inoperable condition at the appointed startup time would create a tremendous burden on the engineer to manually start the needed plant equipment.

To ensure the ability of the FMS to perform its functions, it is prudent to have the system operating status monitored continually. Any number of potential solutions exist to monitor the operating condition of FMS equipment, from autodialers to remote monitoring services. In either event, the intent of the service is to notify the engineering personnel that the FMS is incapable of performing automated startup functions.

Whether this monitoring is performed by on site personnel or remotely, the facility manager must provide emergency operations support in response to the needs of the facility. The criticality of the facilities operations will dictate the level of attentiveness required to ensure the appropriate response. In most instances, it is understood that an emergency is defined as a situation which constitutes an immediate danger to personnel, a threat to damage property, or threat to disrupt occupant operations. Examples include outages in utility systems, clogged drains, broken water pipes, gas leaks, and electrical distribution problems. Incidents such as these require an immediate response to protect the property. Overall responsiveness is then dependent on the extent that the FMS can automatically notify facility management personnel. Key elements of this responsiveness are outlined below:

1. 24 hour per day, 365 days per year emergency call service support in response to emergency conditions. This would include operating the facility during emergency conditions via the FMS. It follows that service calls to restore full FMS operations would be treated as an emergency.

2. The responding technician or craft or trade must be knowledgeable in the operation of the FMS. Using the FMS as the data collection and assessment tool, the responding tradesperson would:

   a. Detect, locate, analyze, report on, and correct, as required, abnormal conditions in the mechanical systems. Typical cases might include freezing temperatures, high winds, snow, heavy rains, and similar adverse weather conditions.

   b. Correct all alarm situations, including emergency shutdown of the facility equipment (neutralizing alarms as necessary prior to correction).

   c. Monitor the automatic restart of the FMS computer and facility equipment after a power failure or any other malfunction. During the emergency situation, provide manual operations, as required, to maintain facility support functions.

Apart from the specific functions noted above, the FMS also provides other forms of support during emergencies. The FMS can be utilized to create an audit or review of the incident. Utilizing trends, point histories, and program analysis, the FMS can provide information on the operating conditions of the equipment and systems prior to the failure. This type of analysis is often very valuable in determining corrective measures to preclude the recurrence of the failure. Used properly and to the full capability of the system, the FMS is an extremely valuable tool in the performance of operations oversight, quality control, and facilities management.

9.3.3 FMS service program

The successful facility manager will maintain all FMS hardware, software, and associated components in accordance with the manufacturer's specifications and sound industry practice. Proper system maintenance will include preventive maintenance (PM), corrective maintenance, software support services, and analysis of declining conditions that warrant further study. All of these elements together represent the requisite attention needed to ensure the successful implementation of the FMS.

Understanding the importance of proper FMS maintenance is demonstrated by the life cycle of the energy management systems and equipment. In use every day as a critical tool in building operations, the FMS service program should include the following features:

1. Daily verification checks on all connected points. Every connected point should be regularly checked for accuracy against portable measurement standards to ensure continued reliability in the sensed values. This is critical to the successful performance of operating programs that base the automatic calculation of setpoints or strategies on the measured values from the field points. Points determined to be "out of tolerance" need to be recalibrated to return the level of credibility requisite of the control strategies.

2. A highly disciplined corrective maintenance program. Central computers, hard disks, printers, and video display terminals should receive monthly inspections to ensure continued reliability. Connected points should be reviewed daily by FMS operating personnel to determine if they are responsive and provide reliable feedback. All points determined to be "suspect" should be tested by onsite O and M personnel in conjunction with the FMS operator. Points that are determined to need corrective action should be either immediately recalibrated or noted on a service and repairs list. This list should be maintained in a database or computerized maintenance management program to ensure adequate oversight. Note that the onus is placed on the O and M personnel to make the initial determination of point trouble to ensure that only truly "bad points" are placed on the point service list. This practice aids in avoiding unnecessary costly repairs by subcontractor personnel to issues that could be easily handled by site personnel.

3. Diligent FMS preventive maintenance services. PM work to be provided should include equipment inspections to ensure proper operation. This includes a schedule of routine maintenance (annual work plan) to be applied consistently for all FMS devices. As with mechanical and electrical systems, PM orders for each task, detailing the work to be performed; required technician skill levels; and any special tools, instruments, and diagnostic programs needed should be generated for all data collection and transmission gear.

4. Service repair records. Dispatched service calls for repairs, calibrations, or parts replacements should require a completed work order for each task completed, reporting on the work accomplished, labor hours, and materials used. This is valuable in assessing recurring issues with specific components of the FMS.

5. System diagnostics. These are essential in detecting early signs of deteriorating performance and serve to predict potential system failures. Potential problem areas should be identified and corrective actions taken, using the most advanced tools, instruments, and diagnostic software.

Much of the tasking elements identified above can be accomplished using the O and M personnel. Conventional FMS service plans tend to require the above functions from the manufacturer's personnel. This practice is very expensive and does not relieve the site personnel from assisting in the process. The facility management engineering personnel, tasked with the proper operation of the mechanical systems, are in the best position to identify problems and the corrective actions needed. Subcontractor personnel will not be as familiar with the nuances of the facility, equipment, or systems—information critical to the proper analysis of system-related issues.

9.3.4   FM operating personnel requirements

This area of facilities management often gets overlooked when developing a comprehensive operating plan. Many facility managers will elect to have system technicians knowledgable in the FMS products and programs perform this function. This is akin to having plumbers perform electrical work or elevator technicians cleaning cooling towers. The FMS is no more than a tool to be utilized in the operations of facilities. The commands to be initiated are no different from timeclocks or manual start stations. The only logical craft to operate facility management systems are operations and maintenance personnel trained and licensed to operate machinery. As such, the licensing provisions requisite of the operating engineers are the same standards that need to be applied to the FMS operating personnel. These standards apply whether the personnel are resident in the facility to be operated or remotely located.

Often facility systems are monitored by personnel that have no training, education, or experience in the systems and equipment that they are monitoring or controlling. While the practice of remote monitoring and control can lead to significant operating cost savings, the methodology employed in selecting these service providers should be no different from a hiring decision for site engineering personnel. Some jurisdictions in the country have sought to develop standards for the remote operations of facilities. What follows is a guideline in developing those standards, reflecting the licensing provisions for the Washington, DC, metropolitan area.

9.3.5   Remote control operating requirements

In establishing licensing requirements for remote monitoring and control facilities, each of the aspects of the operations must be viewed independently to determine the proper methods, staffing levels, training, and communications to protect the property and ensure safe operation. These issues must be evaluated at both the controlling facility and the controlled facility. It is the intent of these standards to define the manner in which remote control and monitoring may be utilized to satisfy the provisions of DC Code Chapter 4, Section 401—Duties of Licensed Operators. The provisions of DC Code Chapter 4, Section 400, apply to the operating engineers utilized in the performance of remote control services.

9.3.5.1 Definitions. The distinction between monitoring and control facilities lies with the fundamental ability to change the operating state of a piece of equipment. Whether an adjustment to a variable position or a change of state (on to off, etc.), the ability to render such an action is the definition of a remote control facility. Monitoring locations, limited to such notices as alarm annunciation and acknowledgment of remotely activated conditions, are not to be considered as remote control facilities.

The term *licensed operating engineer* used here is defined as a District of Columbia 3rd Class Steam Operating license (at a minimum). *Note:* This is the area standard for operating engineers in low-pressure steam plants.

The term *responsible charge* refers to the licensed operating engineer tasked with the operation and oversight of the remote control functions of a facility. This functional responsibility includes actions undertaken by the engineer together with those actions initiated through an automated routine by the remote control system.

Status as defined here refers to the proof of operation of a given device. It is a positive indication that the work the device is designed to perform is being produced. Indicators that fail to positively identify the device as operating and delivering utility resource do not meet the qualifications as a positive status.

Certified remote control facility is defined as an applicant who has been certified to be in compliance with the provisions of the standard.

9.3.5.2 Remote control facility. To be a certified remote control facility, the service provider needs to conform to the following minimum provisions. Conformance includes an initial inspection of the facility, with recertification inspections performed on a recurring basis.

9.3.5.3 Staffing issues

9.3.5.3.1 Licensing requirements of operators. The standard for remote control operating personnel affecting command and/or control initiatives for a property in the District of Columbia is the DC 3rd Class

Steam Engineers license. This measure of the applicant's overall training, experience, and commitment to the trade denotes the minimum licensing of remote operators. Any and all control actions (variable or two-position) are initiated and verified by a licensed operating engineer. The use of personnel that do not meet this requirement renders the operating entity in violation of this standard.

9.3.5.3.2 Staffing levels. A minimum of at least one licensed operating engineer should be present in the remote control facility at any time that remote control services are being performed. This requirement holds for all remote initiated commands and control actions, whether initiated by the operator or by the automated system in use. Command and control actions undertaken by the automated system are deemed to have been made by the licensed engineer with responsible charge of the operation of the particular facility.

9.3.5.3.3 Training and certification. The operating engineer with responsible charge of the remote control functions maintains a valid 3rd Class Steam Engineers license. Any retesting, recertification, or other requirements which become part of the minimum requirements of new applicants for similar licenses are met by the responsible charge engineer. In addition to the license, the engineer with responsible charge possesses a universal CFC license as defined by the Environmental Protection Agency. *It is the intent of this standard that any training or certification required of field operating personnel to maintain a valid local engineering license shall also be required of the responsible charge engineer.*

9.3.5.3.4 Supervision. The minimum licensing provisions for the operating engineer extend to the line supervisors of the remote control functions. A District of Columbia 3rd Class Steam Engineers license is required of the line supervisors.

9.3.5.4   Systems and communications

9.3.5.4.1 Supervised communication formats. Communications systems utilized in the performance of remote control services are supervised. Supervised communication formats include any and all telecommunication services where the availability of the communications path is supervised by the remote control and monitoring system. Examples of these services include dedicated lease lines, ISDN, and dedicated fiber optic networks. Polling systems, whereby the controlled facility only "reports by exception," are not considered supervised unless automatic testing of the availability of the communication path is included with the service. Time-based testing of the availability of the communications service is not to be less than 4 times daily. Evidence of the automatic polling is provided to the reviewing entity in the initial certification inspection together with recertification activities.

9.3.5.4.2 Alarm reporting. Systems which only annunciate alarms in "report by exception" formats are not considered as compliant with these standards and are not certified as remote control facilities. Nothing contained in these standards precludes the ability of a "report by exception" system to be utilized as a monitoring facility only. Systems of this type do not comport with the requirements of remote control operations and are not certified as remote control facilities.

9.3.5.5 Contingent operations

9.3.5.5.1 Loss of communications or control. As remote control services are being provided in lieu of staffed operations, any loss of communications or control by the remote control facility that cannot be restored within 4 hours warrants a return to performance of the operating requirements via direct means.

9.3.5.5.2 Electronic logs. The remote control facility prepares operating logs of system parameters and maintains these records for inspection by the reviewing entity.

9.3.5.6 Fail-safe provisions

9.3.5.6.1 Continuity of service. Redundancy. In the event of a loss of control or communications, the remote control facility must allow for continuity of services through the provision of redundant operations. Redundancy must be in the form of an alternate automatic remote controller or the return to local control.

9.3.5.6.2 Emergency services. The remote control facility must provide for the continuation of operating services in the event of disruption in the normal supply of building services.

9.3.5.6.3 Power. Sufficient emergency power or an uninterruptible power supply must be provided to maintain monitoring and control facilities in the event of a loss of power to the remote control facility. Emergency power/uninterrupted power supply (UPS) should be capable of withstanding a loss of primary utility power for a period of 8 hours.

9.3.5.6.4 Lighting. In addition to the emergency power requirements of the controlling systems, the provision of emergency services extends to the lighting requirements of the remote control facility.

9.3.5.6.5 Air conditioning. Air conditioning and ventilation services needed for the remote control equipment and human comfort should be independent from the base building HVAC systems. The remote control facility should not be dependent on the base building HVAC system for these services.

9.3.5.6.6 Communications. Telecommunications services inclusive of voice-grade services should also be independent of the base facility services.

9.3.5.7   Auditing and inspection

9.3.5.7.1 Initial certification process. The reviewing entity should certify the applicant's compliance with the provisions of this regulation. Certification may include but not be limited to physical inspections of the central monitoring facility, review of emergency services detailed above, and inspection of the qualifications of the operating personnel.

9.3.5.7.2 Recurring compliance inspections. Initial certification notwithstanding, the local reviewing entity retains the right of periodic inspections of the remote control facility to certify continuing compliance with all the provisions of the regulations. Certified applicants found in noncompliance are given 30 days to rectify the deficiency. Pending a reinspection of the identified deficiency, the remote control facility is either recertified as compliant or loses its designation as a certified remote control facility. In accordance with the continuing certification process, the remote control facility provides the reviewing entity with a list of all facilities where the applicant performs remote control services.

9.3.5.8   Controlled facility.
The following provisions relate to the minimum requirements of the facility being controlled by a certified remote control facility. Where variable sensed values are denoted, analog sensors are provided which measure the denoted temperature, pressure, amperage, etc. Where status points are indicated, positive indication is provided.

9.3.5.8.1 Methods. Minimum monitoring points follow.

9.3.5.8.2 Refrigeration equipment. Chillers (200 tons and up). Start/stop, status, evaporator pressure, condenser pressure, compressor amperage (one phase), leaving condenser water temperature, chilled water supply temperature, chilled water return temperature.

Chillers (200 tons and smaller). Start/stop, status, chilled water supply and return temperature, leaving condenser water temperature.

9.3.5.8.3 Boilers. Steam (75 bhp and up). Start/stop, status, flame fail, steam pressure.

Hot water (75 bhp and up). Start/stop, status, flame fail, hot water supply temperature, hot water return temperature.

75 bhp and smaller. No requirement.

9.3.5.8.4 Air-handling equipment. 20,000 cfm and above. Supply fan start/stop, supply fan status, return fan start/stop (if so equipped), return fan status (if so equipped), supply air temperature, return air temperature, mixed air temperature (where equipped with economizers). Supply air static pressure (for variable volume systems).

Below 20,000 cfm. Supply fan start/stop and status, return fan start/stop and status (if so equipped), supply air temperature, return air temperature.

9.3.5.8.5 Pumping systems. Applies to primary hot and chilled water, secondary hot/chilled and condenser water pumping systems, where pumps exceed 25 hp. Start/stop and status of all pumps (except standby devices). Delivery water and return water temperatures.

9.3.5.8.6 Data collection devices. Data collection devices include nonvolatile random access memory for the storage of system parameters. Memory includes the database of connected points, fail-safe positioning, and battery backup to protect the stored information. Data collection devices should be UL listed and installed in accordance with the manufacturer's recommendations. Monitoring and control points may be connected to the data collection devices via hardwired, powerline carrier, or radio-frequency communication methods in accordance with the NEC, NFPA, IEEE, FCC, and any other government agency having jurisdiction.

9.3.5.8.7 Data transmission devices. Data transmission devices (modems, transmitters, active hubs, etc.) should conform to FCC regulations for data transmission devices. All data transmission equipment deployed in support of the remote control services should be dedicated to the remote control system and not be used for voice transmissions or other computer system data communications services.

9.3.5.9 General provisions

9.3.5.9.1 System maintenance. All monitoring and control points utilized in the performance of remote control services should be maintained in proper operating condition in accordance with the manufacturer's recommendations. The requirement extends to the installed monitoring and control points, data collection devices, and data transmission devices located within the controlled facility, together with all data transmission systems, data processing, and storage systems at the remote control facility. The applicant should provide evidence to the reviewing entity of compliance with the provisions of this section.

9.3.5.9.2 Electronic watch tours. All of the provisions of this regulation having been met and certified by the reviewing entity, the applicant may utilize the remote control and monitoring facility to perform engineering watch tours (required by local code) of the connected equipment electronically.

10

Mechanical Equipment and Systems Maintenance Procedures

Bernard T. Lewis

Facilities Management Consultant
Potomac, MD

10.1 Preventive Maintenance Program

A properly conducted preventive maintenance program (PMP) will reduce overall operating costs, aid organization effectiveness, and aid safety practices, and will assure the continued preservation, usefulness, and performance of assets. These benefits are the primary justification for implementation of a sound PMP. Preventive maintenance (PM) should not be considered as a separate matter but as an integral part of the total continuum of facility maintenance.

10.1.1 Purpose of PM

PM carries the connotation that some action must be taken now to prevent a more serious problem at a later date. PM, as the name implies, means taking preventive actions. The purpose is to eliminate the need for radical actions sometime in the future by taking preventive actions now.

10.1.2 Practice of PM

The practice of PM means different things to different people. To most O and M personnel, PM means inspections, adjustments, lubrication, painting, and major overhauls. The term is broadly used, but it essentially represents a philosophy of providing care that will protect and maintain the essential quality of the equipment, system, or facility.

10.1.3   Size of PM activity

The size of the PM activity is immaterial. To an electrical engineer, PM may mean the proper choice and setting of relays in controls to avoid unnecessary downtime. To a mechanical engineer, it may mean a complete overhaul of a pump or compressor.

10.1.4   Definition of PM

To promote a clear understanding of the subject, a simple, practical, and acceptable definition of PM should be established. First, to be considered within the realm of PM, an action must prevent some form of future deterioration or breakdown. The key element of PM consists of actions taken. PM may be defined along the following lines:

> PM consists of periodic inspections, or checking, of existing facilities to uncover conditions leading to breakdowns or harmful depreciation of equipment, and the correction of these conditions while they are still in the minor stage; the preservation, lubrication, replacement of filters and worn parts, and other actions to prevent breakdowns or malfunctions.

10.1.5   Efficiency of PM

PM ultimately reduces the corrective maintenance workload. As PM takes over, the corrective workload is shifted from when you have to do it to when you want to do it. Thus the work can be done more efficiently, at lower cost, and on a schedule, with a substantially decreased likelihood of interfering with program operations.

A rationally planned PM program should include the following components.

10.1.6   Engineering input

The most valuable engineering input is in the original equipment, systems, and buildings design. PM will be facilitated if provisions for its performance are incorporated into the design phase. Designers often overlook the need for maintenance or sacrifice the maintenance aspect in order to emphasize aesthetics or reduce initial costs. Later, the facilities management department (FMD) must modify the system to provide a workable system that can be maintained at a reasonable cost. Obviously such modifications will be costly (such as permanent platforms for accessibility rather than scaffolding).

Some of the responsibility for the frequency with which design and construction shortcomings are encountered must lie with FMD management. In industry, in general, there has not been adequate promotion for better visibility for design flaws during the initial phases of a new project to enable good O and M review.

Inadequate maintenance considerations in engineering design cost industry millions of dollars in unnecessary maintenance costs. The educational process must be more vigorously promoted by FMD management. Presenting proper historical facts can do much to improve the situation.

Designers with good maintenance understanding have contributed greatly to overall reduction of maintenance costs. Sealed motors, modular components, and built-in diagnostic test units are a few of the innovations that have made maintenance an easier task. As the cost for maintenance becomes increasingly significant, designers should produce more such innovations.

10.1.7 Analysis of maintenance needs

When equipment and systems are installed, there is usually no index available for maintenance needs, except manufacturers' recommendations. Further details are acquired during day-to-day operation and maintenance of the equipment and systems. It is advisable to record these details. Analyzing these experiences by using simple measures of statistics or probability will give a sound basis for *what* must be done, *when* it must be done, and *to what extent* it must be accomplished.

10.1.8 Proper use of equipment

A PM program should include means for assuring that equipment and systems are operated properly. This is accomplished by gaining cooperation of operations personnel. It has been determined that approximately 25 percent of all equipment and systems breakdowns are due to improper operations.

10.1.9 Basic controls

A well-engineered PM program, contrasted to the traditional "fire fighting approach," requires skilled personnel; proper tools, materials, and supplies; comprehensive scheduling; and accurate recording. The last two activities, when properly established, become routine administrative tasks. One of the most efficient means available today for sorting, selecting, filing, and recording data is use of a computerized maintenance management system (CMMS) software module for PM control.

10.1.10   Field inspection procedures

Service maintenance manuals issued by equipment and system manufacturers are one of the best sources to facilitate the initial setup of field inspection procedures. They are invaluable guides as to *what* and *when* to inspect, as well as *how* to service and maintain the equipment. The manuals usually contain checklists which itemize all points to be checked and maintained on the equipment and systems, thus providing a uniform and complete inspection. These manuals, coupled with experience and engineering judgment, will normally be sufficient to ensure that used procedures are adequate. However, established procedures need continual refining and updating as experience dictates. During the process of revising manufacturers' recommendations to suit local facility conditions, it is safe to take the most stringent procedures selected from manufacturer's recommendations, engineering judgment, and current FMD procedures to ensure optimum results.

No readymade guide is available on frequencies to perform PM inspections. Engineering analyses of operating cycles, failure history, and maintenance needs will be of assistance in determining these frequencies. In making these determinations, consideration should be given to the following:

- Age, condition, and value (dollar value and value to the organization's mission) of the equipment or system
- Severity of service
- Safety requirements
- Hours of operation
- Susceptibility to wear, exposure to dirt, friction, fatigue, stress, and corrosion
- Susceptibility to damage, vibration, overloading, and abuse

Frequencies should be periodically reviewed and adjusted. New equipment and systems must be checked frequently until they are broken in. PM inspectors should be required to indicate recommended increases or decreases in the PM frequencies, using criteria based on repair statistics. Indications of the need to adjust PM frequencies include:

- No repairs required—overmaintenance is indicated.
- Frequent repairs required—inspectors are not getting to the trouble or cause, or engineering assistance may be required.

10.1.11 Feedback, properly supplied and utilized

Feedback from supervisors, foremen, operating engineers, and mechanics is essential for a PMP to be successful. The first and perhaps one of the most important items needed is *labor hours* and *material costs* expended. The second items needed are records of *completion* of assigned work tasks on the equipment and systems, together with supervisor's verifications, to ensure a measure of productivity and customer satisfaction. The third items needed are descriptions of delays encountered due to lack of transportation, safety permits, materials, tools, supplies, job sites availability, and outage scheduling.

Good engineering judgment should be utilized in a PMP. Techniques such as mean time between failures (MTBF) analyses, technical data accumulation and analyses, and work methods analyses should also be used. Written two-way communications between engineering supervisors and craft and trade foremen should be used to resolve encountered problems.

10.1.12 Labor standards applied to PM

Initially, labor standard times should be used as a basis for estimating labor hour requirements. Section 4.2 describes a set of labor time standards that have proved to be useful to many public- and private-sector organizations. However, as historical data on actual tasks performed are accumulated and then compared with estimates, it is possible to refine the EPS labor time standards to an overall accuracy of approximately ± 5 percent. The identification of labor hours required for the performance of equipment and systems PM would normally be developed by O and M work center craft and trade foremen in conjunction with the work control manager.

The application of labor time standards to repetitive-type work such as PM is highly desirable. Analysis and interpolation of these data for similar but not identical jobs is an excellent guide for PM planning. This approach, coupled with engineering judgment and past experience of qualified personnel, provides the starting point for applying labor hour standard times. Continual evaluation will further assist in refinement of the labor hour standard times.

10.1.13 How PM relates to other routine maintenance

A PMP should be integrated with regular facility maintenance and engineering programs. Supervisors, planners, and schedulers who handle the PM assignments should be the same as those who direct, plan, and schedule regular maintenance work. Mechanics and operating engineers should be rotated periodically so that as many as possible acquire familiarity with the program and its procedures.

10.1.14   Additional PM program techniques

After the basic program has been installed and proved economically beneficial, there are other finer techniques that can be applied. These include the following:

- Protective methods

  Coatings—cathodic and plastic
  Alarms—vibration mounts

- Predictive maintenance—covered in detail in Chap. 6
- Maintenance training
- Standard practice manuals
- Materials research

10.2   Facilities Inspection and Maintenance Program

As facilities become more complex and more resources are required to provide and maintain needed facilities, the question is how can facility management departments keep up with operations and maintenance requirements?

One way would be to hire more craftspeople to take care of both the new and existing requirements. But since needed additional resources are usually not available, this is not a viable option. Any available new resources are usually earmarked for upgrading facilities, equipment, and systems. But this is the way it should be, for any new equipment and system generally makes facilities operations more efficient and cost-effective. So what is the answer? How can the director, facility management department do more with less? A new direction is needed to face this dilemma if you are going to keep up.

One approach is to use a set of programs known as the facilities inspection and maintenance program (FIMP). Although the success or failure of a facility management department (FMD) is hard to ascertain and frequently must be based on empirical evidence, there is evidence that FIMP use has been successful and has helped to meet the changing needs of organizations without placing undo strain on the organization's budget. Generally, the results will be evident in:

- Staff and payroll decreases

- Facilities in better shape
- Square footage requirements, per employee, increased with no ill effect
- Level contract costs

If prevention or decrease in deferred maintenance projects is a criterion for a successful FMD, then FIMP has proved to be a success in many organizations. How will you know if facilities users (customers) are more satisfied? For the following reasons:

- The department's work reception telephone is not constantly ringing with requested work requirements or emergencies.
- Irate telephone calls are very rarely received in the department because the majority of work is done in a timely fashion, and much is done before a customer could even notice that something needed to be done.
- Even though more area is being maintained with fewer people, emergency and service repair tickets volume has not increased.

These results can be achieved using the FIMP programs. However, other alternative solutions should be evaluated also before deciding to use the FIMP. Some alternatives are listed following.

10.2.1 Computers

Computers are good management tools, but you do need the necessary databases to readily determine the required maintenance effort. Computers may help with keeping track of what was done, but a method is needed to provide direction to the required maintenance effort.

10.2.2 Reduction in overall maintenance effort

Reducing the overall maintenance effort, accepting the fact that you cannot get to all requirements anyway, is not the answer because lowered maintenance efforts lead to a downward spiral for a facility's level of maintenance that does not bottom out as equipment and systems age, increase in size, and become more complicated.

10.2.3 Doing only what needs to be done

Another approach is doing only what needs to be done and skipping the rest. This takes care of essential equipment and systems and keeps the facility operating, but invariably the areas where maintenance levels are reduced are the areas that impact on an organization's operations and cause problems.

10.2.4   Plan some maintenance and fire fight the remainder

This could work, but performing maintenance this way means some customers could learn the maintenance management system (MMS) and get all their work done; while others would get almost no work done for they did not know how to use the MMS to their advantage.

From all the reasons listed above, it was deduced that there was a better way. The better way was actually the old ways which were incorporated into FIMP.

Major areas of improvement will result from very minor changes and constant positive reinforcement of old proven procedures. Any FMD should constantly inspect everything to keep abreast of current conditions. Do not hire special inspectors. Use existing personnel; you will be better off. A carpenter or locksmith knows more about maintenance and repair requirements of door hardware than a FMD manager normally would, and a building manager would know more than a user. One particular area where you can achieve considerable success is with regular housekeeping and custodial inspections.

These inspections, conducted by FMD personnel, accompanied by custodial and housekeeping forepersons and building occupants, will spot flaws in custodial and housekeeping routines but ultimately will accomplish much more. The most obvious results will be cleaner, neater buildings. But most importantly, in addition, regular inspections "show the flag" by having FMD personnel, housekeeping personnel, and building occupants dealing with one another. This results in better rapport and better understanding of problems associated with custodial maintenance. Housekeeping inspections are not covered by FIMP but are essential to its success owing to the ease of spotting maintenance requirements when buildings are being cleaned and occupants become accustomed to dealing with FMD personnel. After rediscovering the correct maintenance methods, the next step is getting organized to get the job done.

You begin by determining goals. The goal for maintaining life safety equipment is easy; any condition other than 100 percent is not acceptable. Also, with comparatively reduced financial resources, you cannot afford any emergency work which would have to be deferred. The chosen method would have to allow for an early warning of developing problems. Any system chosen needs to allow for tight labor power scheduling and for rapid labor force movement to concentrate on specific areas, because the window of time availability in some buildings is often quite narrow. Another requirement is to be able to combine usage and visibility in a coherent system; invariably some areas are heavily used and few people see them (kitchens), while some areas are not used hard but everyone sees them (elevator lobbies).

Considering these constraints and objectives, you are able to commence your implementation planning. To keep up with the maintenance requirements, a good place to start is where all concerned can agree on what is needed to determine necessities. As mentioned previously, maintenance requirements for life safety equipment is easy to determine; sprinklers, fire alarms, fire door components, smoke and heat detectors, regular and emergency electrical systems, etc., must be kept in new condition. Other building components need greater or smaller amounts of attention. Determining the correct level of maintenance for any one system requires having a complete, current inventory on hand, research into manufacturer's recommendations, estimates of required labor hours to perform the work, and many educated judgment calls. In this manner, work through each facility's systems and equipment, and then combine the results into a single FIMP for scheduling purposes. This system of integrated programs provides management with historical files for each building. As data is accumulated in each program file for each building, it becomes easier to assess needs and pinpoint areas where problems are developing. Deferred maintenance is thus avoided and labor power is easily adjusted to accurately develop current needs.

FIMP can be originally implemented by drawing together existing programs for centralization. Existing programs form the backbone for FIMP development. One early program to be evaluated is roof inspections. Roof inspections can best be performed by contractor personnel rather than in-house personnel because of the specialty craft trade requirements. It is absolutely essential that roofs be well maintained to ensure watertight building envelopes. Ideally roofs should be inspected twice a year using an in-house inspector who:

- Can perform minor maintenance on the spot
- Knows roofing maintenance procedures in detail
- Can supervise outside contractors to ensure quality workmanship is provided

There should not be too much redundancy in roof inspections. Roof inspections should include nondestructive tests with a nuclear device and infrared spot checks, taking regular core samples of roofing membranes, and twice-a-year thorough visual checks. It is essential to remember that it takes only a few weeks for a roof to go from being maintainable to requiring total replacement. Other programs can be added as required. It is easy to identify areas requiring routine maintenance. Most programs derive from customer needs or complaints.

Other programs will be designed to meet specific needs, such as sprinklers, for insurance purposes. As a need for a particular equipment or system program becomes evident, it is necessary to take the following actions:

- Calculate the frequencies for preventive maintenance visits.

- Conduct an inventory of all similar equipment and systems, including the recording of nameplate data, locations, and current levels of maintenance.

- Develop needed craft labor power requirements to conduct the preventive maintenance visits; including labor hour times for site visits for work, travel, preparation, and delays.

- Develop checklists to control each program, using industry standards, past experience, and engineering judgment.

Each program and its controlling documents should be designed for flexibility to meet FMD, user, and organization needs. For example, manufacturers recommend that emergency diesel power systems should not be started and run without load, but to ensure users that they will start, you can start them once a week anyway and then once a month perform a full-load test.

A third reason to start FIMP programs is for FMD convenience. An example of this type of program is kitchen maintenance. It is easier for FMD to allow the kitchen users (they are the experts) to be responsible for maintenance, but it is more efficient and effective to have the FMD schedule the maintenance on a regular, routine basis. Otherwise craftspeople would be responding to service and emergency calls, and repairing one item at a time.

Programs should be divided into three categories during development:

1. Inspections are programs used to define and identify breakdowns in major equipment or systems such as roofs or structural envelopes, or to look for particular items, such as asbestos during remodeling projects. When a needed repair is identified, a work order will be issued and prioritized. These repairs, controlled by the work order, are either performed according to priority or forwarded to the director for approval, funding, and designation of the performance method—i.e., in-house or by contract.

2. PM programs are also inspections, but with *predetermined maintenance* and *benchmark tests* included. The maintenance and tests

needed for a particular equipment or system are defined and controlled by checklists. Any discovered problems are usually corrected as emergencies.

3. Find/fix are programs that use scheduled teams of craftspeople to systematically inspect for and then perform contingency maintenance on a particular system or equipment or a complete building. During a find/fix procedure virtually any problem discovered is repaired immediately.

In each of these categories the individual programs are broad in scope yet detailed in application. Program examples are preventive maintenance of major switchgear, and find/fix interiors.

Maintenance of major switchgear is critical for the usefulness of a building for obvious reasons. How does a FMD ensure that all switchgear components are in design ranges when the work must be performed by specialty contractors not familiar to FMD personnel? To manage this type of program, the FMD must have complete inventories of all equipment and systems in each building, and a complete set of specifications. These specifications are available from many sources, including the requirements from the appropriate governmental (federal, state, county, and/or city) jurisdiction. The specialty contractor should be tasked to provide full reports, based on accurate guidelines, listing the test results on each piece of switchgear equipment. The report should also list further maintenance requirements needed. This report will be the FMD's control over work performance, and the basis for approving invoice payment.

Another program example is the find/fix series. It should be the FIMP cornerstone for the FMD. Normally this program will use approximately one-half of the routinely scheduled labor hours, control the interior maintenance of buildings, and prevent the deterioration of offices, laboratories, conference rooms, etc.

The normal maintenance management approach is to inspect building interiors and then issue corrective maintenance work orders as needed. This method focused maintenance on necessary areas, but the cost of inspectors, long lead time to complete the repairs, and attendant paperwork made the process economically infeasible. Find/fix enables the FMD to deploy craftspeople to find the needed repairs and fix them on the spot.

The find/fix method ensures that large costly repairs will not result from unnoticed small repairs. The central theme is similar to the method used by homeowners: being observant, making small repairs as needed, or ensuring that someone else makes the repairs. The concept is simple in design, but implementation on a large scale, with large crews, requires much supervision and training and is difficult and time-consuming. A workable program schedule development requires the following elements: needed materials inventories; crew sizes calculations; average production in square footage per labor hour; and a schedule of availability of buildings for the program. After the program is documented and scheduled, checklists that included each item thought to need regular attention should be designed and developed for each participating craft.

The keys to successful maintenance practices are shown following: Go back to basics such as find/fix and inspections, forget computers until your programs are in place and well integrated; and inspect to the point of redundancy, and if at all possible make repairs immediately on the spot; and stick to what you do well—minor maintenance and repairs and not general contracting work. *Use remodeling work to fill in the low spots, contractors to level off the high spots, and stick to maintenance.* Use of FIMP will allow you to *routinely* schedule approximately 50 percent of available labor hours, and find and direct another 30 percent through inspections. The use of FIMP scheduling procedures allows you to plan the use of the remaining 20 percent almost year-round. FIMP takes a great deal of planning and top management support to make it work, *but it does work.*

10.2.5 Facilities inspection and maintenance program specifics

Envelopes   Annually, the structural inspector should perform a complete structural envelope inspection using a checklist and a set of instructions.

Roofs   Annually, a roof inspector should perform a complete roof inspection on each building. The roof inspector should submit inspection forms, complete with findings and repair cost estimates to the director, FMD, for approval and subsequent actions.

Moisture Meter   Every other year, moisture meter tests should be performed on each major flat roof. These tests are performed to establish benchmark data and to pinpoint problems before breakdowns.

Sprinkler Inspections   Quarterly, in-house personnel should perform sprinkler inspections entailing complete checks of flow and tamper switches, alarms, etc. The discovered problems should be reported and processed as "emergencies" in the maintenance management system.

Relamping   Performed, as required on a case-by-case basis, using in-house personnel.

Switchboards   Every 3 years, contract personnel perform switchboard inspections on systems over 1000 A. The inspections should be controlled by FMD documents.

Motor Control Centers   Performed every 3 years by contract personnel, testing and maintaining contacts and heat sensors. The testing and maintaining should be controlled by FMD documents.

Emergency Generators   In-house personnel, controlled by checklists, perform the following listed tasks on the diesel generators:

Weekly   Check all gauges, connections, batteries, and fluids.

Monthly   Perform a complete operational check under load.

Emergency Generators   Weekly, in-house personnel, controlled by checklists, perform complete system checks on gasoline generators.

Emergency Battery Packs   Monthly, in-house personnel, controlled by checklists, test and check for voltage and specific gravity.

Trash Chutes   Quarterly, in-house personnel, controlled by checklists, test and repair trash chutes.

Fire Alarm Systems   Annually, in-house personnel, controlled by FMD documents, test and repair fire alarm systems by activating each station and monitoring alarms, bells, and panels. Necessary repairs are performed as follow-ups on an emergency basis.

Elevators   Weekly, biweekly, or monthly, contract personnel, controlled by FMD documents and specifications, perform maintenance and/or repairs as needed.

Burglar Alarm Systems   Monthly, in-house personnel test each burglar system and perform repairs as needed.

Fume Hoods   Annually, contractor personnel test and certify all fume hoods and exhaust systems for proper velocity and pressure.

Smoke Detectors   Semiannually, in-house personnel, controlled by checklists, inspect, test, or repair smoke detectors as required.

Sprinkler Systems   Monthly, in-house personnel, controlled by checklists, perform preventive maintenance on wet systems and water motor gongs. Weekly (twice a week in freezing weather) on dry systems.

Irrigation Systems   Twice a year, in-house personnel, controlled by checklists, inspect and repair irrigation systems. Systems should be charged in the spring and made ready for the watering season and winterized in the fall.

Halon Systems   Annually, contract personnel with in-house assistance maintain Halon systems.

Painting   On a recurring 5-year basis, paint all buildings.

Clocks   In-house personnel reset clocks in April and November to reflect the changes from EST to DST.

Site Drains   Quarterly, using in-house personnel, inspect and clean, as necessary, site drains.

Water Heaters   Annually, in-house personnel perform routine maintenance on large-capacity gas-fired water heaters.

Uninterruptible Power Supplies   As required, contract personnel maintains uninterruptible power supplies in PBX rooms.

Find/Fix Building Interiors   Three times a year, in-house personnel, directed and controlled by checklists, inspect and repair building interiors.

Find/Fix on Eyewashes and Safety Showers   Twice a year, in-house personnel, controlled by checklists, inspect eyewashes and safety showers for pressure and water volume.

Lightning-Suppression Systems   Annually, in-house personnel, controlled by checklists, inspect lightning-suppression systems for ground and continuity.

Acid Tank Find/Fix   Twice a year, in-house personnel, controlled by checklists, inspect and repair acid neutralizing tanks.

10.3   Equipment and Systems Maintenance and Repair Procedures

10.3.1   Preventive maintenance job tasks

Formal written job tasks should be established for each type of equipment and system in the facility's PMP. The tasks should be written so as to ensure that each particular job task can be performed on more than one equipment or system item. Figures 10.1 through 10.50, inclusive, show typical task descriptions for a representative range of equipment and systems. These figures are used with Georgetown University's approval and are credited to Georgetown University.

Each item or step of the job to be performed is listed with a description of the work to be done. The description of the job is laid out in a stepwise manner and written in terms that the mechanic doing the work can understand. The complete task is packaged, given a suitable reference number, and filed and referenced is such a manner so as to be easily retrievable—normally part of the computerized maintenance management system's PMP module.

GEORGETOWN UNIVERSITY PREVENTIVE MAINTENANCE (GUIDE NUMBER) 1

EQUIPMENT: Electric Motors

MANUFACTURER: All

CATEGORY: Electrical   FREQUENCY: Yearly

PREVENTIVE MAINTENANCE DESCRIPTION:

Conduct Ammeter reading of motor electrical load.

SPECIAL INSTRUCTIONS:

- Exercise special caution while working near energized circuits. Test is to be made on motors operating at full load.

PROCEDURE:

1. Conduct ammeter reading of motor electrical load:

    a. Open the cover door of the motor starter/controller unit or the motor disconnect switch (Actuate defeater catch to open).

    b. With the motor running, take ammeter readings of each of the three motor lines at the output terminals of the contacts.

    c. Observe the amperage load of the motor and compare it with the amperage load for that particular motor.

    d. Record readings on preventive maintenance work order.

TOOLS AND MATERIALS: Revision Date: 8/10/88
Standard hand tools, clamp on ammeter

Figure 10.1 PM Guide Number 1—Electric motors. (*From Georgetown University.*)

GEORGETOWN UNIVERSITY PREVENTIVE MAINTENANCE(GUIDE NUMBER)__2__

EQUIPMENT: Rotating AC Machinery

MANUFACTURER: All

CATEGORY: Test   FREQUENCY: Yearly

PREVENTIVE MAINTENANCE DESCRIPTION:

Megger Insulation Resistance Test (All rotating machinery)

SPECIAL INSTRUCTIONS:
OSHA Standard 1910.213 -Open lock and tag motor disconnect switch
and starter before starting work. Check the voltage of all
incoming line terminals. Positively ascertain that the equipment
is deenergized.

PROCEDURE:
    Insulation resistance should be checked once a year. The
test must be done under the same conditions each time in order
for the readings to have comparison value, (motor temperature,
operating time before test is made, etc.). The following formula
will render the minimum resistance value for the equipment. With
the stator windings at their normal operating temperature,
measured at 500 volts d.c., one minute after motor has been
turned off and ready to test.

$$\frac{\text{rated voltage of machine} + 1000}{1000} = \frac{\text{insulation resistance}}{\text{in megohms}}$$

A.   Use a thermometer to check the temperature of the motor. Do
     this at the top center of the motor housing. Record the
     temperature reading on the PM form.
B.   Deenergize all power to the motor that is being tested.
C.   Connect the megger ground lead to a good ground source and
     test the megger. Now proceed to check each motor lead one at
     a time. Record the insulation resistance of each test on the
     PM form.
*Note-In the event the readings do not meet the minimum operating
standards as previously calculated you should report the
deficiency to the shop foreman responsible for the repair or
replacement of the motor.

TOOLS AND MATERIALS: Revision Date: 8/5/88
Voltage Tester, Safety tags and locks, Megger meter, Hand tools
to remove covers, thermometer to read motor temperature.

Figure 10.2   PM Guide Number 2—Rotating AC machinery. (*From Georgetown University.*)

GEORGETOWN UNIVERSITY PREVENTIVE MAINTENANCE(GUIDE NUMBER)___3__

EQUIPMENT: Electric motors(all)_____

MANUFACTURER: All_____

CATEGORY: Cleaning   FREQUENCY: Yearly_____
PREVENTIVE MAINTENANCE DESCRIPTION:

Cleaning and Drying Electric Motors

SPECIAL INSTRUCTIONS:
OSHA Standard 1910.213 -Open lock and tag motor disconnect switch
and starter before starting work. Check the voltage of all
incoming line terminals. Positively ascertain that the equipment
is deenergized. Volatile Solvent cleaning agents are to be used
in well ventilated areas only -following the manufacturer's
recommendations.

PROCEDURE:
The motor exterior should be kept free of oil, dust, dirt, water,
and chemicals. For fan cooled motors, it is particularly
important to keep the air intake opening free of foreign
material. Dirt buildup inside or outside of motors can cause
overheating and shortened insulation life. A motor subjected to a
10°F temperature rise above the maximum operating temperature
will cause premature motor failure.

Cleaning

Clean the motor inside and outside regularly. Use the
following procedures, as they apply:
A.   Wipe off external surfaces of motor. Use a petroleum solvent
     if necessary.
B.   Remove dirt, dust, and other debris from ventilating air
     inlets on enclosed motors. Remove end-plate covering the
     motor fan. Brush and wipe the fan blade clean.
C.   Clean open motors internally by blowing with clean, dry,
     compressed air at 40 psi.
D.   When dirt and dust are solidly packed, or windings are
     coated with oil or greasy grime, disassemble the motor and
     clean with solvent. Use only high flash naphtha, mineral
     spirits, or stoddard solvent. Wipe with solvent dampened
     cloth or use suitable soft bristle brush. Do not saturate.
     Air dry stator for 24 hours. Then oven dry at 150°F. Be sure
     windings are thoroughly dry before reassemble.
E.   Encapsulated stator windings may be rinsed or sprayed with
     solvent and immediately wiped dry with a cloth. These
     windings may also be cleaned with water and a fugitive
     detergent(ammonium oleate). Rinse with clean, clear water to
     remove all detergent. Hot water or low-pressure steam may
     also be used. Wipe excess water from metal surfaces and oven
     dry at 200°F.

Pg 1 of 2

Figure 10.3 PM Guide Number 3—Electric motors (all). (*From Georgetown University.*)

GEORGETOWN UNIVERSITY PREVENTIVE MAINTENANCE(GUIDE NUMBER)____3____
(continued)

Cleaning

F.   Motors may be dried out by heat from a warm air oven, electric strip heaters, heat lamps, or by passing current through the windings. The temperature should not exceed 167°F. A forced circulation type oven rather than a radiant type oven is recommended. Radiant ovens tend to scorch and burn surfaces before the desirable dryness is reached.

When the motor is dried by passing alternating current through the windings with the bearing housings removed, the rotor must be centered in the stator core. Make certain that the air gap is uniform by wedging fiber strips in the lower portion of the gap. A controlled current of the same number of phases and the same or lower frequency is applied to the terminals.

Voltage should not exceed 10% of normal, and it should not cause more than 60% of normal full-load current to pass through the windings. A voltage 15% of normal may be applied after the insulation resistance has reached ½ of the minimum value for the insulation normal operating resistance. The temperature should not exceed 167°F during drying by this procedure.

G.   After cleaning and drying windings check the insulation resistance with a megger test meter. See PM Guide 2 for procedure.

TOOLS AND MATERIALS: Revision Date:8/5/88
Air compressor, solvent cleaner, soft wire brush, rags for cleaning, vacuum, hand tools to remove covers

Pg 2 of 2

Figure 10.3 (*Continued*) PM Guide Number 3—Electric motors (all). (*From Georgetown University.*)

GEORGETOWN UNIVERSITY PREVENTIVE MAINTENANCE(GUIDE NUMBER) __4__

EQUIPMENT: Electric Motors Grease by Disassembly

MANUFACTURER: Louis Allis

CATEGORY: Lubrication   FREQUENCY: 3 Years
PREVENTIVE MAINTENANCE DESCRIPTION:

Remove motor end shields and repack bearing. (Louis Allis)

SPECIAL INSTRUCTIONS:
OSHA Standard 1910.213- Open lock and tag motor disconnect
switch and starter controller before starting work.

PROCEDURE:
1. Re-grease motor bearings by disassembly:

a.   Motor must be disassembled to lubricate the
     bearings on this type of motor.Remove end shields from
     motor, clean grease cavity.

b.   Refill grease cavity three quarters full with No. 2
     Consistency grease with a -10° to 200°F Ambient.

Caution: Bearings and grease must be kept free of dirt.

TOOLS AND MATERIALS: Revision Date:11/7/88
Standard hand tools, grease, rags.

Figure 10.4 PM Guide Number 4—Electric motors grease by disassembly. (*From Georgetown University.*)

GEORGETOWN UNIVERSITY PREVENTIVE MAINTENANCE(GUIDE NUMBER)__6__

EQUIPMENT: Electric Motor, Ball Bearing, Flush Type_____

MANUFACTURER: G.E/Westinghouse/Baldor/Louis Allis_____

CATEGORY: Lubrication_____   FREQUENCY: Yearly___
PREVENTIVE MAINTENANCE DESCRIPTION:

Regrease single shielded motor bearings.

SPECIAL INSTRUCTIONS:
OSHA Standard 1910.213   Open lock and tag motor.   Disconnect switch before starting work.

PROCEDURE:
A.   Regrease single shielded motor bearings with the motor at normal operating temperature.

B.   Wipe the lubrication fitting, bearing casing and grease relief plug clean at both ends of the motor housing to insure no dirt enters bearings.

C.   Remove the relief plugs;if grease has hardened, probe a short distance into grease chamber to break up grease so it will be forced out by the new grease.

D.   Insert new grease through the lube fitting with a low-pressure grease gun until new grease appears at the grease relief holes.

E.   Energize and run the motor for ten minutes so the bearing can expel any excess grease. Probe the new grease from the relief hole to allow for expansion and replace plug.

F.   Always wipe up expelled grease from around relief plug and filler fitting.

TOOLS AND MATERIALS: Revision Date:8/5/88
Standard hand tools, rags, grease gun.

Figure 10.5 PM Guide Number 6—Electric motors, ball bearing, flush type. (*From Georgetown University.*)

GEORGETOWN UNIVERSITY PREVENTIVE MAINTENANCE(GUIDE NUMBER)_9____

EQUIPMENT: Preheat / Reheat pumps / H.W circ. / CH.W Circ_____

MANUFACTURER:_____

CATEGORY: FREQUENCY:
_____

PREVENTIVE MAINTENANCE DESCRIPTION:

Lubricate the pump _motor_ bearings.

_____

SPECIAL INSTRUCTIONS:

Clean any dirt buildup from air intake openings of motor and wipe
motor clean.

_____

PROCEDURE:
1.   Taco Pump Motors:

       a. Add one teaspoon of SAE20 ND oil to each motor bearing.
          Caution: do not over-lubricate motor bearings.

2.   For Thrush Vertical Mounted Pump Motors:

       a. Add approximately 1/4 ounce of NO2 -20° to 300°F Ambient
grease to upper and lower bearings.
          Caution: do not over-grease

_____

TOOLS AND MATERIALS: Revision Date: 11/01/88
Standard Hand Tools, Grease Gun with NO2 -20° to 300°F Ambient,
Grease Oilcan with SAE 20 Non Detergent oil, 6 foot ladder.

Figure 10.6 PM Guide Number 9—Preheat/reheat pumps/H.W. circulation/CH.W. circulation. (*From Georgetown University.*)

GEORGETOWN UNIVERSITY PREVENTIVE MAINTENANCE(GUIDE NUMBER) 15

EQUIPMENT: Coils and Fans

MANUFACTURER: All

CATEGORY: Cleaning   FREQUENCY: Yearly

PREVENTIVE MAINTENANCE DESCRIPTION:

Coil and Fan Cleaning.

SPECIAL INSTRUCTIONS:

Never steam clean closed pressurized systems. Use eye protection at all times. Be familiar with chemical cleaning agents you are using and their hazards. Use gloves and protective clothing.

PROCEDURE:

Fan Blades, Condenser Coils, Chill and Hot Water Coils, Reheat Coils, Preheat Coils.

Where cleaning can be done with water and detergent and only minimum precautions are necessary to prevent the water damaging any surrounding areas this will be the procedure of first choice. In the event water cannot be used compressed air will be blown through the coils after brushing the coil face. This procedure will be repeated on the back of the coil followed by a thorough vacuuming of each side of the coil and compartments.

A steam cleaner is acceptable in lieu of water and detergent but in no case is steam cleaning to be used on coils containing an isolated fluid or gas. The resultant expansion from hot steam on the coil will produce explosive results and severe personal injury.

The complete removal of dust, oil, and grease is the objective. If, during the cleaning process, surfaces loose their protective coatings a suitable means to restore that protection should be applied as soon as the surface is dry.

TOOLS AND MATERIALS:
Gloves, Coveralls, Wet-vac, Hose, Air Compressor, Brushes, Rags, Eye Protection.

Figure 10.7   PM Guide Number 15—Coils and fans. (*From Georgetown University.*)

GEORGETOWN UNIVERSITY PREVENTIVE MAINTENANCE (GUIDE NUMBER) __16__

EQUIPMENT: Reheat Coils

MANUFACTURER: All

CATEGORY: Cleaning   FREQUENCY: Yearly

PREVENTIVE MAINTENANCE DESCRIPTION:

Clean and Check Reheat Coil.

SPECIAL INSTRUCTIONS:

PROCEDURE:

1.   Clean, Inspect, and Check Reheat Coils:
      a.   Open access cover upstream of reheat coil and clean
      coil face of dirt and debris by brushing and vacuuming.

      b.   Inspect coil for damage, bent fans, evidence of
      leaking or corrosion.

      c.   Actuate valve with thermostat to ensure that it
      functions.

TOOLS AND MATERIALS: Revision Date: 8/10/88
Vacuum, Soft wire brush, Drop light, 9' ladder, hand tools to
open covers.

Figure 10.8 PM Guide Number 16—Reheat coils. (*From Georgetown University.*)

GEORGETOWN UNIVERSITY PREVENTIVE MAINTENANCE(GUIDE NUMBER)__17__

EQUIPMENT: Condensing unit(For Walk-In Box)

MANUFACTURER: Punham-Bush/Tecumseh/Copeland/Carrier

CATEGORY:Refrigeration   FREQUENCY: Yearly
PREVENTIVE MAINTENANCE DESCRIPTION:

1. Conduct annual maintenance of walk-in box condensing unit
    compressor.

SPECIAL INSTRUCTIONS:

Exercise caution when working on running equipment.

PROCEDURE:
1. Conduct annual maintenance of walk-in condensing unit:

        a. Check level of compressor oil.

        b. Flush the water regulating valve by inserting a
            screwdriver under the control valve arm and
            lifting the valve a couple of times.

        c. Check tightness of all flare nuts.

        d. Wipe down unit, inspect for any damage, oil or
            freon leaks.

        e. Tighten all electrical wire terminals on
            machine; inspect for burned or worn contact
            points. Check electrical wire terminals on
            starters for tightness.

        f. Lube motor only in accordance with the assigned PM
            guide.

        g. Check refrigeration charge with manifold
            set. Recharge if needed. Report leaks to
            supervisor and note if there is a leak on PM card.

        h. Ensure that all belt guards are in place when
            done.

TOOLS AND MATERIALS: Standard hand tools, AC manifold set,
rags, grease gun, Freon for recharging.

Figure 10.9   PM Guide Number 17—Condensing unit for walk-in box. (*From Georgetown University.*)

GEORGETOWN UNIVERSITY PREVENTIVE MAINTENANCE(GUIDE NUMBER) 19

EQUIPMENT: Air Dryer

MANUFACTURER: Hankison/Johnson

CATEGORY: Air dryers   FREQUENCY: Yearly
PREVENTIVE MAINTENANCE DESCRIPTION:

1. Clean and check air dryer.

SPECIAL INSTRUCTIONS:

Unplug unit or lock out switch before starting work.

PROCEDURE:
1. Clean air dryer refrigeration unit:

    a.   Turn off unit; remove access panel from cabinet.

    b.   Clean dust and dirt from the inside of the cabinet, the condensing unit, and especially from the condenser coil.

    c.   Wipe the fan blades clean and rotate fan by hand to check the condition of the fan bearings.

2. Clean and check filters:

    a.   Check the oil filter if the media indicator has turned pink. Replace the desiccant filter cartridge.

    b.   Automatic drain: operate the float valve manually during inspection. Clean automatic drain if desiccant filter is changed. Use soap and water only.

    c.   Check by-pass valve to make sure all flows are through the condenser and oil filter.

TOOLS AND MATERIALS: Standard hand tools, strap wrench to open filters if needed, rags.

Figure 10.10 PM Guide Number 19—Air dryer. (*From Georgetown University.*)

GEORGETOWN UNIVERSITY PREVENTIVE MAINTENANCE(GUIDE NUMBER)_20__

EQUIPMENT: Humidifier

MANUFACTURER: All

CATEGORY: Heating and AC   FREQUENCY: Yearly
PREVENTIVE MAINTENANCE DESCRIPTION:

Steam Humidifier inspection and cleaning

SPECIAL INSTRUCTIONS:

PROCEDURE:

    Operate the control valve. Check the discharge from the
manifold. If the unit is not working properly do the following.

A.   Check control valve for dirt or foreign material. If
      necessary, clean.

B.   Check the strainer screen. Remove and clean.

C.   Check trap for dirt and foreign material, clean if
      necessary.

D.   Reassemble and test unit.

TOOLS AND MATERIALS:
Two pipe wrenches, screw driver, wire brush, pressure bulb or
other means to operate pneumatic valve.

Figure 10.11 PM Guide Number 20—Humidifier. (*From Georgetown University.*)

GEORGETOWN UNIVERSITY PREVENTIVE MAINTENANCE(GUIDE NUMBER) 21

EQUIPMENT: Fans

MANUFACTURER: All

CATEGORY: Cleaning   FREQUENCY: Yearly

PREVENTIVE MAINTENANCE DESCRIPTION:

Corrosion control (Fans and Fan Housings).

SPECIAL INSTRUCTIONS:

OSHA Standard 1910.213 -Open lock and tag motor disconnect switch
before starting work.

PROCEDURE:

When the fan is equipped with access panels remove them before
starting inspection.

A.   Inspect the fan casing and blades for corrosion and any
      accumulation of dirt.

B.   Use a dry rag to remove dirt.

C.   Use Varsol to remove dirt that does not come off by wiping
      with a dry rag.

D.   Remove corrosion with a wire brush and paint the area with
      spray primer paint. Fan shaft corrosion should be removed
      with emery cloth and the shaft recoated with a stripable
      corrosion resistant coating.

TOOLS AND MATERIALS: Revision Date: 8/10/88
Rags, wire brush, drop light, primer paint, scrapper.

Figure 10.12 PM Guide Number 21—Fans. (*From Georgetown University.*)

GEORGETOWN UNIVERSITY PREVENTIVE MAINTENANCE(GUIDE NUMBER)   23

EQUIPMENT: Air Compressors

MANUFACTURER: Vilbiss Company/Quincy/Johnson/Champion/Penn

CATEGORY: Air Compressors   FREQUENCY: Monthly

PREVENTIVE MAINTENANCE DESCRIPTION:

Inspect air compressor, check oil, PSI switch, and wipe down
unit.

SPECIAL INSTRUCTIONS:

Any needed repairs should be reported for correction at a later
date. When oil level is found to be low, add oil the same day of
inspection.

PROCEDURE:

Inspect compressor unit for cleanliness:

1.   Visually inspect compressor unit for cleanliness: clean the
     fins on the cylinders, heads, intercoolers, and
     aftercoolers. Wipe down compressor, motor, and tank.

2.   Remove cover of pressure regulator switch and examine
     contract points for damage; examine interior of switch for
     cleanliness. Clean if needed. Replace and secure cover.

3.   Drain tank if unit is not equipped with an automatic drain
     on the lowest point of the tank. Blow down automatic control
     condensate traps when unit is equipped for the function.

4.   Operate safety valves to insure they relieve and reset
     without sticking.

5.   Make sure compressor is operating at the proper pressure
     setting(s) and has the proper oil level.

TOOLS AND MATERIALS: Revision Date: 11/07/88
Standard hand tools, rags.

Figure 10.13   PM Guide Number 23—Air compressors. (*From Georgetown University.*)

GEORGETOWN UNIVERSITY PREVENTIVE MAINTENANCE(GUIDE NUMBER)__28__

EQUIPMENT: Air Compressor, Heavy Duty

MANUFACTURER: Pennsylvania Pump & Compressor Company

CATEGORY: Air Compressors   FREQUENCY:
PREVENTIVE MAINTENANCE DESCRIPTION:

Change crank case oil(Pennsylvania Compressors).

SPECIAL INSTRUCTIONS:

Follow OSHA Standard 1910.213- Open lock and tag motor disconnect
switch before starting work.

PROCEDURE:

1. <u>Change crank case oil:</u>   (Gulf Harmony 53;   13 Quarts for air
compressors)

NOTE: If the oil in the frame oil level gauge darkens and takes
on a dirty appearance before scheduled replacement, it is to be
changed at that time.

    a.   Stop the compressor and open the motor disconnect
        switch.

    b.   Remove oil drain plug from end of drain line and drain
        old oil into waste container.

    c.   Remove Air Breather unit, clean, oil, and reassemble it.

    d.   Thoroughly clean inside of frame sump using kerosene or
        a solvent before refilling with new oil.

    e.   After the compressor is running, check the oil level to
        ensure that it is at the halfway mark of the sight gauge
        on the bed.

NOTE: Lubricating Oils(General):
        FRAME: The lubricant for the splash system should be a
        neutral mineral oil and inhibited against oxidation and
        corrosion, corresponding to the SAE 30 or 40 grade.
        Heavy duty oils are also satisfactory lubricants and
        generally will provide results somewhat superior to
        straight mineral oils.

Page 1 of 2

1

Figure 10.14 PM Guide Number 28—Air compressors, heavy-duty. (*From Georgetown University.*)

GEORGETOWN UNIVERSITY PREVENTIVE MAINTENANCE (GUIDE NUMBER)   28

PROCEDURE, continued-

CYLINDER LUBRICATING OILS (GENERAL) FOR AIR COMPRESSORS ONLY:

Air compressors, due to the moisture in the air, require the use of an oil compounded with 3 to 5% of acidless tallow that contains an oxidation and rust inhibitor and has a low carbon content.

A high grade, 100% distilled , solvent refined, straight mineral oil containing an oxidation, corrosion or rust inhibitor and of an SAE 30 or 40 weight will provide satisfactory compressor cylinder lubrication.   It should have good polar characteristics, good wetting ability, high film strength, good chemical stability, be resistant to sludging and should contain as low a carbon content as possible.

TOOLS AND MATERIALS: Revision date: 11/07/88
Standard hand tools, waste oil container, 13 quarts of oil.

Figure 10.14   (*Continued*) PM Guide Number 28—Air compressors, heavy-duty. (*From Georgetown University.*)

GEORGETOWN UNIVERSITY PREVENTIVE MAINTENANCE(GUIDE NUMBER)___32___

EQUIPMENT: Fan Coil Unit; Ceiling-Mounted; Sizes 1 through 7___

MANUFACTURER:_____

CATEGORY: Fan Coil Units_____ FREQUENCY:_____
PREVENTIVE MAINTENANCE DESCRIPTION:

Clean fan coil unit. Re-oil fan motor.

_____

SPECIAL INSTRUCTIONS:

Covers should be securely bolted to unit to prevent tampering by
tenants(Replace missing fasteners).

_____

PROCEDURE:

1. Clean fan coil unit:

    a.   Open front access panel and remove filter.

    b.   Brush(or vacuum) coils and fins.

    c.   Clean the drain pan and check drain line opening for
       debris.

    d.   Wipe the entire casing clean and inspect for rust.

    e.   Replace or clean the filter.

2. Re-oil fan motor:

    a.   Insert 10 to 15 drops of oil into each of the motor
       sleeve bearings through the oil extension lines.

    b.   Wipe motor fan blades and casing clean: manually rotate
       fan and feel for debris.

    c.   Energize motor and check for excessive vibration and
       noise.

    d.   Check wiring for loose or unprotected connections.
       Test operating controls.

_____

TOOLS AND MATERIALS: Oil can, hand tools, rags, brush or vacuum,
spare fasteners.

Figure 10.15   PM Guide Number 32—Fan coil unit, ceiling-mounted. (*From Georgetown University.*)

GEORGETOWN UNIVERSITY PREVENTIVE MAINTENANCE(GUIDE NUMBER)_33___

EQUIPMENT: Fan Coil Units_____

MANUFACTURER:_____

CATEGORY: FREQUENCY:
_____

PREVENTIVE MAINTENANCE DESCRIPTION:

Change or clean air filters, clean condensate drains and pans, clean fan and coils, oil fan (FCU's)

_____

SPECIAL INSTRUCTIONS:

Turn off power to unit before starting work.

_____

PROCEDURE:

A.   Remove access covers from unit, remove filter. Use a soft brush to clean coil fins and the fan blade(s). Vacuum inside of unit thoroughly.

B.   Clean out any scale or dirt accumulation in drip pan(s).

C.   If the unit has a trap on the condensate line add 4 ounces of drain cleaner to the trap. (Do not use acid type drain cleaners–use enzyme type cleaner such as Mr. Plumber)

D.   Add 6 drops of SAE 20 (ND) oil to each bearing on the fan motor.

E.   Start Fan Motor make sure unit is running smoothly. If there is excessive vibration or noise report it to supervisor so it can be scheduled for repair at a later date.

F.   reinstall clean filter (ensure that filter fits securely and will not fall out).

G.   Replace cover on fan coil unit. Covers should be securely bolted to unit to prevent tampering by tenants. (Replace any missing fasteners.)

_____

TOOLS AND MATERIALS: Revision Date: 10/27/88
Standard Hand Tools, Vacuum, Rags, Filters

Figure 10.16   PM Guide Number 33—Fan coil units. (*From Georgetown University.*)

GEORGETOWN UNIVERSITY PREVENTIVE MAINTENANCE (GUIDE NUMBER)   35

EQUIPMENT: Condensers

MANUFACTURER: All

CATEGORY: Air conditioning   FREQUENCY: Yearly

PREVENTIVE MAINTENANCE DESCRIPTION:

Package airconditioning unit spring check.

SPECIAL INSTRUCTIONS:

OSHA Standard 1910.213 -Open lock and tag motor disconnect switch before starting work.

PROCEDURE:

A. Remove panels. Thoroughly inspect and clean interior and exterior of machine.
B. Clean the condenser by hosing with water. An approved coil cleaner may also be used if necessary. **Do not use hot water or steam.**
C. Carefully inspect all electrical terminals for signs of looseness or overheating.
D. Lubricate motor and fan bearings.
E. Replace belts on condenser fan if so equipped.
F. Start unit and check the refrigerant level. Recharge if needed. Report leaks to supervisor.
G. Check the oil level on compressors equipped with sight glasses.
H. Tighten any fan and compressor motor mounting bolts if found to be loose.
I. Check for proper operation of relays, switches, and safety devices.
J. Check amperages of compressor and fan motors. Compare with nameplate date to see that they are within their proper values.

TOOLS AND MATERIALS: Revision Date: 11/07/88
AC Manifold set, freon, oil-can with SAE 30 ND oil, grease gun, garden hose, hand tools to remove covers.

Figure 10.17 PM Guide Number 35—Condensers. (*From Georgetown University.*)

GEORGETOWN UNIVERSITY PREVENTIVE MAINTENANCE(GUIDE NUMBER) _46_

EQUIPMENT: AIR HANDLER UNIT: SUPPLY (DEHUMIDIFIER)

MANUFACTURER:

CATEGORY: FREQUENCY:

PREVENTIVE MAINTENANCE DESCRIPTION:

1.   Clean and adjust spray type dehumidifier unit.

SPECIAL INSTRUCTIONS:

OSHA Standard 1910.213 -Open lock and tag motor disconnect switch before starting work.

PROCEDURE:

1.   Clean and adjust spray type dehumidifier unit.

  a.   Drain sump tank and scrub clean of any algae, chemical residue, or sediment. Remove and clean spray pump return-line strainer and check for condition.

  b.   Flush out tank thoroughly; close drain valves and fill tank.

  c.   Check operation and condition of float valve. Observe water level when float valve closes; it should be ½" below the tank overflow outlet. Adjust float, if necessary.

  d.   Check spray nozzles for proper functioning; remove and clean, if required.

  e.   Check spray pressure gauge for correct pressure of 10 psi. These units are to be operated continuously.

              263-SB09-023   (AHU-8)
              051-P1-023   (AHU-3)
              051-P1-022   (AHU-4)
              051-P1-024   (AHU-5)

TOOLS AND MATERIALS: Revision Date: 11/07/88
Hose, bucket, hand tools, wire brush, 18" and 24" pipe wrenches.

Figure 10.18 PM Guide Number 46—Air handler unit, supply (dehumidified). (*From Georgetown University.*)

GEORGETOWN UNIVERSITY PREVENTIVE MAINTENANCE (GUIDE NUMBER) _47_

EQUIPMENT: AHU's

MANUFACTURER: All

CATEGORY: Cleaning   FREQUENCY: As Needed

PREVENTIVE MAINTENANCE DESCRIPTION:

Replace throwaway filters

SPECIAL INSTRUCTIONS:

OSHA Standard 1910.213 -Open lock and tag motor disconnect switch before starting work.

PROCEDURE:

A. Turn off air handling unit and remove old filters.

B. Install new filters. Insure that there are no gaps between media.

C. Reinstall any spacers and restart unit.

Note: Two indicators used to determine the need for servicing are a 10% decrease in air flow or an increase in resistance of two to three times the initial resistance. See the table below for some typical velocities and air resistances of various filters.

Clean Filter Operating Data

Type of air cleaner	Nominal velocity through media (fpm)	Resistance through clean filter (in. wg)
Viscous Impingement		
throwaway (2 in.)	300	0.06 - 0.12
renewable (4 in.)	300	0.12 - 0.24
cleanable (2 in.)	300-500	0.04 - 0.12
(4 in.)	300	0.08 - 0.20
automatic self-cleaning	500	0.30 - 0.50

Pg 1 of 2

Figure 10.19 PM Guide Number 47—Air handler units (all). (*From Georgetown University.*)

GEORGETOWN UNIVERSITY PREVENTIVE MAINTENANCE (GUIDE NUMBER) __47__

<u>Clean Filter Operating Data</u> (cont.)

Type of air cleaner	Nominal velocity through media(fpm)	Resistance through clean filter(in. wg)

Dry media
  -cleanable and
    renewable (2 in.)
              (8 in.)
    high efficiency
    renewable
	60
	35
	5-20
		0.08 - 0.13
		0.10 - 0.12
		0.50 - 1.20

Electronic(ionizing)
    plate or cell
    automatic
	300-400
	400-500
		0.15 - 0.30
		0.20 - 0.32

Electronic charged media	35	0.03 - 0.12

TOOLS AND MATERIALS: Revision Date: 8/11/88
Drop light, replacement filters, trash bags, screw driver,
channel locks.

Pg 2 of 2

Figure 10.19 (*Continued*) PM Guide Number 47—Air handler units (all). (*From Georgetown University.*)

GEORGETOWN UNIVERSITY PREVENTIVE MAINTENANCE (GUIDE NUMBER) _51_

EQUIPMENT: Pump Coupling

MANUFACTURER:

CATEGORY: FREQUENCY:

PREVENTIVE MAINTENANCE DESCRIPTION:
1. Check flexible coupling for pump/motor alignment.

2. Check flexible coupling for condition.

SPECIAL INSTRUCTIONS:

PROCEDURE:

1. Check flexible coupling for pump/motor alignment:

   a. Open motor disconnect switch or starter / controller contacts.

   b. To check angular alignment, measure the distance between the coupling flanges at 90° intervals. If this spacing is unequal, this is angular misalignment.

   c. To check for parallel alignment, lay a 6" steel scale across the outside flanges of the coupler halves. If the scale does not lay flat on both flange edges, parallel misalignment exists.

2. Check flexible coupling for condition:

   a. Visually check flexible insert for damage or wear. If necessary, loosen coupling set-screws and slide coupler halves back on shaft; remove flexible part of coupling and inspect it for damage and excessive wear.

   b. Reinstall the flexible part then check coupling hubs for loose shaft keys and set-screws.

TOOLS AND MATERIALS: Revision Date: 11/09/88

Figure 10.20 PM Guide Number 51—Pump coupling. (*From Georgetown University.*)

GEORGETOWN UNIVERSITY PREVENTIVE MAINTENANCE(GUIDE NUMBER)_52__

EQUIPMENT: Centrifugal Pump (not integral with motor)_____

MANUFACTURER:_____

CATEGORY: FREQUENCY:
_____
PREVENTIVE MAINTENANCE DESCRIPTION:
1.   Check operating pump.

2.   Check NON-operating pump.

_____

SPECIAL INSTRUCTIONS:

_____

PROCEDURE:

1.   Check operating pump:
     a.   While pump is in operation, note performance, bearing
           temperature, pressure gauge and operation of any
           installed stuffing boxes. (compare readings against
           previous readings to determine need for gauge
           calibration or repair work.) Conduct static pressure
           check, when possible.

     b.   Check mechanical seal or gland seal leak-off. If gland
           leak-off is excessive and cannot be reduced to a stream
           about the size of a pencil lead, it is necessary that
           the box be repacked.

     c.   Stop and start pump, noting undue vibration, noise,
           pressure, and the action of the check valve.

     d.   Upon completion of test, adjust any gland leak-off to a
           stream about the size of a lead pencil.

     e.   Check the overflow drain hole for foreign matter; clean
           it out as necessary.

     f.   Inspect unit for corrosion or damaged paint work.

Note**-When stuffing box repacking is necessary, close the hand
suction and discharge valves, drain pump casings and note if
valves are holding properly.

Pg 1 of 2

Figure 10.21 PM Guide Number 52—Centrifugal pump (not integral with motor).
(*From Georgetown University.*)

GEORGETOWN UNIVERSITY PREVENTIVE MAINTENANCE (GUIDE NUMBER) __55__

EQUIPMENT: Sump Pump

MANUFACTURER:

CATEGORY: FREQUENCY:

PREVENTIVE MAINTENANCE DESCRIPTION:

1.   Relubricate intermediate and casing sleeve bearings.

2.   Relubricate float-rod felt washer.

SPECIAL INSTRUCTIONS:

PROCEDURE:

1.   <u>Relubricate intermediate and casing sleeve bearings</u>:

   a. Clean off the zerk fittings on the floor plate and attach
      hand grease gun.

   b. Insert Gulfcrown Special grease until a "back pressure"
      is felt, which indicates a filled bearing cavity.

2.   <u>Relubricate float-rod felt washer:</u>

   a. Saturate the felt washer by applying Gulf Harmony 69 oil
      to the float rod and allowing it to drain into the guide
      pipe in the basin cover.

TOOLS AND MATERIALS: Revision Date: 10/20/88

Figure 10.21   (*Continued*) PM Guide Number 52—Centrifugal pump (not integral with
motor). (*From Georgetown University.*)

GEORGETOWN UNIVERSITY PREVENTIVE MAINTENANCE(GUIDE NUMBER)__55____

EQUIPMENT: Sump Pump_____

MANUFACTURER:_____

CATEGORY: FREQUENCY:

PREVENTIVE MAINTENANCE DESCRIPTION:

1.   Relubricate intermediate and casing sleeve bearings.

2.   Relubricate float-rod felt washer.

SPECIAL INSTRUCTIONS:

PROCEDURE:

1.   Relubricate intermediate and casing sleeve bearings:

    a. Clean off the zerk fittings on the floor plate and attach
    hand grease gun.

    b. Insert Gulfcrown Special grease until a "back pressure"
    is felt, which indicates a filled bearing cavity.

2.   Relubricate float-rod felt washer:

    a. Saturate the felt washer by applying Gulf Harmony 69 oil
    to the float rod and allowing it to drain into the guide
    pipe in the basin cover.

TOOLS AND MATERIALS: Revision Date: 10/20/88

Figure 10.22   PM Guide Numbers 55—Sump pump. (*From Georgetown University.*)

GEORGETOWN UNIVERSITY PREVENTIVE MAINTENANCE(GUIDE NUMBER)_56___

EQUIPMENT: Sump Pump

MANUFACTURER:

CATEGORY: FREQUENCY:

PREVENTIVE MAINTENANCE DESCRIPTION:

1.   Relubricate thrust ball bearing.
2.   Check flexible coupling.
3.   Test high water alarm.

SPECIAL INSTRUCTIONS:

PROCEDURE:

1.   Relubricate Pump Thrust Bearings.

      a. Wipe lubrication fitting clean; inject grease until a
back pressure can be felt. Wipe up any excess grease.

2.   Check Flexible Coupling.

      a. Inspect flexible coupling for wear, damage and alignment.

3.   Test High Water Alarm.(test all sumps in the same space
simultaneously)
      a. Open pump motor disconnect switch and allow pump basin to
fill with liquid.
      b. When high water alarm sounds, close disconnect switch and
observe pump and float switch for normal operation.
      c. When no high water alarm exists, lift the float switch
arm manually and observe switch for normal operation. Note: Do
not drop the float switch arm but lower it gently until motor is
de-energized.
      d. Observe the motor to see if it comes quickly up to speed
and maintains a constant rotation rate.
      e. Be alert for unusual or excessive noise. Note that Weil
F2 type pumps have a coarse strainer, be alert for symptoms of
partial clogging.

TOOLS AND MATERIALS: Revision Date: 11/02/88

Figure 10.23 PM Guide Number 56—Sump pump. (*From Georgetown University.*)

GEORGETOWN UNIVERSITY PREVENTIVE MAINTENANCE (GUIDE NUMBER)__58___

EQUIPMENT: Condensate Pump: type HS, Horizontal Centrifugal, V-S

MANUFACTURER:

CATEGORY: FREQUENCY:

PREVENTIVE MAINTENANCE DESCRIPTION:

1. Regrease pump ball bearings.
2. Check float switch and control assembly.

SPECIAL INSTRUCTIONS:

PROCEDURE:

1. Re-grease pump ball bearings.

     a. With the pump stopped, clean off the bearing housing and seal.
     b. Start pump and inject Gulf Crown Grease EP Special, or Equal, into the lubrication fitting until a slight bead of grease appears at the bearing seal. Caution: do not overgrease.
     c. Check metallic packing for proper leak-off.

2. Check float switch and control assembly:

     a. Observe that the float switch opens and closes properly as receiver fills and is emptied by the pump. Check float control linkage for binding.

TOOLS AND MATERIALS: Revision Date: 10/20/88

Figure 10.24   PM Guide Number 58—Condensate pump, type HS, horizontal centrifugal, V-S. (*From Georgetown University.*)

GEORGETOWN UNIVERSITY PREVENTIVE MAINTENANCE(GUIDE NUMBER) _61___

EQUIPMENT: Condensate pump_____

MANUFACTURER: _____

CATEGORY: FREQUENCY: Yearly

PREVENTIVE MAINTENANCE DESCRIPTION:

1.   Drain and flush condensate receiver. Inspect internal parts
of check valve test pump and float switches.

SPECIAL INSTRUCTIONS:
OSHA Standard 1910.213 -Open lock and tag motor disconnect switch
and starter before starting work. Check the voltage of all
incoming line terminals. Positively ascertain that the equipment
is deenergized.

PROCEDURE:

1.   a. Close valves in discharge and return lines; open drain
      and empty receiver, then flush receiver to remove sediment
      and scale.
      b. Check temperature of condensate (it should be
      approximately 30° below steam temperature if traps are not
      leaking).
      c. Open check valve and inspect valve disk and seat for
      cuts, pitting, sediment of scale deposits. Inspect valve
      stem and pin for wear. As an alternate: listen to valve
      operate.
      d. Turn shaft of pumps and see that they rotate freely by
      hand.
      e. Remove float switch cover and examine interior for
      damage, moisture, oil, and dirt. Push down on float switch
      arm to check that it operates pump.
      f. When starting up, open valves in discharge and return
      lines, close drain.
      g. For Reiss Science: Remove strainer clean-out cover from
      side of receiver; remove strainer and clean it.
      h. Check motor bearings for unusual noise.

TOOLS AND MATERIALS: Revision Date: 11/02/88

Figure 10.25 PM Guide Number 61—Condensate pump. (*From Georgetown University.*)

GEORGETOWN UNIVERSITY PREVENTIVE MAINTENANCE (GUIDE NUMBER) __63__

EQUIPMENT: Pump Ball Bearings

MANUFACTURER:

CATEGORY: FREQUENCY: Yearly

PREVENTIVE MAINTENANCE DESCRIPTION:

1.   Add grease to pump bearings.

SPECIAL INSTRUCTIONS:

Excess grease is the most common cause of overheating and bearing damage. Avoid adding too much grease.

PROCEDURE:

1.   Add grease to pump bearings.

a. Clean off the lubricating fittings and inject approximately 1/2 ounce of grease (about a teaspoonful for bearings of small size, and a tablespoonful for larger sizes). The bearing housing is intended to only be 1/4 to 1/3 full of grease.

b. Wipe all visible grease from fittings and bearing housing.

TOOLS AND MATERIALS: Revision Date: 11/02/88
Grease gun, No 2 grease, Rags, Hand tools.

Figure 10.26   PM Guide Number 63—Pump ball bearings. (*From Georgetown University.*)

GEORGETOWN UNIVERSITY PREVENTIVE MAINTENANCE(GUIDE NUMBER) 69

EQUIPMENT: Valves (excluding sprinkler systems)

MANUFACTURER: All

CATEGORY: Lubrication   FREQUENCY: Yearly

PREVENTIVE MAINTENANCE DESCRIPTION:
Inspect and service valve packing and lube stem.

SPECIAL INSTRUCTIONS:
After servicing leave the valve in the position it was found in.

PROCEDURE:

A.   Exercise valve from one limit to the other (open and closed). Lightly lubricate stem with graphite.

B.   Adjust packing gland to stop any leakage.

C.   Any leaks that can not be repaired during this PM whether at the packing or elsewhere is to be noted on the work order and also reported to the shop foreman.

TOOLS AND MATERIALS: Revision Date: 11/09/88
Graphite, Flashlight, 9 ft. ladder.

Figure 10.27   PM Guide Number 66—Radiators, heating. (*From Georgetown University.*)

GEORGETOWN UNIVERSITY PREVENTIVE MAINTENANCE(GUIDE NUMBER) 69

EQUIPMENT: Valves (excluding sprinkler systems)

MANUFACTURER: All

CATEGORY: Lubrication   FREQUENCY: Yearly

PREVENTIVE MAINTENANCE DESCRIPTION:

Inspect and service valve packing and lube stem.

SPECIAL INSTRUCTIONS:

After servicing leave the valve in the position it was found in.

PROCEDURE:

A.   Exercise valve from one limit to the other (open and closed). Lightly lubricate stem with graphite.

B.   Adjust packing gland to stop any leakage.

C.   Any leaks that can not be repaired during this PM whether at the packing or elsewhere is to be noted on the work order and also reported to the shop foreman.

TOOLS AND MATERIALS: **Revision Date:** 11/09/88
Graphite, Flashlight, 9 ft. ladder.

Figure 10.28 PM Guide Number 69—Valves (excluding sprinkler systems). (*From Georgetown University.*)

GEORGETOWN UNIVERSITY PREVENTIVE MAINTENANCE(GUIDE NUMBER)_72___

EQUIPMENT: Belt Drive Alignment_____

MANUFACTURER: Trane_____

CATEGORY: Power Transmission   FREQUENCY: Yearly

PREVENTIVE MAINTENANCE DESCRIPTION:

Belt Drive Alignment Check

SPECIAL INSTRUCTIONS:

OSHA Standard 1910.213 -Open lock and tag motor disconnect switch
before starting work.

PROCEDURE:

A. Sheave Alignment:

Alignment of the belt sheaves is crucial to V-Belt life
and should be checked periodically. Align the fan and motor
sheaves by using a straightedge that is long enough to span
the distance between the outside edges of the sheaves. When
sheaves are aligned the straightedge will touch both sheaves
squarely across their face. For uneven width sheaves a
string drawn tight through the center groove of both sheaves
with an equal measurement made from each end of the string
to an outboard parallel point will also give a good
indication of proper alignment.

When alignment needs to be changed during this PM care
should be taken to ensure that set screws are retorqued to
their recommended values.

Torques for Tightening Set Screws

Set Screw Diameter	Hex Size Across Flats	Recommended Torque Inch Pounds	Foot Pounds
1/4	1/8	66	5.5
5/16	5/32	126	10.5
3/8	3/16	228	19.0
7/16	7/32	348	29.0
1/2	1/4	504	42.0
5/8	5/16	1104	92.0

TOOLS AND MATERIALS:
Straightedge, String, 12 Foot Ruler, Allen Hex Set, Torque Wrench.

Figure 10.29 PM Guide Number 72—Belt drive alignment. (*From Georgetown University.*)

GEORGETOWN UNIVERSITY PREVENTIVE MAINTENANCE(GUIDE NUMBER) __74__

EQUIPMENT: Dampers

MANUFACTURER: All

CATEGORY: Lubrication   FREQUENCY: Yearly

PREVENTIVE MAINTENANCE DESCRIPTION:

Dampers, Linkage, and Actuators care and maintenance

SPECIAL INSTRUCTIONS:

OSHA Standard 1910.213 -Open lock and tag motor disconnect switch before starting work.

PROCEDURE:

A. Dampers and their associated drive components should be inspected periodically for freedom of movement. Also check bolts, clip screws, and locknuts for tightness. Dampers with plastic sleeve type bearings do not require lubrication and should only be cleaned with a dry cloth as conditions dictate. Other types of bushings may be lubricated lightly. Remove any excess lubricant to prevent dirt collecting and binding damper.

B. Check outside air inlet screens during this PM. Clean any debris from screen.

TOOLS AND MATERIALS: Rags for cleaning, Oil Can (SAE 20 [ND] Oil), Hand Tools.

Figure 10.30   PM Guide Number 74—Dampers. (*From Georgetown University.*)

GEORGETOWN UNIVERSITY PREVENTIVE MAINTENANCE(GUIDE NUMBER) 79

EQUIPMENT: Hot Water Heater (heat exchanger)

MANUFACTURER:

CATEGORY: FREQUENCY:

PREVENTIVE MAINTENANCE DESCRIPTION:

1.   Check the temperature of the exit water of the heater.

SPECIAL INSTRUCTIONS:

PROCEDURE:

1.   Check the temperature of the exit water of heater:

        a. If temperature of exit water is 120°F, or less, then
schedule the below descaling procedure for accomplishment: the
rapid descaling method consists of the following procedure.

1.   Close valves in the inlet and outlet water lines. Also drain
        the shell through the blowdown valve while opening the
        shell to atmospheric pressure through the relief valve
        connection.
2.   Close High Temperature return line by means of hand valve.
3.   Inject a small flow of cold water through the open relief
        valve connection, allowing the drain connection to remain
        open at the bottom of the heater.
4.   Shut off cold water flow through the open relief valve
        connection and open high temperature water flow into coils.
        After two minutes interval, close high temperature water
        flow into coils and open cold water flow through open relief
        valve connection. Repeat this procedure for several minutes
        in approximately two minute intervals.
5.   Open the main line of water supply into the heater and allow
        a complete flushing action of the solid particles dislodged
        from the coils to be discharged through the blowdown
        connection.
6.   Restore the heater into service by opening those valves
        which had been closed and closing the relief valve and
        blowdown connection.

Figure 10.31 PM Guide Number 79—Hot water heater (heat exchangers). (*From Georgetown University.*)

GEORGETOWN UNIVERSITY PREVENTIVE MAINTENANCE(GUIDE NUMBER)__79__
-continued

      This method of descaling consists essentially of thermal shocking the coiled heating surfaces by alternately heating and cooling. This causes them to expand and contract freely because of the design of the coil structure and the method of suspension. No damage is done to the internal parts of the heater during this procedure because the heater is designed to permit this continual sudden expansion and contraction.

      It has been proven by experience, that scale formation is very rapid when:

1.   The quantity of water is large (being passed through the heater).

2.   The temperature of the water is elevated beyond 150°F;

3.   High pressures are used in the coils (which is the severe condition in hard water areas); a treatment of a few minutes duration as explained above, has successfully restored heaters to over 90% of their design capacities.

TOOLS AND MATERIALS: Revision Date: 10/21/88

Pg 2 of 2

Figure 10.31 *(Continued)* PM Guide Number 79—Hot water heater (heat exchangers). *(From Georgetown University.)*

GEORGETOWN UNIVERSITY PREVENTIVE MAINTENANCE (GUIDE NUMBER) _80_

EQUIPMENT: Steam Converters

MANUFACTURER:

CATEGORY: FREQUENCY:

PREVENTIVE MAINTENANCE DESCRIPTION:
1.   Inspect converter for leakage.
2.   Clean steam strainer.
3.   Check pressure relief valve.

SPECIAL INSTRUCTIONS:

PROCEDURE:

1.   Inspect converter for leakage.

   a. Inspect external areas and piping connections for leakage.

   b. Check for evidence of internal leakage between tubes and shell: Water leaking into steam system causes <u>continuous</u> flow of water through the steam trap.

   c. When exit water temperature drops to about 150°F, schedule converter for chemical cleaning of the water sides <u>only</u>.

   d. Inspect gauges and thermometers for leaks, broken glass, damaged thermometer bodies, inaccurate readings or faulty operation due to defective parts.

2.   Clean Steam strainer:

   a. remove basket from strainer and clean.

   b. Check operation of relief valve by lifting lever and releasing it. Observe valve for positive seating.

TOOLS AND MATERIALS: Revision Date: 10/21/88

Figure 10.32   PM Guide Number 80—Steam converters. (*From Georgetown University.*)

GEORGETOWN UNIVERSITY PREVENTIVE MAINTENANCE(GUIDE NUMBER) _81_

EQUIPMENT: Domestic Hot Water Heater

MANUFACTURER:

CATEGORY: FREQUENCY:

PREVENTIVE MAINTENANCE DESCRIPTION:

1.   Flush and inspect Hot Water Heater Tank.
2.   Clean Hot Water Tank.

SPECIAL INSTRUCTIONS:

PROCEDURE:

1.   Flush and inspect Hot Water Heater Tank:

     a. Actuate lever of pressure relief valve to test for operation.

     b. Close steam and water valves, release pressure on water side of unit and drain water out of vessel.

     c. Remove manhole cover, flush tank out and inspect the heating surface for fouling and buildup of scale or mineral deposits; mechanically clean heating surfaces or schedule chemical cleaning, if required.

     d. Inspect internal shell for resting or pitting.

     e. Remove anode (if installed) and clean by flaking off decomposed layer with light metal object. Schedule anode for replacement if more than 50% has decomposed.

     f. Clean steam strainer.

     g. Replace manhole gasket; reinstall manhole cover, refill tank and check for leaks.

2.   Shock Procedure:

     -Subject unit to a shock treatment as prescribed by the foreman.

TOOLS AND MATERIALS: Revision Date: 10/21/88

Figure 10.33 PM Guide Number 81—Domestic hot water heater. (*From Georgetown University.*)

GEORGETOWN UNIVERSITY PREVENTIVE MAINTENANCE (GUIDE NUMBER) _82_

EQUIPMENT: Steam Safety Relief Valves

MANUFACTURER: All

CATEGORY: Life Safety   FREQUENCY: Semi-annual

PREVENTIVE MAINTENANCE DESCRIPTION:

Test and inspect steam pressure relief valve.

SPECIAL INSTRUCTIONS:

PROCEDURE:

Lift the operating lever on the safety valve. Check for the following:

A.   That the valve opens and resets properly.

B.   The valve closes tightly without chattering.

C.   That the valve is tightly closed and there is no leakage of steam.

D.   Examine the valve visually. See that there has been no tampering.

E.   Check the atmospheric discharge piping, see that there is proper drainage.

Note-   Atmospheric pipe drainage prevents a column of water from building up over the safety relief valve disk and thus increasing the popping pressure. Water can also produce an explosive-like slug of water being fired out of the valve.

TOOLS AND MATERIALS: Revision Date: 8/15/88
Flashlight, ladder(for certain locations).

Figure 10.34 PM Guide Number 82—Steam safety relief valves. (*From Georgetown University.*)

GEORGETOWN UNIVERSITY PREVENTIVE MAINTENANCE(GUIDE NUMBER) 89

EQUIPMENT: Cooling tower

MANUFACTURER:

CATEGORY: FREQUENCY:

PREVENTIVE MAINTENANCE DESCRIPTION:

1.   Add lubricating oil to fan bearing.
2.   Check operating components.

SPECIAL INSTRUCTIONS:

PROCEDURE:

1.   Add lubricating oil to fan bearing:

  a. Basic Science; Stop fan motor; fill oil cups of fan bearings
      with Gulf Harmony 97 oil, or equal.

2.   Check operating components:

  a. Check for correct water level in basin while pump is
      operating.

  b. Check pump suction strainer-screen for cleanliness.

  c. Check spray nozzle operation (if applicable).

  d. Check belt condition.

TOOLS AND MATERIALS: Revision Date: 10/27/88

Figure 10.35 PM Guide Number 84—Air compressor, heavy-duty, horizontal, reciprocating, 9-in stroke. (*From Georgetown University.*)

GEORGETOWN UNIVERSITY PREVENTIVE MAINTENANCE(GUIDE NUMBER)_89___

EQUIPMENT: Cooling tower_____

MANUFACTURER:_____

CATEGORY: FREQUENCY:
_____

PREVENTIVE MAINTENANCE DESCRIPTION:

1.   Add lubricating oil to fan bearing.
2.   Check operating components.

_____

SPECIAL INSTRUCTIONS:

_____

PROCEDURE:

1.   Add lubricating oil to fan bearing:

  a. Basic Science; Stop fan motor; fill oil cups of fan bearings
      with Gulf Harmony 97 oil, or equal.

2.   Check operating components:

  a. Check for correct water level in basin while pump is
      operating.

  b. Check pump suction strainer-screen for cleanliness.

  c. Check spray nozzle operation (if applicable).

  d. Check belt condition.

_____

TOOLS AND MATERIALS: Revision Date: 10/27/88

Figure 10.36 PM Guide Numbers 89—Cooling tower. (*From Georgetown University.*)

GEORGETOWN UNIVERSITY PREVENTIVE MAINTENANCE(GUIDE NUMBER)_93___

EQUIPMENT: Fire Extinguishers_____

MANUFACTURER:_____

CATEGORY: FREQUENCY:
_____
PREVENTIVE MAINTENANCE DESCRIPTION:

1.   Conduct monthly fire extinguisher inspection.
_____

SPECIAL INSTRUCTIONS:

_____

PROCEDURE:
1.   Conduct monthly fire extinguisher inspection:
  a. Use the inventory list of fire extinguisher locations in the
      designated building and conduct this inspection on each
      extinguisher located at the stations.
  b. Check each extinguisher location to ensure that the correct
      type of fire extinguisher is at the station. (Extinguisher
      type is indicated on the Location List).
  c. Inspect extinguisher to ensure that it has not been used or
      tampered with: water and air- check air pressure; soda-acid
      check nozzle for chemical residue; Co2-check wire seal to
      ensure that it is not broken.
  d. Inspect for any obvious physical damage, corrosion or hose
      impairment.
  e. Check cabinets for accessibility, operable handle, broken
      glass, and trash.
Note- The tag attached to the extinguisher is not to be
initialed for a monthly "inspection". This tag is for recording
the yearly "maintenance action" or a "recharge date" and is to be
initialed at those times.
      NFPA No 10A -1970 Paragraph 1200 is quoted to further define
an "Inspection":
      "An inspection is a 'quick check' that an extinguisher
      is available and will operate. It is intended to give
      reasonable assurance that the extinguisher is fully
      charged and operable. This is done by seeing that it is
      in its designated place, that it has not been actuated
      or tampered with, and that there is no obvious physical
      damage or condition to prevent operation."
-Paragraph 1230: "The whole intent of an inspection is to find
out quickly if something is wrong so that proper corrective
action can be taken."
_____
TOOLS AND MATERIALS: Revision Date:10/28/88

Figure 10.37   PM Guide Number 93—Fire extinguishers. (*From Georgetown University.*)

GEORGETOWN UNIVERSITY PREVENTIVE MAINTENANCE (GUIDE NUMBER) 94

EQUIPMENT: Fire Extinguisher

MANUFACTURER:

CATEGORY: FREQUENCY:

PREVENTIVE MAINTENANCE DESCRIPTION:
1.   Conduct annual fire extinguisher maintenance.

2.   Verify fire extinguisher serial number and hydrostatic test
     equipment.

SPECIAL INSTRUCTIONS:

PROCEDURE:

1.   Conduct monthly fire extinguisher maintenance:

a.   Check for physical damage, dents, corrosion, defective
     nozzle/horn or hose.
b.   Check to ensure that the lettering on instructions is easy
     to read.
c.   Check extinguisher hanger for firm mounting, or that
     extinguisher cabinet door is operable.
d.   Sign the attached inspection tag upon completion of
     maintenance.

-FOR WATER AND AIR EXTINGUISHERS:

a.   Check gauge reading to see that it is in proper range.

-FOR Co2 EXTINGUISHERS:

a.   Check wire seal to ensure that it is not broken.
b.   Weigh extinguisher and compare to amount stamped on bottle.

-FOR SODA AND ACID EXTINGUISHER:

a.   Remove, disassemble, and recharge according to
     manufacturer's instructions.

Caution- Do not invert (pressurize) any extinguisher that shows
signs of mechanical damage or corrosion.

Pg 1 of 2

Figure 10.38 PM Guide Number 94—Fire extinguishers. (*From Georgetown University.*)

GEORGETOWN UNIVERSITY PREVENTIVE MAINTENANCE(GUIDE NUMBER)_94___
-continued

2.   <u>Verify fire extinguisher serial number and hydrostatic test
     equipment</u>:

a.   Verify the serial number on the PM Card against the
     extinguisher serial number; if they are different,fill out a
     "Change Form" for computer updating of correct number.

b.   Check the last hydrostatic test date; if it is more than
     four years ago, schedule the extinguisher for a retest to
     coincide with the 5th year date.

c.   Do not hydrostatic test Soda Acid extinguishers. At time of
     five year testing cycle, replace the extinguishers with
     stored pressure type.

TOOLS AND MATERIALS: Revision Date: 10/27/88

Figure 10.38   *(Continued)* PM Guide Number 94—Fire extinguishers. (*From
Georgetown University.*)

GEORGETOWN UNIVERSITY PREVENTIVE MAINTENANCE(GUIDE NUMBER)___96___

EQUIPMENT: Sprinkler Valves Wet System

MANUFACTURER: OS+Y VALVES, MILWAUKEE

CATEGORY: Life Safety   FREQUENCY:Monthly
PREVENTIVE MAINTENANCE DESCRIPTION:

Inspection of Fire Suppression System (valves and water pressure)monthly.

SPECIAL INSTRUCTIONS:
In the event that the system valve is found unlocked in the open or closed position, report the condition immediately to the plumbing shop foreman(after work hours report it to the Security Department). Note location of the valve and the name of the person to whom the report was given on the PM work order.

PROCEDURE:

A.   Check each valve listed on the PM work order.
     Determine if the valve is fully open and locked.

B.   If in doubt about the valve's condition,
     physically manipulate the valve to be sure it is open.

C.   Check the gauges on the wet-pipe sprinkler system to ensure that normal water supply pressure is being maintained. A pressure reading on the gauge on the system side of an alarm valve in excess of the pressure recorded on the gauge on the supply side of the valve is normal, as the highest pressure from the supply side will get trapped in the system.
     On systems without booster pumps, an equal gauge reading could indicate a leak. If there are no visible leaks, it is possible there has been recent test run on the sprinkler system.

TOOLS AND MATERIALS: Revision Date 8/4/88
Flashlight.

Figure 10.39   PM Guide Number 96—Sprinkler valves—wet system. (*From Georgetown University.*)

GEORGETOWN UNIVERSITY PREVENTIVE MAINTENANCE(GUIDE NUMBER) 97

EQUIPMENT: Sprinkler Valves Dry System

MANUFACTURER: OS+Y VALVES, MILWAUKEE

CATEGORY: Life Safety   FREQUENCY:Monthly
PREVENTIVE MAINTENANCE DESCRIPTION:

Inspection of Fire Suppression System(dry pipe valves, air pressure, freeze protection) monthly.

SPECIAL INSTRUCTIONS:
In the event of that the system is found unlocked in the open or closed position, or the pressure gauges indicate the system is flooded with water, report the condition immediately to the plumbing shop foreman(after work hours report it to the Security Department). Note location of the valve and the name of the person to whom the report was given on the PM work order.

PROCEDURE:

A.   Be sure that all system/control valves are in the open position and are locked.

B.   If in doubt about the valve's condition, physically manipulate the valve to be sure it is open.

C.   Check air and water pressure gauges to be certain that the required air pressure is being applied to the system.

D.   During the winter months, check the sprinkler valve room heater, during each valve inspection to make sure it is operating and providing sufficient heat to prevent freezing.

Air Pressure Chart for Clapper Valves with a
Six to One Operating Differential

Water Pressure Maximum	Air Pressure Minimum	Air Pressure Maximum
50	15	25
75	25	35
100	35	45
125	40	50

TOOLS AND MATERIALS: Revision Date: 8/4/88
Flashlight.

Figure 10.40 PM Guide Number 97—Sprinkler valves—dry system. (*From Georgetown University.*)

GEORGETOWN UNIVERSITY PREVENTIVE MAINTENANCE(GUIDE NUMBER) 108

EQUIPMENT: Sprinkler Systems

MANUFACTURER: All

CATEGORY: Safety   FREQUENCY: Quarterly

PREVENTIVE MAINTENANCE DESCRIPTION:

Sprinkler system main drain flow test.

SPECIAL INSTRUCTIONS:

In the event that the test should prove to be unsatisfactory, report the condition immediately to the plumbing shop foreman. Note location of the test on the PM work order.

PROCEDURE:

A. Open the 2-inch(50-mm) drain valve, watch the supply water pressure gauge. The pressure reading will drop and stabilize. Now close the valve. The pressure reading should return to its former pressure reading quickly.

B. If the pressure reading returns slowly or not at all, this indicates there is a valve closed or some other type of obstruction is present.

C. Observe the discharged water. Watch for any heavy discoloration of the water, or obstructive material flowing from the water outlet. Discoloration or foreign material show a need for investigation to ensure there is a clear unobstructed water supply in an emergency.

TOOLS AND MATERIALS: Revision Date: 11/10/88
Flashlight, channel locks.

Figure 10.41 PM Guide Number 100—Panelboards rated at 600 V or less. (*From Georgetown University.*)

GEORGETOWN UNIVERSITY PREVENTIVE MAINTENANCE (GUIDE NUMBER) __108__

EQUIPMENT: Sprinkler Systems

MANUFACTURER: All

CATEGORY: Safety   FREQUENCY: Quarterly

PREVENTIVE MAINTENANCE DESCRIPTION:

Sprinkler system main drain flow test.

SPECIAL INSTRUCTIONS:

In the event that the test should prove to be unsatisfactory, report the condition immediately to the plumbing shop foreman. Note location of the test on the PM work order.

PROCEDURE:

A.   Open the 2-inch(50-mm) drain valve, watch the supply water pressure gauge. The pressure reading will drop and stabilize. Now close the valve. The pressure reading should return to its former pressure reading quickly.

B.   If the pressure reading returns slowly or not at all, this indicates there is a valve closed or some other type of obstruction is present.

C.   Observe the discharged water. Watch for any heavy discoloration of the water, or obstructive material flowing from the water outlet. Discoloration or foreign material show a need for investigation to ensure there is a clear unobstructed water supply in an emergency.

TOOLS AND MATERIALS: Revision Date: 11/10/88
Flashlight, channel locks.

Figure 10.42   PM Guide Number 108—Sprinkler system. (*From Georgetown University.*)

GEORGETOWN UNIVERSITY PREVENTIVE MAINTENANCE (GUIDE NUMBER) __109__

EQUIPMENT: Fire Pumps

MANUFACTURER: Firetrol

CATEGORY: Safety   FREQUENCY:
PREVENTIVE MAINTENANCE DESCRIPTION:

Fire pump inspection.

SPECIAL INSTRUCTIONS:

Note all deficiencies on PM work order.

PROCEDURE:

A.   Check city water pressure and compare it to the system
      pressure. The system pressure should read 20 PSI higher
      than the city pressure.

B.   Check the pump controllers. Ensure that all switches are in
      the energized position and set for automatic operation.

C.   Report any indicator lights that are not operating to Low
      Voltage Shop foreman for replacement.

D.   Perform (B) and (C) on the booster pump that serves the fire
      system.

TOOLS AND MATERIALS: Revision Date: 6/29/88
Flashlight, screwdriver.

Figure 10.43 PM Guide Number 109—Fire pumps. (*From Georgetown University.*)

GEORGETOWN UNIVERSITY PREVENTIVE MAINTENANCE(GUIDE NUMBER)__111__

EQUIPMENT: Air Compressor (centrifuge) Low Pressure_____

MANUFACTURER:_____

CATEGORY: FREQUENCY:

PREVENTIVE MAINTENANCE DESCRIPTION:

1.   Conduct annual maintenance check on LP centrifugal air compressor.

SPECIAL INSTRUCTIONS:

PROCEDURE:

1.   Lift pressure relief valve of air receiver tank and hold open. Observe pressure gauge to see of pump cuts in at the prescribed pressure and then pumps up tank under usage of air. Check cut-out pressure; release valve and check for proper seating. Check relief valve of separator by hand for proper operation.

      Reiss Science: Lead cut-in at 9 psi; Lead cut-out 18 psi
                     Lag cut-in 8 psi; Lag cut-out 18 psi
      White Gravenor:
Check with operating personnel to ensure that pumps are alternated Lead-Lag monthly.

2.   If pump capacity drops off during the above pump-up test, check for proper amount of water seal and that the compressor suction is unobstructed. (Seal water should be supplied in sufficient quantity so that the unit runs slightly warm to the touch - never hot.) If necessary, schedule compressor for adjustment of the clearance between the rotor and cones.

3.   Listen to both check valves at the pump to determine if air leakage exists. Check condition of motor bearings.

4.   Check line connections for air leaks.

5.   Remove float valve assembly from separator; inspect and clean it. Reinstall and check water level on gauge glass; it should be 1/3 of the glass with the compressor operating.

Page 1 of 2

Figure 10.44 PM Guide Number 111—Air compressor (centrifugal), low-pressure. (*From Georgetown University.*)

GEORGETOWN UNIVERSITY PREVENTIVE MAINTENANCE (GUIDE NUMBER)__111__
-continued

6.   Clean the strainers in seal water line of each pump.

7.   Check packing gland to ensure that leak off is only a slight
      drip. If necessary, replace packing seven rings of graphite
      impregnated packing, 5/16". Ensure that leak-off drain line
      is unobstructed.

TOOLS AND MATERIALS: Revision Date: 11/3/88

Figure 10.44   (*Continued*) PM Guide Number 111—Air compressor (centrifugal), low-
pressure. (*From Georgetown University.*)

GEORGETOWN UNIVERSITY PREVENTIVE MAINTENANCE(GUIDE NUMBER) _112_

EQUIPMENT: Vacuum-Heating / Condensate Pump   (St. Mary's)

MANUFACTURER:

CATEGORY: FREQUENCY:

PREVENTIVE MAINTENANCE DESCRIPTION:

1.   Conduct annual maintenance of Vacuum Heating Pump.

SPECIAL INSTRUCTIONS:

PROCEDURE:

1.   Check contacts of float switches and ensure that switches
      operate condensate pumps. Check "lead" and "lag" operation
      of switches. Check motor bearings.

2.   Check to see if vacuum pump is handling air properly: hold
      the vacuum relief valve in by hand, or remove the pipe plug
      in the vacuum control line; air should be sucked in by the
      pump. Ensure vacuum relief valve seats properly.

3.   Check separator float valve and vacuum pump seal water
      orifice for condition by observing drain line opening. When
      pump is not running, nothing should come out. When pump is
      running, only air, with no water, should be coming out.

4.   Remove strainer flange and remove strainer for inspection;
      clean, if necessary. If stainer contains sludge that cannot
      be removed otherwise, use solvent for removal; do not damage
      strainer by hammering in order to clean it.

5.   See that the air discharge is open to the atmosphere and
      that the check valve in the suction line to each vacuum pump
      is working properly. Access to the check valve is obtained
      by removing the check valve cover on the side of the tank
      adjacent to vacuum pump.

6.   Lube motors.

TOOLS AND MATERIALS: Revision Date: 11/3/88

Figure 10.45 PM Guide Number 112—Vacuum hearting (condensate pump). (*From
Georgetown University.*)

GEORGETOWN UNIVERSITY PREVENTIVE MAINTENANCE(GUIDE NUMBER) 113

EQUIPMENT: Air Compressor (Centrifugal) High Pressure

MANUFACTURER:

CATEGORY: FREQUENCY:

PREVENTIVE MAINTENANCE DESCRIPTION:

1.   Conduct annual maintenance check on HP Nash air compressor.

SPECIAL INSTRUCTIONS:

PROCEDURE:

1.   Lift pressure relief valve of air receiver tank and hold
open. Observe pressure gauge to see of pump cuts in at the
prescribed pressure and then pumps up tank under usage of air.
Check cut-out pressure; release valve and check for proper
seating. Check relief valve of separator by hand for proper
operation. Prescribed cut-in / cut-out pressures:
        Reiss Science: Lead cut-in at 65 psi; Lead cut-out 80 psi
                       Lag cut-in psi; Lag cut-out psi
When pump alternating capability exists, ensure monthly lead-lag
change.
2.   If pump capacity is observed to drop off during the above
pump-up test, check for proper amount of water seal and that the
compressor suction is unobstructed. (Seal water should be
supplied in sufficient quantity that the unit runs slightly warm
to the touch - never hot.) If necessary, schedule compressor for
adjustment of the clearance between the rotor and cones.
3.   Listen to both check valves at the pump to determine if air
leakage exists. Check condition of pump and motor bearings.
4.   Check line connections for air leaks.
5.   Remove float valve assembly from separator; inspect and
clean it. Reinstall and check water level on gauge glass; it
should be 1/3 of the glass with the compressor operating.
6.   Clean the strainers in seal water line of each pump and the
recirculating line to the separator.
7.   Check mechanical seal for excessive leak-off.
8.   Replenish oil supply of the drive-end bearing as needed with
oil.

TOOLS AND MATERIALS: Revision Date: 11/03/88

Figure 10.46 PM Guide Number 113—Air compressor (centrifugal), high-pressure.
(*From Georgetown University.*)

GEORGETOWN UNIVERSITY PREVENTIVE MAINTENANCE(GUIDE NUMBER)__116__

EQUIPMENT: Fire System Control Valves

MANUFACTURER: All

CATEGORY: Safety   FREQUENCY: Yearly
PREVENTIVE MAINTENANCE DESCRIPTION:

Lubricate valve stems.(Fire Suppression Systems)

SPECIAL INSTRUCTIONS:
Always lock the valve in the open position before leaving the
work site. (For any length of time!)

PROCEDURE:

A.   Oil or grease valve stem completely. Close and reopen the
      valve to distribute the lubricant. Lock or seal the valve in
      the open position.

B.   Check and if necessary adjust valve packing gland to stop
      any leakage. Report any leaks that cannot be repaired during
      this PM.

TOOLS AND MATERIALS:
Lubricant for valve stem. Hand tools to adjust packing gland.

Figure 10.47 PM Guide Number 116—Fire system control valves. (*From Georgetown University.*)

GEORGETOWN UNIVERSITY PREVENTIVE MAINTENANCE(GUIDE NUMBER) 119

EQUIPMENT: Life Line T AC Motors 143T-449T

MANUFACTURER: Westinghouse

CATEGORY: Electric Motors   FREQUENCY: Yearly

PREVENTIVE MAINTENANCE DESCRIPTION:

Westinghouse AC Motor Maintenance

SPECIAL INSTRUCTIONS:

OSHA Standard 1910.213 -Open lock and tag motor disconnect switch before starting work.

PROCEDURE:

Motor Location

Open Drip Proof motors are intended for relatively clean dry atmospheres of 104F ambient temperature or less. There are several options offered by Westinghouse that allow use of the motor at higher temperatures. See the modification Data sheets in the back of the PM Guide Book.

Electrical

Check insulation resistance periodically. Any approved method of measuring insulation resistance may be used provided the voltage across the insulation is at a safe value for the type and condition of the insulation. A hand cranked megger of not over 500 volts is the most convenient and safest method.

The recommended insulation resistance of stator windings tested at operating temperature should not be less than:

$$\frac{\text{rated voltage of machine} + 1000}{1000} = \begin{array}{l}\text{Insulation resistance}\\ \text{in megohms}\end{array}$$

If the insulation fails to meet the calculated value then it will have to be dried out.

Motor Lubrication

Lubricating grease should be suitable for an operating ambient temperature of -15F to 130F. All Westinghouse motors are designed to be greased with the relief plug removed. Run the motor for 10 minutes before replacing plug.

Pg 1 of 2

Figure 10.48 PM Guide Number 119—Life line T AC motors, 143T–449T. (*From Georgetown University.*)

GEORGETOWN UNIVERSITY PREVENTIVE MAINTENANCE(GUIDE NUMBER) _119_

Type of Enclosure	Insulation	Frame Size 143 to 215T	254 to 326	364 to 449T
open-DP	B	2 Yrs	18 Mo.	1 Yr
enclosed-FC	B			
open-DP	F	18 Mo.	1 Yr	9 Mo.
enclosed-NV	B			
enclosed-FC	F			
open-DP	H	1 Yr.	9 Mo.	6 Mo.
enclosed-Lint free-FC	B			
enclosed-NV	F			
enclosed-FC	H	9 Mo.	6 Mo.	3 Mo.
enclosed-Lint free-FC	F			

NOTES* -For motors over 1800 rpm use 1/2 of tabled period
       -For heavy duty-dusty locations use 1/2 of tabled period
       -For severe duty-high vibration, shock use 1/3 of table.

Volume-Reference Table
shaft diameter
(at face of bracket)   amount of grease to add
3/4"to 1.25"   .1 oz
1.25" to 1-7/8"   .2 oz
1-7/8" to 2-3/8"   .6 oz
2-3/8" to 3-3/8"   1.6 oz

Oil Lubricated Sleeve Bearings

Before starting the motor, fill both reservoirs through the filler plug with best quality, clean motor oil. The oil should have a viscosity of from 180 to 200SSU (equivalent to SAE #10). During operation, no oil should be added until it drops below the full level. Do not flood the bearing. At about 2 year intervals, dismantle and thoroughly wash out the bearing housing, using hot kerosene oil.

TOOLS AND MATERIALS: Revision Date: 8/4/88
#2 grease with an operating ambient temperature of -15F to 130F, grease gun, hand tools, rags for clean up.

pg 2 of 2

Figure 10.48   (*Continued*) PM Guide Number 119—Life line T AC motors, 143T–449T. (*From Georgetown University.*)

GEORGETOWN UNIVERSITY PREVENTIVE MAINTENANCE(GUIDE NUMBER) 121

EQUIPMENT: Tri Clad 55 143 to 215 143T to 215T

MANUFACTURER: General Electric (G.E.)

CATEGORY: Electric Motors   FREQUENCY: Yearly

PREVENTIVE MAINTENANCE DESCRIPTION:
General Electric AC motor maintenance

SPECIAL INSTRUCTIONS:
   OSHA Standard 1910.213 -Open lock and tag motor disconnect
switch and starter before starting work. Check the voltage of all
incoming line terminals. Positively ascertain that the equipment
is deenergized.

PROCEDURE: INSTALLATION
1.   Location.
a.Dripproof Motors are designed for installation in a well
ventilated place where the atmosphere is reasonably free of
dirt and moisture.
b.Standard Enclosed Motors are designed for installation where
motor may be exposed to dirt, moisture and most outdoor
conditions.
2.   Belts.
a.Sheave ratios greater than 5:1 and center-to-center distances
less than the diameter of the large sheave should be referred to
the Company.
b.Tighten belts only enough to prevent slippage. Belt speed
should not exceed 5000 ft. per min.
c.V-Belt Sheave Pitch Diameters should not be less than the
following values:

	Horsepower		V-Belt Sheave, Min. Dia.		
Synchronous Speed, Rpm				Conventional* A and B	Super♠ 3V
3600	1800	1200	900	Pitch Dia.	Outside Dia.
1½	1	3/4	½	2.2	2.2
2-3	1½-2	1	3/4	2.4	2.4
---	3	1½	1	2.4	2.4
---	---	2	1½	2.4	2.4
5	---	---	---	2.6	2.4
7½	5	---	---	3.0	3.0
10	7½	3	2	3.0	3.0
---	---	5	3	3.0	3.0
15	10	---	---	3.8	3.8

*Max sheave width=2 (N-W)-¼"
♠Max sheave width=N.W.

Figure 10.49   PM Guide Number 121—Tri clad 55, 143T–215T (electric motors). (*From Georgetown University.*)

GEORGETOWN UNIVERSITY PREVENTIVE MAINTENANCE(GUIDE NUMBER) _121_
(continued)

Frame Size		Min
1 Phase	3 Phase	Sheave Diameter
182	182	2¼"
184,213,215	184,213	2½"
	215	3"

3.   Motor Windings
a. To clean, use a soft brush and, if necessary, a slow acting solvent in a well ventilated room. In drying, do not exceed 85°C(185°F).

4.   Inspection
a. Inspect motor at regular intervals. Keep motor clean and ventilating openings clear.

5.   Lubrication
a. **Ball-Bearing Motors** are adequately lubricated at the factory. Relubrication at intervals consistent with the type of service (see table) will provide maximum bearing life. Excessive or too frequent lubrication may damage the motor.

b. **Motors having pipe plugs or grease fittings in bearing housings** should be relubricated while warm and at stand-still. Replace one pipe plug on each end shield with 1/8" inch pipe thread lubrication fitting. Remove the other plug for grease relief. Be sure fittings are clean and free from dirt. Using a low-pressure grease gun, pump in the recommended grease until new grease appears at grease-relief hole. After relubricating, allow motor to run for 10 minutes before replacing relief plugs.

c. **Motors not having pipe plugs or grease fittings in bearing housings** can be relubricated by removing end shields from motor, cleaning grease cavity and refilling 3/4 of circumference of cavity with recommended grease.

Caution: Bearings and grease must be kept free of dirt.

Pg 2 of 3

Figure 10.49 *(Continued)* PM Guide Number 121—Tri clad 55, 143T–215T (electric motors). (*From Georgetown University.*)

GEORGETOWN UNIVERSITY PREVENTIVE MAINTENANCE (GUIDE NUMBER) 121
(cont)

Type of Service	Typical Examples	Hp Range	Relubrication Interval
Easy	Valves; Door Openers; portable floor sanders, motors operating infrequently	½-7½ 10-40 50-150 200-250	10 Years 7 Years 4 Years 3 Years
Standard	Machine tools; air-conditioning apparatus; conveyors; garage compressors; refrigeration + laundry machines; water pumps, woodworking	1½-7½ 10-40 50-150 200-250	7 Years 4 Years 1½ Years 1 Year
Severe	Motors for fans; M-G sets; (running 24 hours/day, 365 days/year; motors subject to severe vibration, etc.	1½-7½ 10-40 50-150 200-250	4 Years 1½ Years 9 Months 6 Months
Very Severe	Dirty, vibrating applications; where end of shaft is hot (pumps and fans); high ambient	1½-7½ 10-150 200-250	9 Months 4 Months 3 Months

TOOLS AND MATERIALS: Revision Date:8/16/88
#2 grease with an operating ambient temperature of -15F to 130F, grease gun, hand tools, rags for clean up.

Pg 3 of 3

Figure 10.49 (*Continued*) PM Guide Number 121—Tri clad 55, 143T–215T (electric motors). (*From Georgetown University.*)

GEORGETOWN UNIVERSITY PREVENTIVE MAINTENANCE(GUIDE NUMBER)__125__

EQUIPMENT: Rotary Vacuum Pump

MANUFACTURER: Leiman Bros. / ITT

CATEGORY: Lubrication   FREQUENCY: Monthly

PREVENTIVE MAINTENANCE DESCRIPTION:

Change Oil on Rotary Vacuum Pump (Leiman/ITT)

SPECIAL INSTRUCTIONS:

    OSHA Standard 1910.213 -Open lock and tag motor disconnect switch before starting work.

PROCEDURE:

A)   Drain old oil and flush tank with new oil.

B)   Refill the automatic oil feed tank to the top of the oil fill elbow with SAE 30 ND oil.

C)   With pump running at required vacuum adjust oil valve to 2 to 3 drops per minute on all pumps except 100 and 107 which should be 6 drops per minute.

D)   Make sure that gasket to oil tank is sealed properly and there are no air leaks.

TOOLS AND MATERIALS: Revision Date: 8/23/88
SAE 30 non detergent oil, container to discard old oil, hand tools to remove covers.

Figure 10.50   PM Guide Number 1251—Rotary vacuum pump. (*From Georgetown University.*)

10.3.2 Publications

The following listed publications are recommended regulations, standards, and codes for accomplishing equipment and systems maintenance and repairs. However, they are not absolute, and other approaches should also be considered based on job site conditions.

Publication	Title	Date
ASME A17.1	American Society of Mechanical Engineers Safety Code for Elevators and Escalators	1990 Edition w/addendums
ASME Section	Boiler Pressure Codes	December 1992
29 CFR Part 1900	OSHA General Industry Standards	Current
NFPA	National Fire Codes	1992
41 CFR Part 101-20	Management of Buildings and Grounds	July 1990
ASHRAE Standard 15	Safety Code for Mechanical Refrigeration	July 1992
40 CFR	Clean Air Act (Refrigerant Venting)	July 1992
ANSI Z245.1	Mobile Refuse Collection and Compactor Equipment Safety Requirements	January 1992
Public Law 94-580	Resource Conservation and Recovery Act of 1976	1976

10.3.3 Personnel requirements

Refrigerants. Personnel who handle refrigerants containing chlorofluorocarbon refrigerants (CFC) must pass an EPA-approved examination to achieve a level IV (universal) certification.

Energy management control system (EMCS). Personnel who operate, maintain, and repair the EMCS should have completed the following: (1) a training course provided by the EMCS manufacturer with a certificate of completion and (2) computer programming experience with the EMCS.

Craftspeople. Trade specialists should be licensed and/or certified by the appropriate local governmental jurisdiction—city, county, or state.

10.4   Energy Management Control System (EMCS)

The EMCS is used to control mechanical and electrical systems that provide required environmental interior temperatures in buildings.

The EMCS is a computer-based system featuring a microprocessor that starts, stops, and monitors mechanical and electrical systems and their individual components throughout the building on a continuous or scheduled basis.

Operation. The EMCS should be operated from the FMD's central control center. This operation will include surveillance of the building rooms, areas, and mechanical systems for adherence to prescribed environmental temperatures. The EMCS operator should take corrective actions to maintain the necessary environmental temperatures within buildings by making any needed EMCS adjustments on a continuous daily basis.

Emergency or unscheduled EMCS shutdown. The FMD should plan to have on call additional temporary personnel or contractor support necessary to maintain full and acceptable performance from all building systems during any EMCS failure. This plan should become effective immediately and at such time as building environmental temperatures cannot be maintained using manual adjustments to the building systems. This plan should remain in effect until such time as the EMCS becomes fully functional as originally designed.

10.5 Equipment and System Warranties

It is advisable to develop a computerized maintenance management system (CMMS) program to track and manage all building equipment and system components that are under current valid warranties. This will assist in ensuring that required repairs or corrections, during the warranty periods, will be made by the installing contractors. Should there be a dispute as to the responsibility for the repair and adjustment, the FMD director will make a determination as to the responsible party and issue directions on how to proceed. The FMD should provide operational service call response (and corrective action) and preventive maintenance functions to the deficient item or system even though the item is under a manufacturer's warranty and may be awaiting repairs or adjustments by others.

11

Electrical Equipment and System Maintenance Procedures

Dana L. Green

Substation Test Company
Forestville, MD

11.1 Introduction

Electricity is a very versatile servant, providing energy to perform a vast array of operations and control functions in relative safety, so long as it is properly respected, used, and maintained. Today's codes, designs, and installations include margins and features to assure public safety. Electrical equipment is always expensive to initially install or to replace. But the total cost of an unscheduled shutdown is significantly higher, perhaps many times higher, than the cost of the equipment repair or replacement alone. Facilities and manufacturing plants require value for cost incurred, in order to financially survive. In order to be cost-effective, expensive equipment must provide reliable, safe performance for a relatively long life, with minimal downtime and repair cost. Adequate maintenance is an absolute requirement. Poorly maintained or nonmaintained equipment becomes unreliable, unsafe, and shorter-lived, at best. At worst, such conditions may lead to extraordinarily expensive fires, shutdowns, injuries, and deaths.

To be effective, an electrical maintenance program must identify the various types of equipment to be serviced, its characteristics, and the affecting conditions of its use and environment, and must address all needs identified. One chapter does not allow inclusion of all types of electrical power and control equipment, or a more expanded definition of equipment needs and maintenance practices, or address all safety concerns. And requirements at one location may not be adequate or appropriate for another location.

This chapter does not address day-to-day maintenance activities but is oriented to scheduled shutdown activities. It will identify for the facility manager and program designer, in very brief form, some of the characteristics and general maintenance requirements of many of the basic types of power equipment and protective devices installed in facility electrical systems.

Development and performance of a complete maintenance program is often performed by the joint efforts of in-house personnel and supporting contractors, and begins with identification of the electrical apparatus to be maintained.

11.2   Electrical Safety

Electrical safety is an integral part of overall plant safety, and general plant safety concerns and rules apply to all electrical workers. However, all work actions include safety concerns peculiar to that particular type of work. And electrical maintenance work is certainly no exception. A few prominent safety reminders or advisements follow.

Ladders should be of nonconductive materials such as wood or fiberglass, even though these materials contribute to weight. Wooden ladders should be sealed with a clear sealer.

Temporary extension cords and drop lights should always include ground fault protection.

Lockout-tagout is more than an OSHA requirement; it may be a matter of life and death. It is necessary to adhere to sound lockout and equipment tagging procedures, developed and performed by knowledgeable, well-trained individuals. All maintenance and operating personnel must understand and comply with the procedures implemented. Lockout-tagout is discussed in OSHA 1926.269(d).

Electrical switching to deenergize equipment and make it safe for hands-on maintenance or repair work includes recognition of potential hazards and inclusion of safety measures that are not ordinarily necessary during normal equipment operations. All potential sources of energy of any type must be recognized and considered.

Testing for dead is a procedure to determine if electrical equipment is deenergized and is performed prior to hands-on maintenance. An appropriate voltage detector, rated for the system voltage to be detected and used in accordance with the manufacturer's instructions, is necessary. Voltage detectors usually provide audible, visual, and/or a vibrating indication of energization, either by contact with or coming into proximity of an energized conductor. When using a voltage detector, it is necessary to first check the detector, by intentionally testing a known energized line to prove the detector is working, and to recheck the detector immediately after testing the intended deenergized equipment to assure the detector did not itself fail during the testing of the equipment thought to be deenergized.

Temporary grounding of electrical equipment or conductors after testing for dead but prior to hands-on maintenance of normally energized conductors is often required. Temporary grounds must be placed between the source of electrical energy and the work site. The ground cables and connectors must be sized to handle the maximum fault current available for time required to assure operation of the upstream protective device. Ground cables for power lines may be no smaller than No. 2 AWG copper and may have to be larger. Grounding system connection to earth, temporary or permanent, should have a resistance low enough to eliminate personnel hazard and allow prompt deenergization by protective devices. Equipment must be locked out, tagged out, and proved deenergized prior to placement of protective grounds.

Confined spaces are found throughout most plants and facilities. All workers must be trained to recognize and properly deal with confined spaces and permit required spaces. Access holes, pits (even outdoor pits entirely open at the top), tanks, pipelines, tunnels, crawl spaces, and excavations are a few examples of confined spaces. Confined spaces require evaluation of potential hazards. Testing of oxygen content, flammable gases, and other potentially toxic air contaminants is required. Adequate forced air ventilation or self-contained breathing apparatus is required. The worker entering an electrical access hole, tank, or other confined space should be connected to a rescue (retrieval) system that can be operated by an appropriately trained attendant. Typical rescue equipment includes a tripod, rope or cable, winch, and harness worn by the worker entering the confined space. Rescue persons entering such a space must be similarly harnessed and attached to retrieval equipment. Permitted confined spaces require completion of an approved permit prior to entry. OSHA 1910.269(e) further describes confined space concerns and requirements.

Personal protective equipment might include eye and face protection, respiratory protection, head protection, work gloves, foot protection, hearing protection, body belts, etc., that are appropriate for conditions present. Clothing should be of a less flammable material, such as wool or cotton, or materials specially made and maintained to be fire resistant. Many popular synthetic materials that serve well under other conditions are extremely dangerous when exposed to fire or electrical arcing and may substantially contribute to the wearer's injury or death. Fire-retardant suits or coveralls are available for use by electrical personnel during switching operations.

Insulating rubber gloves, sleeves, blankets, mats, footwear, and disconnect (hot) sticks are available at most facilities. Proper care and use of the needed protective goods is vital to safety. Unfortunately, too often the user is uninformed, and the protective equipment itself is unsafe to use. Many facilities are not aware that all these items, including aerial lifts, require scheduled testing. Periodic electrical testing of protective rubber goods and insulating tools and apparatus should comply with the appropriate OSHA and ASTM test standards. Testing of these protective goods requires special training and test equipment.

Maintenance testing of electrical equipment can be very dangerous. Some tests require the equipment be in its normal energized state, and other tests may be performed only on deenergized equipment. Many valuable tests of electric insulation and equipment necessarily develop lethal test voltages. Improper connections often result in damage to expensive power equipment, serious injury, or worse. Test technicians must be well trained and very knowledgeable in that art.

The above safety discussion does not begin to address all concerns of electrical safety but merely touches on a few issues. Electrical safety demands study, evaluation, and development of a sound, thorough, safety program, which must include recognition of when outside expertise is either required or provides better risk control.

11.3   Insulation Resistance Measurements

Insulation resistance measurements are one of the most useful, commonly performed field tests and are applicable to most but not all low-, medium-, and high-voltage apparatus. These tests are reasonably easy to perform and are a very valuable indication of insulation system or component condition.

Caution: Do not attempt to perform this test with equipment energized under normal power or electrically charged.

An insulation resistance test is performed by applying a known direct-current voltage across the insulation system under test, measuring the minute ampere flow, and converting the readout to resistance. The test instrument has become known as a *megohmmeter.* Present-day instruments may provide a test potential from 250 to 5000 V in switched increments or fully variable control, and may provide readouts from 0 ohms to 100 gigaohms. Megohmmeters that produce 10,000 to 15,000-V test potentials are also available. The instrument may be hand cranked or electrically operated.

While relatively easy to perform, evaluation of the test results may be complex, requiring a broad understanding of conditions present, combined with experience. The test is temperature-sensitive, and measurements must be temperature corrected to 20°C. Insulation resistances decrease with internal and external moisture, internal or surface insulation deterioration, and other environmental conditions. Lowered resistance on a wet day may be acceptable, but the same value on a dry day may be judged questionable.

11.4   Winding Turns Ratio Measurements

These tests measure the ratio between windings of current or voltage transformers, i.e., the number of turns in one winding to that of the other winding sharing the same core. Therefore, the winding turns ratio of a transformer with high-voltage windings rated 13,800 V and low-voltage windings rated 115 V is 120:1. A current transformer with primary rating of 150 A and secondary rating of 5 A has a ratio of 30:1. Winding turns ratio tests are performed to assure that intended transformer output with respect to input complies with applicable accuracy standards. These tests are performed when the transformer is installed and as a routine maintenance procedure. The ratio of each tap of a multitapped transformer winding is measured. Polyphase transformers may be tested one set of windings at a time. With polyphase (multiphase) transformers, the ratio between the high-voltage winding and low-voltage winding wound around the same area of iron core is measured. Results are recorded and compared to previous measurements to determine if turn-to-turn winding shorts have developed.

Caution: These tests are not to be performed with the transformer energized under normal power. Be certain that the transformer under test is safely deenergized, locked out, tagged out, and safe.

Ratio can be measured by applying an appropriate current or voltage to one winding and measuring output of the other winding. *Example:* If 50 A is passed through the primary of a 250:5 current transformer, 1.0 A should flow in the secondary. Obviously, a means of developing a controlled, metered test current of 50 A, and of measuring 1.0 A, is required.

Voltage transformers are often ratio tested by use of a winding turns ratio test set, available from several test equipment manufacturers. These sets are used to test single and multiphase transformers and usually provide a readout of the ratio measured, to at least the third decimal place. A ratio test set might include a line or hand-powered generator, standard or reference autotransformer, a voltmeter to measure the autotransformer output, an ammeter to measure generator output, null detector, various switches or dials to adjust the autotransformer ratio, and test leads. Connected and operated per the manufacturer's instructions, a winding turns ratio test set provides a relatively easy, accurate means of voltage transformer winding turns ratio measurements.

Voltage turns ratio errors should not exceed 0.5 percent. Therefore, if the primary rating is 13,800 V, and secondary 120 V, the intended ratio is 115:1, and the ratio should measure 115.000 ±0.5 percent.

$$\text{Percent error} = \frac{\text{difference between design and measured ratio} \times 100}{\text{design ratio}}$$

Therefore, if the ratio of the above transformer is measured to be 115.115, the percent error would be

$$\frac{0.115}{115} \times 100 = 0.1\%$$

Ratio tests are often more involved than the above simple examples indicate, and there are other methods of performance.

11.5 Power Factor and Dielectric-Loss Measurements

The ac dielectric-loss and power-factor test is recognized as one of the single most effective methods for locating defective insulation. In the jargon of the electrical utility maintenance engineer, the dielectric-loss and power-factor test is often referred to as the Doble test, primarily because of the extensive use and unique capabilities of Doble field test equipment and the orderly test methods which have been developed by the Doble Engineering Company in cooperation with its client group.

The basic principle of nondestructive tests such as dielectric loss and power factor is the detection of a change in the measurable characteristics of an insulation which can be associated with the effect of moisture, heat, ionization (corona), lightning, and other destructive agents which reduce the breakdown strength of insulation.

11.5.1 Dielectric loss

All commercial solid and liquid insulations have some measurable dielectric loss at normal operating voltage and frequency, since the term "nonconductor" is but a relative one. However, these losses are usually very small and vary approximately as the square of the applied voltage. An appreciable increase in the normal dielectric loss is one of the first indications of deterioration and operating hazards.

Gaseous insulations such as air do not have a measurable loss until they become overstressed and ionized. Solid insulation, unless thoroughly impregnated or immersed in liquid insulation, usually has voids containing air. If these voids become overstressed and ionized during the application of a test voltage, the resulting dielectric losses in the voids will be added to those in the solid insulation; hence the losses will increase at some power higher than the square of the voltage. The presence of voids that become ionized at normal line-to-neutral operating voltage may cause carbonization or radio interference. Apparatus-insulation designers endeavor to eliminate voids where feasible.

The dielectric losses of most insulations increase as the temperature of the insulation increases. In some cases, insulations have failed owing to the cumulative effect of temperature; i.e., a rise in temperature causes a rise in dielectric loss which in turn causes a further rise in temperature.

11.5.2 Power factor

The power factor of an insulation is the cosine of the angle between the charging current vector and the impressed voltage vector. In other words, it is a measure of the energy component of the charging current. The amount of charging volt-amperes and the dielectric loss in watts, at a given voltage, increase with the amount of insulation being tested. However, the ratio (power factor) between the charging volt-amperes and watts loss remains the same regardless of the amount of insulation tested, assuming that the insulation is of a uniform quality. This basic relation eliminates the effect of the size of electrical apparatus in establishing "normal" insulation values and thus simplifies the problem for the test engineer. Power-factor insulation test equipment measures the charging current and watts loss from which the power factor, capacitance, and ac resistance can be easily computed at a given test voltage.

The 60-cycle power factor and dielectric constant of a few typical insulating materials (also for water and ice) are approximately as follows:

Material	%PF at 20°C	Dielectric constant
Air	0	1.0
Oil	0.1	2.1
Paper	0.5	2.0
Porcelain	2.0	7.0
Rubber	4.0	3.6
Varnished cambric	4.0–8.0	4.5
Water	100.0	81.0
Ice	1.0(est.)	86.4

Note: Ice has a volumetric resistivity approximately 144 times that of water, which illustrates why tests for the presence of moisture in an insulation should not be made when apparatus temperatures are at freezing or below.

The normal power factors of assembled apparatus insulation are approximately as follows:

	%PF at 20°C
Bushings, condenser and oil-filled	0.5
Bushings, compound-filled	2.5
Transformers, oil-filled	1.0–2.0 (new transformer 0.5)
Cables, paper-insulated	0.3
Cables, rubber-insulated	4.0–5.0

Normal test values for various types of apparatus insulation have been obtained by testing thousands of similar pieces of apparatus in the field and in the factory.

Frequently, the abnormal test values obtained during the initial tests are due to accumulations of carbon deposit or bad oil which may be cleaned up without permanent injury to the main insulations. These conditions may represent serious operating hazards which require immediate attention. The ability of the test to detect this type of operating hazard before a visual inspection is made of the interior of the apparatus facilitates a more efficient use of maintenance labor power and savings in maintenance costs.

11.6 Motion Analysis of Circuit Breakers

Timing and travel analysis is often performed on outdoor medium- and high-voltage circuit breakers to evaluate the operating mechanism. These tests measure the time it takes the contacts to part after the trip coil has been energized or to make contact after the close coil has been energized. Travel motion is typically analyzed through use of a transducer that converts the motion of the operating rod to an electrical signal. The test equipment required to perform these tests is very specialized and expensive.

The results of these tests will be in graphical form and can be used to calculate opening and closing velocity.

11.7 Insulation and Insulators, General

Materials whose properties include high resistance to the flow of electricity are used as electrical insulators. Depending upon the intended use, the form of electrical insulating materials may be flexible or rigid solids, liquids, or gases.

Manufacturer-applied insulation over cable may be laminated layers of varnished cambric, dacron fiber, oil-impregnated paper, or other types of tape insulation. Alternately, cables are insulated by extrusion of an ethylene-propylene rubber or a cross-linked polyethylene compound. For 600-V class cables, a thermoplastic or thermosetting compound is commonly used. Asphalt-base and resin-base insulating compounds are commonly used to fill cable termination chambers and some splice compartments. Insulating boards are typically a phenolic molding with a fiberglass, dacron, linen, or cotton base. Solid support insulators are often ceramic, molded phenolic, or glass. Liquid insulation may be mineral oil, a paraffin-based compound, silicone, perchloroethylene, or one of several synthetic compounds, and sulfur hexafluoride (SF_6) gas may be used to insulate medium- and high-voltage switch chambers, large generators, and some electrical bus conductors.

Some cable insulations are designed for use in submersed applications, such as found in underground access hole and duct installations, and in some direct buried cable installations. Because of the materials and construction used, these cables are rated to hold up well under wet conditions. However, a tiny flaw in the extrusion process or a jacket puncture during installation will result in deterioration of insulating quality until failure occurs.

Dust and/or airborne contaminates may cling to the surface of insulating support members and to insulators installed in switchgear, motor control centers, dry-type transformers, and the like. The air in a potato chip plant can be laden with oil, or in a newspaper plant laden with ink, to create a film on the insulator surface that attracts conductive dirt and soot. The oil-laden air in a metalworking plant may also carry minute particles of conductive metal to add to the surface contamination.

Solid insulation must maintain high resistance to the flow of electricity both through the insulating medium and across its surfaces.

Visual inspection should include a check of mounting arrangements, and look for cracking, charring, water tracks from overhead leaks or condensation, discoloration, or chips.

Medium- and high-voltage insulation is subject to tracking which may etch or discolor the surface and develop burned patterns in treeing design, and charring at areas of excess corona discharge or surface contamination. Tracking evidences current flow across or through the insulation, and future failure will occur if corrective measures are not undertaken.

Inspection and hands-on cleaning of solid insulation is performed by qualified persons, with the equipment to be serviced deenergized, locked out, tagged out, tested for dead, and the conductors properly grounded. Cleaning solid insulation may be as easy as dusting with a lint-free cloth or vacuuming with a nonconductive vacuum brush and wand. Surfaces with a contaminating film may require application of an electrical-grade solvent by wiping cloth or low-pressure spray. Some insulating surfaces are greatly benefited by waxing and elbow grease, especially indoor and outdoor types of stand-off or support insulators.

Caution: High-pressure air or solvent spray may drive contaminants further into insulated windings.

Electrical insulating material is very negatively affected by contamination and by the presence of moisture. In general, *Keep electrical insulation and insulators clean and dry.*

11.8   Insulating Liquids

Various liquids are used for cooling and insulating purposes. Liquid-filled oil circuit breakers, switches, transformers, capacitors, and cable assemblies are in use throughout most electrical power systems. Mineral oils are least expensive, nontoxic, easy to handle, and easy to maintain. Mineral oil is used extensively in outdoor apparatus. Oil-filled circuit breakers and switches are filled with mineral oil. Transformers and capacitors may be filled with liquids offering high fire-point qualities, such as silicone, R-Temp, and askarel (PCB), especially in indoor applications. Askarel is highly regulated by the Environmental Protection Agency, is no longer manufactured, and is gradually disappearing.

No one test of an insulating liquid provides sufficient information. Depending upon the type of fluid and its use, a screen of tests may be selected. The values received from one test may point to need of another test, in order to better determine what is happening or not happening within the oil-filled device. These tests will aid in determination of oil degradation due to contamination, overheating, sludging, and chemical changes, and may identify carbon, water entry, internal discharge or arcing, and deterioration of other materials.

Reclamation and reconditioning of insulating liquids requires special procedures, experience, and equipment and is best provided by parties with such expertise.

11.8.1   Sampling

Sampling must be performed with great care. The electrical apparatus must have a positive pressure when drawing a sample. When drawing a bottom sample especially, do not allow an air bubble to enter the tank, and which could possibly drift into an area of close clearances. Drawing samples from apparatus having only a small volume of liquid requires extreme caution. Never sample with the apparatus under a negative pressure, especially with the apparatus energized. Samples should be drawn from sample devices provided or from valves. Do not sample from a drain plug. The nameplate of some instrument transformers advises that drawing a sample will void all warranty. Samples should not be drawn from an energized instrument transformer.

Sampling is further described in the standard ASTM D923.

11.8.2 Sample containers

Typical sample containers may be amber-colored glass, high-density polyethylene, and aluminum- or tin-plated cans. Rubber stoppers or cap seals are not used. Clear glass containers are frequently used in order to facilitate visual inspection but must be protected from light. For dissolved gas analysis, samples are drawn into a glass syringe.

The sample container chosen must be absolutely clean and dry.

11.8.3 Sampling location

Determine the correct location for sampling. Liquids with specific gravity less than 1 are sampled at the bottom of the tank. For liquids having a specific gravity greater than 1, samples are drawn from the top of the tank, at the 25° level. A sample device or valve should exist at the appropriate location.

11.8.4 Sampling procedures

Assure that a positive pressure exists at the sample device by placing a slug of insulating oil in a clear plastic tubing that is connected to the sample device so that liquid in the tubing is against the device. Crack the sample valve and note the direction the oil slug moves. If the oil slug moves toward the electrical apparatus, a negative pressure exists, and the sample should not be drawn.

If the sample device is connected to a drain valve, clean the device and the valve.

Remove the drain valve plug and drain at least 2 quarts of fluid from the valve. For very large apparatus, it may be more appropriate to drain several gallons of fluid. Replace the security plug in the drain valve.

Flush the sample device by draining 1 quart of fluid through it.

Collect the sample by holding the sample container so the fluid will run down its sides with minimum aeration. Partially fill the container several times, slosh the fluid, and discard. Liquid flow should be gentle but not interrupted from the start of the flushing action until completion of the sampling.

Fill the sample container to overflowing if a dissolved gas analysis is to be performed. Otherwise leave a $\frac{1}{2}$-in expansion space, especially if the fluid is cold.

If sampling for a dissolved gas analysis (DGA), preferably use a syringe. Connect the syringe to the sample device using tygon tubing. Flush the tubing by passing fluid through the syringe flushing port until no air is present. Then withdraw the syringe plunger to fill the syringe with oil. The filled syringe should include no air.

11.8.5   Periodic tests

Visual Examination ASTM D1524: Field test using the Hellige odor comparator to detect colloidal particles and cloudiness. The test may indicate moisture or sludge. The oil should be clear and bright.

Dielectric Breakdown Strength, ASTM D877 or D1816: Determines dielectric strength by measuring the liquid's ability to withstand electrical stress without failure. It indicates presence of particulates, moisture, fibers, or other conducting particles in the oil. The ASTM D877 method is utilized for apparatus rated 230 kV or less, and the ASTM D1816 method is for higher-voltage-rated apparatus.

Neutralization Number ASTM D974: Determines acid and base number by color titration. It indicates relative changes occurring in the oil during use under oxidizing conditions. A high neutralization number indicates the oil is either oxidized or contaminated.

Specific Gravity, ASTM D1298: Determines the density, relative density (specific gravity), or API gravity of petroleum products. Specific gravity of a fluid affects heat transfer rates. Unusual specific gravity values for a known liquid often indicate contaminating influences.

Interfacial Tension, ASTM D971 and D2285: Method D971 measures the interfacial tension of mineral oils against water, expressed in the dynes per centimeter required to rupture an oil film existing on a body of water. The test uses a tensionmeter, rings of fine platinum wire, and a glass beaker having a minimum diameter of 45 cm. This is often a laboratory test.

The ASTM D2285 test measures interfacial tension of insulating oils against water by the drop-weight method and is a field test. It requires a different style of tensionmeter that injects a measured quantity of water into the oil sample and is not considered as accurate as the D971 test. Interfacial tension is a valuable indication of oil degradation.

Water Content, ASTM D1533: Measures the water content in ppm, by the Karl Fischer reaction method. Excessive water content affects electrical strength, dielectric losses, corrosion, and life expectancy.

Power Factor at 25°C, ASTM D924: The test is a measurement of dielectric losses in insulating liquids. High power factor indicates the presence of contaminants and changes in quality and deterioration in service.

Dissolved Gas Analysis (DGA) ASTM 3612: The single most important test performed on oils from transformers is for dissolved gases. The insulating materials of a transformer break down from excessive thermal or electrical stress, and gaseous by-products are formed which are characteristic of the type of incipient-fault condition, the materials involved, and the severity of the condition. Indeed it is the ability to detect such a variety of problems that makes this test such a powerful tool to detect incipient fault conditions or for investigations after failures have occurred. Dissolved gases can be detected in low concentrations, which usually permits early intervention before failure occurs and allows for planned replacement power. Typical gases generated from a mineral oil–cellulose paper and pressboard insulated transformer include

Acetylene

Carbon monoxide

Carbon dioxide

Ethane

Hydrogen

Methane

Ethylene

In addition there will always be some oxygen and nitrogen present from residual air and the blanket gas.

The composition of gases generated provides information about the type of incipient-fault condition present. For example, four broad categories of fault conditions can be described and characterized by key gases.

Thermal oil	Methane, ethane, ethylene, traces of acetylene
Partial discharge	Hydrogen, methane, acetylene, and ethane
Sustained arcing	Hydrogen, acetylene, ethylene
Thermal—paper	Carbon monoxide and carbon dioxide

Electrical discharges in the paper insulation will cause it to overheat, generating carbon oxide gases. Further refinement of the diagnosis can be provided by examining the relative composition or ratios of gases present.

The severity of a detected incipient fault condition is ascertained by the total amount of combustible gases and their rate of generation. Generally transformers retain a large portion of the gases generated and therefore there is a cumulative history of the degradation of the insulating materials. This is an important attribute for detecting developing problems between tests but also means that some care is needed in interpreting values for the first-time analysis on service-aged transformers more than several years old which could have a history. Some gas generation is expected from the normal aging of transformer insulation. Therefore, it is important to differentiate between normal gassing rates and those which are excessive. Normal aging or gas generation will vary with transformer design, loading, and type of insulating materials. Generally normal gassing rates for all transformers are used to define abnormal behavior. Specific information for a family of transformers can be used when sufficient dissolved gas-in-oil data is available.

11.8.6   Other tests

When appropriate, other useful tests include:

Color, ASTM D1500

Cloud Point, ASTM D97

Flash and Fire Point, ASTM D92

Gas Content, Total, ASTM D2945, D1827, D3612

Oxidation Inhibitor Content, ASTM D1473, D2668

Metals in Oil:

Oxidation Stability (acid/sludge), ASTM D2440, D2112

PCB Content, ASTM D4059

Pour Point, ASTM D97

Viscosity, ASTM D88, D445, D446

11.9   Switchgear

The general term *switchgear* applies to power switching and interrupting apparatus, typically consisting of switches or circuit breakers and associated protection and control equipment.

An open switchgear assembly does not have an overall enclosure. Enclosed switchgear usually includes a metal frame and housing and may be indoor- or outdoor-rated. Enclosed outdoor switchgear assemblies may be walk-in or non-walk-in construction.

Metal-enclosed switchgear typically includes switches and/or circuit breakers, bare bus, instrument and control transformers, relays, metering, and associated wiring. The interrupting devices may be draw-out or bolt-in connected. Metal-enclosed switchgear is available at voltages up to 34.5 kV.

Metal-clad switchgear is a form of metal-enclosed switchgear that provides certain ANSI/IEEE defined features, without exception. To be classified as metal-clad switchgear, the assembly must include a draw-out (removable) style main switch or breaker, metal enclosure around all primary components, insulated primary conductors, safety interlocks, and compartmental construction to separate the main busses, each breaker or switch cubicle, each line and load side set of bus taps, potential transformers, and control transformers. Breaker or switch cubicles include automatic shutters to block access to the energized bus or load connections when the switch device is removed. Protective and control devices may be mounted on switching device cubicle doors. Metal-clad switchgear is rated above 1000 V.

Note: Switchgear bus insulation is not intended to protect against electrocution or electrical shock. Do not touch energized insulated switchgear busses or connections.

11.9.1 Maintenance

First it is necessary to identify how the subject switchgear is constructed and the different components making up the subject switchgear assembly. Review manufacturer's instructions and drawings thoroughly.

Identify and record the nameplate information and ratings of the switchboard assembly and individual components, and compare the information gathered to the most current one-line diagram available.

Inspect the general condition, noting any dented panels, broken hinges, rodent infiltration, access locking devices, paint deterioration, reused areas needing repairs, and any loose or missing hardware. Be attentive to moisture entry. This is especially important with outdoor assemblies whose gasketed seams may have failed. Check anchorage. Inspect, clean, or replace ventilation filters. Confirm cubicle heaters are operational. Confirm correct operation of all mechanical and electrical interlock devices. Assure that fuse and/or circuit breaker sizes and ratings are as shown in the short-circuit coordination study. Check operation of the automatic shutters found in draw-out breaker cubicles. Check grounding and ground connections.

Inspect insulators and insulating barriers for evidence of electrical tracking, burning, discoloration, cracking, contamination, or other abnormalities. Check the tightness of all bus bar and main conductor connections by use of a torque wrench, a low-resistance reading ohmmeter, or infrared thermographic scan. Clean the switchgear thoroughly, using a vacuum cleaner, lint-free cloths, and an electrical solvent where necessary. Lubricate moving parts, slides, or joints per manufacturer's directions. Do not overlubricate, as greases may become grimy over time, and heavily oiled areas collect dust and dirt.

In medium-voltage switchgear, voltage instrument transformers and control transformers, and/or the primary fuses of such transformers are often mounted in a drawer arrangement. When the drawer is opened, the transformers and/or fuses within are disconnected from primary and secondary conductors, and the exposed primary terminals are grounded. Verify the correct operation of these draw-out components, and assure the safety ground feature functions properly.

11.9.2   Tests

Perform ground resistance tests as applicable.

Measure insulation resistances of each set of main busses and of line and load busses, phase to phase and phase to ground. Compare readings to prior test values measured.

Apply the appropriate overpotential between each bus and the grounded bus sections not under test. This is a 1 min go–no-go test, and preferably an ac test voltage. ANSI/IEEE standards do not recognize dc overpotential tests as an acceptable alternative.

Measure insulation resistance of all control wiring at 1000 V dc, excepting any solid-state components.

Measure insulation resistance of voltage instrument transformers and control power transformers, winding to winding and winding to ground.

Confirm ratio and correct operation of instrument transformers and control transformers.

Confirm correct operation of all control and protective devices.

11.10   Protective Relays

11.10.1   General

This brief description of protective relay maintenance will focus on electric power system protection commonly found in manufacturing plants and facilities, to include generator protection.

IEEE 100 defines the function of a protective relay as being "to detect defective lines or apparatus or other power system conditions of an abnormal or dangerous nature and to initiate appropriate control circuit action." Relays accomplish this by monitoring electrical values, i.e., current, voltage, watts, vars, impedance, time, angular, and other relationships, and when certain preset values are met, operating to complete control or trip circuits that in turn initiate operation of circuit breakers or other devices. Protective relays are part of a larger protection system, which may include other control devices, instrument transformers, and circuitry. Protective relays may be factory set, without adjustment. But most relays provide one or more ranges of adjustment or setting levels.

11.10.2 Construction

Relays may be open construction, i.e., without a case, for installation within apparatus housings, or may include a case to provide mechanical and dust protection. Usually, either the case is removable or it includes a removable cover. Typically, most relays can be withdrawn from their case.

The draw-out style cases of well-designed relays include current and trip circuit test switches or a removable electrical connection plug in order to more safely interrupt current, potential, and trip control circuits to the relay at the time of maintenance testing. These relay circuit interrupting devices should include built-in means to automatically short-current circuits such that energized current transformer secondaries are not open-circuited when the test switches are opened or the connection plug is removed. Some relay and case designs provide no test switches or connection plugs, and no circuit shorting means, thereby providing increased risk at the time of maintenance testing.

Protective relays are of either electromechanical or solid-state construction. Both types are reliable if properly maintained.

Electromechanical relays typically operate by principles of electromagnetic attraction or by electromagnetic induction. The attraction relay operates on ac or dc voltage and is used for instantaneous or high-speed operations. Electromagnetic induction relays use torque induced in a rotor, as seen in many time overcurrent and voltage relays having a rotating disk, while many directional, differential, and distance relays use a rotating shaft with a cylinder (cup).

Electromechanical relays have been successfully used for many years. They are relatively easier to adjust but, having moving parts, may more often drift off the intended setting level. Solid-state relays do not include many (or any) moving parts and have earned the reputation of being more accurate and more reliable, especially in areas with a lot of vibration. However, solid-state relays may be more subject to damage from voltage disturbances. While less likely to require routine adjustment, component failure often significantly distorts or disables the intended relay function. Solid-state relays are less likely to be field repairable.

11.10.3   Types and device numbers

Relays are often shown on drawings by ANSI/IEEE recognized device numbers to designate the type of relay employed.

Relay type	Device
Time delay starting or closing	2
Checking or interlocking	3
Distance	21
Synchronizing or synchronous check	25
Undervoltage	27
Annunciator	30
Directional power	32
Undercurrent or under power	37
Field	40
Unit sequence starting	44
Reverse phase or phase balance current	46
Phase sequence voltage	47
Incomplete sequence	48
Machine or transformer thermal	49
Instantaneous overcurrent or rate of rise	50
AC time overcurrent	51
Exciter of dc generator	53
Power factor	55
Field application	56
Rectification failure	58
Overvoltage	59
Voltage or current balance	61
Time-delay stopping or opening	62
Ground protective	64
AC directional overcurrent	67
Blocking	68
Alarm	74
DC overcurrent	76
Phase angle measuring or out of step	78
AC reclosing	79
Frequency	81
DC reclosing	82
Automatic selective control or transfer	83
Carrier or pilot wire	85
Locking out	86
Differential protective	87
Voltage directional	91
Voltage and power directional	92
Tripping or trip-free	94

11.10.4   Maintenance

Inspect the relay and case for evidence of physical damage or corrosion. Inspect the case cover for cracked glass, functioning cover screws, and proper gasket seal. Clean the cover and case top before removing the cover. Visually inspect the relay for burned wires, loose connections, overheating, dust, or dirt. Check tightness of electrical and mechanical connections. Assure all movable parts are free and operate without unintended restriction. Check contacts for burning. As applicable, perform a zero check. Verify target reset. Check disk clearance and spring convolutions. Verify settings are as shown in the coordination study or setting records provided.

11.10.5   Tests

Perform insulation resistance test on each non-solid-state circuit to the relay frame. The manufacture may not allow this test for solid-state or microprocessor relays.

Verify pickup and dropout of electromechanical targets. Check proper operation of light-emitting diode indicators.

Apply the electrical parameters necessary to cause the relay to perform all functions listed in the manufacturer's instruction booklet, by using protective relay test sets and associated apparatus.

Confirm that relay operations result in the control or trip operation intended. Relay testing should include proving out the integrity of the entire protective circuit to assure the intended protective feature is functioning.

Note: Protective relays are best calibrated as mounted in their case during normal operation, if possible. Secondary injection testing provides more accurate calibration. Primary injection testing requires more expensive test equipment.

11.10.6   Cautions

Opening of energized current transformer secondary circuits must be avoided to prevent possible introduction of hazardous high voltage to those circuits, endangering personnel and equipment.

The relay test technician must be fully knowledgeable of all control and protection circuits, devices, and schemes to be tested.

Removing the cover of an in-service relay can result in inadvertent relay operation.

11.11   Ground Fault Protection

Ground fault protection (GFP) is designed to sense abnormal current flow between energized conductors and ground, and to operate to disconnect the circuit for the purpose of limiting equipment damage but without specific concern for electrocution.

Ground fault protection is installed in a grounded distribution system. This protection was commonly applied to medium- and high-voltage circuits. However, for several decades ground fault protection was ignored in 600-V class apparatus, with the erroneous assumption that faults to ground would either be seen by phase overcurrent devices or would develop into phase-to-phase faults, to then be seen by standard phase overcurrent devices. Eventually it became apparent that 480-V apparatus could sustain an arcing fault to ground, with resistance of the arc maintaining fault current levels high enough to melt copper and steel but less than the upstream circuit breaker trip or fuse melting level.

The writer has been present (not responsible) when a 480-V, 4000-A rated fuse-switch switchgear lineup burned with intermittent roaring and booms for 24 min, pending power company arrival to disconnect their network power source. Several hours later, after the smoke cleared, inspection revealed large sections of bus bars melted and gone, or drooped and disfigured, with many feeder cable connections melted and broken by the heat and mechanical stresses present. The 4000-A rated bus bars burned in two and partially melted the source 4000-A fuse connections, but the 4000-A fuses had not operated. Thermal destruction was complete. There was no protective device operation, because the switchgear did not include ground fault protection.

The NEC requires ground fault protection for all new equipment connected to solidly grounded wye services operating at more than 150 V to ground and rated 1000 A or more. However, apparatus built prior to this NEC requirement does not include GFP. Many 480-V switchgear assemblies in service today are not adequately protected for arcing faults to ground, and it is important to recognize whether your assembly includes such protection.

To measure current flow between circuit conductors and ground, the neutral must remain insulated from the ground plane; therefore, the neutral is connected to ground at one location only, for each separately desired system, as required for the GFP scheme employed. The GFP system includes current sensors (cts), the protective device feature, a breaker or switch circuit interrupting means, and associated wiring and connections.

The GFP may be an independent device mounted in a switchboard or switchgear assembly at each switch or breaker rated 1000 A or greater. Most circuit breakers with solid-state trip devices include integral ground fault protection with the other protections provided. GFP systems may measure all phase and neutral currents, such that the imbalance represents ground fault flow. Other GFP systems may measure ground current flow by placing a sensor in the once connecting link between ground and neutral conductors, usually a bus bar connection within the switchboard or switchgear assembly.

Maintenance of a GFP system includes inspection and cleaning of all components. Check connections for tightness. Verify connections are correct. Assure that phase and neutral sensor polarities are correct when all are measured. Confirm that the ground conductor is solidly grounded. If a self-test panel exists, assure it operates correctly. However, do not assume that a self-test button or panel properly determines that a GFP system is operational, even though the circuit breaker or switch opens. These self-test features check part of the overall GFP system, which may have failures among the unchecked components. Where multilayered ground fault protections exist, such as applied to branch circuit breakers, tie breakers, and main breakers in a switchgear assembly, ground fault protection may be zone interlocked to prevent the upstream breaker from tripping until the downstream breaker has had opportunity to clear. If zone interlocked, the GFP or a branch circuit breaker sensing a downstream ground fault condition may send a blocking signal to prevent the upstream tie or main breaker from tripping unnecessarily.

11.11.1   Electrical tests

With the connection between neutral and ground temporarily broken, measure the insulation resistance between the neutral and ground conductor(s). Reconnect the one neutral to ground connection. Compare the resistance value to similar connections.

Electrically confirm that the GFP relay does not operate at 90 percent of its setting and that it does pick up at less than 175 percent of its setting, not to exceed 1200 A.

When GFP systems utilize phase and neutral sensors, it is necessary to pass test current through each phase sensor and the neutral sensor to assure relative polarities at the relay are correct. The relay should operate when the test current passes in the same direction, with regard to sensor polarity marks, of the two sensors being tested. The relay should not operate if the test current passes through the phase sensor in one direction and simultaneously through the neutral sensor in the opposite direction, with regard to sensor polarity marks.

If the GFP relay provides time delay, measure the time delay at a suitable multiple of the setting level, typically 150 to 200 percent. Compare results to the manufacturer-published time current curve.

Verify the power breaker or switch will trip open at 55 percent of the normal ac control voltage or at 80 percent of the normal dc control voltage.

Assure the zone interlock features operate correctly, if existing.

Note: It is the opinion of this writer that GFP relays are more accurately calibrated by most secondary injection test methods. However, primary injection tests are necessary to prove that the GFP system functions properly.

11.12   Ground Fault Circuit Interrupters (GFCI)

These are personnel protective devices intending to prevent electrocution when a person's body, or other object, bridges between an energized 115/230 V conductor and ground. The GFCI interrupts the circuit when current flow to ground exceeds 4 to 6 mA. GFCI protection is available in the form of a GFCI molded case circuit breaker or a GFCI receptacle, and may be fixed in a panel, box, or as part of a portable device. These devices are factory sealed and not adjustable. The GFCI molded case breaker provides overload and short circuit protection for the circuit, as well as ground fault interruption. The GFCI receptacle provides ground fault interruption only. The National Electrical Code specifies use of GFCI devices in several applications. Portable GFCI devices are used at facilities and plants to protect workers when using portable extension cords, drop lights, and electric tools.

Maintenance should be performed per the manufacturer's instructions and should include:

Visual and mechanical inspection. Check tightness of connections.

Verify proper installation and connection.

Exercise the device by operating the test button provided to cause the breaker to trip to the open position. Perform this test as often as the manufacturer recommends.

Replace the GFCI if it fails to trip.

11.13   Low-Voltage Molded (Insulated) Case Circuit Breakers

Molded case breakers are the most commonly used circuit breaker, found in residential electrical panels, commercial buildings of all sizes, and institutional-grade facilities. They are available for single-pole, two-pole, or three-pole use, throughout a wide range of ampere ratings, and for 250-V class and 600-V class installations. Many molded case breakers are factory sealed, with noninterchangeable trip units. Smaller-size molded case breakers include a thermal-style overload trip, utilizing a bimetallic element, which operates in increasingly less time as ampere load increases beyond the minimum operating value. A magnetic trip feature is added to provide a more "instantaneous" trip when short circuits occur. Larger and more expensive molded case breakers may provide interchangeable trip units or replaceable trip units having a variable long-time (overload) trip and variable instantaneous trip features. The trip units of most larger-sized molded case breakers manufactured today are of solid-state design, offering an array of overload and short circuit settings and time-current curve characteristics, to better protect against phase-to-phase and phase-to-ground overcurrent conditions that may occur. For certain uses, a molded case breaker may provide magnetic (instantaneous) trip only, as often used in motor circuits, where the starter provides overload protection. Smaller-sized molded case breakers may plug (snap) onto the panel bus, such that its spring clip conductor connection also serves to mechanically hold the breaker in place. Plug-in breakers are typically rated 15 to 125 A for 120/240-V branch circuit use, at 10,000 to 65,000-A asymmetrical interrupting current (AIC) duty, although a few higher-rated AIC plug-on breakers are available.

Larger and/or better-grade molded case breakers are typically bolt mounted to the switchboard or panel frame, bolt connected to the bus bars, and typically rated 15 to 2500 A, for 600-V class service, and up to 200,000 AIC. The higher available interrupting ratings are accomplished by including current-limiting fuses with the breaker. Breakers having current-limiting fuses should include an anti-single-phase feature to assure that when one fuse opens, all poles of the breaker will open, so that a single-phase condition will not be seen by three-phase motors. Molded case breakers may include an optional shunt trip feature and/or an electrical closing feature, to facilitate breaker operation initiated by a remote device, such as a control switch, undervoltage relay, or other control device.

11.13.1 Electrical tests, energized and carrying load

Measure and record ampere load on all phases.

Perform an infrared thermographic scan.

11.13.2 Electrical tests, deenergized

Measure insulation resistances pole to pole and pole to ground, with breaker closed.

Measure insulation resistance across each set of contacts, with the breaker open.

Measure the resistance of closed contacts, using a low-resistance reading ohmmeter.

Measure insulation resistances of control wiring and associated devices.

Caution: Solid-state devices may be damaged by insulation resistance tests. Review manufacturer's instructions.

Pass a test current that is a multiple (usually 300 percent) of the long-time delay ampere setting through each breaker pole. Verify the breaker trips within the published manufacturer time-current characteristic and tolerances. It may be necessary to series connect two breaker poles in order to eliminate ground fault operation, with certain designs.

Measure the pickup and time delay values of the short-time trip feature, if any.

Measure the pickup of the instantaneous trip feature.

Measure pickup and time delay of the ground fault protection feature.

Confirm correct operation of undervoltage, shunt trip, zone interlocking, trip device operation indicators, electrical close, or other features that may be present.

11.14   Low-Voltage Air Circuit Breakers

Unlike low-voltage molded case circuit breakers, 600-V class air circuit breakers are not housed in an insulating case. Rather, the operating assembly is built into a steel frame. These breakers may be fixed mounted by bolts into a switchgear assembly, but more often each breaker and switchgear breaker cubicle includes a racking assembly so that the breaker can be readily withdrawn from the switchgear. These breakers include a mechanical mechanism to close and open the breaker, arcing and main contacts, and arc chutes placed to break up and extinguish the arc generated by opening while energized, and carrying normal load or high fault currents. These breakers may include electrical operating controls and/or a stored energy closing system. The breaker operating mechanism and contact structure are necessarily hung on a insulating back board, usually premolded and shaped specifically for that breaker make, type, and size. Other components may include interphase insulating barriers, other insulating components, coils, control devices and wiring, and in-line current-limiting fuses. These breakers may be one-, two-, or three-pole assemblies. Multiple-pole breakers are gang-operated. The main conductors of fixed breakers are bolt connected to the switchgear bus bars at the rear of the breaker. The main conductors of draw-out breakers include projecting bus bar stabs each with a female spring tensioned primary contact assembly that mates with a matching male contact within the switchgear cubicle that is hard connected to the switchgear bus. A three-pole draw-out breaker necessarily includes three line side and three load side primary connection stabs. To accomplish automatic operation to the open position at time of overload or downstream fault, a trip unit or relay with auxiliary devices is provided. A flag or other device indicates the breaker position, i.e., open or closed.

Modern arc chutes (interrupters) include ceramic or other nonorganic construction in areas exposed to the arc. Older arc chutes may be made with asbestos, which may release asbestos fibers to place operating and maintenance personnel at risk. Arc chutes containing asbestos should be rebuilt or replaced.

Air breakers include a set of main and arcing contacts on each pole. The arcing contacts close before and open after the main contacts, to limit arc erosion of the main contacts. Consequently, arcing contacts more often require dressing or replacement. When fully closed, the main contacts carry most of the load current. In closing, both sets of contacts require a positive mating to obtain relatively low resistance. Abnormal resistance at the main contacts, when closed, will result in abnormal current flow through the lesser-rated, smaller-sized, arcing contacts, which may overheat. Contact surfaces include a silver alloy to limit melting, burning, and surface resistance. Contact surfaces should be kept clean and bright.

Typical low-voltage power breakers may be 600- to 4000-A frame size, with interrupting rating ranges from 25,000 to 200,000 A. As with molded case breakers, higher-level interrupting capability is achieved by inclusion of high interrupting fuses in series with each pole, and a protective feature to initiate opening the breaker if a fuse blows.

Older-model trip units may be of electromechanical design, incorporating oil dash pots or air bellow devices for time delay, a series trip coil to sense current level and produce inductive energy, and other mechanical apparatus to provide an adjustable pickup point and to cause the breaker to trip. Various makes, designs, and vintages are to be found. Typically, these trip units offer adjustable time delay features, long-time pickup, short-time pickup, and instantaneous trip protections.

Today's trip units are typically solid-state and provide the same protections plus ground fault. Rather than a series trip coil, sensors (ct's) are used to measure line current flow. Sensors for a particular trip device provide a primary ampere rating that equates to the 100 percent (maximum) setting point of the trip device long-time trip protection. The sensor secondary rating is usually 0.5 or 1.0 A. Solid-state trip units may offer many features, ratings, and setting requirements. Each feature should be proved by appropriate testing at the time of maintenance inspection. These trip devices may provide zone interlocking and time-current curve-shaping options. Optional features may include an LED display of measurements of circuit voltage, amperes, power factor, watts, and vars, and the ability to communicate with remote supervision or load control systems.

11.14.1 Maintenance

Review manufacturer instruction books, and become familiar with the particular design, operation, and features of each breaker. A drawout-style breaker should be removed from its switchgear cubicle using the manufacturer-provided breaker lift apparatus. Bolt-in-type breakers are usually serviced without removal from the switchgear assembly.

Caution: Discharge the stored energy system, if any, prior to working on the breaker. Failure to recognize and make safe this feature may result in serious worker injury.

Maintenance includes a thorough cleaning, using a vacuum cleaner with plastic attachments, blower, lint-free cloths, and use of an electrical solvent as necessary to remove films, hardened lubrications, and the like. Remove interphase barriers and clean. Inspect insulating components for cracking, burn marks, tracking, discoloration, or any other abnormality. Do not unnecessarily disturb asbestos arc chutes. Carefully remove nonasbestos-type arc chutes and clean. Remove arc residues, dirt, or other debris, using cloths or a light nonconductive abrasive paper. The insulating quality of an arc chute is especially important at the time of breaker opening, action, and when an energized breaker is in the open position.

Inspect the arcing and main contacts. These contacts should be cleaned (burnished) to remove film or burn marks. A plastic buffing pad, similar to those used to clean pots and pans, but not soap-impregnated and not very abrasive, works great. Burrs may be removed by careful dressing. Check contact wipe, alignment, and pressure. Inadequate contact pressure may result in overheating when the breaker is heavily loaded. Contact lubricant may be lightly applied to cleaned contact surfaces and to movable conductor joints.

The operating mechanism should be cleaned. It should be free of corrosion and grime. Shafts and bearings may be judiciously lubricated with a light nongumming machine oil. Check for loose or missing clips, springs, keepers, hardware, or other parts. Check tightness of all connections. Breaker operations to the close and open positions should be quick and positive, not sluggish or binding. During operational tests, keep hands, fingers, and clothing clear and remain cognizant of a stored energy capability.

Check all breaker auxiliary devices, including coils, shunt trips, interlock components, and solenoids. Pay particular attention to the breaker primary connection stabs and contact clusters and to all secondary wiring breaker to switchgear connections and contacts. Check the condition and alignment of the frame grounding point. Confirm correct operation of the open-close position indicator.

Check the racking mechanism and lubricate as applicable. Confirm alignment with the breaker inserted in its cell or cubicle.

Check tightness of bolted conductor connections with a torque wrench or low-resistance reading ohmmeter.

11.14.2 Tests

Measure insulation resistance pole to pole and pole to ground with breaker closed, and across the open contacts with breaker open. Test with arc chutes and interphase barriers in place.

Measure insulation resistance of control wiring that does not include connection to solid-state apparatus.

Measure contact resistance with a low-resistance ohmmeter.

Typical trip device tests include

1. Long-time feature pickup level

2. Long-time time delay

3. Short-time feature pickup level

4. Short-time time delay

5. Ground fault feature pickup level

6. Ground fault time delay

7. Instantaneous feature pickup level

Review manufacturer instructions and time current curves. Confirm proper trip device and time current characteristics exist. Primary injection method testing will check and provide all of protective systems components to assure that the desired protection exists. Secondary injection calibration tests using manufacturer-provided test kits often test the trip unit functions only. However, secondary injection test apparatus usually provides more accurate calibration capability.

Assure proper operation of all auxiliary devices, to include shunt trips, undervoltage trips, anti-single-phase, electrical interlocks, etc.

Perform complete breaker operational tests.

Perform an infrared thermographic scan.

11.15   Medium-Voltage Breakers

Medium-voltage breakers are three-pole devices that interrupt the circuit with contacts in air, oil, or vacuum medium. Typical breakers may include a mechanical mechanism, a stored energy charging motor, arc chutes, interphase insulating barriers, coils, switch devices, control wiring, main contacts, arcing contacts, three line side and three load side main conductor connection stabs, an assembly of secondary wiring connection stabs, a ground connection, frame, and insulators. These breakers may be designed for indoor or outdoor installations. Most of these breakers are installed within indoor switchgear assemblies, with provision for either a vertical lift or horizontal racking means of connection to the switchgear bus.

11.15.1   Air breakers

Typical air breakers include main contacts which open first, to be followed by the opening of arcing contacts. The arc formed upon parting of the arcing contacts is magnetically drawn and/or air blown into the arc chutes, which by design elongate, cool, and extinguish the arc.

Depending on make and type, all air breakers require a series of maintenance inspections and adjustments in order to provide reliable service when called upon to function.

Caution: All air breakers operate at high speed and very forcefully. Additionally, most air breakers include stored energy for opening and/or closing. It is necessary to understand each particular mechanism design and to safely block or discharge all stored energy sources before working on the breaker.

11.15.2   Maintenance

Review the manufacturer's instruction book for the specific type, model number, and ratings of the breaker to be serviced. The breaker must be withdrawn from its switchgear cubicle for service. Preferably, medium-voltage breakers are serviced at a nearby breaker test station which provides a ready means of operational testing with the breaker removed from the switchgear.

Perform breaker maintenance recommended by the manufacturer. Typical maintenance includes removal, inspection, and cleaning of interphase insulating barriers and arc chutes. Check for cracks, physical damage, charring, erosion of ceramics or other components, and dirt. Inspect contacts for misalignment, burning, pitting, wipe, and wear. Contacts may be smoothed by judicious use of a fine file.

Inspect secondary control contacts, and the six primary stab contacts with disconnecting finger clusters, for alignment, damage, wipe, and evidence of any arcing. Inspect and clean the operating and stored energy mechanisms in accordance with manufacturer provisions. Lubricate moving metal surfaces and joints. Assure bearings, shafts, and slide points are not gummed and operate freely. Apply an approved contact lubricant to the primary stab finger clusters and other current-carrying movable parts or joints only as necessary or indicated desirable. Do not apply excess lubrication. Check blow coils, interlocks, wiring, insulating blocks and insulators, charging motor, control relays and devices, springs, and gears, as applicable. Inspect puffer operation. Check all bolted connections for tightness. As applicable, perform measurements and adjustments of opening and closing speeds, contact wipes, alignments, trip latch clearances, prop clearances, contact gaps, trip latch wipes, release latch settings, interlocks and auxiliary switch alignments, pawls, plungers, limit switches, etc. Verify fit and alignment of the breaker and racking mechanism in the switchgear cubicle.

Record as found and as left operation counter readings.

11.15.3   Tests

Measure contact resistance using a low-reading ohmmeter.

Measure insulation resistance each pole to ground, pole to pole, and across the open contacts.

Perform insulation resistance measurements of control wiring and devices, at 1000 V dc.

Measure resistance of the blowout coil.

With the breaker at the test station or in the cubicle test position (not connected to the energized bus), perform the following operational tests:

1. Trip and close by control switch.

2. Trip breaker by its protective relays and devices.

3. Confirm breaker trip free and antipump features function as intended.

Optional tests as applicable or desired or especially after overhaul or significant repair may include:

1. Power factor (dissipation factor) tests.

2. AC over potential (voltage withstand) tests.

3. Measure minimum pickup voltage of the breaker trip and close coils, and record.

11.15.4   Vacuum breakers

Similar to air breakers, vacuum breakers also include main pole conductors, steel frame, high-speed high-energy operating mechanism, line and load side connection stabs each with finger cluster connectors, cubicle racking and withdrawal provisions for insertion into metal-clad switchgear assemblies, interlocks, an operating mechanism, auxiliary switches, stored energy capability, motor, and control systems.

The principal difference, however, is a significant one. The main breaker contacts are enclosed in a sealed vacuum interrupter, and there are no arc chutes.

Advantages of a vacuum breaker are that it is an ideal dielectric, arc products are not vented into the atmosphere, and it has relative quiet operation, fewer parts, and predicted longer life. Additionally, vacuum breakers are more compact and may be stacked two high in switchgear to conserve floor space in the plant. A disadvantage is that should a vacuum interrupter bottle lose vacuum, it becomes unsafe to operate. To confirm vacuum integrity it is necessary to perform a vacuum interrupter integrity test, which is the application of a high test voltage across the open interrupter contacts, at the test voltage and test method recommended by the manufacturer. This test, as are all switchgear and breaker over potential tests, is preferably performed using appropriate ac test equipment. Use of dc high potential test equipment is acceptable in some applications, and by some manufacturers, but ac is preferred. A vacuum integrity test station should be available for occasional use by operating or test personnel if at all possible.

Cautions: Applying abnormally high voltage across a pair of contacts in a vacuum may produce hazardous x-radiation Also, dc high potential test sets with half or partial wave rectifications are not suitable to test vacuum interrupters, because their peak voltages may be much greater than the metered voltage level. Only test a correctly adjusted breaker, and always comply with manufacturer's directions and test voltage limitations.

11.15.5   Maintenance

Review manufacturer's instructions thoroughly. Follow all manufacturer safety precautions regarding discharging or blocking stored energy features before working on the breaker. Inspect the breaker mechanisms, housing, interphase barriers, insulators, auxiliary devices, and control wiring for evidence of damage, corrosion, overheating, or insulation tracking. Interphase barriers can be removed as necessary for better access. Inspect all bolted connections for tightness or for high resistance. Clean all components. Check and adjust as necessary all interlocks, control switches, auxiliary switches, latches, plungers, ground contact fingers, and other adjustment points as applicable. Apply lubrication sparingly, as the instructions direct. Perform slow close inspection. Measure all critical distances. It is especially important to inspect for contact erosion, contact wipe, and contact gap. Inspect and service breaker racking mechanism. Record as found and as left operation counter readings.

11.15.6 Tests

Measure contact resistance using a low-resistance ohmmeter.
Perform a complete set of breaker trip tests.
Verify trip, close, trip free, and antipump functions.
Measure insulation resistance each pole to ground and pole to pole.
Measure insulation resistance of control wiring.
Perform vacuum integrity test across each opened vacuum interrupter.
Perform an ac voltage withstand (hipot) test pole to pole and pole to ground, with the breaker in closed mode.
Optional tests might include breaker time and velocity analysis, minimum pickup of close and trip coils, and power factor tests.

11.15.7 Oil breakers

Oil circuit breakers (and switches) are much different from air and vacuum breakers. Their main and arcing contacts are immersed in a tank of insulating mineral oil. Oil circuit breakers are referred to as OCBs. Newer oil breakers are for outdoor installation; however, many indoor 5-kV and 15-kV class OCBs exist. Older 5- to 15-kV breakers were typically solenoid-operated and included three line and load breaker connection stabs and a vertical lift mechanism for connection to the switchgear bus. These breakers can be bolt-in connected also. New outdoor OCBs are typically 5-kV to 230-kV class rated, freestanding, and include a pneumatic- or hydraulic-powered operating mechanism, interrupting devices, tail springs, shock absorbers, and a one-tank or gang-operated three-tank design. The bushings of higher-rated units may be oil-filled or condenser type and pass through current transformers installed inside the tank.

Indoor oil breakers provide a risk of oil fire that today is considered unacceptable. However, if properly maintained and operated within their ratings, oil breakers have been a dependable workhorse of the electrical power industry.

The contact arcing that occurs under the oil develops gases, including explosive gases, that must be vented to the atmosphere, so OCB tanks are not sealed. Arcing in oil develops carbon which disperses through mineral oil and collects along internal tank components and the tank floor. Outdoor breakers breathe owing to temperature variations, and tend to collect internal moisture. Most free water collects in the bottom of the tank. Internal spacings take this moisture into account, but excess free water and moisture absorbed in the oil can develop and result in failure.

11.15.8   Maintenance

Study the manufacturer's instructions.

Inspect and service the solenoid, pneumatic, or hydraulic operating mechanism as directed. Check for leaks, operating pressure levels, pressure reliefs, pressure switches, fluid levels, pump actions, mechanical linkages, and connections. Lubricate bearings, slides, and joints as directed. Inspect anchorage, alignment, and grounding. Complete all checks, measurements, and adjustments to the operating mechanism and appurtenances. Check oil levels in each breaker tank, and any oil-filled bushings. Confirm breather vents are clear.

Internal inspection of each tank is performed by either lowering the oil level or completely draining the tank. Some larger breakers require complete draining and provide internal access through a gasketed side-located access hole. Oil handling and temporary storage equipment must be clean, uncontaminated, and dry. Oil should be filtered when transferred between breaker and storage vessels.

Inner tank liners, insulators, and apparatus should be cleaned and inspected thoroughly. Inspection should include contacts, lift rods, resistors, toggle assemblies, interrupters, coil springs, shock absorbers, rollers, internal bushings, and insulators. Replace disturbed gaskets. Check tightness of all connections. Perform all manufacturer-recommended measurements, tolerances, adjustments, and alignments. Slow close the breaker and check for binding contact alignment, penetration, and overtravel. Restore the tank and refill with clean, dry oil.

Caution: Follow all manufacturer warnings and safety directions completely. Follow confined space procedures. Breaker actions are extremely fast and powerful. Keep body and limbs clear.

11.15.9   Tests

Measure contact resistances.

Perform insulation resistance tests pole to pole, pole to ground, and across the open breaker contacts.

Perform insulation resistance test of control wiring and devices.

Trip breaker by each protective relay or device.

Draw an oil sample and perform dielectric breakdown, interfacial tension, visual inspection, color, and power factor tests.

Perform trip, close, trip free, and antipump operational tests.

For larger breakers, perform power factor (dissipation factor) tests on each pole with breaker open, and on each phase with breaker closed. Check tank loss index.

Perform power factor (dissipation factor) tests on each bushing.

Perform time travel tests and assure blade velocity, travel, and overtravel measurements are within tolerances. Check the relative position of significant auxiliary contacts.

11.16   Switches

Disconnect switches are used to connect or disconnect circuits or apparatus and for isolating equipment or circuits. As these switches do not provide interrupting capability and are operated under non-load-break conditions, interlocks are often provided to prevent an unintentional load interruption.

Interrupter switches include provisions to break load at rated voltage and continuous current ratings. These switches have fast close and open actions and may provide close and latch ratings. The fast actions should be independent of the speed of operation provided by the operator. Most of these switches are manually operated. Spring torque is developed by the operating handle, to be released as the operating shaft passes a predetermined point, so that the switch blade snaps open or closed. Interrupting switches may include arc chutes and/or a spring-loaded auxiliary arcing blade. As the main current-carrying switch blade opens, the arcing blade continues to complete the circuit momentarily while sufficient opening of the main blade contacts occurs, and opening spring tension at the arcing blade builds. Then the smaller arcing blade whips open at high speed to extinguish the arc. Interrupter switches may be fused or nonfused type.

Disconnect and interrupting switches are available at all voltage levels and may be of open, free air construction or housed in an enclosure.

Safety switches are commonly used in 600-V class systems. These switches are enclosed and have external operating handles, and the hinged access door is interlocked such that it cannot be opened unless the switch is open. These switches may include fuses.

Bolted pressure switches are a form of safety switch that applies bolted pressure to the movable and stationary contacts through its mechanism. Bolted pressure switches may include electrical tripping, fuses, ground fault, and/or loss of phase protection, especially when applied as main interrupting devices. These switches can offer contact-interrupting ratings 12 times the continuous duty rating.

Double-throw automatic transfer switches are used for 600-V class switching from the normal power source to an alternate or emergency source and do not generally include overcurrent protection. These transfer switches are electrically operated and include voltage sensing of both the normal and alternate source. Time delay features may be provided. Typically, the closed switch is mechanically held and has replaceable contacts. Automatic transfer switches are commonly used in emergency generator systems found in large buildings, hospitals, services to computer centers, and various commercial and industrial facilities.

All of the above switches are available for indoor or outdoor application. Switches may be single-pole or multipole.

11.16.1   Maintenance

Check the enclosure for damage and for secure mounting or anchorage. Note evidence of corrosion or water entry and eliminate the source of any moisture. Check paint condition, clean, prime, and paint as necessary. Assure access door latches, locks, and interlocks operate satisfactorily. Interlocks should prevent opening the access door when the switch is closed. Verify equipment grounding. Check cubicle heaters. Assure any air vents are clear and that dust filters are clean. Thoroughly clean the interior and exterior. Perform operational tests. Inspect all contacts. Confirm correct blade alignment, blade penetration, stops, and contact wipe. Contact lubricant can be judiciously applied to movable current-carrying components. Check bolted connections for tightness by use of a torque wrench, low-resistance ohmmeter, or thermographic scan. Verify fuses are properly rated. Assure fuse holders are mechanically supported and that fuse cartridges are secure in the holders.

11.16.2   Tests

Measure insulation resistance of each pole phase to phase and phase to ground with the switch closed, and across each pole with the switch open.

Measure resistance across each pole of the closed switch, and across each fuse holder, using a low-resistance-reading ohmmeter. Compare values of each pole to the other and to prior recorded readings.

Measure fuse resistances and compare.

For medium-voltage switches, apply an overpotential test voltage each pole to ground, with the other poles grounded and the switch closed.

Test any ground faults, loss of phase, undervoltage, or other protective relays associated with the switch.

Perform a thermographic scan, if possible.

11.17   Fuses

Fuses may be of the current-limiting or non-current-limiting type. When current-limiting fuses melt, the resulting arc is extinguished within a half cycle, to limit current flow. Non-current-limiting fuses may melt in less than half a cycle, but the resulting arc may persist longer than a half cycle, allowing much higher short circuit current until the arc is extinguished. Fuses provide various time current characteristics but are not variable. Fuses are available in an array of voltage and current ratings. The interrupting rating of fuses is an important application concern. Most fuses are one-time, requiring replacement after operation; however, some medium-voltage fuses are available with replaceable fuse elements. Some-low voltage and most medium-voltage makes and types of fuses are sized specific to their holders. Fuses are found in panels, switchboards, starters, controllers, circuit breakers, transformers, and transformer compartments.

11.17.1   Maintenance

Identify fuse locations, ratings, and descriptions, and maintain this list. Verify fuse ratings are correct for each application. Assure that a 100-A fuse has not been placed where a 60-A fuse was required. Inspect the fuse and fuse holder for evidence of overheating, burning, cracked insulation, or abnormal discoloration. Assure the fuse to fuse holder connections are positive and tight. Verify the tightness of electrical connections, and that insulating barriers or covers are not missing.

Maintain your own stock of all fuses, especially large power fuses and all medium-voltage fuses. In one instance we worked around the clock to rebuild and make ready for service a faulted 15-kV switchgear assembly, only to wait the next 4 days deenergized to locate and obtain three specific 15-kV fuses. Some large fuses may cost $1000 to $2000 each, and they may never be needed. But the cost of one otherwise unnecessary plant shutdown of 8 to 24 h may make a once in 30 years fuse inventory purchase a relatively small expense.

11.17.2   Tests

Smaller-sized less valuable fuses are not usually tested.

Measure and record the resistance of very large low-voltage and medium-voltage fuses. Comparison of fuse resistances can reveal a weakened internal element and allow replacement during a planned shutdown.

Perform insulation resistance measurements.

11.18   Motor Control Centers and Starters

Motor control centers are an assembly of motor control units (starters) built by modular design in order to centralize. This grouping is more efficient with regard to power supply, overall space required, installation cost, and operator convenience. Motor control centers are used to supply electric service and to control motors. Motor control centers and starters are dead front assemblies, available in voltage levels up to 7200 V. Motors operating in the 15-kV class level are not often seen and are generally serviced by 15-kV circuit breakers having motor starting protection and controls.

Motor control centers (MCC) may include the housing with cubicles, starter assemblies, bus bar conductors, and insulators. They may also include fuse assemblies, instrument transformers, metering, circuit breakers, switches, cable termination points, and/or a control power transformer. The starters within MCC are almost always draw-out style.

Starters and controllers are assemblies that typically include the control and protective devices necessary to serve a motor. Starters may be free-standing on the floor, wall-mounted, or part of a motor control center. Starters typically include one or more contactors, a primary disconnect means, control fuses, an overload relay assembly, various control devices, line fuses or circuit breakers, and line and load connections. An internal breaker may also serve as the disconnect means. Starters may be single-phase or three-phase. Low-voltage starters are provided in five sizes, size 5 being the largest. Starters may include motor circuit protectors (MCP), which are molded case breakers that include magnetic (instantaneous) protection only, as the starter also includes an overload relay. The more usually seen 208-V or 480-V starter overload relay includes an overload device referred to as an overload heater which thermally senses the load current and opens at a predetermined "overload" current level for that motor. Low-voltage overload relays and heaters are available in a wide variety of operating ranges and time current characteristics. Good motor protection requires selection and use of correctly rated protective devices. A medium-voltage or a very large low-voltage starter may include a reduced voltage starter contactor that opens at a predetermined point after the motor has begun rotating to then allow application of full voltage by closing of a run contactor. Many motor starters or controllers include a variable-frequency drive (VFD) system that controls motor speed by changing frequency and voltage. Additionally, solid-state control, protection, and metering apparatus is regularly included in today's starters. Protection is provided by various protective relays. Medium-voltage contactors may include vacuum interrupters for circuit switching.

11.18.1   Maintenance, low-voltage units

Visually and mechanically inspect for damage, corrosion, and evidence of arcing or overheating. Thoroughly clean the interior and exterior. Check all bolted electrical connections for tightness or high resistance by use of a low-reading ohmmeter or torque wrench. Inspect contactors for contact burning or pitting, alignment, and contact wipe. Replace badly pitted contacts. Insulating supports, barriers, and insulators should be checked for cracking, charring, tracking, or overheating. Check all interlocks and safety devices that exist.

Compare the overload element rating with the motor full-load current rating, including the effect of any existing power factor correction capacitors. If fuses exist, check fuse ratings. Lubrication should be applied only as the manufacturer directs. Check all control devices, such as pilot lights, auxiliary switches, control wiring, and metering present. Check starter disconnecting, racking, and draw-out provisions.

11.18.2   Tests, low-voltage units

Perform phase-to-phase and phase-to-ground insulation resistance measurements of each starter with contacts closed and protective device open.

Perform insulation resistance testing of control wiring and devices. However, do not include solid-state devices.

Measure resistance of bolted electrical connections, using a low-resistance-reading ohmmeter.

Perform breaker and instrument transformer tests as applicable.

Test overload relays by application of an appropriate primary injection test current, measuring the time to trip. Compare to manufacturer's published data. Perform this test with all poles in series for the time test and each individual pole for comparative tests. Test of individual devices will often provide trip times longer than expected.

Operationally test all devices and controls.

Measure and record motor run currents if possible.

11.18.3   Maintenance, medium-voltage units

Review the manufacturer's instructions and descriptive bulletins carefully. Perform a thorough visual and mechanical inspection.

Check for general condition, corrosion, water entry, physical damage, or overheating. Clean thoroughly and carefully. Inspect insulators, insulating supports and barriers for cracks, tracking, charring, overheating, or other abnormality. Check mounting, anchorage, and alignments. Inspect tightness of bolted electrical connections, using a low-resistance ohmmeter or torque wrench. Check operation of shutters, interlocks, racking mechanisms, doors, and other mechanical safety features. Inspect contactor operating mechanisms. Check contacts for burning, pitting, overheating, or excessive wear. Minor contact surface irregularities can be smoothed with a fine file (not sandpaper). Replace worn or damaged contacts. Check contact wipe, alignment, gap, and pressure, as applicable. Apply lubrication only as the manufacturer recommends. Check bearings for gumminess. Compare overload ratings with motor full-load current rating. Check fuses and fuse holder connections. Inspect service switches, instrument transformers, primary and secondary contactor connection points and stabs, control devices, and wiring. Check the wear indicators of vacuum interrupters. Assure gap and other contact parameters are especially correct in vacuum interrupters.

11.18.4   Tests, medium-voltage units

Measure contactor insulation resistance pole to pole and phase to ground, with the contactor closed, and across the open contacts of each phase.

Measure insulation resistance of control wiring and electrical devices, with the exception of solid-state devices.

Measure contact resistances using a low-resistance-reading ohmmeter.

Measure resistance of bolted electrical connections, using a low-resistance-reading ohmmeter.

Perform vacuum interrupter integrity tests per manufacturer directions, preferably using ac rather than dc high-potential test equipment. Do not use dc high-potential test equipment that operates as an unfiltered half-wave rectifier.

Perform ac applied voltage withstand (high potential) tests, phase to phase and phase to ground, with the contactor closed. Apply this test across the open contacts if not a vacuum interrupter-type contactor.

Measure, record, and compare blowout coil circuit resistances.

Perform transformer, circuit breaker, and switch tests as applicable.

Electrically test all electrical safety, trip, and control devices and features.

Measure power fuse resistances.

Test overload and other protective relays as applicable.
Perform operational tests.

11.19   Batteries and Battery Chargers

Batteries are used in electrical switchgear applications, in uninterruptable power supply systems, and for emergency lighting. When part of an electrical switchgear installation, batteries play a vital role in system protection.

When an electrical fault in a medium- or high-voltage power system occurs, the upstream circuit breaker(s) must operate to the open position in order to deenergize and isolate the faulty conductor or equipment. To operate, these circuit breakers require a reliable control voltage and source of operating energy. At time of severe fault, power system voltage may significantly decrease. If the circuit breaker controls were directly served by the same ac source, the control voltage necessary to allow circuit breaker operation may not be present. Therefore, most medium- and high-voltage circuit breakers and protective system devices rely on a dc control source provided by rechargeable batteries.

Again, switchgear batteries are critical to power system protection. Yet they are often unappreciated, and the most neglected component. This writer is aware of many major, even catastrophic electrical fires and losses that grew to such size solely because dc tripping energy was not in existence at the time of a fault. Battery life expectancy and reliability is usually lost by failure to perform simple maintenance tasks on a weekly and monthly schedule.

11.19.1   Batteries

Lead acid and nickel cadmium–type batteries are typically used in electrical substation and switchgear applications. Typical battery banks operate at 24, 48, and 130 V.

There are several different types of lead acid batteries. Lead acid batteries of pasted or "flat" plate with antimony grid construction are the least expensive. But they provide the shortest life and should not be used in switchgear applications. Other types of lead acid batteries, including the lead calcium type, provide an expected life of 20 to 25 + years.

Nickel cadmium batteries can provide relatively long life. They are slightly easier to maintain, which is more important when inspection and maintenance service is less seldom performed. However, they are more expensive to install and to dispose of.

It is important to identify the specific type of switchgear battery in use, its charging voltage and other requirements, by review of the manufacturer's instruction pamphlet.

As applicable, also review the following publications:

ANSI/IEEE 450: *Recommended Practice for Maintenance, Testing and Replacement of Large Lead Storage Batteries for Generating Stations and Substations.*

ANSI/IEEE 1106: *Recommended Practice for Maintenance, Testing and Replacement of Nickel-Cadmium Storage Batteries for Generating Stations and Substations.*

11.19.2 Battery chargers

The battery charger for switchgear batteries should be the regulated type, with characteristics to meet the specific needs of the batteries to be served. The charger must satisfy the dc system load requirements including battery losses, to maintain a "float" charge rate at the battery. Over a period of time, the charged state of several cells in a bank of batteries may become higher or lower than neighbor cells. The battery charger should provide for application of a higher charging voltage referred to as an equalizing voltage level, which application over a specific period of time will tend to equalize the charged state of one cell to the other. Different battery types and dc system characteristics require different equalizing needs. The charger must provide adjustable float and equalize voltage levels. During the time that batteries are subjected to an equalizing voltage, they require much more attention. Chargers that provide an automatic means to cut back from an equalizing charge level to a float charge level are preferable.

11.19.3 Maintenance of batteries

Assure the battery room or area is adequately ventilated and that signs warning of possible presence of explosive gases are present. Note the existence and location of eyewash equipment.

Perform a visual and mechanical inspection. Check liquid levels and add distilled water where needed. Clean the battery surfaces and all terminals. Inspect bolted electrical connections for evidence of corrosion, oxidation, or other abnormalities. Apply an oxidation inhibitor to all battery connections. Check the condition of wires and cables, battery racks, mountings, and anchorage.

11.19.4 Tests

Measure resistance of bolted electrical connections, using a low-resistance ohmmeter. Perform this test after terminals have been cleaned and inspected, but prior to application of an oxidation inhibitor.

Measure and record specific gravity and temperature of each cell.

Measure and record cell voltages and battery voltages with the charger on line and in the float mode.

Measure intercell connection resistances.

A load test may be performed in accordance with manufacturer instructions or the appropriate ANSI/IEEE standard.

Monitor temperatures, cell voltages, liquid levels, and specific gravities if an equalizing charge is applied.

11.19.5 Maintenance of chargers

Visually and mechanically inspect. Note any damage, presence of moisture, and evidence of overheating, corrosion, or other abnormalities. Identify the features provided by the charger. Check all bolted connections. Check mounting and anchorage. Clean carefully.

11.19.6 Tests

Measure resistance of bolted electrical connections, using a low-resistance ohmmeter.

Verify float and equalize voltage levels.

Assure accuracy of all meters.

Check operation of any alarms.

Verify all charger functions.

11.19.7 Cautions

When charging, batteries emit highly explosive hydrogen gas. Open flames and sparks must not be allowed in battery areas or rooms. Adequate ventilation is required.

When working around electric storage batteries do not cause or allow short circuits or arcs to occur.

Protect against battery acids by wearing protective gloves, goggles or face shield, and aprons.

11.20 Transformers

Transformers do not generate or create electrical energy but do alter the form of electrical energy.

Transformers are devices that convert or transform electrical energy from one voltage or current level to another level, with relatively insignificant energy loss. Properly installed, maintained, and utilized, transformers are relatively long-lived devices, or should be.

Transformers of one type and size or another are found in almost every electrical building, including residences.

11.20.1   Power and distribution transformers

At industrial and commercial facilities having large electrical power needs, incoming utility service voltage levels may range from 33,000 to 230,000 V. Large power transformers may be used to step the utility service voltage level down to a 4160- or 13,800-V distribution voltage level, which is then "distributed" to various locations within the plant or building. Other transformers may further step down the plant distribution voltage to provide for 2300- and 480-V motor utilization or to serve 480- and 120-V utilization needs. Transformers may be used to step voltage up to a higher level in order to satisfy load requirements or to more economically transmit electrical power an extended distance. Transformer size is rated in kVA (kilovolt-amperes). Power transformers may be single-phase or three-phase constructed.

11.20.2   Instrument transformers

Instrument transformers are commonly used in electrical substations and switchgear.

These transformers are used in measurement and protective device circuits. The primary of an instrument transformer senses the relatively higher level of voltage or current flow of a power or distribution circuit. The secondary (output) windings provide, by predetermined ratio, a reduced voltage or current level. An instrument transformer is categorized as being a potential transformer or a current transformer.

11.20.3   Specialty transformers

Many deviations from the more usually seen transformer designs exist in order to meet the needs of different industries and applications. Although in less common use, some of these transformers are in regular manufacture while some are custom built to specifications. Facility operating and maintenance personnel should identify the special characteristics of such transformers.

Less usual transformers may be designed to include nonstandard ratings, unusual configurations, accessible tertiary windings, nonaccessible tertiary windings, zigzag winding configurations, unusual voltage taps, submersibility, or autotransformer connections.

Isolation transformers often do not provide voltage transformation but include internal electrostatic shielding and the ability to suppress high-frequency signals between source and ground in order to protect very sensitive loads.

Rectifier transformers include rectification apparatus in order to serve dc loads.

K-factor transformers are designed for use where high current non-sinusoidal requirements exist, and usually provide 200 percent rated neutrals, electrostatic shielding, and reduced core flux. These transformers are necessary where nonlinear loads are served and significant harmonic currents exist.

Arc furnace transformers are specially built to provide high amperage and low voltage, heavier bracing, and extra insulation, to meet the special demands of that application.

11.20.4 Control transformers

Control transformers are principally voltage transformation devices with a wide variety of uses. Larger units may be found in electrical substations or within medium-voltage switchgear to provide low-voltage control power and may be 1.0 to 10 kVA sized. Most control transformers are smaller-sized. They are found throughout various control circuits and plant apparatus. Control transformers may be used to step down or step up voltage levels.

11.20.5 Construction

The simplest transformer consists of two conductor windings of insulated wire wrapped around a common iron core. The winding receiving source energy is considered the primary winding, and the output winding is considered the secondary winding. Confusingly, the electrical industry, including this writer, habitually refers to the higher-voltage winding as the primary, as voltage step-down is the more usual application. In a step-up transformer, the primary winding is the lower-voltage winding, and the higher-voltage winding is the secondary. The number of primary winding turns around the iron core, compared to the number of secondary turns, determines the ratio of transformation.

11.20.6 Dry-type transformers

Dry-type transformers include solid insulation between winding turns, between each winding and another, and between conductors and ground. Dry transformers are usually air-cooled by self-ventilation or by forced air ventilation. Cast coil transformers are a variation of dry transformers and have an epoxy resin casting over each set of windings. Self-cooled units are designated class AA, and the forced air designation is FA. Forced air cooling usually adds 33 percent to the self-cooled kVA rating. With air cooling, the steel enclosure of dry transformers must include ventilation grilles. Most dry transformers are designed for indoor application. Outdoor units are available but not very popular. Generally, dry transformers are cheaper to construct and provide shorter life expectancies.

11.20.7   Liquid-filled transformers

The coil and core assemblies of these transformers are totally immersed in an insulating liquid, all of which is usually enclosed in a sealed tank. The liquid circulates through the windings and through cooling tubes or fins projecting from the tank, which in turn radiate internal heat to the external atmosphere. Liquid circulation by internal convection provides self-cooling and is designated class OA. If external forced air cooling of the transformer tank is also provided, the transformer becomes class OA/FA. If pumps are added to provide forced oil circulation, the transformer cooling class designation becomes OA/FA/FOA. Water-cooled transformers use internal tubes, with circulating cold water flowing, to extract heat, and are designated cooling class FOW. Water-cooled transformers are not often encountered. Increased cooling increases the kVA (power) rating of the transformer. Liquid-filled transformers must include an air or gas expansion space within the tank itself or within an overhead conservator or expansion tank.

The tank of a liquid-filled transformer is often sealed. Prior to sealing, internal air is removed by purging with an inert gas, usually dry nitrogen, which provides an inert gas blanket over the liquid. Larger, more valuable power transformers may include arrangements to maintain a positive inert gas (nitrogen) pressure above the liquid. Tank operating pressure range should be shown on the transformer nameplate.

Liquid-filled transformers may be designed for indoor or outdoor application. Most outdoor transformers are of liquid-filled and sealed construction. Liquid-filled transformers are almost exclusively medium-voltage rated and above.

11.20.8   Transformer insulating liquids

Several types of insulating liquid have been used in transformers.

11.20.9   Mineral oil

Mineral oils have been in use for more than a century. Mineral oil is relatively nontoxic, less expensive, easy to work with, and provides excellent insulating and cooling properties. Used oil may be reclaimed. Principal drawbacks include its flammability level and possible sludge buildup within the transformer. Concern with sludging has decreased with the addition of antioxidant inhibitors, increased understanding, and adequate maintenance. The flammability issue remains, even though correct installation and use, good maintenance, and adequate protection greatly reduce risk of a transformer oil fire. Mineral oil remains the principal filler for outdoor transformers.

11.20.10 Askarel (PCBs)

Askarel is a generic term for the family of polychlorinated biphenyls (PCB), fluids marketed under various trade names. Transformer askarels are a mixture of chlorinated biphenyls and chlorinated benzene. Pure askarel transformer fluids include PCB concentration levels of several hundred thousand parts per million (ppm). First installed in transformers during the early 1930s, askarel soon became the preferred filler for indoor units, even though it was much more expensive. Transformer maintenance and repair workers disliked askarel because of its strong odor and fumes, and because it could be a skin irritant. Askarel inadvertently wiped into or in the immediate vicinity of an eye became especially troublesome. Additionally it was learned that even though PCBs will not continue to burn at room temperatures, they do burn in presence of electrical arcing and may generate very toxic dioxins and furans, which are known carcinogens. Concern that PCBs posed long-term threats to the environment resulted in a federal EPA manufacturing ban in 1977, followed by many regulations regarding continued use, storing, record keeping, marking, contamination, spill cleanup, maintenance, repair activities, transporting, and disposal of these substances. PCBs are regulated under the federal Toxic Substance Control Act (TSCA). Spill cleanups have become extremely expensive, and most owners have elected to become PCB-free.

The federal EPA recognizes the following three PCB content classifications:

Non-PCB	< 50 ppm
PCB contaminated	≥ 50 to < 500 ppm
PCB substance	≥ 500 ppm

Twenty years after the EPA ban on PCBs, transformer maintenance and repair workers regularly encounter PCB contaminated and PCB substance transformers and must be knowledgeable of the special concerns and regulations that apply. Many transformers that were originally PCB filled or contaminated were retrofilled and reclassified non-PCB, only to later become recontaminated as PCBs continued to leach from internal absorbent materials.

11.20.11 R-Temp

R-Temp is a high-fire-point fluid used in lieu of askarel for indoor transformer use or wherever concern for transformer oil fire exists. Its toxicity is similar to that of mineral oil and it provides high dielectric breakdown strength. Heat transfer capability is fair, but not as good as that of other liquids in use; therefore, transformers filled with this fluid must include better circulation and heat radiation. This fluid has relatively high viscosity and pour point and may require heating during transfer.

11.20.12 Silicone

Silicone transformer fluid also provides a high-fire-point rating and is used in indoor transformers. Silicone is considered nontoxic and is environmentally acceptable. Silicone fluids provide high dielectric and good heat transfer properties and are more viscous. Silicone fluids are not as affected by heat. This fluid has a propensity to absorb water, which is not readily removed by usual filtration measures, and its cost per gallon is high.

11.20.13 Perchloroethylene

This fluid has been used as a solvent and interim transformer filler during retrofill programs to convert askarel transformers to non-PCB status. A few new transformers were furnished with this fluid. Perchloroethylene provides a high fire point, high dielectric strength, and adequate heat-transfer qualities. Its solvent qualities provide skin irritation and fumes that are an eye irritant. Perchloroethylene is considered a toxic environmental hazard. Transformers previously askarel filled that were retrofilled with perchloroethylene as the interim fluid, then ultimately filled with silicone, may still be both PCB and perchloroethylene contaminated. Perchloroethylene is regulated under the federal Resource Conservation and Recovery Act (RCRA).

11.20.14 Transformer appurtenances

As applicable, power and distribution transformers may include a winding temperature gauge and device, a nonload break tap changing means, cooling fans, and fan controls. Liquid-filled transformers may include drain and filter valves, liquid sample devices, a liquid temperature gauge and device, pressure relief device, nitrogen gas fittings, nitrogen pressure tank, sudden pressure device and relay, liquid pumps and controls, an oil conservator or expansion tank, and dehydrating breather equipment. Large power transformers may include a load breaker tap changer (LTC) with manual and automatic controls.

11.20.15   Maintenance, dry-type transformers, as applicable

With covers removed, inspect physical, mechanical, and electrical condition. Check all insulating components for visible evidence of deterioration, corona tracking, cracks, or other abnormalities. Clean thoroughly, using lint-free cloths, soft brushes, and a vacuum, being especially careful not to damage winding insulation and winding tap connections. Clean housing ventilation openings. Confirm resilient mountings are free and that shipping brackets have been removed. Verify proper grounding of the core, frame, and housing. Assure that temperature controls and cooling fans function as intended.

11.20.16   Tests of dry-type transformers

Measure insulation resistance winding to winding and winding to ground of all transformers.

Measure resistance of bolted connections of distribution and power transformers, using a low-resistance-reading ohmmeter.

Perform winding turns ratio tests of all transformers other than control transformers.

Measure winding resistance of power and distribution transformers rated 150 kVA and greater.

Measure core insulation resistance at 500 V dc if the core is insulated and may be readily isolated.

Perform insulation power factor and dissipation factor tests of distribution and power-type transformers rated >600 V and ≥167 kVA.

11.20.17   Maintenance of liquid-filled transformers

Inspect physical and mechanical condition. Record all as found gauge readings. Verify correct liquid level in all tanks and bushings. Check that pumps, fans, alarm, temperature, liquid level, and other controls operate as intended. Verify grounding. Assure that nitrogen-blanketed transformers maintain positive pressure. Check all bolted connections using a torque wrench, low-resistance ohmmeter, or thermographic scan. Check for evidence of liquid leaks at all tank or compartment entrances, fittings, gaskets, cover plate bolts, and bushings.

11.20.18   Tests for liquid-filled transformers

Measure insulation resistance winding to winding and winding to ground.

Perform dielectric absorption test and calculate polarization index.

Measure resistance of bolted connections using a low-resistance-reading ohmmeter.

Measure winding resistance using a low-resistance-reading ohmmeter.

Perform winding turns ratio tests at the desired tap setting.

Perform power factor and dissipation factor tests on all windings and bushings.

Perform an oxygen content test of the nitrogen blanket.

Draw a sample of the insulating liquid and perform dielectric breakdown, acid neutralization, interfacial tension, color, water content, visual condition, power factor, and dissolved gas analysis tests.

11.21   Voltage-Regulating Apparatus

11.21.1   Step voltage regulators and load tap changers (LTC)

Step voltage regulators increase or decrease line voltage, usually by plus or minus 10 percent. Load tap changers are a component added to the medium-voltage secondary windings of a three-phase distribution transformer. Voltage regulators are usually single-phase. Step voltage regulators and load tap changers are used to maintain output voltage within preset parameters, in order to meet the needs of connected downstream loads at the owner's premises. The windings of LTC apparatus include multiple taps that pass from the main tank through an insulating and sealing tap board, into a separate, liquid-filled compartment. The tap changing compartment houses a moving and stationary contact assembly that changes winding tap connections, to change the primary to secondary winding turns ratio, thereby changing the output voltage. Load tap changers (LTC) commonly include 16 raise and 16 lower positions, separated by a center neutral position, to allow plus or minus 10 percent voltage regulation. The LTC contact assembly operates such that tap changing occurs with the transformer energized and under load, in a make-before-break fashion so that load is not interrupted and contact arcing is minimal. The tap changing compartment also houses part of the mechanical operating assembly. Present-day LTC contacts are vacuum bottle type. A separate dry chamber houses the remaining drive assembly with motor, a voltage regulating relay, and other control devices. The winding taps and contacts of a step voltage regulator are more often located within the main tank.

11.21.2 Induction-type voltage regulators

An induction-type voltage regulator does not use winding taps and contacts, usually consists of one main tank, and is usually single phase. It consists of a rotatable primary (shunt) winding connected across the high-voltage line, and a stationary secondary or series winding that is connected in series with the line. The shunt winding is rotated on an axis. When the shunt winding is rotated in one direction from the center neutral position, the voltage induced in the series winding increases (boosts) the feeder voltage. When the shunt winding is rotated in the opposite direction from the center neutral position, the voltage induced in the series winding reduces (bucks) the output voltage. The shunt coil is rotated by a motor, shaft, and gear drive assembly, and is controlled by a voltage-regulating relay system.

11.21.3 Voltage-regulating relay system

This system is usually fed by a voltage and current instrument transformer(s). The current transformer is usually located within the main tank. The system includes a voltage-regulating relay and other control devices. The voltage-regulating relay provides variable voltage, time delay, and line drop compensation setting capabilities, and is sensitive to voltage and phase angle changes on the power line. Older-style relays are the electromechanical type. Present-day voltage-regulation relays are solid-state. These relays include provision for testing and adjusting with the transformer energized and in service.

11.21.4 Maintenance

Review all applicable manufacturer instruction booklets and develop a thorough knowledge of the type of apparatus at hand, its specifications and requirements.

Determine which inspections and tests may be or should be performed with the regulator or LTC transformer energized, and which inspections and tests absolutely require safe deenergization, lockout, and tagout.

During visual inspection, and as applicable, record position indicator, liquid-level gauge, and operation counter readings. Compare the new readings to prior readings. Look for leaks at gaskets, valves, sample devices, packing glands, and welded seams, and repair as needed. Operating mechanisms require cleaning and lubricating per manufacturer directions, including motor bearings, gears, and any grease fittings. Inspect physical and mechanical condition. Check for corrosion, clean rust spots, prime, and paint as needed. Check operation of heaters. Verify equipment grounding. Check motor and driving apparatus for correct operation. Verify motor cutoff at extreme positions. Verify tightness of bolted connections. Inspect all control devices, relays, wiring, and connections. Inspect vacuum bottle wear or erosion indicators, as applicable.

If performing an internal inspection, remove the oil, preferably by a filtration process, into clean, dry storage containers. Remove or open the access cover plate, and remove internal carbon or debris that may be present. Check tightness of bolted connections with a low-resistance-reading ohmmeter or torque wrench. Inspect the tap changer terminal board and other insulating components for evidence of moisture, tracking, cracks, or other abnormalities. Inspect contacts for wear, pitting, wipe, alignment, and secure mounting. Electrically operate the tap changer to all tap positions, observing contact and drive system operations. Replace gaskets as necessary, replace the compartment cover plate, and refill with filtered clean oil or new oil.

11.21.5   Electrical tests

Identify equipment grounding, measure connection resistances with a low-reading ohmmeter, and measure resistances between earth and the ground system or grid.

Perform insulation resistance measurements of each winding to ground and between windings, as applicable. A 10-min absorption test will allow calculation of polarization index values. Test step voltage regulators in an off neutral position. Correct values to 20°C.

Measure winding resistances at all tap positions, using a low-resistance ohmmeter.

Perform insulation power factor (dissipation factor) tests on bushings and windings, as applicable, and correct values measured to 20°C.

Measure turns ratios on each tap position of the step voltage regulator, or LTC.

If the compartment or tank is intended to be sealed and nitrogen-blanketed, measure the percent oxygen content within.

Perform voltage-regulating relay tests to assure proper operation and accuracy of bandwidth, time delay, voltage, and line drop compensation characteristics.

As applicable, vacuum bottle integrity tests (over potential) should be performed, per manufacturer instructions.

Insulating liquid samples drawn from step voltage regulator tap changer and load tap changer compartments are tested for dielectric-breakdown strength, visual condition, and color. Dissolved gas analysis tests are also performed, with results compared to other identical tap changer compartments.

Liquid samples drawn from induction regulator tanks usually

receive dielectric breakdown, acidity, visual, color, power factor, water content, interfacial tension, and dissolved gas analysis tests. Even though there are no ANSI standards for dissolved gas values in a regulator or load tap changer compartment, DGA values from one unit can be compared to the test values of similar units.

11.22   Cable and Wire

These conductors are found throughout the power and control systems of all buildings, plants, facilities, and distribution systems. Cable is an assembly of multistranded wire, or of conductor(s) and insulation or of several insulated wires in a group or within a common outer jacket. Cables and wires are installed in various types of raceways, such as conduits, pull and junction boxes, cable trays, underground ducts, and access holes. Cables are also installed direct buried in the earth or suspended aerially.

Regardless of make, type of construction, materials used, or ratings, most medium- and high-voltage power cables include a metallic shield in the form of evenly spaced copper wire strands or a thin copper tape, spiraled over the conductor insulation, but under the outer jacket. This shield serves to maintain an even distribution of electrostatic stress, to avoid the concentration of stress that would occur with every bend of the cable, which would reduce insulation life at that area. Wherever two medium-voltage cable ends are joined (spliced) the shield of each cable must be continuous across the splice area, brought out of the splice in a waterproof manner and connected to ground. At each medium-voltage cable end or termination, a means of stress relief and shield termination must be installed. There are many types and designs of medium-voltage cable terminations today, but all include or require connecting the cable shield to ground. Grounding of cable shielding at splices and terminations is important. Continuation or replacement of internal cable shielding, as in a splice, is accomplished by using a woven braid or mesh tape of fine copper strands, often referred to as tinsel braid. However, if this tinsel braid is brought out of the splice or termination and used to connect the cable shield to ground, the exposed braid may lose conductivity due to corrosion or melt and part at a time of excess current flow. Cable shielding that is not maintained at ground potential may become energized at the main conductor voltage, leading to cable failure at best. Therefore, the wire that connects cable shielding in splices and terminations, exits the cable, and connects to a system ground must be at least No. 6 AWG copper size.

Cable and cable splices are often installed in underground access holes or in cable vaults, which are confined spaces, which must be dealt with accordingly. Underground access holes are also inherently very damp or submersed areas. Cables should be installed along their walls, not across the center. Cable supports in access holes are generally a vertically mounted rack (bracket), with adjustable steps, and a porcelain seat or insulator. These cable racks and steps are usually constructed of heavily galvanized steel or of fiberglass. Where medium-voltage cables of one feeder or circuit share a common access hole or junction box space with cables of another circuit, each set of feeder cables should be wrapped with an arcproofing tape to assure that when one feeder cable faults it will not burn into an adjacent feeder cable to compound the problem. We once responded to an extensive unscheduled outage at a large naval facility to find that one 3/C 15-kV cable had faulted. Arcing had also occurred between many 3/C 15-kV feeder cables in the same crowded access hole to seriously damage or fault six different 3/C feeder (circuit) cables in the same hole, greatly extending the downtime, loss of mission, and repair cost. Application of cable arcproofing would have restricted the damage to the originally faulted cable. Beware of asbestos arcproofing, which can release harmful asbestos fibers. If asbestos must be disturbed or is otherwise able to release fibers, it must be abated. Asbestos utilized to arcproof cables is an unnecessary hazard, and removal is recommended.

11.22.1 Maintenance

Cable raceways are inspected for corrosion, mounting and support, peeled paint, and physical damage, as applicable. Coverplate gaskets are inspected for deterioration, especially if outdoors. Accessible sections of cables are inspected for evidence of jacket or insulation damage, such as nicks, cuts, or cracking, and for overheating or discoloration. Where cables lie or break over an edge, an indentation occurs which may lead to failure, especially if several heavy cables are stacked at that point. Assure that cables are installed and supported so that their weight does not hang on a connection point and so that the cable will not appreciably move at the time of sudden electrical or mechanical stress associated with starting currents or surges. Check that cable racks are securely fastened to the wall and that cables are secured to the racks.

Cable connectors should be checked for proper application, i.e., metal, voltage, ampere, size, and temperature rating. It is important that connection devices for aluminum conductors be marked AL and for copper conductors be marked CU. Connections should be checked for tightness, cracking, corrosion, or overheating. Assure that wire strands are not nicked or cut where the insulation was cut and removed at the termination. Compression connectors are inspected to assure the connector was sufficiently compressed or crimped, that the connector did not split when compressed, and that the crimp did not occur at the end of the connector barrel to result in broken wire strands. Mechanical connectors and connecting bolts can loosen, the result of temperature changes, vibration, dissimilarity of metals, and conductor movement.

Assure that splices are adequately supported on both sides and that grounding connections are in good condition.

Assure that shield grounding is intact as the splice or termination manufacturer intended.

Assure that splice and termination shield grounds are equivalent to No. 6 AWG copper for 5- to 25-kV cables 250 kcmil or smaller, and that shielding ground connections are in good condition. Inspect splices and terminations for evidence of water entry, overheating, or tracking. Look for compound or filler materials that might be leaking from poured or filled splices, potheads, or terminal chambers.

Assure that arcproofing tapes are in place and secured by glass tape binders.

Access holes should be cleaned of debris and dirt, including the internal sump or drain, if any.

11.22.2 Testing

Perform an infrared thermographic scan of cables, wires, splices, and terminations.

Perform insulation resistance measurements, phase to phase and phase to ground. The tests should not include solid-state equipment without consulting the manufacturer.

For medium-voltage and other shielded cable, this test is applied between each main cable conductor and its grounded shield, with all other cables and shields grounded.

The dc test voltage for 600-V class cables should generally be 1000 V, and for medium-voltage class cables 2500 or 5000 V.

Check tightness of bolted electrical connections, using a low-resistance-reading ohmmeter or a torque wrench.

A dc high-voltage test is applied to all medium-voltage cables, between each cable conductor and its grounded shield, with all other cable conductors and shields connected to ground. Prior to performance, the cable should be disconnected and isolated from surge arresters, switches, transformers, starters, motors, and the like. It is important to assure that each cable end is known, safely guarded from human contact, adequately clear of adjacent objects, and protected from excess corona discharge (leakage) prior to application of the test voltage. Test parameters and precautions specified in the appropriate ICEA standard for that cable should be adhered to. The maximum maintenance test voltage to be applied should not exceed that listed in Fig. 11.7 or the manufacturer's published limits, whichever is lower.

The test voltage should be applied in at least six equal increments, holding at each step for 1 min or until the current level decays and stabilizes. Leakage currents are read and plotted at each step, and the test is halted immediately if the leakage current slope becomes nonlinear or if voltage or current values become unstable, to indicate impending specimen failure. The intended maximum test voltage should be held no longer than 5 min. Properly performed, these tests will not harm good insulation. Comparison of leakage current values and plots can reveal cable deterioration and allow repair or replacement during a planned shutdown. These tests should be performed only by well-trained, fully qualified cable test personnel.

Note: Harm or failure of defective insulation during a planned shutdown and test under controlled conditions has less negative effect than an unscheduled shutdown caused by an unexpected cable fault.

Caution: When dc voltage is applied, power cables become capacitively charged and may contain lethal or damaging energy after the source of energy is removed. Provision to safely bleed down or discharge this stored energy is necessary. Application of a full ground to a highly charged cable may damage the cable, or worse. Cable testing is best and safer performed by a trained, experienced professional.

11.23 Rotating Machines

11.23.1 Introduction

Today's electrical machines, motors, and generators have been and still are a workhorse of modern society. They are very reliable but can be very expensive to purchase and/or repair. Therefore, it is of great interest to those concerned with electrical systems that the machines keep running and do not require repair or replacement.

Electrical maintenance and tests are conducted at the completion of a motor or generator installation and periodically over the life of the machine. This ensures good electrical insulation and operation as intended by the manufacturer. Also, the recorded results can be tracked and analyzed to determine when repairs are warranted.

This brief description of rotating machines focuses on the maintenance required of generators and motors found in plants and facilities. Emphasis is on generator maintenance and testing.

11.23.2 Safety

Observe lockout-tagout procedures carefully.

Approved blocking and grounding, as appropriate, should be in effect before proceeding with any maintenance or tests.

Cleaning solvents may deteriorate varnished insulation on machine coils.

Do not use trichloroethylene or carbon tetrachloride when cleaning electrical machines; these have been proved harmful to health.

Never wear loose clothing or jewelry that can get caught on motor fans, pulleys, or shafts.

11.23.3 Maintenance

Generator and motor cleaning should take place in conjunction with prime mover overhaul or as scheduled for periodic maintenance. Low-pressure compressed air in addition to a cleaning solvent will provide needed results. Metal tools must not be used to poke or scrape dirt from the windings, laminations, or connections.

Vibration causes structural-insulation failure. Insulation is weakened. Banding, blocking, wedging, and tying become loose, and coils are allowed to become loose. Second, vibrations cause sparking at current-collecting devices; this sparking produces burning at commutator or collector rings. Vibrations also tend to embrittle copper, which leads to eventual breakage. Last, vibrations can cause premature bearing failure. Antifriction bearings may crack, sleeve bearings are pounded out of round, or a babbitt loosens in the shell and the shell wears the frame housing. Vibrations may be caused by an electrical or mechanical unbalance in the machine, a misalignment between couplings, an unbalanced load, or a poor foundation.

> Maximum allowable peak-to-peak vibration amplitude varies with rated speed of machine. Parameters for vibration amplitude commonly accepted at 3000 rpm and above are 0.001, at 1500 to 2999 rpm are 0.002, at 1000 to 1499 rpm are 0.0025, and at 999 rpm and below are 0.003.

Bearing insulation test: Variations in the magnetic flux passing through a shaft may cause a current flow in the shaft-bearing-frame circuit. Arcing across the thin film of oil between the shaft and bearing causes pitting. This pitting eventually causes friction, heating, and premature bearing failure. This electrical path is broken by insulating one of the bearings. Some machines having two insulating layers and a so-called high point in between can be tested with a megohmmeter (minimum 20 megohms).

11.23.4   Commutator, slip rings, and brush maintenance

Runout adjustment for brushes: Collectors on large low-speed machines can tolerate eccentricity of up to $\frac{1}{32}$ in. On higher-speed machines the manufacturer's specifications should be checked. The rings should be clean and free from any shipping coatings. They should be wiped clean with dry canvas or cleaned with cleaning stones. This commutator maintenance seems to be the most overlooked. A canvas wiper can be easily homemade by wrapping 8 to 12 layers of 6- or 8-oz canvas over the end of a hardwood board and then bolting in place. As a layer of canvas becomes saturated with carbon, gum, and smut, cut away a layer of canvas, exposing a new one. Never use a lubricant on rings or commutators. Brushes must be seated with full contact.

Commutator brushes can be seated using a seating stone. Collector brushes can be seated with a stone or sandpaper. Carborundum or emery paper should not be used because it can be embedded in the brush and wear the collector rings during operation. To seat brushes using a stone, increase the brush tension. While the machine is in operation, press the stone against the ring or commutator directly so the stone's power will carry under the brush and seat it. After seating, vacuum all brush particles and then reset brush tension.

Many different rigging systems are used to set brush tension. Consult the specific manufacturer's literature for determining proper tension. Brush alignment is critical for current transfer and limiting sparking. Variations of more than $\frac{1}{16}$ in should be corrected. In order to promote uniform wear of collector rings the polarity of the brushes should be changed periodically.

11.23.5   Alignment

Shafts of the machine and the driven load must be aligned horizontally, or they have parallel misalignment. The centerlines of the shafts must not intersect, or they will have angular misalignment. Coupling separation must also be set in accordance with the manufacturer. Detailed instructions of aligning shafts is beyond the scope of the maintenance electrician. For details see manufacturer's literature.

11.23.6   Air gap measurement

The air gap between the rotor and the stator of any rotating machine should not be more than 10 percent with the measurements taken 90 percent apart. The air gap more specifically is the radial iron-to-iron clearance between the center of a pole face and the face of a core tooth. The air gap must be checked after the rotor shaft has been aligned with the prime mover of a generator or the driven load of a motor. Measurements should be taken at a minimum of four equally spaced locations around the circumference at both ends of the rotor. Some equipment cannot be accessed for direct measurement. In those cases, the manufacturer should have provided machined surfaces and punch marks. In either case, always read the manufacturer's instructions when measuring alignment or air gaps. Periodic measurements can be used to determine bearing wear in the machine. It follows that the alignment process and the air gap setting are part and parcel of the same exercise. If one is adjusted the other must be checked.

11.23.7 Grounding straps

Grounding straps provide a path of low impedance for fault currents that flow when insulation breaks down. When grounding straps are improperly connected, stray currents may pass through engine components with resultant damage. Bearings may be damaged. Electrical discharge damage can be identified in its early states as random streaks on the crankshaft journals, and the bearing shells will be completely worn. This electrical damage can be prevented by maintaining straps in good order, i.e., clean, dry, and tight.

11.23.8 Tests

Measure insulation resistances per ANSI/IEEE Standard 43. For motors 200 hp or greater, perform a dielectric absorption test and calculate the polarization index value.

Perform vibration tests. Plot amplitude versus frequency of motors>200 hp. Perform vibration amplitude test for small motors.

Measure run current and compare to motor rating.

Measure resistance of bolted electrical connections using a low-resistance-reading ohmmeter.

Apply a dc overpotential test by the step voltage method on motors 1000 hp and greater, and on all medium-voltage class motors.

Verify resistance temperature detector (RTD) circuits and devices, as applicable.

Optional tests include surge comparison and insulation power factor.

11.23.9 Synchronous motor tests

Measure voltage drop on salient poles.

Insulation resistance measurements of the following windings: main rotating field, exciting armature, and exciter field, per ANSI Standard 43.

Resistance measurements of motor field windings, exciter-stator windings, exciter-rotor windings, and field discharge resistors.

Check or adjust the exciter-field current to nameplate value.

Test setting and operation of all timers and other control devices present.

Optional tests might include high potential testing of the excitation system, high-speed recording during acceleration to measure stator voltage and current and field current, a plot of stator current vs. excitation current at 50 percent load, and testing of any power factor relay.

11.24   Surge Arresters

Surge arresters are available in all voltage classes, and there are many types. They are applied in both indoor and outdoor installations, and commonly are found on overhead lines, in transformer compartments, in switchgear assemblies, within medium-voltage switch cabinets, and at motors and generators. Surge protectors are also used to protect sensitive solid-state equipment such as computers, copiers, control devices, and the like. Surge arresters are connected between line and ground and in certain applications phase to phase. Under normal voltage conditions, the surge arrester is seen as an insulator and allows no intentional current flow through it. Should line voltage exceed a predetermined level, as might occur from a lightning strike, switching surge, fault condition, transient spike, or other means, the arrester becomes conductive so that stress of the abnormal high-voltage surge is relieved. As soon as the surge voltage collapses, current flow through the arrester ceases, and it reverts to its insulating mode. Most medium-voltage surge arresters include a porcelain housing.

11.24.1   Maintenance

Clean the arrester thoroughly. Inspect the arrester for physical damage. Look for cracks or chips in the porcelain or evidence of external flashover. Check bolted connections for tightness by using a torque wrench or low-resistance-reading ohmmeter. Check mounting and conductor clearances. Confirm the grounding conductor is securely connected to a ground bus or electrode.

It is not reasonably possible to field test and assure that a surge arrester operates as intended. However, the normally present insulating quality may be tested.

11.24.2   Tests

Measure insulation resistance across the arrester, and compare reading to like arresters or to manufacturer listings.

Test the grounding connection.

Measure watts loss by using power factor–dissipation factor test equipment. This test is more usually applied to larger medium-voltage and all high-voltage surge arresters. Compare measurements to those of like units.

11.25   Grounding Systems and Equipment Grounding

Article 100 of the 1996 National Electrical Code (NEC) provides the following definitions.

Ground: A conducting connection, whether intentional or accidental, between an electric circuit or equipment and the earth or to some conducting body that serves in place of earth.

Grounded: Connected to earth or to some conducting body that serves in place of earth.

Grounded effectively: Intentionally connected to earth through a ground connection or connections of sufficient current-carrying capacity to prevent the buildup of voltage that may result in undue hazards to connected equipment or persons.

Grounded conductor: A system or circuit conductor that is intentionally grounded.

Article 250 of the NEC provides general requirements of grounding or bonding of electrical installations.

11.25.1   Grounding systems

In order of frequency of use, electric power systems may be classified as being:

Solidly grounded: Having a direct, solid connection between neutral and earth.

Ungrounded: Having no intentional connection between conductors and ground.

Resistance grounded: Having an intentional resistance connected between the neutral conductor and ground.

Reactance grounded: Having an intentional reactance element connected between the neutral and ground.

11.25.2   Equipment grounding

With rare exceptions for special apparatus having its structure insulated from ground, the external conductive housing or surface and internal frames of electrical raceways and equipment are maintained at ground potential through a bonding system and solid ground connection. Ground system design and integrity are an important component of personnel safety and electrical protective device operation.

Connections to earth may include one or more driven ground rods interconnected to form a grid. Ground grids are typically installed at transformer and switchgear locations. Ground grids between various transformer and switchgear locations should be interconnected. Single rods are typically found in access holes and at the base of power poles, especially if the pole includes switches, surge arresters, or cable terminations. The grounding electrode system should include connection to the interior metal water pipe at a point no more than 5 ft from the entrance into the building, and to the building structural-steel frame. The NEC describes other types of grounding electrodes that may be installed. Buried or otherwise nonaccessible grounding connections should be exothermic welded. Mechanical connections must be accessible for maintenance and repair.

The fall of potential method of measuring the resistance between a ground electrode system and earth employs a reference electrode placed a sufficient distance from the ground electrode being tested, and a known current is passed between these two electrodes. Another reference electrode is placed in line and between the test electrode and the outer reference electrode, at 62 percent of the distance between the test electrode and the outer reference electrode, and the voltage drop between test electrode and the 62 percent reference point is measured. A ground resistance test set allows easy performance and provides a direct resistance reading in ohms.

11.25.3   Maintenance

Identify the type(s) of grounding systems and connections utilized throughout the plant. Inspect accessible grounding connections for tightness, corrosion, and cleanliness.

Warning: It is not safe to interrupt ground connections when the associated electrical system is energized.

11.25.4   Tests

Measure resistances between the main grounding system and the frames of major electrical equipment. Resistance should not exceed 0.5 ohm.

Measure the resistance between the main grounding electrode system and earth by the fall of potential method, using a ground resistance test set. For commercial or industrial installations this resistance should be 5.0 ohms or less, and for generating or transmission substations not more than 1.0 ohm.

11.26 Frequency of Performance

How frequently various maintenance activities are required depends on the type of equipment, its characteristics, operating demands, environment, and any significant occurrences. Excess dirt, surface contaminants, heavy demand on moving parts, or unusual mechanical or electrical stresses might be cause for more frequent service. Industrial plants often place harsher demands on equipment than university or government usage might impose. Some contactors, breakers, and switches operate much less frequently than others, so wear patterns will differ.

All electrical apparatus should be visually inspected while in service at least weekly, if not daily.

Batteries require weekly inspection, monthly service, and annual complete service and testing.

All low-voltage (600 V class) drawout style breakers and molded case breakers ≥400 A size should be serviced and tested at least every 3 years. Under harsher environments, yearly service may be necessary. Longer intervals may allow breakers with failing trip units to remain on line unnecessarily, leaving the appearance of safety where none exists. Critical or life safety breakers should be tested annually.

Most low-voltage apparatus requires inspection, service, and testing at least every 2 to 3 years.

Regardless of voltage rating, contactors, breakers, and switches that are operated frequently to energize and deenergize loads may require annual or even more frequent service in order to monitor contact wear and other deterioration.

Medium-voltage insulation can deteriorate to flashover in much less than 2 to 3 years. Medium-voltage apparatus operates at greater electrical stress and can intentionally or undesirably deliver much more electrical energy. All medium-voltage apparatus, including protection and control devices, should be inspected, serviced, and tested annually.

In general, motor inspection and service does not always include significant disassembly unless tests indicate a need to do so.

Insulating liquids should be drawn and tested annually.

11.27 Personnel Qualifications

Performance usually centers around the facilities electrical staff, whose knowledge of the electrical system and equipment at that facility is invaluable. These are the people who day to day maintain operation, perform ongoing maintenance and repair, and perform or oversee system and control modifications and additions. Depending upon their availability and technical training, much electrical maintenance work can be completed by in-house personnel. At time of a major shutdown, however, the facility staff is often concerned with other downtime tasks, which limits available time for completion of a thorough maintenance program. Few facilities employ electrical personnel with in depth knowledge of the mechanical measurements and clearance adjustments necessary, or the ability to properly provide good testing performance.

A reliable electrical contractor may well provide electricians who are trained and experienced in industrial apparatus and controls, and who may provide strong support in many areas of shutdown maintenance, especially with regard to cleaning and tightening work.

Inspection, testing, calibration, and evaluation is best performed by an independent, fully qualified electrical testing service company that provides in-house electrical engineering staff and certified test technicians. Independent test companies routinely work on all makes and models of all electrical apparatus, to provide evaluating opinions and recommend actions that are completely unbiased by manufacturer, supplier, or installing contractor influences. The testing company's in-house engineering staff specializes in performance of acceptance and maintenance inspections and tests, helps shape and technically define the program, reviews test procedures and data collected, and the test report, most of which cannot be provided by an after-the-fact cursory review by a non-full-time engineer whose principal engineering duties are design-oriented.

Test technicians performing this work should hold current certification by International Electrical Testing Association (NETA), Certified Technician Level III or Certified Senior Technician Level IV, or by the National Institute for Certification in Engineering Technologies (NICET), in the field of Electrical Testing Engineering Technology, Level III or IV. NETA and NICET are both nonprofit certification sources, completely independent of influence from training institutions, equipment manufacturers, and installers. Technician certification by either NETA or NICET requires written verification of specialized experiences and passing a very comprehensive, proctored, written examination.

11.28   Tables of Values

Figures 11.1–11.19 are reprinted with permission from MTS-1997, NETA Electrical Maintenance Testing Specifications for Electrical Power Distribution Equipment and Systems, Copyright 1997, with permission of and by courtesy of the International Electrical Testing Association, Morrison, CO 80465. Telephone 303-697-8441. All rights reserved.

For purposes of economy of space in this text, the NETA order of appearance has been changed and an appropriate test number provided, while retaining the NETA title and table content completely.

(NETA TABLE 10.1)
Insulation Resistance Tests on Electrical Apparatus and Systems

Maximum Rating of Equipment in Volts	Minimum Test Voltage, dc in Volts	Recommended Minimum Insulation Resistance in Megohms
250	500	25
600	1,000	100
5,000	2,500	1,000
8,000	2,500	2,000
15,000	2,500	5,000
25,000	5,000	20,000
35,000	15,000	100,000
46,000	15,000	100,000
69,000	15,000	100,000

In the absence of consensus standards dealing with insulation-resistance tests, the NETA Technical Committee suggests the above representative values.

See Table 10.14 for temperature correction factors.

Actual test results are dependent on the length of the conductor being tested, the temperature of the insulating material, and the humidity of the surrounding environment at the time of the test. In addition, insulation resistance tests are performed to establish a trending pattern and a deviation from the baseline information obtained during maintenance testing enabling the evaluation of the insulation for confined use.

Figure 11.1 Insulation resistance tests on electrical apparatus and systems.

(NETA TABLE 10.2)
Switchgear Low-Frequency Withstand Test Voltages

Type of Switchgear	Rated Maximum Voltage (kV) (rms)	Maximum Test Voltage kV	
		ac	dc
LV (Low-Voltage Power Circuit Breaker Switchgear)	.254/.508/.635	1.6	2.3
MC (Metal-Clad Switchgear)	4.76	14.0	20.0
	8.25	27.0	37.0
	15.0	27.0	37.0
	38.0	60.0	+
SC (Station-Type Cubicle Switchgear)	15.5	37.0	+
	38.0	60.0	+
	72.5	120.0	+
MEI (Metal-Enclosed Interrupter Switchgear)	4.76	14.0	20.0
	8.25	19.0	27.0
	15.0	27.0	37.0
	15.5	37.0	52.0
	25.8	45.0	+
	38.0	60.0	+

Derived from ANSI/IEEE C37.20.1-1993, Paragraph 5.5, *Standard for Metal-Enclosed Low-Voltage Power Circuit-Breaker Switchgear*, C37.20.2-1993, Paragraph 5.5, *Standard for Metal-Clad and Station-Type Cubicle Switchgear* and C37.20.3-1993, Paragraph 5.5, *Standard for Metal-Enclosed Interrupter Switchgear*, and includes 0.75 multiplier with fraction rounded down.

The column headed "DC" is given as a reference only for those using dc tests to verify the integrity of connected cable installations without disconnecting the cables from the switchgear. It represents values believed to be appropriate and approximately equivalent to the corresponding power frequency withstand test values specified for voltage rating of switchgear. The presence of this column in no way implies any requirement for a dc withstand test on ac equipment or that a dc withstand test represents an acceptable alternative to the low-frequency withstand tests specified in this specification, either for design tests, production tests, conformance tests, or field tests. When making dc tests, the voltage should be raised to the test value in discrete steps and held for a period of one minute.

Because of the variable voltage distribution encountered when making dc withstand tests, the manufacturer should be contacted for recommendations before applying dc withstand tests to the switchgear. Voltage transformers above 34.5kV should be disconnected when testing with dc. Refer to ANSI/IEEE C57-13-1978 (R1987) *IEEE Standard Requirements for Instrument Transformers* [10], Section 8 and, in particular 8.8.2, (the last paragraph) which reads "Periodic kenotron tests should not be applied to transformers of higher than 34.5 kV voltage rating."
+ Consult Manufacturer

Figure 11.2 Switchgear low-frequency withstand test voltages.

(NETA TABLE 10.5)
Transformer Insulation-Resistance

Transformer Coil Rating Type in Volts	Minimum dc Test Voltage	Recommended Minimum Insulation Resistance in Megohms	
		Liquid Filled	Dry
0 - 600	1000	100	500
601 - 5000	2500	1000	5000
5001 - 15000	5000	5000	25000

In the absence of consensus standards, the NETA Technical Committee suggests the above representative values.

NOTE: Since insulation resistance depends on insulation rating (kV) and winding capacity (kVA), values obtained should be compared to manufacturer's published data.

Figure 11.3 Transformer insulation resistance.

(NETA TABLE 10.14)
Insulation Resistance Conversion Factors
For Conversion of Test Temperature to 20°C

Temperature		Multiplier	
°C	°F	Apparatus Containing Immersed Oil Insulations	Apparatus Containing Solid Insulations
0	32	0.25	0.40
5	41	0.36	0.45
10	50	0.50	0.50
15	59	0.75	0.75
20	68	1.00	1.00
25	77	1.40	1.30
30	86	1.98	1.60
35	95	2.80	2.05
40	104	3.95	2.50
45	113	5.60	3.25
50	122	7.85	4.00
55	131	11.20	5.20
60	140	15.85	6.40
65	149	22.40	8.70
70	158	31.75	10.00
75	167	44.70	13.00
80	176	63.50	16.00

Figure 11.4 Insulation resistance conversion factors for conversion of test temperature to 20°C.

(NETA TABLE 10.9)
Instrument Transformer Dielectric Tests
Field Maintenance

Nominal System (kV)	BIL (kV)	Periodic Dielectric Withstand Test Field Test Voltage (kV)	
		ac	dc
0.6	10	2.6	4
1.1	30	6.5	10
2.4	45	9.7	15
4.8	60	12.3	19
8.32	75	16.9	26
13.8	95	22.1	34
13.8	110	22.1	34
25	125	26.0	40
25	150	32.5	50
34.5	150	32.5	50
34.5	200	45.5	70
46	250	61.7	+
69	350	91.0	+
115	450	120.0	+
115	550	149.0	+
138	550	149.0	+
138	650	178.0	+
161	650	178.0	+
161	750	211.0	+
230	900	256.0	+
230	1050	299.0	+

Table 10.9 is derived from Paragraph 8.8.2 and Tables 2 and 7 of ANSI/IEEE C57.13, "Standard Requirements for Instrument Transformers."

+ Periodic dc potential tests are not recommended for transformers rated higher than 34.5 kV.

* Under some conditions transformers may be subjected to periodic insulation test using direct voltage from kenotron sets. In such cases the test direct voltage should not exceed the original factory test rms alternating voltage. Periodic kenotron tests should not be applied to (instrument) transformers of higher than 34.5 kV voltage rating.

Figure 11.5 Instrument transformer dielectric tests field maintenance.

(NETA Table 10.17)

Metal-Enclosed Bus Dielectric Withstand Test Voltages

Type of Bus	Rated kV	Maximum Test Voltage, kV	
		ac	dc
Isolated Phase for Generator Leads	24.5	37.0	52.0
	29.5	45.0	--
	34.5	60.0	--
Isolated Phase for Other than Generator Leads	15.5	37.0	52.0
	25.8	45.0	--
	38.0	60.0	--
Nonsegregated Phase	0.635	1.6	2.3
	4.76	14.2	20.0
	15.0	27.0	37.0
	25.8	45.0	63.0
	38.0	60.0	--
Segregated Phase	15.5	37.0	52.0
	25.8	45.0	63.0
	38.0	60.0	--
DC Bus Duct	0.3	1.6	2.3
	0.8	2.7	3.9
	1.2	3.4	4.8
	1.6	4.0	5.7
	3.2	6.6	9.3

Derived from ANSI-IEEE C37.23-1987, Tables 3A, 3B, 3C, 3D and paragraph 6.4.2. The table includes a 0.75 multiplier with fractions rounded down.

Note:

The presence of the column headed "dc" does not imply any requirement for a dc withstand test on ac equipment. This column is given as a reference only for those using dc tests and represents values believed to be appropriate and approximately equivalent to the corresponding power frequency withstand test values specified for each class of bus. Direct current withstand tests are recommended for flexible bus to avoid the loss of insulation life that may result from the dielectric heating that occurs with rated frequency withstand testing.

Because of the variable voltage distribution encountered when making dc withstand tests and variances in leakage currents associated with various insulation systems, the manufacturer should be consulted for recommendations before applying dc withstand tests to this equipment.

Figure 11.6 Metal-enclosed bus dielectric withstand test voltages.

(NETA TABLE 10.6)
Medium-Voltage Cables
Maximum Maintenance Test Voltages (kV, dc)

Insulation Type	Rated Cable Voltage	Insulation Level	Test Voltage kV, dc
Elastomeric:	5 kV	100%	19
Butyl and Oil Base	5 kV	133%	19
	15 kV	100%	41
	15 kV	133%	49
	25 kV	100%	60
Elastomeric:	5 kV	100%	19
EPR	5 kV	133%	19
	8 kV	100%	26
	8 kV	133%	26
	15 kV	100%	41
	15 kV	133%	49
	25 kV	100%	60
	25 kV	133%	75
	28 kV	100%	64
	35 kV	100%	75
Polyethylene	5 kV	100%	19
(see Note 4)	5 kV	133%	19
	8 kV	100%	26
	8 kV	133%	26
	15 kV	100%	41
	15 kV	133%	49
	25 kV	100%	60
	25 kV	133%	75
	35 kV	100%	75

Derived from ANSI/IEEE Standard 141-1993 Table 12-9 and by factoring the applicable
ICEA/NEMA Standards by 75% as recommended in Section 18-9.2.4 of NFPA 70B, 1994 Edition
Electrical Equipment Maintenance.
Refer to notes on the following page.

Figure 11.7   Medium-voltage cables maximum maintenance test voltages (kV, dc).

(NETA TABLE 10.6 - NOTES)

NOTE 1:   Selection of test voltage for in-service cables depends on many factors. The owner should be consulted and/or informed of the intended test voltage prior to performing the test. Caution should be used in selecting the maximum test voltage and performing the test since cable failure during the test will require repair or replacement prior to re-energizing.

NOTE 2:   AEIC C55 and C56 list test voltages approximately 20 percent higher than the ICEA values for the first five years of service. These values are based on 65 percent of the factory test voltages. A reduction to 40 percent is recommended for a cable in service longer than five years.

NOTE 3:   ANSI/IEEE 400-1991 specifies much higher voltages than either the ICEA or the AEIC. These voltages overstress cables and are intended to find marginal cable during shutdown to avoid in-service failures. These test voltages should not be used without the concurrence of the owner. If the cable is still in warranty, the cable manufacturer should be consulted for their concurrence. (See the Standard for a discussion of the pros and cons of high direct-voltage tests.)

NOTE 4:   See Electric Power Research Institute Report, EPRI TR-101245, "Effect of DC Testing on Extruded Cross-Linked Polyethylene Insulated Cables." DC high potential testing of aged XLPE-insulated cable in wet locations may reduce remaining life.

Figure 11.7 *(Continued)* Medium-voltage cables maximum maintenance test voltages (kV, dc).

(NETA TABLE 10.19)
Overpotential Test Voltages for Electrical Apparatus
Other than Inductive Equipment

Nominal System (Line) Voltage[1] (kV)	Insulation Class	AC Factory Test (kV)	Maximum Field Applied AC Test (kV)	Maximum Field Applied DC Test (kV)
1.2	1.2	10	6.0	8.5
2.4	2.5	15	9.0	12.7
4.8	5.0	19	11.4	16.1
8.3	8.7	26	15.6	22.1
14.4	15.0	34	20.4	28.8
18.0	18.0	40	24.0	33.9
25.0	25.0	50	30.0	42.4
34.5	35.0	70	42.0	59.4
46.0	46.0	95	57.0	80.6
69.0	69.0	140	84.0	118.8

In the absence of consensus standards, the NETA Technical Committee suggests the above representative values.

[1] Intermediate voltage ratings are placed in the next higher insulation class.

Figure 11.8 Overpotential test voltages for electrical apparatus other than inductive equipment.

(NETA TABLE 10.16)
High-Potential Test Voltage
for Periodic Test of Line Sectionalizers

Nominal Voltage Class kV	Maximum Voltage kV	Rated Impulse Withstand Voltage kV	Maximum Field Test Voltage kVAC	DC 15 Minute Withstand (kV)
14.4 (1Ø)	15.0	95	26.2	39
14.4 (1Ø)	15.0	125	31.5	39
14.4 (3Ø)	15.5	110	37.5	39
24.9 (1Ø)	27.0	125	45.0	58
34.5 (3Ø)	38.0	150	52.5	77

Derived from ANSI/IEEE C37.63-1984(R1990) Table 2 (*Standard Requirements for Overhead, Pad-Mounted, Dry-Vault, and Submersible Automatic Line Sectionalizers of AC Systems*).

The table includes a 0.75 multiplier with fractions rounded down.

In the absence of consensus standards, the NETA Technical Committee suggests the above representative values.

NOTE: Values of ac voltage given are dry test one minute factory test values.

Figure 11.9 High-potential test voltage for periodic test of line sectionalizers.

(NETA TABLE 10.15)
High-Potential Test Voltage
for Automatic Circuit Reclosers

Nominal Voltage Class, kV	Maximum Voltage, kV	Rated Impulse Withstand Voltage, kV	Maximum Field Test Voltage, kVac
14.4 (1Ø and 3Ø)	15.0	95	26.2
14.4 (1Ø and 3Ø)	15.5	110	37.5
24.9 (1Ø and 3Ø)	27.0	150	45.0
34.5 (1Ø and 3Ø)	38.0	150	52.5
46.0 (3Ø)	48.3	250	78.7
69.0 (3Ø)	72.5	350	120.0

Derived from ANSI/IEEE C37.61-1973(R1993) (*Standard Guide for the Application, Operation, and Maintenance of Automatic Circuit Reclosers*), C37.60-1981(R1993) (*Standard Requirements for Overhead, Pad-Mounted, Dry-Vault, and Submersible Automatic Circuit Reclosers and Fault Interrupters for AC Systems*).

Figure 11.10 High-potential test voltage for automatic circuit reclosers.

(NETA TABLE 10.7)

Molded-Case Circuit Breakers Values for Inverse Time Trip Test
(At 300% of Rated Continuous Current of Circuit Breaker)

Range of Rated Continuous Current Amperes	Maximum Trip Time in Seconds For Each Maximum Frame Rating[1]	
	<250V	251 – 600V
0-30	50	70
31-50	80	100
51-100	140	160
101-150	200	250
151-225	230	275
226-400	300	350
401-600	------------	450
601-800	------------	500
801-1000	------------	600
1001-1200	------------	700
1201-1600	------------	775
1601-2000	------------	800
2001-2500	------------	850
2501-5000	------------	900

Reproduction of Table 5-3 from NEMA Standard AB4-1991.

[1] For integrally-fused circuit breakers, trip times may be substantially longer if tested with the fuses replaced by solid links (shorting bars).

Figure 11.11   Molded case circuit breaker values for inverse time trip test.

(NETA TABLE 10.8)

Instantaneous Trip Setting Tolerances for Field Testing
of Marked Adjustable Trip Circuit Breakers

Ampere Rating	Tolerances of High and Low Settings	
	High	Low
<250	+40% -25%	+40% -30%
>250	±25%	±30%

Reproduction of Table 5-4 from NEMA publication AB4-1991.

For circuit breakers with nonadjustable instantaneous trips, tolerances apply to the manufacturer's published trip range, i.e., +40 percent on high side, -30 percent on low side.

Figure 11.12   Instantaneous trip setting tolerances for field testing of marked adjustable trip circuit breakers.

(NETA TABLE 10.3)
Recommended Dissipation Factor/Power Factor
of Liquid-Filled Transformers

	Oil Maximum	Silicone Maximum	Tetrachloro-ethylene Maximum	High Fire Point Hydrocarbon Maximum
Power Transformers	2.0%	0.5%	3.0%	2.0%
Distribution Transformers	3.0%	0.5%	3.0%	3.0%

In the absence of consensus standards dealing with transformer dissipation factor/power factor values, the NETA Technical Committee suggests the above representative values.

Figure 11.13 Recommended dissipation factor/power factor of liquid-filled transformers.

(NETA Table 10.13)
SF$_6$ Gas Tests

Test	Test Limits
Moisture by hygrometer method	Per manufacturer or investigate greater than 200 ppm [1]
SF$_6$ decomposition byproducts by ASTM D2685	Greater than 500 ppm [2]
Air by ASTM D-2685	Greater than 5000 ppm [3]
Dielectric Breakdown using hemispherical contacts at 0.10 inch gap at atmospheric pressure	11.5 – 13.5 kV [4]

[1] According to some manufacturers.
[2] In the absence of consensus standards dealing with SF$_6$ circuit breaker gas tests, the NETA Technical Committee suggests the above representative values.
[3] Dominelli, N. and Wylie, L., Analysis of SF$_6$ Gas as a Diagnostic Technique for GIS, Electric Power Research Institute, Substation Equipment Diagnostics Conference IV, February 1996.
[4] Per Even, F.E., and Mani, G. "Sulfur Fluorides", Kirk-Othmer Encyclopedia of Chemical Technology, 4th ed., 11,428, 1994.

Figure 11.14 SF$_6$ gas tests.

(NETA TABLE 10.4)
Suggested Limits for Service-Aged Insulating Fluids

Mineral Oil*				
Test	ASTM Method	69 kV and Below	Above 69 kV through 288 kV	345 kV and Above
Dielectric breakdown, kV minimum	D877	26	26	26
Dielectric breakdown, kV minimum @ 0.04 gap	D1816	23	26	26
Dielectric breakdown, kV minimum @ 0.08 gap	D1816	34	45	45
Interfacial tension, mN/m minimum	D971	24	26	30
Neutralization number, mg KOH/g maximum	D974	0.2	0.2	0.1
Water content, ppm maximum	D1533	35	25	20
Power factor at 25°C, %	D924	1.0***	1.0****	1.0****
Power factor at 100°C, %	D924	1.0***	1.0****	1.0****

Test	ASTM Method	Silicone**	Less Flammable Hydrocarbon ***
Dielectric Breakdown, kV minimum	D877	25	24
Visual	D2129	Colorless, clear, free of particles	N/A
Water Content, ppm maximum	D1533	100	45
Dissipation factor, % max. @ 25°C	D924	0.2	1.0
Viscosity, cSt @ 25°C	D445	47.5 - 52.5	N/A
Fire Point, °C, minimum	D92	340	300
Neutralization number, mg KOH/g max	D974	0.2	N/A
Neutralization number, mg KOH/g max	D664	N/A	0.25
Interfacial Tension, mN/m minimum @ 25°C	D971	N/A	22

*   IEEE C57.106-1991 *Guide for Acceptance and Maintenance of Insulating Oil in Equipment*, Table 5.
**   IEEE C57.111-1989 *Guide for Acceptance of Silicone Insulating Fluid and Its Maintenance in Transformers*, Table 3.
***   IEEE C57.121-1988 *Guide for Acceptance and Maintenance of Less Flammable Hydrocarbon Fluid in Transformers*, Table 3.
****   IEEE Standard. 637-1985 *IEEE Guide for the Reclamation of Insulating Oil and Criteria for Its Use.*

Figure 11.15 Suggested limits for service-aged insulating fluids.

(NETA TABLE 10.18)
Thermographic Survey'
Suggested Actions Based on Temperature Rise

Temperature difference (DT) based on comparisons between similar components under similar loading.	Temperature difference (DT) based upon comparisons between component and ambient air temperatures.	Recommended Action
1°C - 3°C	0°C - 10°C	Possible deficiency; warrants investigation
4°C - 15°C	11°C - 20°C	Indicates probable deficiency; repair as time permits
-- -- --	22°C - 40°C	Monitor continuously until corrective measures can be accomplished
> 16°C	> 40°C	Major discrepancy; repair immediately

Temperature specifications vary depending on the exact type of equipment. Even in the same class of equipment (i.e., cables) there are various temperature ratings. Heating is generally related to the square of the current; therefore, the load current will have a major impact on ▲T. In the absence of consensus standards for ▲T, the values in this table will provide reasonable guidelines.

Figure 11.16 Thermographic survey.

(NETA TABLE 10.10)
Maximum Allowable Vibration Amplitude

Speed - RPM	Amplitude - Inches Peak to Peak
3000 and above	0.001
1500 - 2999	0.002
1000 - 1499	0.0025
999 and below	0.003

Derived from NEMA publication MG 1-1993, Sections 20.53, 21.54, 22.54, 23.52, and 24.50.

Figure 11.17 Maximum allowable vibration amplitude.

(NETA TABLE 10.11)
Periodic Electrical Test Values for Insulating Aerial Devices
Insulating Aerial Devices with a Lower Test Electrode System
(Category A and Category B)

Unit Rating	60 Hertz (rms) Test			Direct Current Test		
	Voltage kV (rms)	Maximum Allowable Current Microamperes	Time	Voltage kV	Maximum Allowable Current Microamperes	Time
46 kV & below	40	40	1 minute	56	28	3 minutes
69 kV	60	60	1 minute	84	42	3 minutes
138 kV	120	120	1 minute	168	84	3 minutes
230 kV	200	200	1 minute	240	120	3 minutes
345 kV	300	300	1 minute	360	180	3 minutes
500 kV	430	430	1 minute	602	301	3 minutes
765 kV	660	660	1 minute	924	462	3 minutes

Insulating Aerial Devices without Lower Test Electrode System
(Category C)

Error! Bookmark not defined. Unit Rating	60 Hertz (rms) Test			Direct Current Test		
	Voltage kV (rms)	Maximum Allowable Current Microamperes	Time	Voltage kV	Maximum Allowable Current Microamperes	Time
46 kV & below	40	400	1 minute	56	56	3 minutes

Figure 11.18   Periodic electrical test values for insulating aerial devices.

(NETA TABLE 10.11 (cont.))

Insulating Aerial Ladders and Insulating Vertical Aerial Towers

Unit Rating	60 Hertz (rms) Test			Direct Current Test		
	Voltage kV (rms)	Maximum Allowable Current Microamperes	Time	Voltage kV	Maximum Allowable Current Microamperes	Time
46 kV & below	40	400	1 minute	56	56	3 minutes
20 kV & below	20	200	1 minute	28	28	3 minutes

Chassis Insulating Systems
Lower Insulated Booms

60 Hertz (rms) Test			Direct Current Test		
Voltage kV (rms)	Maximum Allowable Current Milliamperes	Time	Voltage kV	Maximum Allowable Current Microamperes	Time
35	3.0	3 minutes	50	50	3 minutes

NOTE:
1. Derived from ANSI/SIA A92-2-1990.
2. A method of calculating test voltages for units rated
    other than those tabulated here is as follows:
    1. The 60 Hz test values are equal to line to ground at
    the unit rating value time 1.5.
    2. Multiply the 60 Hz test values times 1.4 to arrive at
    the direct current values.

Figure 11.18 (*Continued*) Periodic electrical test values for insulating aerial devices.

US Standard Bolt Torques for Bus Connections
Heat-Treated Steel - Cadmium or Zinc Plated

Grade	SAE 1 & 2	SAE 5	SAE 7	SAE 8
Minimum Tensile (P.S.I.)	64K	105K	133K	150K
Bolt Diameter In Inches	Torque (Foot Pounds)			
1/4	4.0	5.6	8.0	8.4
5/16	7.2	11.2	15.2	17.6
3/8	12.0	20.0	27.2	29.6
7/16	19.2	32.0	44.0	48.0
1/2	29.6	48.0	68.0	73.6
9/16	42.4	70.4	96.0	105.6
5/8	59.2	96.0	133.6	144.0
3/4	96.0	160.0	224.0	236.8
7/8	152.0	241.6	352.0	378.4
1.0	225.6	372.8	528.0	571.2

Bolt Torques for Bus Connections Silicon Bronze Fasteners[1]
Torque (Foot-Pounds)

Bolt Diameter in Inches	Nonlubricated	Lubricated
5/16	15	10
3/8	20	14
1/2	40	25
5/8	55	40
3/4	70	60

[1] Bronze alloy bolts shall have a minimum tensile strength of 70,000 pounds per square inch.

Aluminum Alloy Fasteners[2]
Torque (Foot Pounds)

Bolt Diameter in Inches	Lubricated
5/16	8.0
3/8	11.2
1/2	20.0
5/8	32.0
3/4	48.0

Figure 11.19 U.S. standard bolt torques for bus connections.

[2] Aluminum alloy bolts shall have a minimum tensile strength of 55,000 pounds per square inch.

Bolt Torques for Bus Connections
Stainless Steel Fasteners[3]
Torque (Foot Pounds)

Bolt Diameter in Inches	Uncoated
5/16	14
3/8	25
1/2	45
5/8	60
3/4	90

[3] Bolts, cap screws, nuts, flat washers, locknuts: 18-8 alloy. Belleville washers: 302 alloy.

Figure 11.19 (*Continued*) U.S. standard bolt torques for bus connections.

Bibliography

ASTM, *1994 Annual Book of Standards—Electrical Insulation and Electronics,* vol. 10.3, American Society for Testing and Materials, Philadelphia, PA, 1994.

James G. Biddle Co., *Getting Down to Earth,* Blue Bell, PA, 1981.

T. Craft and W. Summers, *American Electricians Handbook,* McGraw-Hill, New York, 1992.

Doble Engineering Co., *Reference Book on Insulating Liquids and Gases,* Doble Engineering Co., Watertown, MA, 1992.

Doble Engineering Co., Various Published Papers, Watertown, MA.

A. S. Gill, *Electric Equipment Testing and Maintenance,* Reston Publishing Co., Reston, VA, 1982.

IEEE, *Recommended Practice for Electric Power Distribution for Industrial Plants,* IEEE/John Wiley & Sons, Inc., New York, 1976.

International Electrical Testing Association, Maintenance Testing Specifications, NETA, 1997.

Meyers, Kelly, and Parrish, *A Guide to Transformer Maintenance,* Transformer Maintenance Institute, Akron, OH, 1981.

National Fire Protective Association, National Electrical Code 1996, NFPA, Quincy, MA, 1994.

National Fire Protective Association, NFPA 70B—Recommended Practice for Electrical Equipment Maintenance, NFPA, Quincy, MA, 1994.

Anthony J. Pansini, *Electrical Transformers and Power Equipment,* Prentice Hall, Englewood Cliffs, NJ, 1988.

James G. Stallcut, OSHA Electrical Regulations Simplified, Gray Boy Publishing, Ft. Worth, TX, 1992.

Various Electrical Manufacturer Equipment Instruction Books, Guides, and Catalogs.

Westinghouse, *Electric Maintenance Hints,* Westinghouse Electric Corporation, Forbes Rd., Trafford, PA, 1984.

12

Outsourcing Considerations

Bernard T. Lewis

Facilities Management Consultant
Potomac, MD

Paul S. Lewis

Attorney-at-Law
Potomac, MD

12.1 Recommended Tasks for Outsourcing

A facility's use of contract operations and maintenance can vary from a single operations or maintenance service contract to a complete facility's operations and maintenance contract. A facility might contract its custodial, pest control, recycling, and trash removal services; electrical distribution system maintenance and testing; landscaping; etc.; or 100 percent of its operations and maintenance functions; including supervision and engineering. In some cases, facilities use contract operations and maintenance as a way of obtaining contractors' skills and knowledge. Other functional operations and maintenance service items that are usually readily available, and useful to the facility manager, as outsourcing service tasks items are shown following: water treatment; custodial, pest control, trash disposal, and recycling; backflow preventers preventive maintenance; electrical distribution system preventive maintenance and testing; fired and unfired pressure vessel inspection and testing; landscaping, exterior and interior; elevator and escalator maintenance and repair; energy management control systems—preventive maintenance, servicing, and programming; window washing; security; eddy current testing; and fire alarms, fire extinguishers, and sprinkler systems maintenance and testing.

For a facility to have this full range of potential outsourcing operations and maintenance service tasks, it must be in a favorable geographical location, where such contractor capabilities exist to furnish these services. There are a growing number of such locations nationwide today.

More rather than less involvement appears to be most attractive to new facilities or ones that have undergone major renovations, since contracting all or a significant portion of their operations and maintenance services would not include changing existing systems. However, an established facility would not have a problem in this regard for it may be a useful facilities engineering and management project to evaluate usefulness of existing systems to determine the need for change. Under these circumstances, the facility manager can develop procedures during operations phase-in for new facilities or during continued normal operations for existing facilities. From a practical standpoint, contract operations and maintenance services, in general, are more applicable to existing facilities, particularly if the facility manager is thinking in terms of total contract operations and maintenance services.

The facility manager's outsourcing approach can use the following alternative methods:

1. The facility manager can have in-house operations and maintenance workforce perform all operations and maintenance work—both day-to-day requirements and peak loads—with some craft specialty requirements contracted. Until recently, this approach was used by the majority of facilities in this country.

2. The facility manager can use in-house operations and maintenance workforce for performing all day-to-day normal requirements and bring in contractors' forces to handle major peaks—that is, emergencies, turnarounds, overhauls, and the like. In this case, the contractors' forces would be building trades craftspeople performing architectural and structural alterations, improvements, maintenance, or repair tasks; and/or HVAC contractors operations and maintenance craftspeople performing specific HVAC operations, maintenance, and repair tasks.

3. The facility manager can have a minimum in-house operations and maintenance workforce perform only part of the day-to-day requirements and use contractors' operations and maintenance services to perform the balance of the day-to-day work and to handle peak requirements.

4. Facility managers can elect to have no operations and maintenance working personnel assigned to their facility and rely entirely on contractor forces to provide all operations and maintenance personnel for normal and peak operations and maintenance needs. This should be the exception rather than the rule, for it has been found to be more efficient, effective, and economically beneficial to have a small number of in-house trades and craft personnel to provide HVAC operations and maintenance services, service calls performance, preventive and predictive maintenance procedures, and energy management control system operations and programming.

12.2 Other Considerations

To avoid possible labor union problems, the facility manager should give the facility's in-house workforce all the work it can handle, including overtime, before using contractors. The facility manager should review the labor union contract(s) the facility has with its own in-house operations and maintenance workforce. By union contractual agreement, some facilities may not use contractors except for construction of major new facilities or major overhauls and turnarounds; others may have other specific restrictions that must be met regarding outsourcing of operations and maintenance service work.

The main objective, in this regard, is to accomplish the facility's required task or tasks on schedule within cost limits and without any disruption from the in-house or contractors' workforces. Every effort should be made to eliminate or forestall any disputes by notifying facility personnel and the various union committees about the work planned to be performed by an outside contractor and why. Such communication can minimize labor disputes, improve operations and maintenance personnel work attitude, and reduce friction when the contractor's personnel move into the facility to do their job.

12.2.1 Can you motivate contract maintenance workers?

When the facility manager elects to employ an outside operations and maintenance service contractor, the final contractor selection is usually based on reputation, experience, performance record, scope of services required, labor relations, and of course, price. Too often neglected areas that should also be considered in evaluating a contractor's ability to perform are methods of motivation and productivity measurement.

For contract operations and maintenance services to be profitable, the contractor must operate and maintain a client's facility more efficiently and more economically than can the client by using the in-house workforce. To accomplish this objective, contractors must strive for the optimum utilization of their workforce, which largely depends upon proper motivation of the tradespeople and craftspeople, and measurement of their productivity, to ensure optimum output in accordance with the facility manager's contract performance work specifications.

12.3   The Advantages and Disadvantages of Outsourcing

12.3.1   Advantages

Outsourcing advantages are shown following:

1. The ability to fluctuate worker use to exactly meet facility day-to-day needs permitting expenditure of fewer worker-hours per year to perform a given amount of work. This method eliminates employment of full-time personnel to handle peaks and reduces the tendency for built-in featherbedding. It permits scheduling of seasonal work requirements to take advantage of seasonally good weather.

2. Seniority is not part of contract operations and maintenance services. Contractor tradespeople and craftspeople must compete with fellow workers to retain their jobs. In normal times, a job in the permanent nucleus of a contractor's operation and maintenance force is considered to be a choice job by most tradespeople and craftspeople. They usually get 40 hours per week of work, some overtime work and, of course, area wage rates. They must produce to hold their jobs. In normal times, some facility managers feel that a facility can develop a more productive workforce with contract labor than under average "facility" conditions using in-house labor.

3. A facility can meet the extraordinary labor needs for startup without overmanning for normal operations and should work less overtime during startup.

4. Since contractor tradespeople and craftspeople are already trade and craft trained, they require less training than the typical facility operations and maintenance employee to reach a given level of work performance efficiency.

5. Unlimited worker availability permits scheduling of turn-arounds, overhauls, alterations, etc., to best fit overall facility economics and need. There is no need to attempt to fit these requirements to the availability of a given number of facility in-house employees. Equipment and system downtime can also be shortened by using more personnel—to the extent practical.

6. The facility manager can pass on to the contractor the responsibility for essentially all or a significant part of the facility's operations

and maintenance services. This can leave more of the facilities' key personnel with more time to concentrate on process operations—the making of more and better products, or providing better organization services. This does not mean that operations and maintenance services will be overlooked; because contractors performing the services have a real economic incentive to do a better job, for operation or maintenance service contracts are usually an integral part of their annual income.

7. The facility manager can draw on the contractor for additional operations and maintenance services over and above normal facility operations and maintenance tasks. Contractors—if they are the right ones—can provide additional supervision as needed for the provided services. The contractor can also provide intermittent supervision in specialty areas such as rigging, critical equipment inspection, and repair, etc. Facility managers can use additional assistance from time to time for construction drawing takeoffs, jobs planning and estimating, jobs sketching, equipment tests, etc., and still keep their own staff at a constant level.

8. Another advantage often discussed is that the facility manager can reduce the investment in facilities, buildings, tools, and equipment. This is not a significant advantage. Over a period of time it has been observed that shops, storerooms, offices, etc., for contractors' personnel approach the norm for the facility manager's in-house personnel. Shops of various types can be kept minimal. Larger, infrequently used equipment, is available for lease by the facility manager from the contractor or others, regardless of "contract" versus "in-house" operations and maintenance requirements. There may be some savings on investment in tools if the contractor supplies them, but this does not present a significant savings.

9. The facility manager can shift essentially all responsibility for labor relations for operations and maintenance personnel to contractors, rather than be responsible for in-house personnel.

10. Facility managers can start with contract operations and maintenance and if not satisfied with the results can phase into a facility in-house personnel performed operations and maintenance organization. If they start out with an in-house workforce, there will be great difficulties in making changes later on. If the facility manager is not satisfied with a contractor's performance, the contractor can be changed. This happens occasionally.

Summary

To summarize, the prime advantage of contract operations and maintenance services is that the facility manager has a "wide degree of flexibility," substantially more than that enjoyed with a typical facility in-house operations and maintenance workforce.

12.3.2   The disadvantages of outsourcing

Facility managers should not examine only one side of any issue. Question what are the pitfalls—the problems?

1. Some facility labor relations conditions permit the use of in-house mechanics who are not bound by trade or craft union jurisdictional lines in performing operations and maintenance work. Contractors, by and large, follow trade and craft union jurisdictional lines, except in emergency situations or where there are "right to work" laws in the state where the work is performed.
   (a) Trade and craft performance restrictions cannot be held as tight on operations and maintenance work as on construction work. On the other hand, unnecessary disregard for trade and craft performance restrictions should be avoided. A contractor needs to use good judgment and be consistent in trade and craft work assignments.
   (b) The degree of rigidity on trade and craft performance restrictions will vary from contractor to contractor, and from project to project. This is because different people are involved—different project managers, supervisors, craftspeople, tradespeople, etc. This is not the fault of the issued contract but results largely from a lack of thorough training, good judgment, communication, and leadership on the part of certain contractor personnel.
   (c) With trade and craft performance restrictions, detailed work tasks planning is an absolute necessity. A facility manager who has all of the right people (the right trades and crafts), materials, tools, and equipment in the right place at the right time can have tradespeople and craftspeople working at their specialty without jurisdictional delays or squabbles.
   On occasions, the "skilled" craftspeople and tradespeople are not so skilled—particularly when a geographical area is faced with an extremely high level of construction and facilities operations and maintenance opportunities. In some cases, facilities operations and maintenance personnel will be attracted back to construction work by substantial overtime opportunities, by offers of supervisory positions, and by special deals or incentives on the part of construction contractors. Facility managers contribute to this agenda by not controlling overtime and their contracts.
   A related problem has been that of staffing for substantial peaks. During periods of heavy construction work in a geographi-

cal area, normally the only way a contractor can attract substantial numbers of personnel for facilities work for relatively short periods of time is to work overtime—enough to meet or exceed overtime that is being worked on construction projects, and/or pay higher than normal union or area hourly wage rates. Although the contractor and facility manager, for other considerations, often elect to work some overtime on turnarounds or overhauls, some feel that during high construction periods in their geographic areas, they work more overtime than normal to attract sufficient capable personnel to meet job demands. Thus there is a conflict between construction, and contract operations and maintenance, owing to unusual boom conditions in a geographic area. This situation will return to normal conditions, as the area construction level tapers off and as training programs produce more skilled trade and craft personnel.

Some facility managers feel that they cannot use contract operations and maintenance performance for certain specialty areas because contractors and cognizant unions cannot provide sufficient skills supervision, and tradespeople and craftspeople, with reasonable support backup. This has been true for instrumentation, particularly electronic; HVAC mechanics; and licensed stationary operating engineers. This, of course, does not preclude contract operations and maintenance use at a facility, but it has been a deterrent. Accordingly, many facility managers have elected to develop their own workforce with the necessary technical skills and cognizant licenses and certifications. Many contractors, and the unions concerned, are aware of this possibility and similar problems and are taking steps to provide these services.

Contract operations and maintenance wage rates are usually matched to construction rates in an area. On the average, construction wages have risen substantially faster than facility operations and maintenance wages, including fringe benefits. Most facility managers monitor this development carefully. They like the flexibility that goes with contract operations and maintenance; but, if the cost of contract labor gets too high in relation to facility in-house labor rates, they cannot or will not be able to afford the need for contracting operations and maintenance functions.

12.4 Contract Services

Contractors can provide facility support services in many ways. They can offer general operation and maintenance services that include a review and audit of the facility department's organization; facility operations and maintenance work performance standards and procedures; custodial, pest control, trash removal, and recycling operations; response to emergency situations; procedures relating to the installation, operation, repair, and maintenance of equipment and systems; facility inspection techniques, scheduling, record keeping, and reporting; operations, maintenance, and testing procedures for HVAC equipment and systems; and electrical distribution systems testing, maintenance, and repair. Facility inspections, as noted above, may extend to a complete facility structural examination, including rebar detection in concrete, underwater examinations, examination of timber and wood construction, load test certification, and corrosion studies. Inspection methods may include radiographic and ultrasonic, magnetic particle, and liquid penetrant testing, and visual examination.

Many contractors' shops are equipped with all the required tools and equipment to place the facility's equipment and systems back in service in the shortest possible time. There are, however, times when the equipment requiring repairs is so large that it would be difficult to move it out for repair. When the size of the equipment, its weight, or its stationary position prohibits such removal, on-site maintenance and/or repair may be the only option. In such cases, the contractor will dispatch personnel to the facility to evaluate the problem and discuss with the facility manager possible maintenance or repair procedures. When the facility manager has selected the most practical maintenance or repair option, the contractor will provide the necessary supervision, labor, materials, parts, equipment, and tools to complete the maintenance or repair at the facility.

A temporary contract arrangement, or one on a scheduled basis, can provide the additional personnel the facility manager requires to handle emergencies or service breakdowns, of organization mission impacting equipment; i.e., if the equipment or system fails, the organization has to halt production. Such an arrangement permits the inhouse operations and maintenance workforce to meet the facility's daily needs without concern for peak load conditions.

Facility managers must carefully evaluate the abilities of their personnel to determine whether they can properly perform the equipment operations and maintenance tasks involved under normal conditions, and during possible emergencies, without any serious production losses. In the evaluation, the facility manager should also compare the costs of doing the work in-house to the cost of having it performed by a contractor.

12.5 Quality Control

Conclusions on work quality by contractors are listed following:

1. Contract operations and maintenance workforce performance offers the facility manager an opportunity to obtain a workforce that produces good-quality work. When the facility manager takes the opportunity to evaluate and eliminate contractors who normally perform less than optimum quality work, the provided work quality will usually be of excellent quality. Where no screening is performed, work quality provided will generally be less than satisfactory.

2. Work quality provided depends on the quality of supervision provided. Good supervision is critical to obtaining quality work, yet this is an important factor rarely considered. However, the quality of project supervision, if less adequate than the contractors had promised, will normally result in low-quality work.

3. Contract operations and maintenance work performance is no guarantee of delivered quality work. Tradespeople and craftspeople hired by contractors are not better than those recruited for the facility's in-house workforce. They lack the specific experience that the regular employees acquire in time at the facility, but they often compensate for this by bringing a greater breadth of experience to the job.

12.6 Flexibility

The biggest advantage of outsourcing is flexibility. The ability to recruit and dissolve a large workforce quickly is most often cited as the contract operations and maintenance major advantage. Other advantages are:

1. Outsourcing of operations and maintenance work is applicable to more than rebuilds, turnaround, or alterations.

2. It is difficult to institute a reduction in force for an in-house workforce.

12.7 Reduced Capital Expenditures

In these competitive and rapidly changing times, private and public sector organizations are all deeply concerned with either maximizing profits or making a major improvement in services delivery, i.e., making as high a return on investment as possible. When an organization for profit (private sector) is reduced to its simplest operating concept—profit—profit can be increased as shown following. Public sector organizations' simplest operating concept is to provide required quality services—optimum delivered services—at the least overall cost:

1. *Private Sector*   Higher production, so higher sales, and higher average prices for products or services; thus, more income.

2. *Public Sector*   Meet required services delivery specifications with the required services quality.

3. *Public and Private Sectors*   Hold the line on, or reduce operations costs of any and all kinds—less operation costs.

Facility operations and maintenance performance can have significant effects on the above items. Facilities kept in good condition can provide higher product output per unit of time and can also enjoy a higher service or use factor—both adding to increased production and in turn higher sales, therefore more income. Efficient and effective operations and maintenance work performance can reduce direct operations and maintenance costs and therefore reduce the organization's overall operations costs.

Private and public sector facility managers, in seeking new ideas for improving the quality of operations and maintenance tasks performance and reducing operations and maintenance costs, have designed and developed the concept of a comprehensive outsourcing (contract) operations and maintenance program to achieve these objectives. From a cost-benefit comparison viewpoint, it would be difficult to prove that contract operations and maintenance usage is less expensive than in-house performance owing to the lack of good cost data in most facility organizations.

Engineering judgment and a review of many facility outsourcing operations and maintenance work experiences reveal that:

1. There is little "make work" in outsourcing operations and maintenance services performance.

2. Contractor's profits were not too high.

3. Less money is spent on "frills" than what an in-house workforce spends.

4. Contractor's wage rates are not too high, indicating a supposition that other available savings will offset the wage differential between in-house and contractor employees.

12.8   Employing Specialty Contractors

What are the procedures that should be used for outsourcing operations and maintenance services? By this is meant what actions and thought processes should facility managers use from the time they decide they want to consider outsourcing operations and maintenance services support until a contract is awarded and the contractor is on-board performing the services. In logical sequence these actions are:

1. Most facility managers have background and experiences for managing in-house operations and maintenance work performance but need to gather comparable information and data on past outsourcing operations and maintenance work performance. The best source for this information are other facility managers currently using operations and maintenance contractors.

2. After as much information and data as possible is gathered, a careful analysis and evaluation will determine what outsourcing task is really applicable in the individual facility's case.

3. Next, facility managers must make a comparison of the advantages offered by outsourcing operations and maintenance with using their own in-house workforces. However, caution and complete objectivity must be exercised because intangibles have to be considered and assumptions made. The analysis made must consider the overall effect on the facility, as follows, during the contract: startup (phase-in) period, operational period, and the close-down (phase-out) period after the contract has expired or has not been renewed.

4. If, after an overall evaluation of costs and benefits has been made, it is concluded that outsourcing operations and maintenance services offer distinct advantages, facility managers should then further refine the evaluation and decide on the procedure to be followed. They should also decide who will perform work scheduling and planning, the extent the contractor will be involved in the facility's cost accounting services, whether the contractor will have to provide and staff support services such as purchasing, drafting, reporting, record keeping, etc.

The defined tentative contractor operational procedures should be established so that when taking bids or negotiating with a contractor, the general scope of work, and performance work statement can be described. It will also be useful to consider the type and size of contractor and in-house organizations required to manage the contract.

Contractor selection should be made based on a projected long-term contractual arrangement. The contractor and the facility manager must work as a team in executing the awarded contract. The following major points should also be considered in this regard:

1. The contractor's integrity.
2. The contractor's experience. Has operations and maintenance work been performed with the staff, or have they been acting as a labor broker with facility managers shouldering the responsibility? What are the extent and depth of the contractor's experience?
3. What does the contractor offer in terms of supervision, and trade and craft personnel skills? This is a most important consideration.

The following criteria must be carefully reviewed as regards the contractor's capability.

(a) Is the staff adequate to support project workforces? Are these personnel experienced on contract operations and maintenance projects?

(b) Who is offered as the project manager? Has the person been in the contractor's employ prior to this contract for at least 3 years? Does the contractor really know the proposed project manager's capabilities? Is the manager experienced and skilled in supervising operations and maintenance work?

(c) Can the contractor offer sufficient skilled assistant supervisors to manage the job? What is their experience and background? Will unknowns have to be hired to staff the project? Can the contractor offer adequate supplemental and/or specialty supervisors?

(d) What is the contractor's situation with regard to any unions involved—particularly at the local level?

(e) Does the contractor have complete knowledge of the individual tradespeople and craftspeople basic skills, special skills, and capabilities?

(f) Are the company and key people really respected by the various unions and their members?

(g) Is the contractor inclined to manage in short-term expediencies or to adhere to the formal labor agreements? Does the contractor obtain what is specified under the labor agreements but still deal fairly with union members?

(h) How complete is the contractor's service in the specialty? Do the services match your needs? If not, are you willing to split work responsibility? Does the contractor have necessary equipment, tools, and facilities to support your operation?

It is difficult for the facility manager to obtain complete and detailed answers to these questions. Some contractors are prone to exaggerate when discussing their capabilities. The facility manager needs to probe deeply and take a "show me" attitude. Be sure to obtain the true story before drawing any conclusions.

One good measure of a contractor is overall effectiveness, consistently performing repeat work for facility managers in the area, receiving a fair share of the area's contract work, and making good profits on contracted work. Facility managers should request financial information from the contractor, not only to be sure the contractor is fiscally sound but also to measure the company's progress in terms of profits. The ability to consistently make good profits is conclusive evidence of a well-managed organization.

12.9   Selecting an Outsourcing Source

Several approaches are open to facility managers in contractor selection:

1. One approach is to check with other facility managers in the area to determine what contractors they use and the degree of success they are having. Or they may obtain information of experiences of other facility managers at monthly meetings, seminars, etc.

2. Another approach is to contact the manufacturer of some of the more sophisticated equipment currently operating in the facility. The manufacturer may provide maintenance and/or repair services for equipment or may be able to suggest a contractor for that service. Many manufacturers of electrical equipment (motors, generators, transformers, and switchgear), boilers, air-conditioning units, and temperature controls (calibration) have service shops throughout the country and can provide such a service on a contract basis or on an "as needed" arrangement. Any production or facility equipment that would cause the facility to shut down would be of primary concern to the facility manager. If such equipment is beyond the skills and abilities of in-house personnel, the facility manager should determine the types of services that are provided by the equipment manufacturer.

Many engineering and construction firms provide an operations and maintenance contract service for a facility's operating equipment. Such services are offered mainly for refineries, chemical plants, and power generating stations, but they are also available for any type of public or private sector facility. These firms can supply specialists in planning and scheduling, rotating equipment, piping, rigging, and electrical equipment. They are capable of handling all equipment operations, maintenance, emergency repairs, capital improvements, overhauls, turnarounds, and other required services. Some of their systems incorporate sophisticated record keeping and maintenance management (CMMS) procedures to include preventive and predictive maintenance programs, and emergency work procedural techniques, to ensure that all work is performed logically and efficiently.

One other approach would be to examine a directory containing the names of various contractors' associations to determine those with specific specialties interest. The associations will provide the names of contractors you are interested in. Also look for contractors' names in the local telephone book. Follow-up telephone calls can determine if further correspondence or contact with those companies is warranted.

12.10   Judging Abilities
of an Outsourcing Firm

How can the facility manager judge a contractor's work quality and technical ability? There are eight important characteristics to look for prior to obtaining services of an operations and maintenance contractor: technical expertise of personnel, local support availability, service reputation, pricing policy, business experience, financial stability, equipment repair facilities (including availability of tools and materials), and speed of response to service calls.

Besides the facility manager, the operations supervisor, maintenance supervisor, engineering supervisor, and/or electrical and mechanical engineers should be involved in the contractor selection process.

Annual preventive maintenance agreements, a 2 or more years preventive maintenance agreement, and service on an as needed basis are the types of contracts normally selected. Other types would be for special frequency on specific equipment and for emergencies.

It is important that the facility manager include in the service contract all details related to the equipment and systems being serviced. The facility manager and the contractor must agree to all performance work statement specifications and the contract costs.

One major consideration, to be evaluated before signing a service contract, is to ensure that the contractor has adequate insurance coverage and that the insurance information and the amount of coverage involved becomes part of the contract. Accidents that occur in the facility as a result of the contractor's personnel actions should be covered by insurance. Contractors must carry ample insurance in the following areas so that the facility is not held responsible for any accidents to the contractor's or company's employees or damage to facility equipment: workmen's compensation and occupational disease and protective liability insurance.

It may be advisable for the facility manager to have a protective liability insurance policy to protect the facility from subcontractors that may be used but are not covered under the prime contractor's insurance policies.

The contract awarded should guarantee for the life of the contract that the contractor remedy failures, at no additional cost to the facility manager, due to poor workmanship and/or defective materials furnished by the contractor.

A penalty clause in the contract, to protect the company, is advisable if an equipment or system installation has to be completed by a deadline. The penalty clause can serve as an incentive and also protect the company against costly delays. However, it may be detrimental in pushing the contractor to rush the installation and may result in poor workmanship being provided. The final decision as to whether or not to include or exclude a penalty clause should be based on the contractor's reputation and the criticality of the project.

No matter what type of service contract is selected, the facility manager should assign one of the engineering or supervisory staff to review, periodically, all contractor submitted reports. With such an arrangement, it is possible to monitor the progress of the work being performed and to call attention quickly to tasks, work performance, or costs that may be deviating from the contracted norms.

12.11   Responsibility for Outsource Firm Administration and Control

Since, as shown above, the "contract document" is considered a key to contractor administration and control, it is necessary to ensure that this document provides details on every aspect of the work to be performed or material to be provided and the standards to which it will be performed. In addition to the controls provided by a well-defined contract, the supervisor of each section which will use contractor services should be delegated operational control and tasked to complete a monthly evaluation report such as that shown in Fig. 12.1 for each facility contractor.

12.12   Outsource Contract Administration

Viable and effective outsource agreements require an extremely high degree of specificity in every phase. Certainly, that concept may be applicable to any form of contract. However, the very nature of the outsource agreement, i.e., reliance on "outside" maintenance and operations personnel, not employed on a full-time basis, and thereby beyond routine controls, makes it critical that every phase of the contract be specific enough so as to avoid any misunderstandings. Nonetheless, the onus of such restrictions should not deter the facility manager from seeking outside contracting. Again, various mechanical and technical disciplines simply cannot be delegated "in-house" without sacrificing safety and efficiency.

12.12.1   Negotiating the outsource contract

The initial step in negotiating the contract occurs well before the final contractual arrangement is put to paper. More specifically, facility managers must attempt to articulate their needs before initiating negotiations with any contractors. In fact, it would be good practice to detail such requirements in a written prospectus, whether in the form of a bid proposal or by way of a request for contractual services.

Monthly Evaluation Report for Contractors			
Performance Factor	Raw Score 0-100	Weight Factor	Weighted Score (points)
1. Contractor has demonstrated flexibility and responsiveness to overall support.	_____	X 0.10	_____
2. Contractor has provided required planning and scheduling of resources to meet needs.	_____	X 0.10	_____
3. Contractor's organization is adequately staffed and properly supervised.	_____	X 0.10	_____
4. Contractor has met all reporting requirements in timely and accurate manner.	_____	X 0.10	_____
5. Contractor has identified potential problems and provided alternative solutions.	_____	X 0.10	_____
6. Contractor's personnel are qualified for assigned positions.	_____	X 0.10	_____
7. Contractor has provided efficient utilization of equipment and facilities.	_____	X 0.10	_____
8. Contractor has responded to previously noted problems or deficiencies.	_____	X 0.10	_____
Total Weighted Score			_____

Remarks: _____

_____

_____

A Raw Score of 70 or above = **Acceptable Performance**
A total Weighted Score of 49 or below = **Unacceptable Performance**
(Maximum Weighted Score Possible is 80 points)

Figure 12.1 Monthly evaluation report for contractors.

However, in either case, the key element is the provision of complete notice, to the prospective contractor, of the facility manager's expectations. Further, any such document must ultimately be incorporated into the final contract, either as an addendum or in some other form.

In determining pricing requirements, the facility manager should always look to costs for comparable in-house services, as a baseline.

Obviously, specialists, such as elevator maintenance personnel, must be able to perform such services less expensively, or at least be able to accomplish such services more efficiently than in-house personnel.

The prospectus, or bid proposal, must require references and suggested maintenance and operation plans by each bidder. These written plans are necessary so that accurate comparisons can be effected between bidders. Again, such written submissions must be included in the written contract.

12.12.2   Specification of services; responsibility for communications; supplies, materials, equipment, and utilities

Once the facility manager has determined the particular services that must be supplied by a contractor, those services must be memorialized in a *support services agreement*. Accordingly, it becomes of paramount importance to allocate responsibility between the parties. In other words, there will be a confirmable understanding as to each party's duties and responsibilities, e.g., who is expected to supply materials, equipment, and utilities; which costs are included in the agreement; and the extent of labor included.

12.12.3   Certified outsource firm personnel to be used

It is not enough for the facility manager to rely on a contractor's reputation with regard to a particular field of endeavor. Better practice is to include language in the contract that requires that the nature and/or level of certification for each employee that will be utilized on-site reach a certain level in accordance with the laws, codes, and regulations of the jurisdiction in which the work is to be performed. Moreover, it is not unreasonable to require that only those individuals qualified or certified to perform the *particular services specified in the outsource agreement* be utilized.

12.12.4   Scheduling work and/or reporting requirements

Every outsource agreement must include a particular basic maintenance and operation schedule, along with a system of reporting the level that maintenance and operations have been accomplished, per the agreement. A variety of methods are available for such reporting requirements. However, it is suggested that the operative phrase should be "the simpler the better."

Accordingly, the outsource agreement should include a "check-off" list and/or schedule of required contacts with appropriate supervisory personnel to suffice as reporting requirements.

It is also important to specify in any contract what is *not* to be accomplished. In other words, the basic scope of any outsource agreement is invariably, as stated above, a listing of what services the facility manager expects to be performed, for the fees to be paid. However, to avoid any misconceptions in terminology or expectations, it is good practice to specify what services will *not* be performed, what parts and/or supplies will *not* be provided by the contractor (e.g., whether worn-out parts are included in the contract), and/or what assistance or notification will *not* be required of the facility manager.

12.12.5 Supervision of outsourcing agreement

In any outsource agreement the supervision of contract execution must be relegated in-house. While this statement is simple in nature and, perhaps, logical in essence, nevertheless it is good practice to specify the nature and scope of supervision to avoid any potential problems.

By delegating the facility manager's responsibility for supervision of *each* particular contract duty, you ensure that a manager or management team will develop sufficient expertise to understand and evaluate the technical services. By the same token, this provides the contractor with a contact person or committee for emergency actions and with respect to any modification on the provided services.

12.12.6 Conclusion

The traditional adage "good fences make good neighbors" certainly applies in the outsource contract agreements. Certainly, any contract may have to be modified over its life. However, by crafting a specific and detailed agreement, you avoid communication problems, failure of service, and, hopefully, litigation. (See Figs. 12.2 and 12.3 for samples of well-written outsource agreements.)

Bibliography

Czegel, Barbara, *Running an Effective Help Desk; Planning, Implementing, Advertising, Improving, Outsourcing,* New York, Wiley/QED Publication, 1994.

Hicks, Donald K., Williamson, John H., Lewis, Carl G., *Guide to Preparing Work Descriptions for Performance Work Statements for Contracted Maintenance Activities for Army Installation Directorates of Logistics,* Champaign, IL, U.S. Army Corps of Engineers, Construction Engineering Research Laboratory, 1991.

Rothery, Brian, and Robertson, Ian, *The Truth about Outsourcing,* Brookfield, VT, Gower, 1995.

SUPPORT SERVICES AGREEMENT

TO:

PROPOSAL No.   01MP95009
DATE SUBMITTED: August 14, 1994
PHONE No.:

AGREEMENT No.:

Location of Agreement: 11510 Falls Road

Siebe Environmental Controls agrees to provide the services described in the attached schedules in accordance with the following terms and conditions.

Schedules Attached	Type of Service
A	Temperature Control
B	Mechanical System
E	Filter Service
G	Additional Services
H	List of Equipment

Terms and Payment:

We agree to furnish the services as described on the schedules checked above for the annual price of $7,440.00
SEVEN THOUSAND FOUR HUNDRED FORTY DOLLARS**

This Siebe Environmental Controls Service Agreement shall begin on the 15 day of August, 1994 and shall continue for a period of one year(s) and from year to year thereafter until terminated. Either party may terminate this agreement upon thirty days written notice prior to the anniversary date of the agreement.

Invoices will be issued Monthly (XX) Quarterly ( ) Semi-Annually ( ) Annually ( ) Special ( ) as agreed. Payment will be made in 30 days.

The contract price shall be subject to adjustment yearly to recognize any changes in costs. Notice of proposed adjustments to the annual price will be provided to you at least sixty days prior to agreement renewal date.

SIEBE ENVIRONMENTAL CONTROLS

Company: _____   Submitted By: Doug Wiggins _____

Accepted By: _____   Accepted By: _____

Title: _____   Title: _____

Date: _____   Date _____

Figure 12.2   Support services agreement.

12.20   Equipment and Systems Operations and Maintenance

General Conditions:

Preventive maintenance will be performed during normal working hours and are defined as _____7:30 AM_____ _____4:00 PM_____ Monday through _____Friday_____ inclusive, excluding holidays.

Reasonable means of access to the equipment being maintained shall be provided by the owner. Our service does not include the normal operation of your system such as starting, stopping, or resetting of the equipment described. However, Siebe Environmental Controls shall be permitted to start and stop all equipment necessary to perform the herein agreed services.

Siebe Environmental Controls shall not be liable for any loss, delay, injury or damage that may be caused by circumstances beyond its control including, but not restricted to acts of God, war, civil commotion, acts of government, fire theft, corrosion, electrolytic action, floods, lightning, freeze-ups, strikes, lock-outs, differences with other trades, riots, explosions, quarantine restrictions, delays in transportation, shortage of vehicles, fuel, labor materials or malicious mischief.

Siebe Environmental Controls' responsibility for injury or damage to persons or property that may be caused by or arise through the maintenance service, or use of the system(s) shall be limited to injury or damage caused directly by our negligence in performing or failing to perform our obligations under this agreement. Siebe Environmental Controls shall not be responsible for the failure of furnishing labor or material caused by reason of strikes or labor troubles affecting our employees who perform the service called for herein, or by unusual delays beyond our reasonable control in procuring supplies or for any other cause beyond our reasonable control. In no event shall Siebe Environmental Controls be liable for business interruption losses or consequential or speculative damages.

Siebe Environmental Controls will present a service report for your signature at the completion of each visit.

We will not be required to make safety tests, install new attachments or appurtenances, add additional controls, and/or revamp or renovate existing system with devices of a different design or function to satisfy conditions established by insurance companies, laboratories, governmental agencies, etc.

In the event the system is altered, modified, changed or moved, Siebe Environmental Controls reserves the right to terminate or renegotiate the agreement based on the condition of the system after the changes have been made.

If emergency service is included in this agreement and is requested at a time other than when we would have made a scheduled preventive maintenance call, and inspection does not reveal any defect required to be serviced under this agreement, we reserve the right to charge you at our prevailing service rates.

Siebe Environmental Controls reserves the right to discontinue this maintenance service agreement at any time, without notice, unless all payments under this contract shall have been made as agreed.

If replacement of parts are included in this agreement, it is understood that Siebe Environmental Controls will not be responsible for the replacement or repair of boiler tubes, boiler sections, boiler refractory, chimney, breeching, refrigeration evaporators, refrigeration condensers, water coils, steam coils, concealed air lines, fan housings, ductwork, water balancing, decorative casings, equipment painting, disconnect switches, electrical power wiring, water, steam, and condensate piping or other structural or non-moving parts of the heating, ventilation, and air conditioning systems. Replacement control valves and dampers, when in our judgment they are required, are covered by this agreement. However, removal from and reinstallation to pipes and ductwork is not included. Siebe Environmental Controls will not be required to make replacements or repairs necessitated by reasons of negligence, misuse or other causes beyond our control except ordinary wear and tear.

If equipment becomes non-repairable due to unavailability of replacement parts, Siebe Environmental Controls will no longer be required to maintain or service such equipment as a part of this agreement. However, Siebe Environmental Controls will assist the owner in replacing the equipment at prevailing service rates.

It is agreed that the equipment, piping, ductwork, controls, etc., have been installed basically as shown on the contract drawings for this building and that the installation and performance of these systems is acceptable to the owner.

It is further understood that the equipment covered under this agreement is in maintainable condition and eligible for a maintenance agreement. If at the time of start-up or on the first inspection, repairs are found necessary, such repair charges will be submitted for the owner's approval. If these charges are declined, those items will be eliminated from the agreement and the price of the agreement will be adjusted in accordance with equipment covered.

Figure 12.2   *(Continued)* Support services agreement.

AUTOMATIC TEMPERATURE CONTROLS

Scheduled maintenance inspections shall be performed during normal working hours

Frequency: Coverage:

0   Weekly   0   Monthly   0   Scheduled Maintenance Only
(x) Quarterly   0   Semi-Annually   (x) Scheduled+Unscheduled Maintenance
0   Annually   0   Repair Materials
0   Other _____   (x) 24-Hour Response

Equipment Covered: All electric and pneumatic controls _____
_____

Air Compressor
Drain Tank and Inspect
Note Condition of Supply Air
Check Run Time
Check Pulley and Belt Tension
Check PRV and Filter Station
Change Oil (Annually)

Air Handling Units
Review Sequence of Operations
Inspect and Lubricate Dampers and Valves
    (Annually)
Check Pilot Positioners
Check Panel Mounted Controls
Check Transmitters and Gauges
Calibrate as Needed
Clean Panel Face
Check Auxiliary Devices (PE's EP's)
Check Safety Devices (Freeze, Fire)

Air Drier
Check Traps
Clean Condenser
Check Expansion Valve and Refrigerant

Time Clocks
Check Time Settings and Operations

Terminal Equipment
Check Room Thermostats (Annually)
Calibrate as Needed
Check Valves; Damper Actuators

Water Systems
Check Master/Submaster Controls
Calibrate as Needed
Check Valves, PE's, Step Controllers
Check Safety Controls

Figure 12.2 *(Continued)* Support services agreement.

CONDENSING UNITS

Scheduled maintenance inspections shall be performed during normal working hours.

Frequency: Coverage:

0	Weekly	(X)	Monthly	
0	Quarterly	0	Semi-Annually	
0	Annually			
0	Other			

0	Scheduled Maintenance Only
(X)	Scheduled+Unscheduled Maintenance
0	Repair Materials
(X)	24-Hour Response

Equipment Covered: See Equipment Schedule H

Air Cooled - Startup Inspection

1. Review manufacturer's recommendation for startup.
2. Energize crank case heater per manufacturer's recommendation for warmup.
3. Remove all debris from within and around unit.
4. Visually inspect for leaks.
5. Check belts, pulleys and mounts. Replace and adjust as required.
6. Lubricate fan and motor bearings per manufacturer's recommendation.
7. Inspect electrical connections, contactors, relays and operating/safety controls
8. Check motor operating conditions
9. Check and clean fan blades as required.
10. Check and clean coil. Straighten fins as required.
11. Check vibration eliminators. Replace or adjust as required.
12. Check compressor oil level, acid test oil and meg hermetic motor. Change oil and refrigerant filter drier as required.
13. Check and test all operating and safety controls
14. Check operating conditions. Adjust as required.

Mid-Season Inspection

1. Visually inspect for leaks.
2. Lubricate fan bearings per manufacturer's recommendation.
3. Lubricate motor bearings per manufacturer's recommendation.
4. Check belts and sheaves. Replace and adjust as required.
5. Clean and straighten fins as required.
6. Check operating conditions. Adjust as required.

Air Cooled Seasonal Shutdown

1. Review manufacturer's recommendation for shutdown.
2. Verify economizer operation (if applicable).
3. Visual inspection of leaks.
4. Verify auxiliary heater operation (if applicable).
5. Inspect belts and filters.
6. Test all safety controls

Optional to Above Procedures

0 1. Clean coil with chemical and high pressure spray. (See Water Treatment)

Figure 12.2 (*Continued*) Support services agreement.

<u>Water Cooled - Startup Inspection</u>

1. Review manufacturer's recommendation for startup.
2. Energize crank case heater per manufacturer's recommendation for warmup.
3. Visually inspect for leaks.
4. Vent system of trapped air.
5. Inspect electrical connections, contactors, relays and operating/safety controls.
6. Check vibration eliminators. Replace or adjust as required.
7. Check compressor oil level, acid test oil and meg hermetic motor. Change oil and refrigerant filter drier as required.
8. Check and test all operating and safety controls
9. Check operating conditions. Adjust as required.

<u>Mid-Season Inspection</u>

1. Visually inspect for leaks.
2. Check operating conditions. Adjust as required.

<u>Air Cooled Seasonal Shutdown</u>

1. Review manufacturer's recommendation for shutdown.
2. Verify economizer operation (if applicable).
3. Visual inspection for leaks.
4. Verify auxiliary heater operation (if applicable).
5. Inspect belts and filters.
6. Test all safety controls.

<u>Optional to Above Procedure</u>

0   1. Inspect and punch condenser tubes.

Figure 12.2 (*Continued*) Support services agreement.

<u>Preseason Inspection</u>

1. Inspect fireside of boiler and record condition.
2. Brush and vacuum soot and dirt from flues and combustion chamber.
3. Inspect firebrick and refractory for defects. Patch and coat as required.
4. Visually inspect boiler pressure vessel for possible leaks and record condition.
5. Disassembled, inspect and clean low-water cutoff.
6. Check hand valves and automatic feed equipment. Repack and adjust as required.
7. Inspect, clean and lubricate the burner and combustion control equipment.
8. Reassemble boiler.
9. Check burner sequence of operation and combustion air equipment.
10. Check fuel piping for leaks and proper support.

<u>Seasonal Startup</u>

1. Review manufacturer's recommendations for boiler and burner startup.
2. Check fuel supply.
3. Check auxiliary equipment operation.
4. Inspect burner, boiler and controls prior to startup.
5. Start burner, check operating controls. Test safety controls and pressure relief valve.
6. Perform combustion test and adjust burner for maximum efficiency.
7. Log all operating conditions.
8. Review operating procedures and owner's log with boiler operator.

<u>Season Shutdown</u>

1. Review owner's log. Log all operating conditions.
2. Shut off burner and open electrical disconnect.
3. Close fuel supply valves.
4. Review boiler operation with boiler operator.

<u>Monthly Operating Inspection</u>

1. Review owner's log. log all operating conditions.
2. Inspect boiler and burner and make adjustments as required.
3. Test low water cutoff and pressure relief valve.
4. Check operating and safety controls.
5. Review boiler operation with boiler operator.

Figure 12.2 (*Continued*) Support services agreement.

FANS & CENTRAL FAN SYSTEMS

Scheduled maintenance inspections shall be performed during normal working hours.

Frequency: Coverage:

0	Weekly	(X)	Monthly	
0	Quarterly	0	Semi-Annually	
0	Annually			
0	Other			

Coverage:

0   Scheduled Maintenance Only
(X) Scheduled+Unscheduled Maintenance
0   Repair Materials
(X) 24-Hour Response

Equipment Covered: See Equipment Schedule H

CENTRAL FAN SYSTEMS - ANNUAL INSPECTION

1. Check and clean fan assembly.
2. Lubricate fan bearings per manufacturer's recommendations.
3. Lubricate motor bearings per manufacturer's recommendations.
4. Check belts and sheaves. (Replace and adjust as required.)
5. Tighten all nuts and bolts.
6. Check motor mounts and vibration pads. (Replace and adjust as required.)
7. Check motor operating conditions.
8. Inspect electrical connections and contactors.
9. Lubricate and adjust associated dampers and linkage.
10. Check fan operation.
11. Clean outside air intake screen.
12. Check and clean drains and drain pans.
13. Check and clean strainers, check steam traps and hand valves.
14. Check filter advancing mechanism. Lubricate and adjust as required.
15. Inspect filters.
16. Check heating and cooling coils.
17. Inspect humidifier.

SEMI-ANNUAL INSPECTION

1. Lubricate fan bearings per manufacturer's recommendations.
2. Lubricate motor bearings per manufacturer's recommendations.
3. Check belts and sheaves. (Replace and adjust as required.)
4. Clean outside air intake screen.
5. Check filter advancing mechanism. Lubricate and adjust as required.
6. Inspect filters.
7. Check heating and cooling coils.
8. Check humidifier.

Figure 12.2 *(Continued)* Support services agreement.

PUMPS

Scheduled maintenance inspections shall be performed during normal working hours.

Frequency: Coverage:

0	Weekly	(X) Monthly	0	Scheduled Maintenance Only
0	Quarterly	0 Semi-Annually	(X)	Scheduled+Unscheduled Maintenance
0	Annually		0	Repair Materials
0	Other _____		(X)	24-Hour Response

Equipment Covered: See Equipment Schedule H _____

ANNUAL INSPECTION

1. Lubricate pump bearings per manufacturer's recommendations.
2. Lubricate motor bearings per manufacturer's recommendations.
3. Tighten all nuts and bolts. Check motor mounts and vibration pads. (Replace and adjust as required.)
4. Visually check pump alignment and coupling.
5. Check motor operating conditions.
6. Inspect electrical connections and contactors.
7. Check and clean strainers and check hand valves.
8. Inspect mechanical seals. Replace as required OR inspect pump packing. Replace and adjust as required.
9. Verify gauges for accuracy.

SEMI-ANNUAL INSPECTION

1. Lubricate pump bearings per manufacturer's recommendations.
2. Lubricate motor bearings per manufacturer's recommendations.
3. Check suction and discharge pressures.
4. Check packing or mechanical seal.

Figure 12.2 (*Continued*) Support services agreement.

PACKAGE UNITS

Scheduled maintenance inspections shall be performed during normal working hours.

Frequency: Coverage:

0	Weekly	(x) Monthly	0	Scheduled Maintenance Only
0	Quarterly	0 Semi-Annually	(x)	Scheduled+Unscheduled Maintenance
0	Annually		0	Repair Materials
0	Other	_____	(x)	24-Hour Response

Equipment Covered: See equipment schedule H _____

_____

COMPRESSOR ASSEMBLY

Air Cooled - Startup Inspection

1. Review manufacturer's recommendation for startup.
2. Energize crank case heater per manufacturer's recommendation for warmup.
3. Remove all debris from within and around unit.
4. Visually inspect for leaks.
5. Check belts, pulleys and mounts. Replace and adjust as required.
6. Lubricate fan and motor bearings per manufacturer's recommendation.
7. Inspect electrical connections, contactors, relays and operating/safety controls
8. Check motor operating conditions
9. Check and clean fan blades as required.
10. Check and clean coil. Straighten fins as required.
11. Check vibration eliminators. Replace or adjust as required.
12. Check compressor oil level, acid test oil and meg hermetic motor. Change oil and refrigerant filter drier as required.
13. Check and test all operating and safety controls
14. Check operating conditions. Adjust as required.

Mid-Season Inspection

1. Visually inspect for leaks.
2. Lubricate fan bearings per manufacturer's recommendation.
3. Lubricate motor bearings per manufacturer's recommendation.
4. Check belts and sheaves. Replace and adjust as required.
5. Clean and straighten fins as required.
6. Check operating conditions. Adjust as required.

Air Cooled Seasonal Shutdown

1. Review manufacturer's recommendation for shutdown.
2. Verify economizer operation (if applicable).
3. Visual inspection of leaks.
4. Verify auxiliary heater operation (if applicable).
5. Inspect belts and filters.
6. Test all safety controls

Optional to Above Procedures

0   1. Clean coil with chemical and high pressure spray. (See Water Treatment)

Figure 12.2   *(Continued)* Support services agreement.

Water Cooled - Startup Inspection

1. Review manufacturer's recommendation for startup.
2. Energize crank case heater per manufacturer's recommendation for warmup.
3. Visually inspect for leaks.
4. Vent system of trapped air.
5. Inspect electrical connections, contactors, relays and operating/safety controls.
6. Check vibration eliminators. Replace or adjust as required.
7. Check compressor oil level, acid test oil and meg hermetic motor. Change oil and refrigerant filter drier as required.
8. Check and test all operating and safety controls
9. Check operating conditions. Adjust as required.

Mid-Season Inspection

1. Visually inspect for leaks.
2. Check operating conditions. Adjust as required.

Air Cooled Seasonal Shutdown

1. Review manufacturer's recommendation for shutdown.
2. Verify economizer operation (if applicable).
3. Visual inspection for leaks.
4. Verify auxiliary heater operation (if applicable).
5. Inspect belts and filters.
6. Test all safety controls.

Optional to Above Procedure

0   1. Inspect and punch condenser tubes.

Fan Assembly - Annual Inspection

1. Check and clean fan assembly.
2. Lubricate fan bearings per manufacturer's recommendations.
3. Lubricate motor bearings per manufacturer's recommendations.
4. Check belts and sheaves. (Replace and adjust as required.)
5. Tighten all nuts and bolts.
6. Check motor mounts and vibration pads. (Replace and adjust as required.)
7. Check motor operating conditions.
8. Inspect electrical connections and contactors.
9. Lubricate and adjust associated dampers and linkage.
10. Check fan operation.
11. Clean outside air intake screen.
12. Check and clean drains and drain pans.
13. Check and clean strainers, check steam traps and hand valves.
14. Check filter advancing mechanism. Lubricate and adjust as required.
15. Inspect filters.
16. Check heating and cooling coils.
17. Inspect humidifier.

Semi-Annual Inspection

1. Lubricate fan bearings per manufacturer's recommendations.
2. Lubricate motor bearings per manufacturer's recommendations.
3. Check belts and sheaves. (Replace and adjust as required.)
4. Clean outside air intake screen.
5. Check filter advancing mechanism. Lubricate and adjust as required.
6. Inspect filters.
7. Check heating and cooling coils.
8. Check humidifier.

Figure 12.2 (*Continued*) Support services agreement.

AIR FILTER SERVICE

Siebe will furnish and install replacement media for the following air handling units:

SEE SCHEDULE H FOR EQUIPMENT COVERED

A. H. Units	# Filters	Size	Changes/Year	Type Filter

***AIR FILTERS ON CONVECTORS TO BE CHANGED ONCE PER YEAR.

***AIR FILTERS ON ALL OTHER UNITS TO BE CHANGED QUARTERLY.

If additional changes are required, they will be made with the mutual consent of the owner in regard to cost for this additional service.

Figure 12.2 *(Continued)* Support services agreement.

ADDITIONAL SERVICES

In addition to the services listed in Schedules A through F, Siebe will furnish the additional services and/or special routines as listed below:

REPLACEMENT OF WORN PARTS OR COMPONENTS: (From Schedules A&B)

Replacement of worn parts or components identified as being part of the mechanical equipment listed on Schedule H is not included in this agreement, with the exception of refrigerant for the air conditioning system.

Refrigerant will be provided when necessary at no additional cost. Robertshaw also agrees to make required repairs and replacements to the equipment listed on Schedule H with the owner's authorization and will bill the owner for the parts and components at the prevailing material rates.

Replacement of major equipment and/or components such as AC units, boilers, compressors, etc., are not included. In such instances, Robertshaw will either provide a proposal to do the work or will make the replacement with the owner's authorization and bill the owner for the parts and labor at the prevailing rates.

Figure 12.2 *(Continued)* Support services agreement.

Master
Maintenance Agreement

AGREEMENT FOR
    MASTER MAINTENANCE SERVICE

TO: _____
(Purchaser - herein called You)

    11510 Falls Road

    Potomac, Maryland 20854

BUILDING LOCATION _____

    11510 Falls Road

    Potomac, Maryland

    . Company (herein called We) will provide   MASTER MAINTENANCE SERVICE on the elevator equipment in the above building and described below (herein called the equipment) on the terms and conditions set forth herein.

No. Elevators and Type	Manufacturer	Serial No.
One Hydraulic Passenger		E-90263

EXTENT OF COVERAGE

    We will:

    Regularly and systematically examine, adjust, lubricate and, whenever required by the wear and tear of normal elevator usage, repair or replace the equipment (except for the items stated hereafter), using trained personnel directly employed and supervised by us to maintain the equipment in proper operating condition.

    Furnish all parts, tools, equipment, lubricants, cleaning compounds and cleaning equipment.

    Relamp all signals as required during regular examinations only.

    Periodically examine and test the hydraulic system and/or governor, safeties and buffers on the equipment, at our expense, as outlined in the American National Standard Safety Code For Elevators and Escalators, A.N.S.I. A17.1, current edition as of the date this agreement is submitted. It is expressly understood and agreed that we will not be liable for any damage to the building structure occasioned by these tests.

ITEMS NOT COVERED

    We assume no responsibility for the following items, which are not included in this agreement:

The cleaning, refinishing, repair or replacement of
- Any component of the car enclosure including removable panels, door panels, sills, car gates, plenum chambers, hung ceilings, light diffusers, light fixtures, tubes and bulbs, handrails, mirrors, car flooring and floor covering.
- Hoistway enclosure, hoistway gates, door panels, frames and sills.
- Cover plates for signal fixtures and operating stations.
- Intercommunication systems used in conjunction with the equipment.
- Main line power switches, breakers and feeders to controller.
- Emergency power plant and associated contactors.
- Emergency car light and all batteries, including those for emergency lowering.
- Smoke and fire sensors and related control equipment not specifically a part of the elevator controls.
- Jack unit cylinder, buried piping and buried conduit.

Figure 12.3   Maintenance master agreement.

PRORATED ITEMS

The items listed on the schedule below show wear and will have to be replaced in the future. To provide you with the maximum of service from these items, we are accepting them in their present condition with the understanding that you agree to pay, in addition to the base amount of this agreement, an extra at the time the items listed are first replaced by us. Your cost for the replacements will be determined by prorating the total charge of replacing the individual items. You agree to pay for that portion of the life of the items used prior to the date of this agreement, and we agree to pay for that portion used since the date of this agreement.

SCHEDULE OF PARTS TO BE PRORATED

NAME OF PART   DATE INSTALLED

-NONE-

HOURS OF SERVICE

We will perform all work hereunder during regular working hours of our regular working days, unless otherwise specified. We include emergency minor adjustment callback service during regular working hours of our regular working days.

If overtime work is not included and we are requested by you to perform work outside of our regular working hours, you agree to pay us for the difference between regular and overtime labor at our regular billing rates.

PURCHASER'S RESPONSIBILITIES

- Possession or control of the equipment shall remain exclusively yours as owner, lessee, possessor or custodian.
- Your responsibility includes, but is not limited to, instructing or warning passengers in the proper use of the equipment, taking the equipment out of service when it becomes unsafe or operates in a manner that might cause injury to a user, promptly reporting to us any accidents or any condition which may need attention and maintaining surveillance of the equipment for such purposes.
- You will provide us unrestricted access to the equipment, and a safe workplace for our employees.
- You will keep the pits and machine rooms clear and free of water and trash and not permit them to be used for storage.
- You agree that you will not permit others to make changes, adjustments, additions, repairs or replacements to the equipment.

TERM

This agreement is effective as of _____August 28_____, 19 88 (the anniversary date) and will continue thereafter until terminated as provided herein. Either party may terminate this agreement at the end of the first five years or at the end of any subsequent five-year period by giving the other party at least ninety (90) days prior written notice.

This agreement may not be assigned without our prior consent in writing.

Figure 12.3 *(Continued)* Maintenance master agreement.

CONDITIONS OF SERVICE

No work, service or liability on the part of   , other than that specifically mentioned herein, is included or intended.

The parties hereto recognize that with the passage of time, equipment technology and designs will change. We shall not be required to install new attachments or improve the equipment or operation from those conditions existing as of the effective date of this agreement. We have the responsibility to make only those adjustments, repairs or replacements required under this agreement which are due to ordinary wear and tear and are disclosed to be reasonably necessary by our examination. You agree to accept our judgement as to the means and methods to be used for any corrective work. We shall not be required to make adjustments, repairs or replacements necessitated by any other cause including but not limited to, obsolescence, accidents, vandalism, negligence or misuse of the equipment. If adjustments, repairs, or replacements are required due to such causes, you agree to pay us as an extra to this agreement for such work at our regular billing rates.

We shall not be required to make tests other than those specified in the extent of coverage, nor to install new attachments or devices whether or not recommended or directed by insurance companies or by federal, state, municipal or other authorities, to make changes or modifications in design, or make any replacement with parts of a different design or to perform any other work not specifically covered in this agreement.

It is understood, in consideration of our performance of the service enumerated herein at the price stated, that nothing in this agreement shall be construed to mean that we assume any liability on account of accidents to persons or property except those directly due to negligent acts of Dover Elevator Company or its employees, and that your own responsibility for accidents to persons or properties while riding on or being on or about the aforesaid equipment referred to, is in no way affected by this agreement.

We shall not be held responsible or liable for any loss, damage, detention, or delay resulting from causes beyond our reasonable control, including but not limited to accidents, fire, flood, acts of civil or military authorities, insurrection or riot, labor troubles, including any strike or lockout which interferes with the performance of work at the building site or our ability to obtain parts or equipment used in the performance of this agreement. In the event of delay due to any such cause, our performance under this agreement will be postponed without liability to us by such length of time as may be reasonably necessary to compensate for the delay. In no event will we be responsible for special, indirect, incidental or consequential damages.

PRICE

The price for the service as stated herein shall be _____One Hundred and Fifty_____ _____ Dollars ($_150.00_____) per month, payable monthly in advance upon presentation of invoice. You shall pay as an addition to the price, the amount of any sales, use, excise or any other taxes which may now or hereafter be applicable to the services to be performed under this agreement.

This price shall be adjusted annually and such adjusted price shall become effective as of each anniversary date of the agreement, based on the percentage of change in the straight time hourly labor cost for elevator examiners in the locality where the equipment is to be examined. For purposes of this agreement, "straight time hourly labor cost" shall mean the straight time hourly rate paid to elevator examiners plus fringe benefits which include, but are not limited to, pensions, vacations, paid holidays, group life insurance, sickness and accident insurance, and hospitalization insurance. The straight time hourly labor cost applicable to this agreement is $_24.035_ of which $_7.35_ constitutes fringe benefits.

A service charge of 1½% per month, or the highest legal rate, whichever is less, shall apply to delinquent accounts. In the event of any default of the payment provisions herein, you agree to pay, in addition to any defaulted amount, all our attorney fees, collection costs or court costs in connection therewith.

Figure 12.3 (*Continued*) Maintenance master agreement.

SPECIAL CONDITIONS

ADDITIONAL PROVISIONS

This instrument contains the entire agreement between the parties hereto and is submitted for acceptance within 30 days from the date executed by us, after which time it is subject to change. All prior negotiations or representations, whether written or verbal, not incorporated herein are superseded. No changes in or additions to this agreement will be recognized unless made in writing and signed by both parties.

No agent or employee shall have the authority to waive or modify any of the terms of this agreement.

We reserve the right to terminate this agreement at any time by notice in writing should payments not be made in accordance with the terms herein.

Should your acceptance be in the form of a purchase order or similar document, the provisions, terms and conditions of this agreement will govern in the event of conflict.

ACCEPTANCE BY YOU AND SUBSEQUENT APPROVAL BY AN EXECUTIVE OFFICER OF DOVER ELEVATOR COMPANY WILL BE REQUIRED BEFORE THIS AGREEMENT BECOMES EFFECTIVE.

Accepted:_____
      (Full Legal Company Name or Individual Purchaser)

By: _____
      (Signature of Authorized Official)

_____
      (Type or Print Name)

Title _____Executive Director_____
      (Type or Print)

Date Signed: ___September 23, 1988___

BILLING ADDRESS:

By: _Chas Stump_____

Date Signed: _____

APPROVED:

By:_____

Title:_____

Date Signed: _____

Figure 12.3 *(Continued)* Maintenance master agreement.

13

General Cleaning in Today's Modern Facility Environment

Steve Sadler and Paul Zeski

Pritchard Industries, Inc.
New Haven, CT

13.1 General

The term "general cleaning" in its basic form leaves much to the interpretation of the parties responsible for getting a facility "clean." Therefore, the gray area that is produced regarding interpretation of the term "general cleaning" can only be made clear through precise instructions by the facility manager. It is the intention of this chapter to help dissolve any misconceptions that can occur between the facility manager and the cleaning vendor during the scope of their contractual relationship and to offer insight into the mind-set of the contractor.

13.2 Types of Cleaning

As an effective manager, it is your responsibility to define the types of cleaning that will be required and expected within the facility. This can only be accomplished by getting an understanding of the property and its needs through an investigation of the following criteria:

The types of areas requiring service:

The parallel (or "in common") requirements of the inhabitants

The uncommon needs of the tenants

The budget guidelines for the project

With a clear understanding of these concepts, you should have the basis for building a cleaning program that will best serve to protect the assets of the property and to keep the facility inhabitants happy.

13.3   Cleaning Program

The first step in creating a clear understanding between management and vendor is to pre-prepare a list of *specifications*. In the cleaning industry, we like to call this your list of "specific *actions*." From this offering, the cleaning contractor will decide (1) the number of employees required on a daily basis to clean the facility (2) the amounts and types of supplies and equipment that will need to be present on site to accomplish scheduled goals, and ultimately, (3) what costs will be charged to the facility to justify the contract. So, as you can see, your attention to detail in writing the proper specifications will determine your "effectiveness-to-budget" ratio.

In order to get you started, let's begin with the fundamental educational process within the cleaning industry. Every cleaning contractor begins by referring to the "Five Steps of Basic Cleaning." These steps (or categories) are as follows:

Step 1: Trash   Functions that must occur within the scope of the contract that will pertain to the collection, removal, recycling, compacting, breakdown, and hauling of rubbish.

Step 2: Ash   Actions that will occur within the parameters of the cleaning agreement that will address the collection, cleaning, and responsible removal of all refuse produced by the act of smoking (whether within the general office environment or within the confines of smoking areas).

Step 3: Dust   All dusting tasks that must occur within the assigned workspace to minimize the collection of dust particles on horizontal and vertical surfaces.

Step 4: Spot Cleaning   Referring to damp-cleaning glass and counter surfaces (both horizontal and vertical) to remove the presence of stains, spills, and handprints. Note that this does not fall under the category of dusting because the nature of the action involved requires specific technique and solutions to accomplish this task.

Step 5: Floor Work   This category encompasses all tasks that would be required to maintain any and all floor types within the serviced property. For example, carpeted floor areas and the different pile types, rugs, vinyl composition floors, stone floors (marble, granite, terrazzite, etc.), ceramic tile, wooden floors, and raised floorings would all need specific maintenance requirements.

It is from this reference guide that you should build your facility "specific *actions.*"

Okay, you've done a terrific job of listing the tasks that will need to be performed within the facility to make sure that everything is clean and maintained properly. Now it is time to address the concept of "frequency." Each task on your list will require a times per year total to determine how often the action will need to be performed. It is important to separate which tasks will require more frequent occurrence from the tasks that will require less attention. By determining this, you will appropriate the funds within your budget with greatest efficiency. Keep in mind that some of the tasks that will require the lowest occurrence can be the most costly to perform each time due to labor burden.

Frequency terminology:

Daily   Specifies that this task will be performed every day that the property will be open for business. Most facilities provide "full service cleaning" for their tenants 5 days per week either Monday through Friday or Sunday through Thursday. (Of course, porter and matron services are typically on a Monday through Friday schedule regardless.) The term daily indicates the number of days during the year in which a facility will be open for business, typically 260 days per year for 5 day per week service or 365 days per year for 7 day per week service.

Weekly   Indicates that a task is to occur 1 time per week throughout the areas in which it is specified. A weekly frequency representation can also be indicated.

Monthly   Tasks that are specified to be performed monthly will generally occur 1 time every 30 days and may be listed as having a frequency of 12 times per year.

Quarterly   A quarterly frequency task is scheduled to be accomplished 4 times per year or 1 time every 3 months. Certain specification programs may describe this task and then simply list the number 4 next to the description, indicating 4 times per year as the occurrence schedule.

To quickly touch on the remaining, most common, frequency descriptions without belaboring the point, we offer the following:

3 times per week = 156 times per year

4 times per week = 208 times per year

Biweekly = 26 times per year

Bimonthly = 6 times per year

Semiannual = 2 times per year

Annual = 1 time per year

It is important to identify frequency of service as an important part of your thought process when building the "specific *actions*" for your facility, as this will be the guideline to which the contractor will adhere for the duration of the agreement. The tasks that are typically underestimated in frequency mainly deal with the upkeep of floor surfaces. Whether it be carpet extraction, hard surface floor maintenance (i.e., stripping and refinishing, spray buffing, stone polishing, etc.), or restroom ceramic and grout care, it is important to consider the traffic that will cause the deterioration of these areas and to address them through the specifications accordingly.

The next step in determining the amounts of staff needed to maintain your facility is to touch on the subject of "productivity rates." These rates are directly derived from the combination of "specification" and "frequency." Each task that will be accomplished within your property will have (by industry studies and hands-on knowledge) a productivity rate individually assigned to it based upon the average taken from having repeatedly performed each task and assessing a productivity value to each duty. This productivity rate is generally based upon "the amount of square feet that can be covered by an individual task in one hour." For example, the industry productivity rate for collecting trash in a typical office environment is 20,000 ft² per hour according to the Daniels Associates time and motion study for performing this "general cleaning" function (see Bibliography). And just as the task "trash collection" has an industry productivity standard assigned as a general rule, so does each task that will be performed within your facility. Some floor finishing tasks will be accomplished at a rate as low as 200 to 300 ft² per hour based upon industry standards.

Once each task has been assigned a productivity value based upon "square feet coverage rate per hour," an overall productivity rate can be determined that best suits the specific needs of the facility.

13.4 Staffing

And it is from this "overall production rate" for completing all daily, weekly, monthly, semiannual, and annual duties that the cleaning contractor will most efficiently staff your property. Without exposing any preconceived thoughts about productivity, as each facility requires different and specific production needs, we offer the following hypothetical observation:

Let's say that ABC Corporation is the single occupant of a 300,000 (cleanable space) ft^2 building. This building is broken down into the following categories: executive office space, general office space, hallway and corridor space, restroom space, deli space, dining room area, kitchen area, delivery dock(s), fitness area, elevators, main lobby areas, etc. Now, once these areas have been identified, a set of specifications will need to be written to address the specific cleaning needs of each space. It will not be unusual to find that many of the cleaning processes found within the five basic steps of cleaning (trash, ash, dust, spot cleaning, and floor work) will apply to all of the areas at first writing, but it is when we begin to break down the special requirements of each area that we discover vast differences in cleaning application. It is the responsibility of those composing the facility specifications to thoughtfully apply the correct cleaning tasks to each area in order that the proportionate amount of time will be allocated to each space. Once this has been accomplished, each separate area will be once again combined to equal the facility cleanable square footage, and a production rate will now be exposed for the facility. The following illustration will substantiate this process.

Area	Space utilized, ft^2
Executive office space	50,000
General office space	210,455
Hallway and corridor space	21,000
Restroom space	5,000
Deli space	500
Dining room area	2,800
Kitchen area	1,200
Delivery dock(s)	1,000
Fitness area	2,300
Elevator(s)	245
Main lobby area(s)	5,500
Combined cleanable area total	300,000

As you can see, each area has been assigned individual square footage for the amounts of space utilized. As discussed, each area has its own individuality with regard to cleaning process, so we have chosen the proper set of "specific *actions*" for each area and have decided what frequency these specifications will need to be performed throughout the contract. Each individual area has its own productivity rate based upon the combination of specification and frequency. When each area is melded into the total, an overall productivity rate is determined for the entire facility.

From this rate the general cleaning contractor proposes staffing for the project. We hope that the above information has proved itself worthy insight and illustration to determining productivity levels within your facility. The next logical course of action is to further explore the arena of staffing as this pertains to identifying job positions for the staff of workers that will be responsible for cleaning your building on a daily basis. To illustrate this we once again refer to our hypothetical 300,000 ft^2 building, occupied by ABC Corporation.

ABC Corporation is a company that produces widgets and sells them on the world market. Like many companies, ABC inhabits an office structure with many areas hosting a list of amenities that produce a comfortable working environment for the employees and visiting guests. As previously mentioned, ABC Corporation has provided its facility with each of the different areas as listed above. The property management responsible for contracting cleaning services for the building has provided specifications for each different area, along with a desired frequency schedule, and has allowed the prospective janitorial provider to submit a proposal that outlines staffing for the project. From this list of "specific *actions*" the contractor has applied industry-accepted productivity standards and has established a cleaning rate of 3750 ft^2 per hour. Keep in mind that this proposal has been tailored specifically to the requirements set forth by ABC management and directly answers the criteria presented. (In other words, production rates vary from facility to facility.) With this knowledge in mind, we offer the following formula:

ABC Corporation building cleanable square footage = 300,000 ft^2

Determined overall productivity rate = 3750 ft^2 per hour

300,000/3750 = 80 h per day

The general cleaning contractor has proposed 80 h as being the minimum requirement for performing up to the standards set by ABC management. The next phase of this process is to determine how long a cleaning shift will be available (or required) by facility management. We can look at this two ways for illustration purposes.

Full-time shift: 80 h per day/8 h (or full-time) shift = 10 available positions

Part-time shift: 80 h per day/4 h (or part-time) shift = 20 available positions

As you can see, both scenarios utilize the same number of hours to clean the facility, but based upon the length of the shift, the number of available positions (to accomplish the exact amount of work) is different. This can become confusing if you are accepting competitive bids and one contractor proposes a full-time shift and one opts for part-time. This can easily be clarified by simply comparing cleaning hours. Do not be confused by numbers of personnel, because as demonstrated above, the contractor proposing 20 persons is providing the same number of cleaning hours as the contractor proposing 10 people. (Which is heavier? 10 lb of iron? or 10 lb of feathers?)

13.5 Job Descriptions

We have now reached the part of the process where job descriptions need to be assigned to the proposed staff. At this point, we have a facility that needs to have general cleaning services, and a staff of 20 persons which will be utilized to perform the custodial duties. The next step is to create "position descriptions" for each of the cleaning staff and to direct them to individual work areas to execute these duties. To simplify this process, we refer again to the "five basic steps of cleaning," for these job assignments will be created from the list of "specific *actions*" provided by facility management. Before moving into specific job categories by position, consider the following job position titles:

General cleaner [general service worker (GSW), maid or matron]

Restroom cleaner

Floorwork personnel (floor technicians)

Utility cleaners (utility support cleaners, zone cleaners)

Stairway and stairwell cleaner

Elevator and escalator cleaner

Periodic cleaner

Window and glass technician

Trash removal persons (trash and recycle technicians)

Day porter (day person, restroom restocker)

Day matron (day person, restroom restocker)

As you can see, the facility can have a number of different job positions which will be allocated to the personnel employed within the general cleaning staff. These staff members perform specific functions based upon assignment to make sure that the custodial contractor is within compliance of the job functions or specifications laid out by facility management. To further illustrate this, we once again refer back to our hypothetical building occupied by ABC Corporation.

The facilities manager of the ABC Corporation has accepted a number of competitive bids from the most reputable janitorial service providers in the vicinity of the ABC property. The initial qualifying criteria required for a service organization to be selected as a prospective bidder included proof of financial stability, years in the janitorial service industry, references of "like accounts" in square foot size and build-out, the ability to provide insurance coverage as required by the ABC Corporation, and demonstration of competent management personnel. Once all bids were submitted, a comparison of hours to be worked and shift length was performed as well as a comparison of other financial data that was pertinent to each group of prices. From this comparison the ABC Corporation selected the commercial cleaning company offering its services at the most competitive rate. ABC Corporation did not have an obligation to select the lowest or highest bidder but should choose the bidding contractor that offered experience, competitive price, managerial know-how, insurance coverage, and financial strength. The chosen cleaning organization proposed a part-time shift utilizing 20 personnel at 4 h each or 80 h total dedicated to performing janitorial services per night. (Refer to the productivity illustration above.) From this identification of numbers of persons required for the job position descriptions can be assigned.

Position 1: General service worker. Responsible for performing all daily housekeeping and security tasks (i.e., locking tenant doors, turning off lights upon job completion) as stipulated within the specifications and directed by management and direct supervision. Duties include trash collection and relining of receptacles with plastic liners as needed, general dusting to include daily and periodic high (objects over 60 in) and low (objects under 24 in) dusting, spot cleaning of horizontal and vertical glass and laminant surfaces with a clean damp cloth, and general vacuuming of carpets and rugs.

In other words, the general service worker (GSW) position is responsible for most of the duties found within the "trash, ash, dust and spot cleaning" categories. Therefore, when the position breakouts are demonstrated, the GSW position will consume the majority of the 20 staff members proposed because, as shown previously, the ABC Corporation consists mainly of executive office space, general office space, corridors, and hallways (281,455 ft^2).

Position 2: Restroom cleaner. Responsible for performing all daily restroom cleaning tasks as required by observation of building specifications and directed by facility management and direct supervision. Duties to include restroom trash collection and relining of receptacles with plastic liners daily, damp disinfecting of all fixtures and sinks, restocking supplies (towels, bathroom tissue, feminine products dispensers, soap dispensers, etc.), damp disinfecting of all horizontal and vertical surfaces to include partitions and mirrors, sweeping and mopping (with a germicidal solution) floor surfaces.

Position 3: Floorwork personnel (floor technician). The floorwork position is held by an individual experienced in maintaining various floor coverings and surfaces throughout the facility and has the skills necessary to operate the equipment required to perform this work. A knowledgeable floorwork person will have insight and background in shampooing, extracting, and spot cleaning carpeted floors as well as being able to perform hard surface floor care tasks such as polishing stone floors, stripping and waxing vinyl composition floors, spray buffing (the technique utilized to spray a fine layer of floor finish onto a waxed floor and buff in order to maintain the luster of that floor on a daily or weekly basis without having to strip the surface), dust mopping (or sweeping), and damp mopping. These duties are generally performed under direct guidance by facility supervision in accordance with property specifications and scheduled frequencies.

Position 4: Utility cleaners (utility support cleaners, zone cleaners). Utility cleaners are also utilized and assigned their duties by facility management and supervision in accordance with the areas of the specifications that would not fall under the responsibility of the GSW position. Typically, it is the job of the utility person to maintain the normal day-to-day cleanliness of areas such as facility lobbies, elevators, escalators, retail areas, fire escape stairwells, building entryways, walkways, and plazas, loading docks, freight elevators, skywalks, basement level service hallways, and tunnels. This position can also be designated as "utility support cleaner" because the individuals occupying these positions may be called upon to assist GSWs in accomplishing periodic task work that is performed not on a daily schedule but on a periodic frequency such as high and low dusting. A utility cleaner who is assigned as a support position to accomplish periodic tasks may choose to accomplish these duties by breaking up the assigned work areas into "zones" (thus the title "zone cleaner"), accomplishing the periodic schedule of tasks in these smaller portions of the assigned work area on a daily basis vs. say, all of a task in the entire work area in one day. (Is that as clear as mud?) To clarify, let's say that our utility support cleaner has been assigned to perform the "weekly" "low dusting" task and the "monthly" "high dusting" tasks in a designated work space. This person may choose to break this work area into four equal quarters (or "quadrants") and execute this work utilizing the following sample schedule:

Monday	Perform weekly low dusting in Quadrant 1
Tuesday	Perform weekly low dusting in Quadrant 2
Wednesday	Perform weekly low dusting in Quadrant 3
Thursday	Perform weekly low dusting in Quadrant 4

At this point, all low dusting is 100 percent complete in the entire work area.

Friday   Perform monthly high dusting in Quadrant 1

This monthly high dusting schedule is to be completed in one quadrant every Friday until all quarters have been high dusted for the month. At the beginning of each month this cycle repeats itself.

Position 5: Stairway and stairwell cleaner.   The person or persons assigned to maintaining the cleanliness of the facility stairways (tenant internal suite stairs, carpeted public or common area stairs) and building stairwells (fire escape stairs, generally concrete or vinyl tile used very little or only in emergencies) will do so in accordance with the specifications assigned to these areas. It is not uncommon for the stairway and stairwell position to also fall under the utility person category. This is so because many facilities that have stairs do not provide enough stairway space to call for an individual to spend an entire shift cleaning stairwells.

You will generally find the stairway and stairwell position to be assigned to an individual or individuals in very tall buildings (say, 20 stories or greater) or in a multiple building complex. Other than these two scenarios, stairwell cleaning will fall under the responsibility of the utility position, which will also be subject to accomplishing other assignments.

Position 6: Elevator and escalator cleaner.   Much like the stairway and stairwell cleaner, the elevator and escalator cleaner will typically fall into the utility person role as well. The facility or building would have to provide a tremendous number of elevators and escalators to occupy the entire shift length of one individual. It is possible, but more often than not the elevators and escalators will be cleaned by a utility person. Once again, in the situation where a facility has a tremendous number of elevator cabs, it is possible to create a single position or multiple positions dedicated solely to this type of cleaning. The main responsibility of the elevator and escalator person is to clean the elevator cabs in their entirety to include all walls, ceilings, and fixtures, metal doors, metal railings, door tracks, button push plates, escalator tracks, rubber hand rails, glass, decorative metal, etc.

Position 7: Periodic cleaner.   In custodial terms, "periodic work" or "periodics" are jobs that are outlined within the facility specifications as needing to be accomplished on a frequency schedule less that daily. These duties are generally scheduled on a weekly, monthly, quarterly, biannual, or annual basis. Therefore, many facility custodial managers will create full positions dedicated to accomplishing these tasks on a periodic schedule which complies directly with building specifications. The role of the periodic cleaner can also fall under the utility person category depending on facility size. The periodic cleaner position may have the responsibility of performing duties such as light lens cleaning, which typically happens on a quarterly to annual frequency.

Position 8: Window and glass technician.   Many facilities are built with a skin which consists of windows that are set in sills. This window glass has two sides, the exterior side and the interior side. It is usually the interior side of this glass that is directly addressed by the janitorial contractor per facility specifications. Along with this interior and exterior glass, many buildings have been furnished with many decorative accessories that have either glass or plexiglass panels such as partitions, glassed-in conference areas, glassed-in offices, and computer room areas. These glass panels require cleaning on a daily basis to remove spots and fingerprints, as well as periodic cleaning of the panels in their entirety to remove dust that accumulates on the upper portions that are not subject to fingerprints and smudges. It is therefore the responsibility of the custodial service to address the glass cleaning issue by either assigning a sole position or positions to clean and maintain window glass or to schedule this task to be executed by the utility crew. It is highly unlikely that any part of the cleaning crew will have any responsibility to clean any of the exterior window glass higher than the first story. It is generally common practice for a separate specialized window crew or service to be responsible for maintaining all exterior window glass higher than that of the first story of the structure. This also will remain true for interior skylights, atrium glass, and lobby glass that is higher than the first floor elevation where either scaffoldings, platforms, or lifts are required. In this case, special insurance requirements will also need to be addressed owing to liability.

Position 9: Trash removal persons (trash and recycle technicians).   Trash removal persons (not to be confused with individuals assigned to trash collection) are the people responsible for taking collected and bagged trash to its final destination at the facility for pickup and transporting (or hauling) to the dump or recyclery. These persons can also be classified as "recycling technicians" if the job requires that they become proficient in the sorting of materials that have the potential for recycling such as glass, plastic, paper, and cardboard. Many facilities have a recycling program in place where the building inhabitants have received instructions to place reusable materials in bins which have been labeled for the collection of those materials only. The responsibility of trash collection will generally fall under that of the general service worker (GSW).

The GSW will typically gather all trash in 44 gallon containers, or "brute barrels," which are large plastic collection buckets on wheels for easy transport. These containers are generally lined with a large plastic bag which is tied off when full. The GSW will then place the full bags at a gathering point somewhere in their work area (near an exit door or in the elevator lobby). Then, at some point during the shift, a trash removal person will make rounds with a very large collection receptacle, say a 1.5 yd^3 buggy, and gather the bagged trash. This trash is then moved to the area on site designated as the final collection point before transport to the dump. This area is typically a loading dock or platform which provides a dumpster or trash compactor. In some instances, the facility will provide a number of different variations for trash to be sorted. This might include a number of different dumpsters labeled "Paper Only," "Plastic Only," "Glass Only," or "Cardboard Only." It is therefore the responsibility of the recycle technician to properly sort these items in order to provide the greatest possibility for the most efficient recycling.

At this point in our position descriptions a distinguishable line can be drawn which separates the facility cleaning staff into two entities, the "night staff" and the "day staff." Both of these facets play an integral role to comprise the whole with regard to providing full service within a facility. Whereas the inhabitants of the facility rarely come in contact with the evening general cleaning staff, the opposite is true for the day staff. The day porters and day matrons act as the "glue" to keep the entire cleaning operation together and generally grow to be relied upon by building tenants to provide continuity to the cleaning service throughout the day. It is the important responsibility of the day personnel employed within the property to maintain the flow of service during normal building hours. Without the day staff (and assuming that the facility is of the size where day services are required), it is possible that the efforts of the night staff could fall short in a number of areas such as restroom supply restocking due to daytime usage, cleaning of spills, and in the instance where a large quantity of trash has been produced and is in need of removal. From this information the following positions have come to play an integral role in the success of many facilities.

Positions 10 and 11: Day porter/day matron (day person, restroom restocker). The position of the day person has become vital to the continuous flow of custodial service within many facilities. It is not uncommon for many tenants to rely upon the daytime cleaning crew for a number of crucial services which make inhabitance of the workplace pleasant. For example, the day crew will generally report to their assigned positions at least 1 h prior to normal facility hours, and sometimes earlier. Day people will make sure that common areas and entryways are clean and free of debris. These day positions are also responsible for keeping plazas and walkways safe by removing any standing water, snow, or spills that might be present prior to normal operating hours and that might occur throughout the day. Once a normal rotation has taken place to make common areas and entryways clean and safe, a scheduled check of facility restrooms takes place. This is one reason why day persons might also be titled "restroom restockers." During the morning hours, normal building activity will deplete restroom supplies such as hand towels, bathroom tissue, and hand soap. Therefore, it is the standard schedule of the day person to check and restock these supplies as needed to avoid having any of these items totally diminished. The only real distinguishable difference between the job of the day porter and day matron is that it proves less of an inconvenience to facility tenants when no interruption in restroom services occurs owing to opposite gender restocking.

This is why "day person" has become the industry-accepted title for this position. It is not uncommon to see facility day custodial personnel assisting building management in a number of tasks when not attending to normally scheduled janitorial duties.

At this juncture we summarize the above information and tie up any loose ends left unaddressed.

The ABC Corporation staff consists of 20 general cleaning personnel. Each of these people is given a job assignment in accordance with building specifications. We know from the information above that their job titles could be any one of the categories listed. Each person's eligibility for any of these positions is based on previous job experience. The following shows the building staff by position:

Area	Space, ft^2
Executive office space	50,000
General office space	210,455
Hallway and corridor space	21,000
Deli space	500
Dining room area	2,800
Kitchen area	1,200
Fitness area	2,300
Combined square footage	288,255
Required "general service workers"	15
Restroom space	5,000

Area	Space, ft^2
Total square footage restrooms	5,000
Required "restroom cleaning staff"	1
Main lobby area	5,500
Elevators	245
Delivery dock	1,000
Total "common entryway" area	6,745
Required "elevator, stairway, lobby, dock" cleaning staff	1
The remaining staff will be responsible for the remainder of the work, which is based on task and not square footage.	
Required "trash removal" staff	1
Required "utility cleaner, glass technician, periodic cleaner"	1
Required "floorwork technician"	1
Total cleaning staff	20

As demonstrated, some of the job positions require proficiency in a number of skills to provide the most efficient allocation of staff. The most important factor in the proposed staffing profile is the overall effectiveness of these individuals as they are positioned. This will generally be revealed within the first few weeks of their presence within the facility, so it will be critical to gauge the satisfaction of the tenants during this time. The feedback of the building inhabitants will be a crucial link in allowing the contractor to make the proper adjustments (if needed) with regard to efficient job allocation.

13.6   Quality Control

Quality control (QC) is an aggregate of functions designed to ensure adequate quality in manufactured products by initial critical study of engineering design, materials, processes, equipment, and workmanship followed by periodic inspection and analysis of the inspection to determine causes for defects and by removal of such causes.

Quality control as it pertains to the general cleaning industry is a term that has parameters that encompass a broad scope of smaller entities making up the QC program as a whole. The facets that should comprise a comprehensive quality control program in the cleaning service industry are as follows:

Adherence to specifications + cleanliness + communication + execution + personnel + response + training + follow-up = "Quality"

It has been our experience that any incident within a facility will fall into at least one of the above categories. The factor which will determine the success (or failure) of the janitorial service being provided at your facility will rely solely on the execution of *all of the above*.

With this in mind, we introduce you to the quality rating program implemented at the ABC Corporation facility.

The ABC Corporation is the sole tenant of a building that is about 15 years old. The building has been well maintained mechanically and has been renovated and remodeled recently. It has been the goal of ABC to provide a pleasant working environment for its inhabitants and their guests, and an excellent job has been done to keep the interior and exterior of the building looking neat and clean. The recent renovation fell right into ABC's plan to update carpets, wall coverings, some furniture, and fixtures in order that the building can maintain its class A status. The goal of ABC management is to provide the vehicle for quality by instilling a sense of cleanliness and pride within the facility boundaries. The building managers made quality control a major issue when choosing their janitorial service provider. ABC expects quality from each of its vendors.

During the final selection process, ABC was impressed by the "quality control program" that was outlined by the successful custodial service bidder. It is important to keep in mind that the janitorial service that was awarded the cleaning contract is able to provide quality control in its service effort while maintaining competitive pricing. This program is now in effect at the facility site and is as follows:

Step 1: Communication. It was the objective of the successful janitor company to establish lines of communication with the ABC Corporation management office and the facility tenants from the beginning of the relationship. A number of brief meetings were attended prior to commencement of services to discuss issues ranging from security, keys, special working hours, special facility needs, areas where tenants may work late, heavy traffic areas, possible specific tenant needs, day porter duties (scheduling), and project management and supervisory requirements. A "communications log" will be present at the site to track cleaning requests and comments so that total customer satisfaction is achieved.

Step 2: Training. Once a crew was established to clean the ABC site, a training outline was drafted to demonstrate hazard communication (Hazcom) training, proper use of HEPA (high-efficiency particulate air) filter vacuums, special cleaning requirements within their assigned working areas, an overview of facility specifications which will be performed within individual workstations, the proper use of tools and equipment, the proper use of cleaning chemicals, and correct maintenance practices of all supplies, equipment, and storage areas.

Step 3: Startup. A team was assembled by the successful janitorial vendor to be on-site at the ABC Corporation facility during the first few days of startup. This team consisted of management personnel assigned to a number of individuals on the cleaning crew for the purpose of conducting extensive training to those persons. The members of the start team will also be responsible for checking a specific assigned area for these days so that all specifications have been addressed. Should any deficiencies be detected, further training will be provided.

Step 4: Equipment. An extensive equipment list was provided to demonstrate the types of innovations that will be introduced on the ABC Corporation site. Vacuum cleaners equipped with HEPA filters will be utilized so that dust particles that are removed from carpets are not recycled into the air through vacuum exhaust. All vacuum bags will be contained within compartments featured on this equipment. Equipment will be of sturdy construction by manufacturers with an outstanding reputation in their field. The cleaning contractor will have open lines of communication with equipment representatives who will answer any questions that may arise with regard to proper maintenance and operation.

Step 5: Monitoring. The quality control program will be monitored through a number of site visits each year to gauge the success of the cleaning operation. Adherence to a number of criteria will be judged in order that an overall picture of quality can be produced. The categories graded are to include overall account cleanliness, management satisfaction, equipment maintenance, adequate supply stock, tenant satisfaction, log entries, adherence to specifications, periodic scheduling, accurate billing records, cleaning personnel satisfaction, on-time payroll deliveries, response to service requests, competence of site cleaning management and supervision, and compliance regarding health, safety, and hazard regulations.

Step 6: Reporting. The final step in providing an all-encompassing quality control program is the proper reporting of inspection results. Therefore, the janitorial company has implemented a system where a "report card" will be distributed to facility management personnel, and cleaning management and supervision so that deficiencies may be corrected. As a result of these reports, rewards will be issued to individuals demonstrating outstanding efforts and retraining will be conducted as needed. As you can see, a number of factors comprise an effective quality control effort. Communication is the key in making the entire system work. In all cases, failure to express cleanliness objectives by facility management to the cleaning contractor will expose weaknesses in the program. Also, failure of the janitorial company to respond correctly to customer requests, and within an adequate amount of time, will undermine established quality objectives. Communication is the fuel that drives the quality machine.

13.7   Waste Management Program

Requirements for different types of facilities with regard to establishing a successful "waste management program" within a facility may vary, so we refer to the ABC Corporation building once again as we present our final model. We will try to include a number of scenarios, terminology, and equipment choices which can be utilized to effectively address the management of waste collection and removal from the ABC Corporation site.

The procedure for collection of rubbish on any site is, in essence, very much the same. Generally, the entire idea is to collect, remove, and transport the waste produced on site to the dump. When the specific goals of the facility are addressed and Environmental Protection Agency (EPA) compliance issues are factored in, very distinguishable differences may be brought to light. These differences can be based upon a number of different factors including economics, recycling, and philosophy (with regard to the environment). A knowledgeable cleaning contractor can assist in producing the best waste removal situation for the project. As a first step the contractor listens to the needs of the facility manager, for it is from these initial requirements that the program will be established.

Without getting into specific state and EPA requirements due to the vast scope pertaining to these issues, we assume that the facility manager and the cleaning contractor are aware of specific requirements for the dumping of waste. For additional information see the Bibliography to assist in answering any questions you may have regarding Environmental Protection Agency waste-handling issues. We once again cast ourselves back to the site of the ABC Corporation.

The ABC Corporation has discussed their facility needs for waste disposal at the property. These needs include providing a comprehensive waste management program that will address all issues regarding the collection and disposal of all trash. It is the goal of the ABC Corporation and its inhabitants to recycle as much waste as possible. Between the facility manager and the cleaning contractor an interoffice e-mail has been composed and sent out to each of the building tenants outlining the recycling program. Because the ABC Corporation is a single inhabitant of the building, the e-mail was the most efficient way to inform its employees of the program while saving any paper that might have been utilized for this process. As you can see, ABC is very serious about reaching its recycling goal. The memo lists a number of methods that will be adhered to by its workers to assure that the maximum amount of recyclable materials will be eligible for reuse. Some of the instructions are as follows:

1. Each employee will be provided with two office trash cans at their desk. One receptacle will have the recycling symbol clearly posted. The recycling bin will be for all paper waste that is produced at each desk area. The second waste can will be for any other trash that is not specified as being eligible for the recycling program.

2. Any consumption of food or drink (where drink containers produce waste) is prohibited at employee desks to prevent accidental contamination of recyclable paper. This practice is also exercised to minimize the number of plastic trash can liners that will need to be replaced due to wet trash. (Dry liners that are not soiled or damaged can be used over and over.)

3. Large (95-gallon) containers on wheels will be located in the freight elevator lobbies on each floor. Each employee will be responsible for carrying and disposing of the contents of their recyclable (dry) paper waste containers one time per week into these big containers. When full, these 95-gallon containers will be moved to the area designated as recycle goods storage. These items will then be loaded and hauled to the recyclery.

4. All "wet trash" items are to be disposed of in the proper recycling receptacle in the kitchen and cafeteria areas only. Wet trash is defined as any waste materials produced by the consumption of food products. Absolutely no food packaging is to be disposed of at the desk areas to prevent contamination of eligible recyclable waste.

5. A number of clearly labeled receptacles will be located in the kitchen and cafeteria areas that will direct employees where to dispose of plastic containers, glass containers, and aluminum cans. All other wet trash packaging (i.e., sandwich wrappings, sandwich boxes, Styrofoam meal containers) will be disposed of in one barrel which is also located in the kitchen and cafeteria area. This barrel will not contain any items considered for recycling. It is critical that no wet trash come in contact with the paper waste that will be recycled. This contact could contaminate the paper trash, possibly rendering it useless to the recycling process. All collection bins will be removed when full and the contents will be hauled for reuse.

As you can see, the ABC Corporation is determined in its effort to introduce a successful recycling program which encompasses all eligible goods.

The next part of the waste management process is to provide a list of equipment that will be necessary for accommodating the recycling goal of the ABC Corporation. This equipment is essential in providing to the program the means for proper execution. The waste containers must be of the proper size to hold the amounts of waste that will be produced at the facility. This equipment must also be of sturdy construction and designed well in order that transport of full containers will be manageable. The following is a list of commonly proposed equipment:

1. 95-gallon paper collection containers with wheels and lids. These large containers are constructed of a very durable plastic. The objective is to place them in an area where employees can dump paper wastebaskets from their desks. These containers roll off much like a two-wheel furniture dolly (tilt and roll) for easy transportation to and from the areas in which they are located. They are available in a variety of colors for easy identification.

2. 50-gallon containers with wheels and lids. These medium-large containers are constructed of durable plastic that can be easily cleaned by hosing the entire unit. The objective is to place these receptacles in large kitchen and cafeteria areas where employees can choose the correct receptacle (as labeled) for disposal of glass, plastic, and aluminum packaging. These containers are built with design features that include lids and furniture dolly-style wheels that allow for easy transport to and from the areas in which they are positioned. A number of colors are available.

3. 28-gallon containers with wheels and lids. These square containers are available in a variety of colors and feature a four-wheel base and lids. Constructed of durable plastic, the wheels swivel in order that the units can be easily moved in any direction. These 28-gallon units can be placed in smaller kitchen areas where 50-gallon receptacles would be too large or where smaller amounts of recyclable trash will be produced.

4. 5-gallon "We Recycle" recycle containers. These containers are recommended to be stationed at each employee work area. They are constructed of durable plastic and have the recycling symbol (three arrows forming a continuous circle) clearly posted on them. They are molded in "recycle blue" plastic for easy identification. It is the intention that only paper produced at the desk site be placed in these cans. Once they are full, employees can easily carry the can to one of the 95-gallon paper collection containers located near the work area.

5. 5-gallon waste receptacles. These are positioned at the employee disquiet, adjacent to the "We Recycle" container. A plastic liner will be placed inside this container, as it is the intention of the company for employees to place any nonrecyclable waste in these containers. They are constructed of durable plastic and can be furnished in a variety of

colors. Usually cast in beige or black plastic, these waste cans will be emptied nightly by the cleaning crew and relined as necessary.

6. 44-gallon "brute barrel" containers. Also referred to as "janitor carts," 44-gallon "brutes" are probably the most recognizable piece of custodial equipment in a building. Each staff member assigned to general cleaning receives a brute barrel as standard issue equipment for the collection of disquiet trash in their assigned work area.

These barrels are constructed of durable plastic and are situated on five swivel wheels for easy mobility in all directions. It is standard practice to line these collection carts with large plastic liners similar in size to "leaf and lawn" bags. The "general service workers" (GSW) empty all nonrecyclable disquiet and kitchen trash into these large lined containers nightly for disposal. This trash is hauled directly to the dump (or similar site) owing to the nonreusable nature of this trash.

7. 1.5 cubic yard "tilt truck." Constructed of durable heavy-duty plastic, the 1.5 yd^3 tilt truck (or larger, up to 2.5 yd^3) is a large buggy, usually gray in color. These carts are manufactured with two fixed rear wheels on an axle and two heavy-duty swiveling front wheels for easy maneuverability. Once the GSWs have collected disquiet and kitchen nonrecyclable trash, it is general practice for them to place this bagged and tied trash in a gathering area (typically a freight elevator lobby) for removal from their work area. The person or persons on the custodial crew assigned to "trash removal" load these tilt trucks with the nonrecycle trash collected in the 44-gallon brutes and transport them to the building area containing trash dumpsters or compactors. It is the intention of the janitorial company to be able to move large amounts of rubbish to the trash docks in the large tilt truck containers each trip, creating a most efficient operation. These containers are designed to be tilted at the edge of a dock platform for easy unloading or dumping, thus the word "tilt" in the description. Tilt trucks are not motorized or mechanical and are pushed much like an oversized shopping cart with fixed rolling wheels in the rear and swiveling wheels in the front.

At this point we discuss the removal of recyclable and nonrecyclable trash from the "trash dock" area of the facility to the final dumping or recycling site. This removal procedure is set up on a schedule by the company or companies contracted to haul recyclable and nonrecyclable waste.

Once again, it is highly recommended that the reader refer to the Bibliography regarding the usefulness and value of recyclable products that will be considered for collection and regeneration within your facility.

We now move to the subject of removal of waste items from the

ABC Corporation building. To accomplish this, we place ourselves in the waste collection area of the facility in the loading dock area of the building, which features direct street access for large trucks to pick up and deliver materials at the site. In a conversation between the management company, the janitorial contractor, and the waste collection service, it has been agreed that a number of large metal dumpster containers will be provided to the building for the gathering of waste and recyclable goods. Each dumpster bin is clearly labeled as to what items will be placed in each bin exclusively. To further reinforce this effort, these bins have been color coded for identification of contents as well as posted in a number of languages for the same purpose. The janitorial contractor has implemented a training program that details these criteria and will designate individual persons to place separated materials into these containers. No person will be placed in this job category who has not been sufficiently instructed in the handling of the facility waste materials. It has also been determined that the waste collection service will handle the removal of all facility waste including recyclable goods. Therefore, a pickup and removal schedule has been set up as follows:

1. The waste collection contractor will service the account every day, 5 days per week. Since the amounts of the various types of trash (i.e., paper, aluminum, plastic, glass, and general waste) will be separated by the building tenants and placed in designated cans, the need to haul any one of these categories of waste daily will be eliminated. (There just won't be enough volume of any one item produced for this to be required).

2. The by-product that will accumulate the largest amount of waste will be paper and paper products such as corrugation. This waste will need to stay dry from the ABC Corporation site to the recyclery. Therefore, a sizable vehicle that hauls dry paper goods only will be required to service the account 1 day per week. A number of vehicles can be utilized for this process as long as the loading compartment or trailer provides adequate cover for the paper product within.

3. A second vehicle will be utilized to remove the remaining recyclable products at the site (i.e., aluminum cans, plastic containers, glass containers, and general waste). Since these items typically produce equal amounts of moisture from their contents, it is not necessary to provide separate hauling vehicles (or cover for that matter). However, it will still be important to prevent commingling of these products from the ABC Corporation site to the recycling facility. This can easily be accomplished by placing large (clear) plastic liners in the various collection bins where tenants will be placing their recyclable goods. (The clear plastic liners provide an easy identification method in order that the waste collection contractor can quickly

determine the contents of each load.) Also, when these goods get to the dump site, they can be separated quickly from one another, as the clear liners afford an easy view of their contents. This process can save the trash collection service valuable time and money as individuals are not required to "hand sort" the trash, which has already been accomplished at the building site.

In closing, we hope that we have demonstrated the resourcefulness that has become common practice within the general cleaning industry. Today's largest janitorial contractors employ thousands of people nationally and internationally and can produce annual revenues in the millions (and in some cases, billions) of dollars. It is no longer required, and in most cases impossible, that facility managers become experts in the field of custodial services and waste removal. With ever-changing rules, regulations, and Environmental Protection Agency requirements (as well as Occupational Safety Hazard Administration or OSHA compliance), it has become common practice for large general cleaning operations to assign compliance officers to keep abreast of changes in this field. There are many advantages to contracting with a larger custodial provider that cannot typically be afforded by smaller companies. The largest demonstration of this is the requirement that will be needed by your facility regarding insurance coverage. Many small companies just cannot provide the insurance boundaries needed to adequately protect and "hold harmless" the property management company and the tenants that occupy the building. Another advantage of financial strength will be clearly realized in the equipment provided by the custodial company within your property. It is important to look to the companies that can afford to offer the latest technological discoveries available. One example of this are the HEPA compliance vacuums which have a number of filters that can filter the exhaust air of the vacuum down to 0.3 micrometers at 99.97 percent efficiency. These systems can be very expensive compared to lesser effective commercial vacuums but provide cleaner, healthier air for the inhabitants of the building. They also keep dust particulates from recirculating throughout the building, thus keeping air ventilation systems clean.

Bibliography

Five Basic Steps of Cleaning, as instructed by Pritchard Industries, Inc., *Training Procedures Manual,* Pritchard Industries, Inc., 1120 Avenue of the Americas, New York.
HEPA information based upon performance quotes by vacuum systems manufactured by Castex Incorporated (A Tennant Co.), 12875 Ransom Street, Holland, MI.

Production rates established in database providing accepted custodial industry productivity standards. Database provided by Daniels Associates, Inc., 2930 East Northern Avenue, Phoenix, AZ.

The McGraw-Hill Book on Integrated Solid Waste Management by Tchobanoglous, Theisen, and Vigil.

The McGraw-Hill Recycling Handbook by Herbert F. Lund. Handbook of Environmental Management and Technology by Holmes, Singh, and Theodore, Wiley-Interscience, New York.

# Chapter 14

Landscaping Services

Carleen M. Wood-Thomas

Consolidated Engineering Services, Inc.
Arlington, VA

Professional landscaping gives properties curb appeal and this is what initially attracts potential occupants. The more attractive a property is from the street, the more likely prospective occupants are to come into the property. Professional landscaping greatly enhances the appearance of a building and can set it apart from others in the community. The following information should assist a facility manager in managing the landscape contractor responsible for maintenance of the grounds and assist in understanding the basic principles of landscaping. It is important to note that the information provided is for temperate climates and can only be generally applied to other climates in the country.

This chapter outlines the principal responsibilities that a facility manager should assume in overseeing the design and maintenance of a property, some of the more common challenges normally encountered, and recommended guidelines for specifying contractor responsibilities, including turf maintenance, which includes mowing and trimming, pest and weed control, and fertilization; and maintenance of trees, shrubs, and other plant materials including pruning, fertilization, and watering. Also addressed are recommendations on plant materials, trees, and shrub seasonal color and sample budgeting and maintenance forms that are generally applicable to most facilities.

14.1   General Responsibilities
of the Contractor

It should be the responsibility of the contractor to furnish all labor, equipment, and materials to perform all specified landscape maintenance tasks. All equipment should be of such type and condition as needed to effectively perform the task intended and to avoid any unreasonable hazards or dangers to the properties, occupants, and pedestrians. Equipment should be maintained in good condition and should not produce excessive noise or noxious fumes when operated under normal conditions. Some communities have noise restriction ordinances. These restrictions may apply to mowers, leaf blowers, and other landscape equipment. Check with your local jurisdiction to determine if any restrictions apply.

Personnel employed by the contractor should be thoroughly trained using current and accepted horticultural practices. Trained personnel are less likely to make mistakes and have accidents. Personnel should wear company uniforms at all times as supplied by the contractor. Uniforms help identify employees and add a professional appearance.

There should always be a supervisor or foreman on the property to direct all contracted personnel and maintenance operations. Supervisors should report to the building during each visit to verify their presence and provide an update on the status of the grounds. The supervisor should also leave a work ticket with the facility manager which includes information on tasks completed during the visit. If personnel are not on site, the contractor should leave the work report in a designated location on site or mail the pertinent information.

The landscape contractor should comply with any applicable codes and building regulations. This information should be communicated to the contractor prior to the first visit to avoid unnecessary problems. For instance, many facilities have restrictions on mowing hours. If mowing should not occur before 7:30 A.M. this is important to communicate to the contractor to avoid tenant complaints.

All existing and new plant material should be replaced at the contractor's expense if the death or damage of the plant was caused by negligence or a direct act by the contractor. Plant material should be replaced by the owner if the death or damage occurred by natural phenomena outside of the contractor's control. All replacement plant material should be in excellent health and acclimated to the area. It should be a form and type indicative of the species and should be of the same size, type, and form as the plant being replaced.

14.2   Selecting a Contractor and the Bidding Process

When selecting a landscape contractor, several contractors should be interviewed before a final selection. Cost should not be the only determination in the selection process. Landscape contractor abilities and competence vary widely and it is important to obtain an objective assessment of the experience and skills that a contractor offers.

The company chosen should have the appropriate equipment for the job, should be professional, and should be appropriately trained. Look for a contractor that listens to your ideas and problems. Ask for references, visit the company's offices, and visit properties they are currently maintaining.

A contractor that is not appropriately trained will likely lead to a frustrating situation where both simple and more complicated maintenance tasks fail to be properly performed. Over time, this will degrade the appearance of the property and will invite additional problems due to the contractor's inability to properly assess plant diseases, identify pests, prune properly, and perform other important landscape tasks.

In order to make apples to apples comparisons on different proposals from different contractors, set up guidelines that each contractor must follow for bid submission. Since each property may have different landscape needs and may require different levels of maintenance, it may be best to have a professional examine the property to analyze its maintenance requirements. Based on this analysis, a comprehensive maintenance specification can be developed along with a list of each task necessary to maintain the landscape and the number of occurrences per task (see Fig. 14.1).

Once the maintenance specifications and any other relevant expectations have been identified, a mandatory prebid conference should be called to walk the potential contractors through the property. This meeting should emphasize any special requirements of the property and its boundaries. Any questions regarding the landscape can be addressed at this time. In this way each contractor is bidding on the same exact set of specifications, tasks, and number of occurrences. If this is not done prior to bidding, contractors are bidding on what they think needs to be done, not necessarily what the facility manager wants. By spending the time up front, costly extras and misunderstandings can be avoided.

14.3   Lawn Maintenance

Lawn areas require a consistent maintenance program to stay healthy. If lawn areas are neglected for only a couple of seasons, they

14.4 Equipment and Systems Operations and Maintenance

	NO. OF OCC.	COST/ OCC.	EXTENSION
A. TURF MAINTENANCE			
1. Mowing	28	$_____	$_____
2. Leaf Removal	8	$_____	$_____
3. Edging	28	$_____	$_____
4. Fertilizing	3	$_____	$_____
5. Weed Control			
a. Pre-Emergent	2	$_____	$_____
b. Post-Emergent	2	$_____	$_____
6. Pest Control	2	$_____	$_____
7. Core Aeration	1	$_____	$_____
8. Mechanical Slit-seeding	1	$_____	$_____
9. Lime (if necessary)	1	$_____	$_____
10. Dethatching (if necessary)	1	$_____	$_____
		SUBTOTAL	$_____
B. PEST MANAGEMENT			
1. Monitor and Control All Pests and Diseases	39	$_____	$_____
		SUBTOTAL	$_____
C. MAINTENANCE OF TREES, SHRUBS AND OTHER PLANTINGS			
1. Edging	6	$_____	$_____
2. Mulching	1	$_____	$_____
3. Weeding	28	$_____	$_____
4. Clean-up	40	$_____	$_____
5. Pruning			
Trees	1	$_____	$_____
Shrubs	10	$_____	$_____
Ground cover	10	$_____	$_____
6. Fertilization			
Trees	1	$_____	$_____
Shrubs	2	$_____	$_____
Ground cover	2	$_____	$_____
Annuals and Perennials	4	$_____	$_____
7. Monitor and Adjust Irrigation	19	$_____	$_____
8. Manual Watering	As needed	$_____	$_____
		SUBTOTAL	$_____
		TOTAL	$_____

Figure 14.1 Bid form for apples to apples comparisons from contractors.

will decline. A maintenance program should consist of proper mowing, pest and weed control, fertilization, watering, aerating, and overseeding. This program will encourage the growth of healthy turf, which in turn crowds out unwanted weeds.

There are two basic types of grasses: cool season grasses and warm season grasses. Cool season grasses can withstand cold winters, but most do poorly in hot summers. Cool season grasses are used mainly in northern areas. Warm season grasses are lush during hot weather and go into dormancy when temperatures are below freezing. A reputable contractor, nursery, or local extension service can determine what type of grasses will thrive in your area.

14.3.1 Mowing and trimming

Mowing should begin in March or April at intervals of 5 to 10 days (maximum) between mowings. Mowing should be done frequently enough so that no more than one-third of the leaf area is removed at one time. This will help the turf to develop a more extensive root system and withstand environmental stresses. The optimum height for turf is determined by the type of grass. Horizontally spreading grasses are typically cut shorter than vertically growing grasses. Frequent mowing tends to produce a finer-textured turf, since cutting frequently stimulates new growth. If the turf is neglected and becomes too tall, the growth becomes coarse and may produce seeds. Mowing the turf too short will cause the grass to dry and burn, which will allow weed seeds to germinate.

Mowing should be done in alternate directions at least every four mowings. This will eliminate ruts and a striped or streaked look. Mowers should be well maintained and cutting blades kept sharpened at all times to prevent tearing of the leaf blade.

Litter and debris should be removed from all lawn areas prior to mowing. Clippings can be left in the turf area. This is more cost-effective, since the clippings put nutrients back into the soil. If, owing to long periods of wet weather, the turf becomes excessively long, the clippings should be removed, since they can form a mat on the turf and shade out and kill the grass. In areas where there is concern that clippings are unsightly, such as entrances, clippings can be bagged or raked and removed. Mowing should be done in such a way that clippings are not blown into the shrub beds and tree rings, since this can be unsightly.

Areas around posts, signs, buildings, and trees should be trimmed at the same height as the lawn. Lawnmowers and string trimmers should not be used at the base of trees and shrubs, since they can cause damage to the base of the plants. The contractor should be responsible for replacement and/or repair of plants, irrigation equipment, exterior lights, signs, posts, fences, automobiles, paved surfaces, building exterior, and any other ground structures damaged by their activities while working on the property.

All sidewalks, pathways, and curbs should be blade-edged on a regular basis. Shrub beds and tree rings should be edged to maintain a 2-in vertical edge between turf and mulch areas. Grass and weeds should be continuously removed from cracks and expansion joints in all walks and curbs. Clippings and all natural debris should be cleaned off all paved areas after mowing and edging.

14.3.2   Weed control and pests

Weeds are simply plants that grow in the wrong place. There is no such thing as a weed-free lawn, but with proper control weeds can be minimized. To minimize weeds, the following recommendations are generally applicable:

In early spring, when daytime air temperatures reach 55 to 60°F, a broad-spectrum, preemergent (applied before weeds emerge) herbicide that controls both noxious grasses and broadleaf weeds should be applied to all turf areas in accordance with the manufacturer's recommendations. Additional applications of preemergent weed control may be necessary to effectively control all weeds.

In late spring and again in early fall, when daytime air temperatures are not above 80°, the contractor should apply a broad-spectrum, postemergent (applied after weeds emerge) herbicide to control all weeds. The presence of certain weed species that are difficult to control may require additional applications of herbicides.

The contractor should regularly monitor all turf areas for insect, disease, and weed infestations and treat as needed. The contractor should be responsible to replace all turf areas damaged as a result of pest and disease problems with sod. The sod should match the surrounding healthy turf.

14.3.3   Fertilization

Soil fertility is one of the major considerations in any program of lawn management. A healthy lawn requires a soil that is fertile from year to year. Since grass can quickly deplete soil of essential nutrients, the nutrients should be added into the soil on a regular basis. The essential nutrients for turf areas are nitrogen, phosphorus, and potassium. Nitrogen is critical since it stimulates leaf growth and keeps turf green. Phosphorus is needed for the production of flowers, fruits, and seeds and induces strong root growth. Potassium is valuable in promoting general vigor and increases the resistance to certain diseases. Potassium also plays an important role in sturdy root formation.

Fertilizers contain three numbers such as 10-6-4. The first represents the percentage of nitrogen, the second the percentage of phosphorus, and the third the percentage of potassium. All fertilizers list the essential nutrients in this order.

While these nutrients are critical to a healthy lawn, excessive use and application of fertilizers is unnecessary and damaging to both the lawn and the environment. For this reason, apply only the minimum amount necessary to achieve a healthy lawn area.

Soil pH is critical to growing a healthy stand of turf (or any plants). Soil pH is the acid-alkaline balance of the soil. The pH scale divides the range of alkaline and acid materials into 14 points. The middle value of 7.0 is neutral, marking a balance between acidic and alkaline soil values. Some plants thrive best in neutral conditions, while others prefer a more acid or alkaline soil. Turf grows best at a pH of 6.0 to 7.0; consequently to have a healthy stand of turf it is important to make sure the pH is correct. The pH can be changed by adding lime if the soil is too acid or by adding sulfur if the soil is alkaline.

A soil test will give the pH as well as the level of nutrients available in the soil. From the soil test results a contractor can determine the amount of fertilizer to be applied and if an application of lime or sulfur is necessary.

14.3.4   Dethatching and aeration

Many soils can become compacted over time. When this occurs, nutrients and water have difficulty in penetrating to the root zone. This causes the root zones to become shallow, which in turn causes the plants to dry out quickly. Aeration can help to reduce compaction. Aeration is a method whereby holes are punched into the turf to allow moisture, oxygen, and nutrients to penetrate the soil and reach the root zone. Aeration holes should penetrate 2 to 4 in into the soil.

Thatch is a layer of dead grass and other organic matter that accumulates between the surface of the soil and the grass blades. When the thatch depth reaches $\frac{1}{2}$ in the thatch should be removed. Cool season grasses should be dethatched in the fall. Warm season grasses should be dethatched in the spring.

14.3.5   Overseeding

Once a year turf should be overseeded to reestablish the lawn in bare or thin areas. Overseeding is most successful in the fall. The seed should be planted early enough so that the grass has enough time to get established before the cold weather sets in. The next best time to overseed is in the spring after the last frost and before it gets too hot.

14.4   Maintenance of Trees, Shrubs, and Other Plantings

14.4.1   Mulch and weed control

Mulch regulates soil temperatures, insulates plant roots from temperature extremes, reduces water loss from the soil surface, and minimizes the time and labor required to maintain the garden by minimizing the germination of weed seeds. The most important function of mulch is moisture retention. Mulch allows water to percolate through and protects the soil from the drying effects of the sun.

Mulch is available in a number of forms both organic and inorganic. Organic mulch is typically recommended because it will eventually decompose and will add humus to the soil. This in turn improves the soil composition and texture. In addition, during decomposition nutrients are released, which increases the fertility of the soil. In selecting mulches, there are a number of factors to consider: the availability of the material, the cost, and the appearance. Local nurseries or an extension service should be able to advise you on the type of mulch suitable for your needs.

Prior to mulching, all beds and tree rings should be defined and edged. The edge should be maintained throughout the season to give the landscape a clean and crisp appearance. Edging debris should not be placed in the beds or rings but should be removed from the site since excess soil at the base of plants can be detrimental to their health. All tree rings should be evenly concentric around the tree and all bed edges should be maintained as one smooth and continuous line.

The beds and tree rings should be mulched at a depth of 2 to 3 in in early spring. Any mulch existing from previous years that is in excess of 2 in deep should be removed or worked into the soil before new mulch is applied. Often the mulch is left in the beds and after years it builds up to 8 to 10 in. This is not a case where more is better. Since roots need oxygen to survive, excess mulch will prevent oxygen from penetrating the soil and the plants can suffocate. Excess mulch will also cause roots to grow into the mulch and not into the soil. This causes the plants to be shallow-rooted, which in turn causes the plants to dry out quickly and suffer during droughts.

It is a common practice in many areas to mound mulch around the trunk of a tree. This practice should be avoided for a number of reasons. First, mulch, if kept in contact with the bark, can promote attack by insects and disease. Second, the mulch will keep excess moisture on the trunk or crown of the plantings, which can cause decay.

In midsummer, the mulch should be lightly raked and loosened to break up any water-impermeable layers. A light top-dress application of mulch should be applied in early fall and at any other time during the season to maintain a consistent 2-in layer of mulch in all beds and tree saucers.

All beds, tree rings, and planting areas should be kept weed-free at all times. Weeds should be controlled by hand. A preemergent herbicide may be applied before mulching to reduce weed germination. Postemergent, nonselective contact herbicides should only be used as spot treatments.

14.4.2 Pruning

There are many reasons to prune: to keep plants healthy, to restrict or promote growth, to encourage bloom, to repair damage, to remove structurally weak or otherwise undesirable branches, to clear a building, or to allow light to penetrate to the ground. Trees, shrubs, and ground covers all may require some type of pruning during the growing season to achieve these goals.

14.4.2.1 Shrubs. It is important to remember that different types of shrubs have different growth habits and characteristics. Plants are selected for a particular area based on form, color, and texture. If all the plants on a site are sheared into hedges or as individual balls or squares, the characteristics of the plants are lost. Therefore, it is important to follow proper pruning techniques so that the natural beauty of a plant is recognized.

The best time to prune shrubs depends on their flowering habits. Shrubs which flower on new growth should be pruned in early spring before new growth emerges or during the last weeks of winter. Shrubs which flower on old growth should be pruned directly after flowering. If these shrubs are pruned throughout the growing season, the flower buds will be removed and there will be no flower display. As a general rule of thumb spring-flowering shrubs should be pruned immediately after blooming. Broadleaf evergreen trees and shrubs should be pruned after new growth hardens except for hollies, which should be tip-pruned in early spring. Conifers should be pruned by pruning new growth (candling) and again, only if necessary, after the new growth hardens off. Shrubs that flower in summer should be pruned in late fall to early winter or early spring.

If plants are healthy, pruning should occur to maintain the shape of

Figure 14.2 Proper pruning cuts. (1) A proper pruning cut is achieved by making a clean 45° angled cut away from the bud. The cut should be made about ¼ in above the bud. (2) If a cut is made too close to the bud the bud may die. (3) A long stub will be unsightly and will eventually decay.

the plant and to encourage new growth. Depending on the type of shrub, one-fifth to one-fourth of the old branches should be removed from the ground. This will encourage new growth from the base of the plant and will maintain the natural shape of the shrub. If additional pruning is necessary, hand pruning is appropriate as long as the natural habit of the plant is maintained. Proper pruning cuts are essential to maintaining the health of the plant (see Fig. 14.2).

Plants that are disease- and insect-ridden and unmanageable require rejuvenation pruning. The first step in rejuvenation pruning is to remove old and diseased wood. The second step is to cut back healthy wood to encourage branching. Lastly, the sucker growth should be thinned. The key to rejuvenation pruning is to strive for a well-balanced and uniform plant.

There are occasions where a plant or group of plants is completely overgrown or scraggly and it is appropriate to cut the plant to the ground in the early spring and allow new growth to renew the plant. Although this may seem extreme, certain plants will come back healthy and strong within one to two seasons. Check with your local nursery to see which plants will survive renewal pruning. If a plant cannot be cut to the ground, pruning can occur over a 3-year period, each year cutting back one-third of the oldest canes. Whether a plant is cut to the ground or cut back a third at a time, the plant should be fertilized and watered to encourage new growth.

Hedges should be pruned by hand as necessary to maintain a neat and trim appearance. Hedges should be maintained at an exact and equal height for the entire length of the hedge and should be shaped with the bottom of the hedge slightly wider than the top (see Fig. 14.3).

Figure 14.3 Hedge pruning. Proper pruning for a hedge leaves the bottom of the hedge wider than the top. This allows sunlight to all areas of the shrub.

14.4.2.2 Trees. Trees may require pruning, particularly if they have been neglected for many years. All dead, diseased, weak, and cross branches should be removed to improve the structural integrity of the tree. To avoid having to extensively prune large trees, the trees should be pruned and trained while young. Properly pruned trees will grow into structurally sound trees as they mature. Cross branches should be removed, permanent branches should be carefully selected, and a strong branch structure should be developed (see Fig. 14.4).

Typically, the contractor should be responsible for all pruning that can be reached from the ground (or the bed of a truck on all roadways and parking lots) with an extended pole pruner. If a tree requires climbing into the tree, it should be pruned by a certified arborist.

Any trees or shrubs that are pruned to the point where the aesthetic quality of the plants is severely damaged or the health of the plants is jeopardized should be replaced entirely at the contractor's expense.

Figure 14.4 Tree pruning. Proper pruning of a branch is cut as close to the branch collar as possible. The branch collar should not be removed.

14.4.2.3   Vines and ground covers. Vines and ground covers should be pruned regularly to maintain a neat and manicured appearance but should not be sheared. String trimmers are never to be used to prune ivy or other ground covers. Ground covers should be pruned at the nodes, with the cut hidden. Depending on aesthetic preferences, ground covers should be maintained within or partially overhanging all planters and off all paved surfaces. Ground covers should be maintained 4 to 6 in away from the trunks of trees and shrubs.

14.4.3   Fertilization

Plants need different amounts and proportions of nutrients to stay healthy. In most areas, even if the soil has ample amounts of organic matter, supplemental fertilization will be necessary. Many soils have insufficient amounts of one essential nutrient with an overabundance of another. As discussed earlier, the amount and proportions of fertilizer required should be determined by a soil test.

Assuming that fertilization is required, shrubs and ground covers should be fertilized in early spring. Ericaceous (acid-loving plants) plant material should be fertilized with an acidifying fertilizer. Deciduous and evergreen trees should be fertilized in late fall after leaves have dropped and the danger of forced growth is past.

All trees and shrubs indicating chlorosis (a systemic condition in which new and old growth turns yellow) should be fertilized with chelated iron and other micronutrients as needed. Root-stimulating hormones should be applied to all plants that are in poor condition. Soil drenching is preferred over foliar application to prevent phytotoxicity (burning of the leaves).

14.4.4   Cleanup

All areas, including planting areas, plant materials, lawns, and paved areas, should be kept clean at all times. The contractor should remove and dispose of any and all trash (including cigarette butts), sticks, natural debris (including soil, sand, rocks and gravel, withered flower buds, seed pods, leaves, etc.) from all landscaped areas, including all raised planters, turf, and ground cover beds during each visit.

In autumn, leaves should be raked and removed on a regular basis. All leaves should be removed from all lawn and bed areas before mowing, including leaves and branches that drop throughout the spring and summer months.

14.5 Pest Management

The contractor should be responsible for the detection, monitoring, and control of all pests. The contractor should be aware of the potential pests and should make regular and thorough inspections of all plant material and treat as necessary using products and methods that target the insect pest with minimal residual effects.

It would be difficult to discuss all the potential pests which can affect plantings. What is important to understand is that healthy plants perform well. How do you keep your plants healthy? Locate plants in a location that is suitable for the particular plant, prepare the soil well, water appropriately, and fertilize when necessary. All this is part of the philosophy of integrated pest management (IPM). If plants are in a healthy environment they are less likely to have diseases and pests. If they do have a particular problem only the problem area should be addressed. The practice of spot treatment greatly reduces unnecessary pesticide use. It should never be assumed that if there is a problem in one area it must be widespread and therefore everything should be treated. This is often a waste of time and money. It is also important to realize that a low population of pests does not necessarily mean that chemical control is required. Since no landscape can be kept pest-free, a management program should be developed which maintains pest populations below a damaging level.

If there are beneficial insects (ones that feed on the pests), then no chemicals need to be applied. Applied IPM techniques minimize the need for pesticides to control problems as well as decrease the probability of problems in the first place. If the philosophy of IPM is used, there should be less need for chemical applications, which ultimately lowers the cost for maintenance while enhancing the performance of the landscape.

If there is a need for chemical application, the contractor must adhere to the Department of Agriculture Regulations for Commercial Application of Pesticides. All pesticide applicators should be licensed or directly supervised by a licensed applicator. All licenses should be for commercial application and should be current.

The contractor should inform the owner of the pesticides to be used on the property and should receive approval from the owner before making substitutions on specified chemicals. The contractor should submit copies of product safety sheets of all pesticides used to the facility manager to be filed in the building office.

Upon completion of each pesticide application, the applicator should record all information on a data sheet and file it in the building office (see Fig. 14.5).

PESTICIDE APPLICATION RECORD AND DATA SHEET

Applicator's Name(s): _____

Date and Time of Application: _____ 19_____ . _____ M.

Weather Conditions (Temperature, Wind Direction and Velocity, Precipitation,

etc.): _____

_____

Pesticide Used - Label Name and Formulation: _____

_____

Concentration: _____

Total Amount of Pesticide Mixed: _____   Amount of Pesticide Used: _____

Amount of Pesticide Left Over: _____

Disposal of Leftover Pesticide: _____

Application Equipment Used: _____

_____

Safety Equipment Used: Face Shield ( ) Goggles ( ) Respirator ( ) Gloves ( )

Boots ( ) Headgear ( ) Tyvek Suit ( ) Rainsuit ( )

Other _____

Plants Treated: _____

_____

Number of Plants or Area Treated: _____

_____

Reason for Pesticide Application: _____

_____

Applicator's Comments: _____

_____

_____

_____

_____

(Applicator's Signature)

Figure 14.5   Sample pesticide application and data sheet.

Pesticide applications should not be done as calendar-scheduled or general cover sprays. Pesticides should only be applied as needed, when pests are detected through regular inspections. Pesticides should be applied at a time of day when human activity is at a minimum.

14.6   Watering

The key to watering is to water deeply and infrequently. This helps to develop an extensive, deep root system. Frequent, light waterings encourage roots to stay near the surface. This encourages the plants to be more and more shallow-rooted. Shallow-rooted plants tend to be less vigorous and suffer during drought. It is always preferable to water early in the morning so the sun will burn off the excess moisture. This will decrease the potential for fungus and disease to set in.

If plants are too dry their leaves and flowers will wilt and the plant will eventually wither and die. Plants can also die from too much water, especially if the water accumulates around the roots of the plant. With too much water, the leaves and flowers turn black and fall off and the roots will rot. When the roots are exposed, they will be black with no healthy white roots and the exposed roots will often smell foul. The correct amount of water will vary according to soil type, plant and turf species, climate, and weather. Once the correct amount of water is determined, plants will be healthy and vigorous.

It is best to have one person responsible for the proper and adequate irrigation of all lawn areas, beds, trees, annual flowers, and all other plantings throughout the season. In most situations the watering should be the full responsibility of the landscape contractor, since the contractor is most familiar with the plant needs.

Automatic underground irrigation systems should be used where they are installed and working. Again, the contractor should be solely responsible to coordinate the timing and control of these systems. Irrigation systems should be checked regularly to make sure heads are spraying correctly and lines have not been broken.

Areas without irrigation systems should be watered by hand. If the contractor is responsible for the watering, they should supply hoses, water trucks, and all other equipment needed to properly and adequately irrigate by hand.

14.7 Seasonal Color

Seasonal color is essential for distinguishing one property from the next. For very little money, annuals, perennials, and bulbs can enhance the overall appearance of a property.

Annuals are defined as plants that complete their life cycle in one season. In most areas of the country they will flower throughout the growing season but will need to be removed at the end of the season. With careful planning and design, the display will perform for the entire season.

Annual planting beds are high maintenance, and this should be considered in determining the size of the bed and the location. Annual plantings may need daily watering and will need regular fertilization. Scattered beds throughout a property that do not have access to water will decline quickly and show poorly. On the other hand, a small bed near the front entrance will be noticed by everyone entering the building and will be easier to maintain.

Perennials are plants which will come back each year but will typically have a shorter blooming time. Mixing perennials which bloom at different times will give a constant display of color. Since perennials will come back year after year, it is recommended that a landscape architect or garden designer be involved in the layout and design to assure a successful planting.

Bulbs have a short blooming period but give a beautiful display and are always welcome in the early spring. Many bulbs, such as tulips, are removed after they bloom. Others, such as daffodils, can be left in place for many years as long as the bed is not disturbed.

The seasonal color displays should be unified rather than disjointed groupings throughout the property. To achieve the greatest visual impact, the displays should be planted at key points to a building such as entrance drives and the front door. Keep the designs simple, using a few colors and a limited number of plant types. A single color will have more impact than five or six colors mixed together.

Soil preparation is one of the most important elements in achieving a successful seasonal display. Seasonal plantings need good drainage and fertile soil to look their best. First, the soil should be tested for pH to determine whether it is acid (6.9 and lower) or alkaline (7.1 and higher) and if there are any nutrient deficiencies. For the most accurate results, a soil sample can be sent to a soil testing lab.

Next the area should be checked for drainage. An easy test to assess whether there is proper drainage is to dig a whole 1 foot by 1 foot by 1 foot and fill it with water. If the water percolates out within 1 to 2 h the drainage is adequate. If it takes more than 1 to 2 h, some type of drainage system may be necessary such as drain tiles or a dry well. It is important to note that very few plants can survive in wet conditions, so poor drainage should always be corrected.

Once the drainage is checked and the area is draining adequately, the soil should be amended. Amending soil with organic material will improve the soil quality and improve aeration and may also improve drainage. As the organic matter decomposes it will add nutrients into the soil which then can be taken up by the plants.

The plant bed should be tilled to a depth of approximately 12 in. The amendments should be worked into the soil and the consistency of the soil should be light and airy. Once this is completed, the seasonal display can be installed.

14.8   Quality Control

Quality control should be the responsibility of the landscape contractor. The landscape contractor should employ site supervisors and grounds technicians that accept responsibility for the appearance of the site. Inspections of the property should occur on a regular basis (monthly or quarterly) to establish crew goals for quality improvements as well as for identifying the site's needs. Copies of the inspection reports should be sent to the facility manager for review. The landscape contractor should familiarize all site personnel with the specific requirements of the site and the specifications for maintenance.

The facility manager should also perform quality control inspections with the contractor. Figure 14.6 is a landscape maintenance schedule which can be used as a checklist to assure that all tasks are being performed and that all deficiencies are being addressed and actions are taken to rectify them. The schedule should be modified to meet the specific requirements of the site and the specifications.

Landscape Maintenance Schedule

TASK	Jan	Feb	Mar	Apr	May	Jun	Jul	Aug	Sept	Oct	Nov	Dec
Spring Clean-up				✓								
Mulching				✓								
Mowing			✓	✓	✓	✓	✓	✓	✓	✓	✓	
Remove Litter/Debris in Turf Areas			✓	✓	✓	✓	✓	✓	✓	✓	✓	
Weed Control			✓	✓	✓	✓	✓	✓	✓	✓	✓	
Pre-emergent Herbicide				✓								
Grub Control					✓	✓	✓	✓	✓	✓	✓	
Post-emergent Herbicide				✓					✓			
Fertilize Shrubs/Trees			✓									
Fertilize Turf			✓								✓	
Core Aerate Lawn									✓			
Dethatch Lawn									✓			
Soil Tests									✓			
Over Seed									✓			
Lime											✓	
Dormant Oil			✓									
Other Pests			✓	✓	✓	✓	✓	✓	✓	✓	✓	
Pruning Trees	✓	✓										✓
Pruning Shrubs/Ground Covers				✓	✓	✓		✓	✓	✓		

Figure 14.6 Typical landscape maintenance schedule

Bibliography

American Horticultural Society Encyclopedia of Gardening. New York: Dorling Kindersley, Inc., 1993.

American Standard for Nursery Stock. Washington, D.C.: American Association of Nurserymen, Inc., 1990.

Baumgardt, John P. *How to Prune Almost Everything.* New York: William Morrow & Company, 1968.

Bush-Brown, James and Louise. *America's Garden Book.* Pennsylvania: Charles Scribner's Sons, 1967.

Landscape Specification Guidelines. Maryland: Landscape Contractors Association, 1993.

Lawns & Ground Covers. California: Sunset Publishing Corporation, 1989.

Taylor's Master Guide to Gardening. New York: Houghton Mifflin Company, 1994.

Tree-Pruning Guidelines. Illinois: International Society of Arboriculture, 1992.

15

Elevator and Escalator Equipment and Systems Maintenance and Repair Services

Ronald D. Schloss

SEEC, L.L.C., Columbia, NJ

15.1 Introduction

The vertical transportation equipment must consistently deliver people and freight safely and efficiently throughout your building or complex. Unlike most of your electrical and mechanical equipment, elevators, escalators, dumbwaiters, cartlifts, moving walks, and stage lifts are operated and ridden by *all* of your employees, tenants, and visitors. Therefore, your first responsibility is to assure that operating signage and instructions for passengers are communicated as required. Second, for the equipment to perform as intended day in and day out, periodic maintenance will be required. Periodic tests to help assure safe operation will also be required by the local code-enforcing authorities. It will be your responsibility to establish a competent outside contractor and/or hire the necessary workforce to assure safe and efficient equipment operation. Third, you will need to monitor and document that the equipment is operating as intended, at least as your defense for liability should an accident occur. A central monitoring station, outside consultant, and your own trained eye will be helpful. A good working relationship with your vertical transportation contractor and your elevator maintenance personnel will be essential for assuring that the contract is fulfilled.

This chapter's aim is to provide the building owner or manager with their responsibilities for providing safe and reliable vertical transportation. Section 15.2 explains the necessary steps to help assure that the riding public properly use your building's transportation. Section 15.3 details the required maintenance and explains the contractor's qualifications and responsibilities and the options available to the building owner/manager. Section 15.4 describes monitoring and documentation of equipment performance. Section 15.5 offers methods for monitoring maintenance performance. Section 15.6 presents the necessary steps to help improve safety and limit liability. Section 15.7 describes a program to restore equipment that is approaching obsolescence.

15.2 Operating Instructions

Signs that are required by ASME A17.1-1996 Safety Code for Elevators and Escalators are listed in this section. It should be noted that your local code authorities may have adopted only a portion of this national code or may have made their own revisions. Currently the A17.1 code is being combined or harmonized with the B44 Canadian code and will change many rules when the harmonization efforts become the standard. Also the version of the code in effect at the time your equipment was installed or modernized will govern the rules you must follow except for retroactive rules. Symbols for blind or sighted passengers (American Disability Act—ADAAG) and signage required for firefighter's operations are not shown below, as vintage and local codes dictate the exact requirements. Your local code-enforcing authorities can help with any questions.

15.2.1 Elevators

Blind hoistway and emergency doors must have the following sign (at least 2-in lettering) DANGER, ELEVATOR HOISTWAY posted on the corridor side of the access door.

A photograph as shown in Fig. 15.1 should be posted over each elevator corridor call station. It is 5 in wide and 8 in high.

A plate stating the capacity (pounds and number of passengers) and the rated load should be fastened in a conspicuous place inside the elevator. Lettering should be $\frac{1}{4}$ in or larger.

15.2.2 Freight Elevators

Signs (letter at least $\frac{1}{2}$ in high), in addition to the capacity and data plates that should be permanently and conspicuously posted inside the cab, should include:

5" wide, 8" high

Figure 15.1   Elevator corridor call station pictograph. (*Courtesy of ASME.*)

- For class A freight elevators permitted to carry passengers: CLASS A LOADING. THIS ELEVATOR DESIGNED FOR GENERAL FREIGHT LOADING.

- For class B freight elevators permitted to carry passengers: CLASS B LOADING. THIS ELEVATOR DESIGNED TO TRANSPORT MOTOR VEHICLES HAVING A MAXIMUM GROSS WEIGHT NOT TO EXCEED   LB.

- For class C-1 freight elevators permitted to carry passengers: CLASS C-1 LOADING. THIS ELEVATOR DESIGNED TO TRANSPORT LOADED INDUSTRIAL TRUCK. MAXIMUM COM-BINED WEIGHT OF INDUSTRIAL TRUCK AND LOAD NOT TO EXCEED   LB.

- For class C-2 freight elevators permitted to carry passengers: THIS ELEVATOR DESIGNED FOR LOADING AND UNLOADING BY

INDUSTRIAL TRUCK. MAXIMUM LOADING AND UNLOAD-
ING WEIGHT WHILE PARKED NOT TO EXCEED   LB.
MAXIMUM WEIGHT TRANSPORTED NOT TO EXCEED   LB.

- For class C-3 freight elevators permitted to carry passengers must
  meet the requirements for passenger elevators: CLASS C-3 LOAD-
  ING. THIS ELEVATOR DESIGNED TO TRANSPORT CONCEN-
  TRATED LOADS NOT TO EXCEED   LB.

- For the above freight elevators that are *not* permitted to carry pas-
  sengers, the sign should read: THIS IS NOT A PASSENGER ELE-
  VATOR. NO PERSONS OTHER THAN THE OPERATOR AND
  FREIGHT HANDLERS ARE PERMITTED TO RIDE ON THIS
  ELEVATOR.

15.2.3   Hand Elevators

Hoistway doors must have the following sign (at least 2-in lettering)
DANGER—ELEVATOR—KEEP CLOSED posted on the corridor side
of each door.

15.2.4   Hand-Operated Dumbwaiters

Each hoistway door should have the following sign posted on the corridor side with letters not less than 2 in high: DANGER—DUMB-
WAITER—KEEP CLOSED.

15.2.5   Hand- and Power-Operated Dumbwaiters

A sign stating no RIDERS should be located in the car in letters not
less than $\frac{1}{2}$ in high.

15.2.6   Wheelchair Lifts

A passenger restriction sign should be provided at each landing and on
the platform. It should be securely fastened in a conspicuous place and
state PHYSICALLY DISABLED PERSONS ONLY. NO FREIGHT. The
sign letters should not be less than $\frac{1}{4}$ in high and should include the
international symbol for physically disabled persons.

15.2.7   Escalators and Moving Walks

A caution sign shown in Fig. 15.2 should be located at the ends of the
escalator or walk visible to the boarding passengers. The size should
be 4 in wide and $7\frac{3}{4}$ in high.

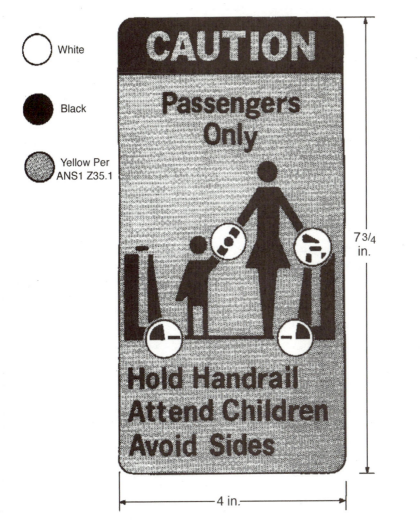

Figure 15.2 Escalator/moving walk sign. (*Courtesy of ASME.*)

■ The ends should contain a red-colored stop button containing the words EMERGENCY STOP.

15.2.8 ADA Signage (American Disabilities Act)

Signage is required to aid the handicapped. While most of the regulations are aimed at braille operating controls placed at wheelchair-accessible levels in the elevator, some signage may be required.

15.2.9   Additional Signage

Caution, warning, or traffic regulation signs may be placed in areas where appropriate to deter equipment misuse or improper operation. While only the signs stated above are required by code, these additional signs can be helpful. An example would be a sign or caution tape placed across an elevator entrance reading ELEVATOR BEING SERVICED. Another example would be a sign on the equipment room door reading DANGER, ELEVATOR EQUIPMENT. Signs on switches and disconnects controlling equipment lighting and power circuits reading ALWAYS LEAVE ON may help to prevent inadvertent shutdowns or entrapments.

15.2.10   Equipment Identification Signs

All vertical transportation equipment within the building should be marked with a unique identification number. Where practical, mark each elevator sequentially from 1 through 24 if there are 24 elevators in the building. All components of the elevator should contain the same number on the control panels, the machine, motor-generator, the governor, the main line power disconnect, the pit equipment, and inside the elevator. This will help ensure that a component is not mistakenly serviced, resulting in personal injury or equipment damage. It also permits passengers to report a misoperation of an elevator or correctly report an entrapment. Floor markings should be placed on the leading edge(s) of the hoistway doors, visible when the door(s) open so that a problem at a specific floor can also be communicated to the repair technician or identify a floor to emergency personnel. Escalators, dumbwaiters, and other unique equipment can be identified with similar markings of elevators, since there would be little chance for confusion.

Operating instructions and signage described above will help assure proper use of the equipment. It does *not* provide the passengers with the best time of day to use the equipment. The design speed and capacity of the equipment were based on the projected traffic flow from 7 A.M. to 7 P.M. on normal working days. The peak periods of passenger usage for most office buildings will occur between 7:30 and 9 A.M. for up-traveling passengers and between 4:30 and 6 P.M. for passengers leaving the building. Lunchtime peaks usually occur between 11:30 A.M. and 1 P.M. The number of units required to prevent "bunching" of passengers (for example, 30 or more people waiting at the lobby floor to board an elevator at 8 A.M.) in the building lobby and other corridors was considered by the designer. For more information on the application of the units see *Vertical Transportation, Elevators and Escalator,* 2d ed., by George Strakosch or *Vertical Transportation*

Standards, 7th ed., by the National Elevator Industry, Inc. Both can be obtained through the Educational Materials Department of *Elevator World Magazine.* To minimize bunching and other long corridor waiting times in your building, and therefore optimize equipment performance, consider one or more of the following actions:

- Keep all available equipment "in service" during peak periods, avoiding planned maintenance shutdowns, deliveries, and other utilization of the equipment that could be postponed to in between or after peak periods.

- Keep the elevator supervisory (dispatching) systems in proper working condition and their time clocks set to the correct time of day.

- Run pairs of escalators in the directions favoring the flow of traffic.

- Where practical, implement staggered starting times and lunch periods for building personnel.

- Encourage building personnel to follow good passenger practices, for example, no registering of both corridor pushbuttons in an effort to get an elevator more quickly.

- Discourage passenger flow opposite to the major traffic flow where possible.

Contact your elevator contractor or consultant if you encounter problems that may only be corrected by the addition of special features.

15.3   Maintenance of Equipment

15.3.1   Definitions and Terminology

Traction elevator. A cab and counterweight system (see Fig. 15.3) connected by a set of several steel ropes. The ropes are routed over the sheave (grooved wheel) of a driving machine. The machine motor can drive the sheave directly (*gearless traction elevator*) or indirectly with worm and gears (*geared traction elevator*). Owing to design requirements, the traction elevator requires more maintenance time when compared to all other types of vertical transportation equipment in the building.

Hydraulic elevator. A cab connected to a jack consisting of a plunger and cylinder (see Fig. 15.4). The most common types of hydraulic elevators have underground jacks with plungers that push the bottom of

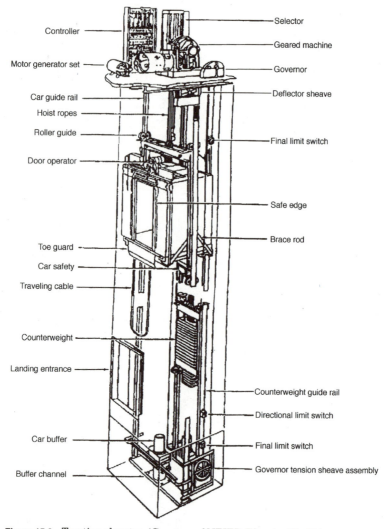

Figure 15.3 Traction elevator. (*Courtesy of NEIEP, Elevator World.*)

the cab upward to the floors (up to 8 or fewer). The jacks reside in the ground to a depth approximately the same as the travel of the cab. An electric motor and pump, using hydraulic oil, propels the plunger and cab upward. The motor and pump is idle in the down direction. Gravity causes the cab and plunger to descend at a speed controlled by a valve in the oil line. The conventional hydraulic elevator is the least complex type of passenger and freight elevator and will require 2 to 4 times less maintenance than the traction elevator. Maintenance costs should reflect this ratio.

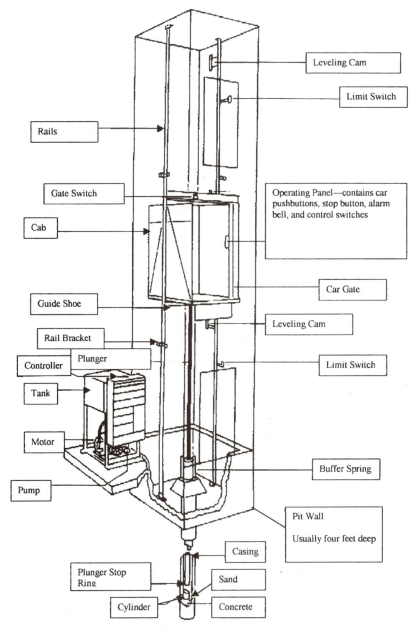

Figure 15.4   Hydraulic elevator. (*Courtesy of Otis Elevator Co., Elevator World.*)

Maintenance tasks. Tasks are the actions performed on the equipment. There are six actions—examining or testing are the first two actions to determine which, if any, subsequent actions are required. These are cleaning, lubricating, adjusting and repairing, or replacing a part or all of a component (or cleaning equipment areas such as the car top, pit, or machine room). The tasks are necessary to help maintain equipment in a safe and efficient operating condition.

Routine maintenance examinations. A set of observations and tasks is performed each time the unit is scheduled for maintenance. The typical routine examination includes equipment ride quality with an emphasis on leveling, door operation, and the signal lights and bells. A machine room check is made of all rotating equipment for oil levels and leaks, brushes and commutators, brake operation, and relay functionality. Escalators and walks will require examinations of the steps, comb teeth, handrails, key switches, and brake stopping distances. Routine examination frequency on elevators and escalators will vary owing to age, condition, operating environment, and usage. More maintenance time is required for older vintages (with the same condition and usage) unless the equipment has been modernized with "state of the art" components.

Periodic maintenance tasks. The tasks include equipment tests performed quarterly (every 3 months) or at different intervals. Most equipment tests are done periodically. The typical periodic examination is made in three general areas: Machinery room(s), car tops, and pits on elevators; the upper and lower pits; and the incline on escalators and walks. Brakes, moving selectors, controllers, and governors and tests will require periodic maintenance 1 to 4 times annually. Cleaning, car shoes, rope examinations, doors, door operators, and locks will be included in the car top area examination and will be made 1 to 4 times each year. From the pit, the lower car equipment, the buffers, compensation, and governor tension sheaves must be inspected, and this is usually done every 3 months.

Maintenance schedule. This is a list or chart that contains routine and periodic tasks, as well as the interval between the tasks. The intervals can be based on calendar days or on usage of the equipment, e.g., number of stops on an elevator or the hours of operation on an escalator. The schedule directs the technician as to *what* tasks to perform and *when* to perform them. When the tasks are completed, the schedule can be referred to as the maintenance log. The schedule and log can be either a written document that resides at the job site or an electronically transmitted database from the equipment's control processor, stored on a computer.

Maintenance guide. This is a procedure or a group (book, document, or database) of procedures describing in detail *how* a task, exam, or test is to be performed.

Time-based maintenance. A fixed amount of time between scheduled maintenance. For example, a traction elevator may receive *scheduled* maintenance monthly, a hydraulic elevator every 6 weeks, and an escalator every 8 weeks.

Use-based maintenance. These types of scheduled examinations and periodic maintenance tasks are dependent on an event, for example, the number of stops or miles traveled on an elevator, the number of operating hours on an escalator or walk. For example, 8000 stops could schedule a traction elevator routine examination; 4000 stops may schedule an examination on hydraulic elevators; and 600 h may schedule an escalator or walk (depending on vintage, operating environment, and condition of equipment). Microprocessor-based equipment has features that facilitate the acquisition of this information, and in some cases the information can be remotely accessed.

Use- and time-based maintenance. This frequency of maintenance scheduling is a combination of the use-based and time-based maintenance scheduling defined above. Past history may indicate that the elevator, for example, makes 8000 stops in a 6-week period. Then for the next 12 months, the routine examinations would be performed at 6-week intervals. If the number of shutdowns (callbacks) and customer satisfaction inquiries indicate that the interval is adequate, then it may be continued at that frequency. This method of scheduling proves to be more effective on older equipment where stop counts and running hours are not readily available. Where stop counters and time accumulators are installed on the units, the routine exams can be made at fixed intervals of time and the periodic exams can be made based on usage data. This approach is productive when more frequent visits are required.

Interval. The period of time between examinations. A fixed interval implies *time-based* approaches to maintenance. A *use-based* maintenance approach would generally imply a variable or floating interval.

"Callback." This is the term that is used for an *unscheduled* visit by the service technician to an elevator or other type of vertical transportation unit requiring some form of service. Callbacks may occur owing to an equipment misoperation, misuse or abuse, a need for transporting large loads on the elevator under supervision by the elevator technician, or recovery of a lost item inside the elevator hoistway or escalator.

Availability. This indicates the amount of time that the equipment is available to the user, usually indicated in hours or percent. An elevator normally available for passenger use may be down for 34 h in 1 month, therefore only available for 80 percent of the time. Causes for downtime include equipment maintenance, breakdowns, and planned shutdowns ("mothballing") due to passenger inactivity in the building.

Repair work. Any restoring action taking more than 4 h and/or needing two or more technicians.

Openings. The floors where passengers or freight can ingress or egress. There may be one or more openings at each floor that the elevator serves.

15.3.2 Contracted Maintenance

This is an agreement between the building and a maintenance contractor. Contracts include maintaining most equipment components in a safe and efficient operating condition. Some components are excluded, as shown in the "sample contract." Contracts are usually written to cover a period of 1 to 5 years with a renewal clause after the initial term. Cancellation can occur when a party to the agreement gives 30-day notice of a desire to terminate the agreement. While some buildings hire personnel to maintain their equipment, contracted maintenance, originating in the early 1920s remains as most popular.

Companies can be chosen from the firm that originally installed or upgraded (modernized) the equipment or from a company that only performs maintenance and not installation and modernization. At this writing, several U.S. and most European companies are ISO 9000 certified. This or an equivalent certification requires a rigorous documentation of operating procedures by the company as well as periodic compliance inspections by the certifying organization. While many companies may do good maintenance—any need for engineering and extensive modifications can usually be done more efficiently by the companies having access to the original design and application database. Examples would be modifications of speed and/or capacity, or adding an additional unit. Before deciding which company will be awarded the contract, ask the contractors to show you both their maintenance-operating plan and their plan for upgrades, as it would apply to your building. When done correctly, this plan describes *how* the company will execute their contractual requirements. Review the plan carefully with your elevator consultant.

If the equipment in your building has just been installed then traditionally you will receive at least 90 days of new installation service by the installing company. This is done to make further adjustments to the equipment as it gets broken in. At the end of the free service period the installing company and other maintenance companies will as a rule call on you offering a service contract. From the installing company it will be very important that you obtain a complete set of "as built" external wiring diagrams. The set can vary from several to one hundred or more sheets of paper. Without the diagrams, no company can provide proper service to the equipment. Keep the diagrams with your important records and provide a copy to the maintenance company if it is different from that of the installing company. It will be necessary to have these drawings updated when any major changes are made. Keep one copy as a master to reflect these changes. Examine carefully each company's contract for the following points:

- Are the major components covered in full, with parts and labor included if replacement is required? The hoist motor, brakes, motor-generator set (if it exists), all ropes, cab or escalator and walk structure, controllers, governors, selectors, guide shoes, and rails for both the car and counterweights should be included. Some contracts include only oil and grease (lubrication) and charge for all parts needed, while some supply only normal renewal parts such as contacts and motor brushes along with the lubrication.

- Most contracts will *exclude* the following items: cabs, cab finishing and flooring; all doors and door frames and sills; light fixtures and lamps; cover plates for fixtures and operating stations; power switches and breaker feeds to the equipment; smoke detectors; external emergency power equipment feeds; underground piping and cylinders on hydraulic elevators; and the renewal or refinishing of all exposed trim on escalators. This is a fair practice, since most of these items are beyond the elevator maintenance company's control.

- While most maintenance contracts offer overtime callbacks, most repairs are to be done during regular working hours. For a premium, planned repairs on equipment critical to the building are done on overtime or building holidays.

- Insurance and governmental tests should also be included. Some tests, however, can best be done by building personnel during off-hours and should not be included in the contract. These tests include a monthly check of the firefighter's service, emergency power tests, and tests of life support systems. Doing these tests during off-hours can be more productive and much less disruptive to the traffic flow in the building. Smoke detectors should be tested by specialists as prescribed in NFPA-72.

- Hoistway cleaning on elevators and truss oil pan cleaning on escalators should be included to avert a several hundred dollar annual cost.

- Overtime callbacks should be included. These are callbacks other than 8 A.M. to 4:30 P.M. Monday to Friday. A typical contractor will pay the straight time costs, with the building incurring the bonus labor for overtime work.

- Exclusions will almost always include consequential damages caused by the equipment being unavailable for use. Equipment obsolescence creating unavailability of parts and other material will usually be excluded. Have your consultant review this carefully, as it can be critical as equipment age exceeds 20 years.

- Performance bonds which may be in the form of a penalty for not meeting expected goals relating to the number of callbacks, equipment availability (see definitions above), and corridor call waiting times could be included in the contracts. Have an elevator consultant guide you as to how the equipment is to perform, and/or to write the performance bond into the contract.

After the contract is awarded to a maintenance contractor, the following should be established:

- The day of the week or month routine examinations will be performed.

- The building person that will be screening any requests for callbacks (so the company does not respond to bogus calls).

- An in-house log-in, log-out book or system for the elevator technician.

15.3.3 In-House Maintenance

This is the alternative to contracted maintenance. While the up-front costs of staffing, training, and equipping elevator personnel to maintain and repair the vertical transportation in the building can be large, it can have long-term paybacks. Some situations where in-house maintenance should be considered include remote areas where elevator service companies have no personnel; highly diversified and/or unique elevator equipment; high-security buildings where any outside personnel would require accompaniment by building personnel; or dissatisfaction with contracted maintenance performance in the area. Most in-house maintenance of elevator equipment is currently being done only in large complexes such as large real estate firms, universities, and industrial complexes.

Here are some staffing guidelines to follow when considering in-house maintenance and repair of vertical transportation equipment. If the number of units in your building or complex is less than 40 traction elevators (20 each geared and gearless) or 70 mixed units (traction and hydraulic elevators, escalators, etc.) then consider having one maintenance technician. Contract out any major repair work needed with the local maintenance contractor, such as reroping of elevators, large motor and bearing repairs, or any job taking more than 4 h and/or requiring two people rerope. The guideline for the number of units required for two in-house technicians to perform *both* maintenance and repair would be 70 traction elevators or 175 equally mixed escalator and hydraulic elevator units. More units will require more people. Use these numbers as a guide for cost justifying in-house maintenance and the staffing required. However, it is important to keep in mind that the above loading is based on some assumptions. They are that the number of openings on the elevators are 15 or less, that the distance between work assignments is 10 min traveling time or less, that the equipment is operating in a clean and controlled environment, is in good condition, and is less than 20 years old. Reduce the technician(s) workloads for equipment falling short of these assumptions. If in doubt, consult with an elevator field productivity consultant for estimating your workloads, and by all means, monitor equipment performance periodically and adjust the workloads accordingly.

If you are currently performing in-house maintenance of the vertical transportation equipment, you may consider what this author refers to as "maintenance alternatives." Included are "specialized maintenance assignments" and "amplified or dedicated maintenance." Depending on the size of your workforce, consider escalator specialists, hydraulic specialists, or certain traction elevator specialists, doing maintenance and/or repair functions on your vertical transportation equipment. A repair team may only perform major repair and/or adjusting of the equipment, a callback technician may only respond to callbacks, while the remaining technicians perform basic maintenance functions. Alternating the technicians periodically can broaden their expertise. The advantages in addition to the "economies of scale" are that the amount of specialized tools and transportation vehicles would be minimized, resulting in tool cost reduction; also the training time per technician would be minimized.

Equipment shutdowns or callbacks can occur throughout the normal workday or occur on nights and weekends. Each period, night or day, requires different administrative and management decisions and actions.

Elevator management people have perhaps their biggest challenge in orchestrating the workforce to answer overtime callbacks. Callbacks occurring on workdays between the hours of 6:30 and 8 A.M. and between the hours of 4:30 and 6 P.M. are the most difficult to orchestrate. Technicians should use mobile phones and pagers during these hours. Another method would be to utilize and pay overtime rates to standby personnel, if only to answer any callbacks involving entrapment of passengers. A technician standby list would be required. Take actions along these lines if callbacks are occurring during these hours. While costly, the callback response time can be reduced by 1 to 2 h during these peak traffic hours.

The same rationale applies for all other "after hour callbacks." The service company develops a rotational list of technicians that would be called first to answer callbacks. If the person(s) on call is unable to respond to the callback or needs assistance, then other people not on the list would be called until someone is assigned to answer the callback. Lists with primary and secondary (backup) technicians will help. While this procedure may appear haphazard to some, the method is in wide use today.

Will there be enough elevator technicians to provide your large complexes with "after hour" callback coverage? Logistically it takes a technician $2\frac{1}{2}$ to 3 h to answer a single callback. Less time would be required for successive or back-to-back callbacks. Phones, pagers, and/or beepers are a must to schedule successive callbacks. Otherwise you may not have enough technicians to cover the area. Summer vacations and holidays will compound the problem. A log listing the history of your building(s) callbacks is necessary, not only for staffing purposes but also for liability defense if entrapment and/or injury occur.

A maintenance control program (MCP) or maintenance plan describes in detail *what* routine examinations, periodic maintenance tasks, and tests are performed by the maintenance technician, *when* (at what frequency) they are performed, and *how* they are to be performed. The MCP should describe the method used to document maintenance activity on each unit. This documentation is referred to as a maintenance log. The maintenance technician fills out the log while performing the maintenance activities. An example of a simple time-based maintenance log for an eight-car bank (group) of traction elevators is shown in Fig. 15.5. Use-based or combination use-based and time-based logs are more complex but can be more effective in logging maintenance activity. The MCP and the maintenance log

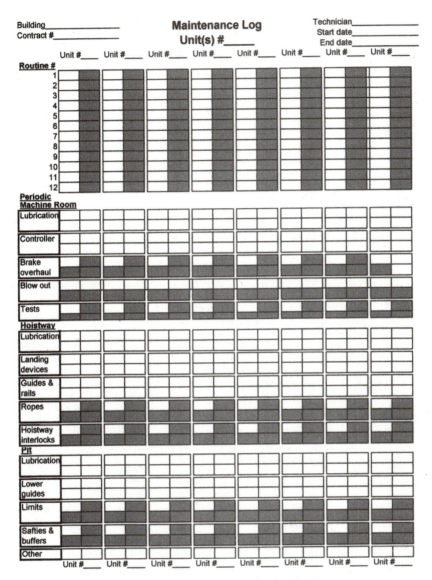

Figure 15.5 Maintenance activity log. (*Courtesy of SeeC, L.L.C.*)

should be customized for each bank of units. A maintenance guide describing how the tasks and tests are to be performed should also be included in the MCP. Building management, the elevator consultant, the maintenance contractor, or a combination of the three can develop the MCP.

15.3.4   Removing Equipment from Service

Your watch engineer or other authorized building personnel will need to remove equipment from normal operation when malfunctions occur. The best place to turn equipment off is inside the elevator or at the top or bottom of the escalator, to avoid entrapments of passengers in the elevator or someone falling on the escalator. Turn the keyed lock in the elevator car station to the off position. Turn off both the "in service" and the "power switch." On some elevators the switches are behind the locked door in the car station. Turning off the cab lighting will deter people from entering the cab. Pull the doors closed if possible. For the escalator, simply make a visual check that no one is riding the escalator or about to board; then press the stop switch located at the top or bottom newel ends. A sign on the key switches will alert other personnel that a malfunction has occurred and to check with the building office prior to returning the escalator to service. The signs can also be placed inside and outside the elevator.

15.4   Monitoring Equipment Performance

15.4.1   What to Monitor

The following items need to be monitored as frequently as monthly, particularly on equipment built in the 1980s or earlier that has not had more recent control modifications. Later equipment has more reliable solid-state and computerized timing circuits; however, local code requirements still may require monthly testing of firefighter's service and the documentation of hydraulic oil additions. Your elevator consultant, elevator contractor, or local code-enforcing representative can provide you with the methods of testing required in your area.

- *Firefighter's service* (monitored monthly by building personnel)— Test the system by recalling all elevators with the lobby switch— all elevators in normal service at that time should return to the lobby. Then test the firefighting operation of at least one of the elevators that returns to the lobby using the switch inside the elevator.

- *Emergency power* (monitored periodically by building personnel)— Test the operation of the equipment that has been designated to operate on the emergency power supply system (diesel generator or secondary supply system) when normal power has failed.

- *Elevator performance* (monitored periodically by elevator technician, witnessed occasionally by building management)—Randomly select one or more elevators, or perhaps an elevator that you sense is running more slowly than the others, and have the elevator technician verify that it is performing within 10 percent of contractual requirements. Included will be the following: *contract speed* measured with a tachometer on the governor rope in the machine room (hydraulic elevators seldom have governors and the speed is measured from the top of the car; therefore, only the elevator technician, for safety reasons, should be checking these speeds); *door opening and closing times* measured with a digital stopwatch (times will vary depending on door system types; however, opening speed will exceed the closing speed which is limited by codes to force and energy maximums, for example, 1.5 s opening time and 2.5 s closing time); *door fully opened (dwell) time* measured with a stopwatch from inside the elevator as it travels in response to corridor calls (time will vary between 3 and 7 s, the setting dependent on the distance from the elevator to the corridor call button, as well as the type of call to which the elevator is responding); *elevator "car start" time* measured from inside the elevator with a stopwatch and defined as the elapsed time between when the door has fully closed in response to a call and when the elevator motion begins (time should be about ½ s on traction elevators but could be 2 or 3 times longer on hydraulic elevators); *elevator movement time* ("brake to brake time") to move one floor measured from inside the elevator using a stopwatch and timing "when the elevator starts moving to when it stops moving" (time will be about 5 s for a traction elevator with floor heights of 12 ft, but at least twice as long for hydraulic elevators). All of the above can be combined into overall performance time, which is the time from the starting of the door closing at a floor and ending at the time the door reaches the fully opened position at the next floor down or up.

- *Elevator group performance* (measured periodically by the elevator technician) indicates how quickly a group or bank of elevators respond to corridor calls and lobby service demands during peak periods as well as off-peak periods. This measurement is done most efficiently using an elevator traffic analyzer. Many types are available on the market today, and some later microprocessor elevator supervisory systems have built-in analyzers. A typical printout is shown in Fig. 15.6. Comparisons from one period should be similar in corridor waiting times when compared to other periods. A rise in long waiting times indicates that one or more of the systems features need to be serviced.

Note: Refer to NEII (National Elevator Industry, Inc.) performance standard supplements for more information. These documents can be obtained through *Elevator World Magazine.*

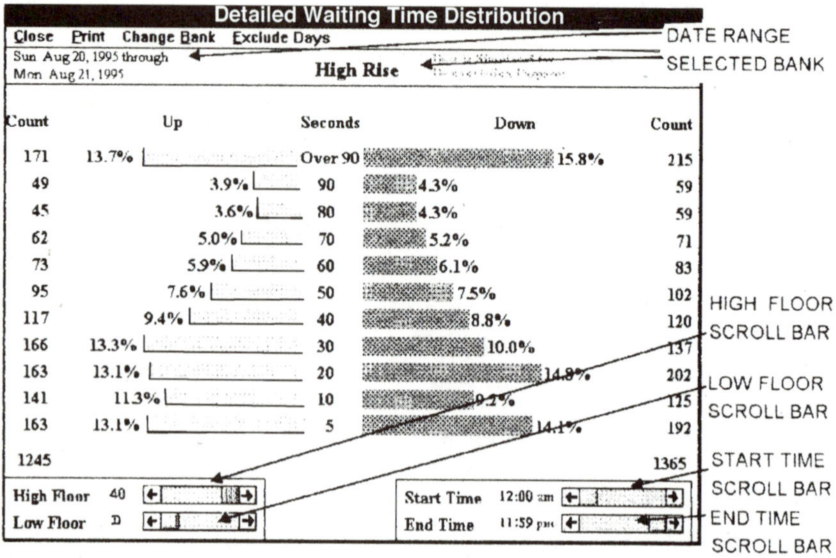

This depiction shows us the overall percentage of answered hall calls. In the illustration above;

- 13.1% of the up hall calls were answered in 5 seconds or less.
- 14.1% of the down hall calls were answered in 5 seconds or less.

Figure 15.6   Distribution detailed. (*Courtesy of Integrated Display Systems, Inc.*)

15.4.2   Callback and Downtime Performance

This measurement reflects maintenance efficiency as well as the overall condition of the equipment. Building personnel must work together developing plans to reduce callbacks. The starting point is to know and communicate the volume and causes of callbacks. Having a callback code to easily identify the most common callback causes will be a necessary step in callback reduction. Codes permit easy entry by the technician and facilitate sorting callback cumulative data. The code should not be so detailed as to discourage the use by technician, dispatcher, or building watch person. It should identify an assembly, not a component. It must be written generically. It must red flag unsafe conditions, e.g., entrapments, mislevels, broken steps, stopped handrail. Callbacks of the same type or causes can then be reduced. Intensified maintenance on affected components is one of numerous methods for reducing or eliminating a recurring callback. An effective coding process will help identify equipment components requiring more intensive maintenance or system upgrades. But, like anything, it's only effective if it's accepted and used.

What is the *acceptable* number of callbacks? What is the *achievable* number of callbacks? Answers to the above questions, while highly circumstantial, can be made with some assumptions.

The *acceptable* number of callbacks for a given period should be better than (fewer callbacks) the previous period, or at least the same in cases where the trend has been highly favorable.

An *achievable* number of callbacks for reliably designed equipment could be as follows:

Escalators: two per year (180 days)

Hydraulic: three per year (120 days)

Traction: four per year (90 days)

Another helpful practice is to log the time that callbacks occur, through the workday or at nights or weekends.

Downtime is the number of hours that the equipment is *not* available for passenger or freight usage. A simple chart tracking callbacks and downtime is shown in Fig. 15.7. Observing the trend monthly will track degradation in callback and downtime performances.

15.5 Monitoring Maintenance Performance

Your building's maintenance control plan (MCP), described in Sec. 15.3.3 in conjunction with the maintenance log, can be the source data for developing a chart that provides the elevator technician's maintenance performance. Certain tasks are to be performed at given frequencies of time and/or usage. Are they? A chart similar to the one shown in Fig. 15.5 will provide the answer. Figure 15.8 logs maintenance activity, and the Pf column computes the ratio of when the tasks were performed and when they were scheduled. A Pf value of 80 or 90 may be acceptable. On the other hand, a Pf value of 50 implies that maintenance tasks in that category are being performed only one half as often as the MCP promises. This indicates that either steps should be taken to accelerate maintenance activity or the MCP should be revised. Reviewing the chart at least every 3 months is suggested. While this is another piece of paper to generate, the document can be invaluable in defending a "negligence of maintenance" lawsuit.

Computerized maintenance monitoring is the most efficient method for documenting maintenance. The maintenance technician displays the task or test on a laptop computer or handheld terminal, then enters the task as completed. The entry documents what maintenance tasks were completed, when, and by whom. The program compares the date due versus the date completed, permitting a performance log. A microprocessor stores the information, perhaps for years

Area _____ Supt. _____
Route _____ Bldg. _____
Bank _____

Unit #1			Unit #2			Unit #3			Unit #4			Unit #5			Unit #6		
Date	Time	Act	Date	Time	Act	Date	Time	Act	Date	Time	Act	Date	Time	Act	Date	Time	Act
10/29/97	10A	RCF	10/29/97	10A	FUS	10/29/97	10A	TCAB	10/29/97	10A	HEAT	10/29/97	10A	TCAB	10/29/97	10A	HEAT
10/29/97	4P	DLM	11/02/97	11A	BRU	10/30/97	9A	LH	11/05/97	2P	COMB	11/05/97	2P	LH	11/05/97	2P	COMB
11/16/97	3P	DLE	11/16/97	10A	OIL	11/16/97	11A	LUB	11/16/97	12N	STEP	11/16/97	1P	LUB	11/16/97	2P	STEP
11/24/97	10A	DOA	11/24/97	10A	WRE	11/24/97	10A	HDWE	11/24/97	10A	H/R	11/24/97	10A	HDWE	11/24/97	10A	H/R
12/28/97	4P	SBP	12/28/97	4P	CADJ	12/28/97	4P	LEAK	12/28/97	4P	ROA	12/28/97	4P	LEAK	12/28/97	4P	ROA
01/30/98	3P	O/S	01/30/98	3P	PCB	01/30/98	3P	NOIS	01/30/98	3P	OTH	01/30/98	3P	NOIS	01/30/98	3P	OTH
01/30/98	4P	MPF	01/30/98	4P	SAFO	01/30/98	4P	LOW	01/30/98	4P	OTH	01/30/98	4P	LOW	01/30/98	4P	OTH
												02/15/98	2P				

Figure 15.7 Callback log. *Note:* ACT is the action performed to correct the problem. (*Courtesy of SEEC, L.L.C.*)

15.22

when battery backups are provided, and when equipped with a modem can transmit the information periodically to a central station.

Remote monitoring systems are readily available in today's market and are the state of the art peripheral for today's vertical transportation. These systems can be adapted to any type of equipment. In addition to providing maintenance documentation and performance information as described above, remote monitors can transmit live data on the current state of the elevator or escalator system, similar to the information on an automobile's dashboard but with much more detail. Repeated misoperations, equipment status, alarms, entrapments as well as system performance information can be transmitted, as they occur, to the central watch station computer. Some monitoring systems have two-way communication to the elevator, helpful for consoling entrapped passengers and deterring equipment misuse. Shown in Fig. 15.9 is a typical remote monitoring system, employing monitors in the security officer's room, at the desk of the lobby attendant, and in the machine rooms for use by the technician. Figures 15.10 through 15.16 provide valuable information to all parties and can be programmed to alert them when certain performance parameters are exceeded. Figures 15.17 through 15.25 illustrate several methods for displaying corridor call waiting times. This information documents how efficiently the elevator dispatching or supervisory system is performing. A comparison can be made with the time of day the elevators may be performing less efficiently and the number of cars in service at that time.

15.6 Improving Safety and Limiting Liability

This section is devoted to reducing possible causes of accidents to the building personnel as well as the riding public. No transportation equipment and its associated machinery will be free of potential dangers, no matter how well maintained, no matter how extensive the training programs and cautionary signage. No matter how compliant with safety codes, accidents will occur. How well you have done your job with respect to the issues addressed in the first sections of this chapter will help improve safety in the building. But accidents will occur, the following being among the most common or serious:

- Trips and falls while entering and exiting the elevator or while riding an escalator.

- Attempting to escape from an elevator that is stopped between floors.

- Being struck by elevator doors.

Area _____ Supt _____
Route _____ Bldg. _____
Bank _____

Maintenance Activity Log

Report Date _____

Unit #1

Date	Time	Act	Pf
10/29/97	10A	RE	82
10/29/97	4P	PP	82
11/16/97	3P	PM	79
11/24/97	10A	RE	82
12/28/97	4P	RE	69
01/30/98	3P	RE	67
01/30/98	4P	PP	62
AVERAGE TOTAL			72

Unit #2

Date	Time	Act	Pf
10/29/97	10A	RE	82
11/02/97	11A	PM	68
11/16/97	10A	PP	79
11/24/97	10A	RE	82
12/28/97	4P	RE	69
01/30/98	3P	RE	67
01/30/98	4P	PP	62
AVERAGE TOTAL			73

Unit #3

Date	Time	Act	Pf
10/29/97	10A	RE	82
10/30/97	9A	PM	73
11/16/97	11A	PP	79
11/24/97	10A	RE	82
12/28/97	4P	RE	69
01/30/98	3P	RE	67
01/30/98	4P	PP	62
AVERAGE TOTAL			73

Unit #4

Date	Time	Act	Pf
10/29/97	10A	RE	82
11/05/97	2P	PM	80
11/16/97	12N	PP	79
11/24/97	10A	RE	82
12/28/97	4P	RE	69
01/30/98	3P	RE	67
01/30/98	4P	PP	62
AVERAGE TOTAL			74

Unit #5

Date	Time	Act	Pf
10/29/97	10A	RE	82
11/05/97	2P	PM	67
11/16/97	1P	PP	79
11/24/97	10A	RE	82
12/28/97	4P	RE	69
01/30/98	3P	RE	67
01/30/98	4P	PP	62
02/15/98	2P	PM	67
AVERAGE TOTAL			72

Unit #6

Date	Time	Act	Pf
10/29/97	10A	RE	82
11/05/97	2P	PM	75
11/16/97	2P	PP	79
11/24/97	10A	RE	82
12/28/97	4P	RE	69
01/30/98	3P	RE	67
01/30/98	4P	PP	62
AVERAGE TOTAL			74

Unit #7

Date	Time	Act	Pf
10/29/97	10A	RE	82
11/05/97	2P	PM	60
11/16/97	3P	PP	79
11/24/97	10A	RE	82
12/28/97	4P	RE	69
01/30/98	3P	RE	67
01/30/98	4P	PP	62
AVERAGE TOTAL			72

Unit #8

Date	Time	Act	Pf
10/29/97	10A	RE	82
11/05/97	2P	PM	75
11/16/97	4P	PP	79
11/24/97	10A	RE	82
12/28/97	4P	RE	69
01/30/98	3P	RE	67
01/30/98	4P	PP	62
AVERAGE TOTAL			74

Figure 15.8 Maintenance activity log. *(Courtesy of SEEC, L.L.C.)*

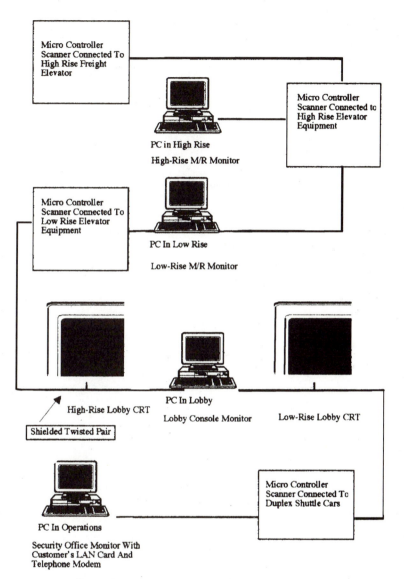

Figure 15.9 Remote monitoring system. (*Courtesy of Integrated Display Systems, Inc.*)

- Being jarred by an elevator making a sudden stop.
- Entrapment on the sides, backs, or front of the escalator steps.
- Falling into escalator steps.
- Falling down an open hoistway.

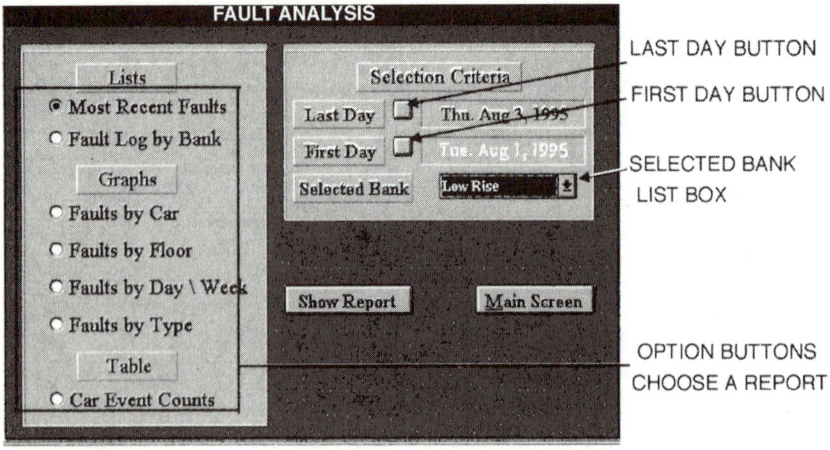

Figure 15.10 Fault report menu. (*Courtesy of Integrated Display Systems, Inc.*)

Lists
Most Recent Faults a list of the most recent faults.
Fault Log by Bank a list of faults within a date range.

Graphs
Faults by Car a graph showing the number and type of faults per car.
Faults by Floor a graph showing the number of faults per floor.
Faults by Day\Week a graph with the number of faults by days or weeks.
Faults by Type a graph with the number of faults per type.

Table
Car Event Counts a table of minor elevator system failures.

Selection Criteria (see pg. 4.9)
Last Day Button choose the last day for the report.
First Day Button choose the starting day for the report.
Selected Bank List Box select a bank to report.

Show Report to bring the selected report on screen.
Main Screen to return to the main screen.

The following sections suggest possible causes for these types of accidents.

Elevator accidents involving tripping and falling while entering or exiting the elevator cab can occur as the elevator approaches the floor or when the elevator stops above or below the corridor floor. Many traction elevators open their doors prior to reaching floor level. This is

10 Most Recent Faults						
Help Close Print Change Bank						
Mon. Aug 7, 1995			Low Mid			Data is Simulated for Demonstration Purposes
Car	Date	Time	Cleared	Floor	Fault	Direction
Car 7	Mon. Aug 7, 1995	19:20:48	19:20:49	4	Lost Power	UP
Car 7	Mon. Aug 7, 1995	17:12:37	17:12:38	4	Lost Power	DOWN
Car 5	Mon. Aug 7, 1995	15:48:27	15:48:27	0	Safety	
Car 4	Mon. Aug 7, 1995	13:39:05	13:39:05	0	Safety	
Car 5	Mon. Aug 7, 1995	11:04:00	11:04:00	18	Lost Power	
Car 8	Mon. Aug 7, 1995	09:35:05	09:35:05	0	Safety	DOWN
Car 4	Mon. Aug 7, 1995	07:45:41	07:45:42	20	Lost Power	
Car 4	Mon. Aug 7, 1995	05:51:08	05:51:09	0	Safety	DOWN
Car 5	Sun. Aug 6, 1995	07:24:49	07:24:49	0	Safety	
Car 7	Sun. Aug 6, 1995	05:09:34	05:09:35	17	Lost Power	DOWN

Switch values for record 1 of 10
Power: Down

SELECTED BANK

HIGHLIGHTED BAR

INPUT STATUS AREA

At the top of the scrolling fault window are the following headings:
- **Car** which identifies the car at fault
- **Date** and **Time** the fault occurred
- **Cleared** the time the fault was corrected
- **Floor** what floor the car was on when the fault occurred
- **Fault** the type of fault
- **Direction** the direction the car was traveling.

Figure 15.11 Most recent faults. (*Courtesy of Integrated Display Systems, Inc.*)

referred to as "preopening the doors on approach" and is done to improve performance times. Where preopening is provided, the doors should be less than 24 in open when the elevator is within ½ in from floor level, i.e., less than the width of a person (24 in wide). Observe the approaches as you use the elevator and report any misoperations to your elevator technician. Tripping and falling can also occur when the elevator stops ½ in beyond floor levels. Point out any of these misoperations to your elevator technician.

Elevators stopping above or below floor level more than 2 ft will not normally open its doors. If they are forced open by an entrapped passenger attempting to exit the elevator, accidents can occur. Falls down the hoistway sometimes occur when passengers jump from the elevator that has stopped 3 ft or more above the floor, lose their balance, and fall into the hoistway. Provide warnings to your tenants. Explain that an entrapped passenger must wait for an authorized person to release them from the elevator, and how to contact that person using the alarm and communication system. The passengers should attempt to move the elevator only with the floor buttons and should stay clear of the doors.

Note: See ANSI A17.4 for more information.

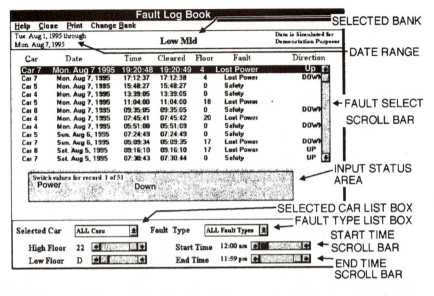

At the top of the scrolling fault window are the following headings:
- **Car** which identifies the car at fault
- **Date** and **Time** the fault occurred
- **Cleared** the time the fault was corrected
- **Floor** what floor the car was on when the fault occurred
- **Fault** the type of fault
- **Direction** the direction the car was traveling.

Figure 15.12   Fault log by bank. (*Courtesy of Integrated Display Systems, Inc.*)

Most elevator passengers are aware that the doors close automatically. Most elevator passengers assume that the door reopening devices, e.g., safety edges and light rays, are working properly. However, these devices are not usually effective on the corridor doors, unbeknownst to the average passenger. Instruct them of this fact. Also do not circumvent any door reopening device. If a device is not working properly, alert your technician. Also have your technician demonstrate how to test the doors to assure that they comply with both the force and energy codes. The force code is written to reduce the squeezing force of the doors to a tolerable level (30 lbf) should the closing doors trap someone. This is measured with a force gauge. The energy code (not to exceed 7 ft-lb) is written to limit the striking force of the door system. Testing is done with a stopwatch. Elapsed door closing time should not be less than a value that would indicate the door speed has reached a maximum (usually 1 ft/s for standard door system weights). For example, a 3-ft-wide single panel door should

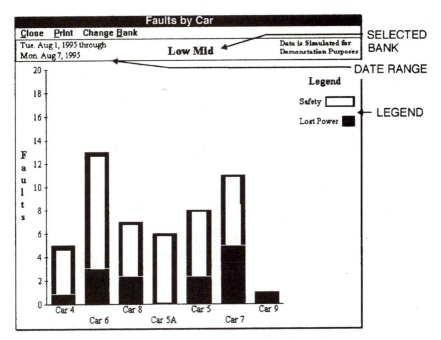

Figure 15.13   Fault graph by car. (*Courtesy of Integrated Display Systems, Inc.*)

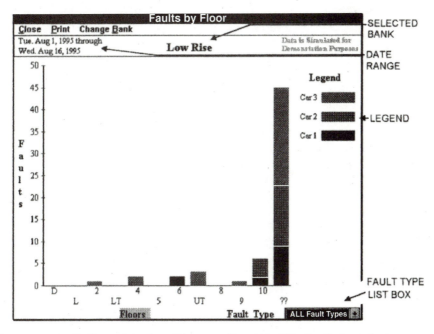

Figure 15.14   Fault graph by floor. (*Courtesy of Integrated Display Systems, Inc.*)

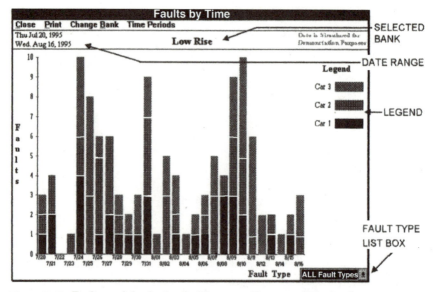

Figure 15.15   Fault graph by day/week. (*Courtesy of Integrated Display Systems, Inc.*)

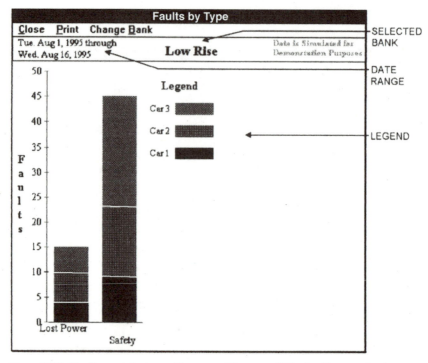

Figure 15.16   Fault graph by type. (*Courtesy of Integrated Display Systems, Inc.*)

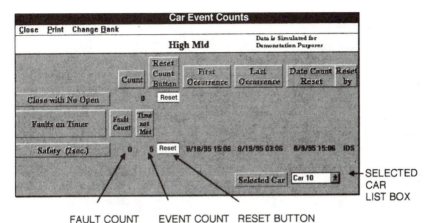

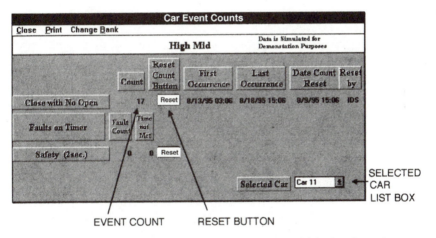

In the example above, Car 10 has had 5 safety circuit faults which were cleared within 2 seconds between 8/18/95 and 8/19/95.

In the example above Car 11 has had 17 events in which the door close limit was active without a corresponding door open limit. This could be due to an intermittent contact or other door malfunction.

Figure 15.17 Car event counts. (*Courtesy of Integrated Display Systems, Inc.*)

take at least 3 s to close, perhaps longer for heavier doors such as ornamental bronze doors.

Falls down elevator hoistways have occurred. They can occur because of circumventing of the electrical and mechanical interlock on the hoistway door, unauthorized personnel using a door interlock releasing tool, broken relating mechanisms on multipaneled hoistway doors, failure of a mechanical door closer, top or bottom door retaining

TRAFFIC ANALYSIS

Hall Button Counts (Graphs)
- By Floor
- By Time of Day

Wait Time (Graphs)
- Average by Floor
- Average by Time of Day
- Distribution by Hourly Interval
- Distribution Detailed
- Longest Wait Times (List)

Selection Criteria

Last Day	Tue. Apr 9, 1996
First Day	Mon. Apr 8, 1996
Primary Bank	Bank #1

Change Selection Criteria

History Available
- Use Selection Criteria
- Use Today's Data

Show Report   Main Screen

SELECTED BANK
LIST BOX
SELECTION CRITERIA
BUTTON

OPTION BUTTONS
USE SELECTED CRITERIA
OR GET AN UP TO THE
MINUTE REPORT ON
TODAY'S TRAFFIC

OPTION BUTTONS –CHOOSE A GRAPH OR LIST

Hall Button Counts (Graphs)
By Floor a breakdown of hall button density per floor.
By Time of Day for a 24 hour timeline of hall button density at 10, 15, 20, 30, or 60 minute intervals.

Wait Times (Graphs)
Average By Floor to see the average waiting time per floor.
Average By Time of Day for a 24 hour timeline of average waiting times at 10, 15, 20, 30, or 60 minute intervals.
Distribution By Hourly Interval to get 24 hour timeline showing the percentage of calls answered in 5, 10, 20 ... 90 seconds or longer.
Distribution Detailed for a more detailed view of the percentages.

Longest Wait Times (List) to produce a list of the floors with the 10 longest waiting times.

Figure 15.18   Traffic report menu. (*Courtesy of Integrated Display Systems, Inc.*)

devices not secure, or simply when excessive force is applied on the corridor side of the hoistway door. To reduce the risk of this happening in your building, frequent and thorough hoistway door maintenance is necessary. Door interlocks, hangers, sills, bottom shoes (gibs), relating cables, and closers must be in good condition. Door interlocks should not be circumvented unless absolutely necessary, and then only by an elevator technician when the elevator is *out* of normal service. In the event that a door is opened and the elevator is at another floor, place a barricade in front of the opening and take the elevator out of service until repaired.

Hydraulic elevators having leaks or unaccounted oil losses are

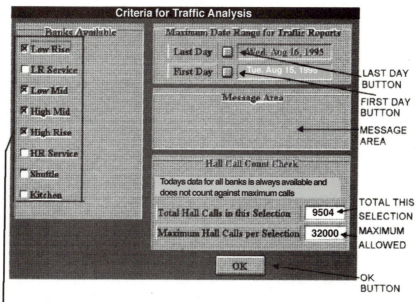

Figure 15.19 Selection criteria menu. (*Courtesy of Integrated Display Systems, Inc.*)

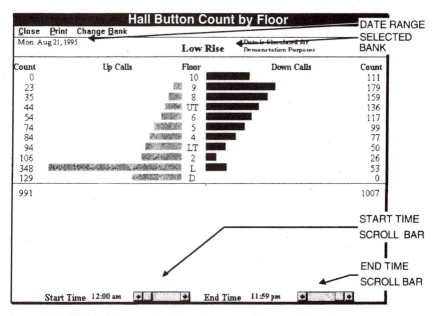

Figure 15.20 Hall button count by floor. (*Courtesy of Integrated Display Systems, Inc.*)

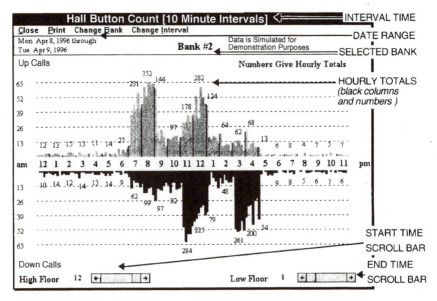

Figure 15.21 Hall button count by time of day. (*Courtesy of Integrated Display Systems, Inc.*)

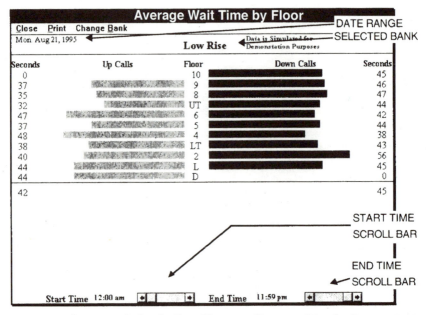

Figure 15.22 Average wait time by floor. (*Courtesy of Integrated Display Systems, Inc.*)

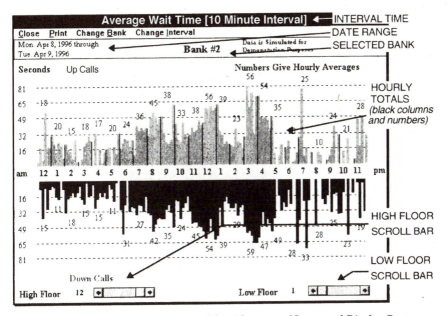

Figure 15.23 Average wait time by time of day. (*Courtesy of Integrated Display Systems, Inc.*)

potentially dangerous. A pipe or cylinder rupture can result in an elevator descending at a rapid speed and contacting the bottom springs with great force. Injury or death is likely. Complete records of any adding of hydraulic oil must be maintained, and unexplained losses investigated.

Falls on escalators can be caused by slower-moving or stopped handrails, sudden stops, transporting of large packages, wheelchairs, carts, and freight, and simply the loss of balance. Instructing passengers to hold the handrails, to avoid the sides of the step, and to step up over the combfingers when exiting the escalator assists in averting accidents. Figure 15.2 of this chapter shows the required signage to emphasize that only passengers are permitted on escalators. Signs should also be placed near an escalator directing wheelchaired passengers and less agile people to the location of the elevator. Instruct the building tenants accordingly. Shut down any escalator having either or both handrails stopped or moving more slowly than the steps. Flaps or guards should be secured over the stop buttons at each end of the escalator to prevent accidental engaging of the stop button. Escalators are permitted by code to travel at speeds up to 125 ft/min. If you feel this speed is too fast for your building's tenants, consider a slower speed. Some escalators by design have a slow and fast speed

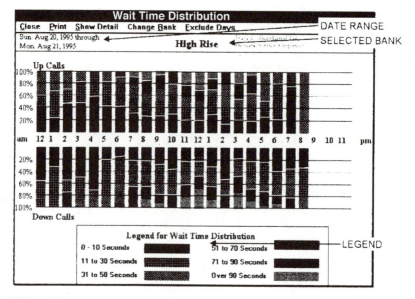

This graph shows a 24 hour time line with the percentage of wait time for every hour.

For example, between the hours of 8pm and 9pm;

- 100% of the up calls were answered in 31 to 50 seconds.

The down calls for that same time period read as follows;

- 36% were answered in 0 to 10 seconds.
- 44% were answered in 11 to 30 seconds.
- 20% were answered in 31 to 50 seconds.

See Distribution Detailed *(pg. 5.7)* for a breakdown of this graph's percentages.

If waiting times are unusually long, check the *FAULT LOG* and the *SERVICE LOG* to see if 1 or more cars were out of service for maintenance or malfunction at the time the waiting was the longest.

Figure 15.24   Wait time distribution. *(Courtesy of Integrated Display Systems, Inc.)*

selection on the starting key switch. Contact the escalator manufacturer to obtain escalator speed-reducing information. They as well have a vested interest in reducing escalator accidents.

Entrapments on escalators occurring around the edges of the step can be reduced if passengers stand in the middle of the step. Yellow markings forming a box around the step help to direct passengers to stand in the middle. Consider having these markings applied to your steps. The entrapments are generally caused by excessive gaps between adjacent steps, between steps and the side skirts, and the

10 Longest Wait Times — DATE RANGE / SELECTED BANK

Close Print Change Bank Exclude Days
Sun. Aug 20, 1995 through
Mon. Aug 21, 1995   **High Rise**   Data is Simulated for Demonstration Purposes

Day	Time	Wait Time	Floor	Up /Down
Mon. Aug 21, 1995	12:53 pm	223	40	DOWN
Mon. Aug 21, 1995	01:03 pm	211	36	DOWN
Mon. Aug 21, 1995	12:48 pm	210	35	DOWN
Mon. Aug 21, 1995	12:16 pm	203	L	UP
Mon. Aug 21, 1995	12:21 pm	202	38	DOWN
Mon. Aug 21, 1995	12:59 pm	202	xx	DOWN
Mon. Aug 21, 1995	12:27 pm	200	L	UP
Mon. Aug 21, 1995	03:26 pm	199	38	DOWN
Mon. Aug 21, 1995	12:16 pm	196	L	UP
Mon. Aug 21, 1995	12:24 pm	196	D	UP

High Floor 40   Low Floor D   Start Time 12:00 am   End Time 11:59 pm

WAIT TIME (SECONDS) / HIGH FLOOR SCROLL BAR / LOW FLOOR SCROLL BAR / START TIME SCROLL BAR / END TIME SCROLL BAR

This report shows the ten longest waiting times of the hall buttons pressed. The date, time, floor, and direction are recorded. This can be an accurate and unbiased answer to the complaint, "I WAITED FOREVER FOR AN ELEVATOR ON THE 29TH FLOOR THIS MORNING".

If waiting times are unusually long, check the *FAULT LOG* and the *SERVICE LOG* to see if 1 or more cars were out of service for maintenance or malfunction at the time the waiting was the longest.

Figure 15.25   Long wait times. (*Courtesy of Integrated Display Systems, Inc.*)

failure of the comb fingers to mesh into the step treads. Excessive friction on the side skirts increases the chance for entrapments. All of the above should be minimized. Side gaps should be limited to $^3/_{16}$ in or less on both sides of the step throughout the escalator travel but especially near the ends, where the steps go from offset to level. If scraping noises are heard while trying to comply with these clearances, applying a plastic-like material to the sides of the steps can also reduce the gap. Contact your elevator parts supplier or your consultant for more information on this product. The side skirts should have less rubbing friction than the surrounding metal surfaces to help reduce entrapments. If not, apply a friction-reducing substance either permanently or periodically to the side skirts, being careful not to get any of the substance on the top of the steps or on the landings. Serious injuries or deaths can occur when step frames and/or rollers are cracked or broken. Annual internal inspections of the steps are a must to help ensure the structural integrity of the steps. Figure 15.26 illustrates typical escalator components.

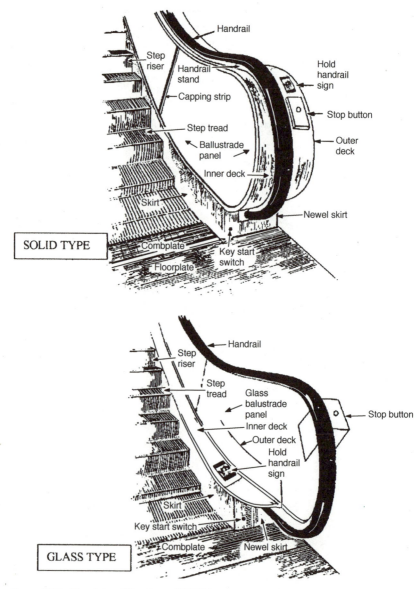

Figure 15.26 Escalator. (*Courtesy of NEII.*)

Exposing unauthorized personnel to vertical transportation machinery rooms can be hazardous. Keep the rooms locked and provide entry to only authorized personnel. Do not use these rooms for storage of equipment other than maintenance equipment and parts needed for vertical transportation. This includes the elevator hoistways and pits. These areas contain dangerously high voltages and many moving parts. Cleaning personnel, painters, and other people who have no operational equipment in these areas should not be allowed to use the areas. In addition to personal injury and death hazards, equipment damage can occur. For example, the storing of large objects in the pit underneath the counterweight area can cause frame damage resulting in the dislodging of the counterweight from their containment rails. If as a result a collision were to occur between the car and counterweight, it could destroy the elevator cab and injure or kill its occupants. This is not a doomsayer's fantasy; these collisions have occurred.

Elevator technician safety can be enhanced with proper lighting in all working areas. Included are the machine rooms, the car tops, and the pits. Yellow striping on low overheads, ledges, and at pinch points will assist the technician in being aware of these dangers. High-voltage signage should also be placed where needed. Hoistway access switches will assist the technician to gain safe access to the hoistway. This feature should be added if it does not already exist. Ask the technician to suggest other safety measures or point out any potentially hazardous conditions.

Failures to warn of any potentially dangerous conditions have resulted in numerous settlements in favor of the plaintiff. If a code-enforcing authority, elevator technician, or consultant informs you of potential dangers, you are obligated to correct the condition or forewarn the users of the equipment. In like manner if you notice a potentially dangerous condition that should have been detected and corrected by a contracted elevator maintenance company, notify them immediately by registered letter or courier of the condition. If in doubt shut down the equipment until the correction is made.

15.7   Modernization of Equipment

The first significant changes to vertical transportation equipment in the last 50 years occurred in the 1980s. Microprocessor technology was introduced permitting programmable features on both the driving and supervisory systems. Prior to this era, features and options required to optimize equipment performance when a building traffic pattern changed were costly to add and required extensive downtime.

System performance information and error messages so valuable to the technician for troubleshooting equipment malfunctions were virtually nonexistent. Reliability was poor. Maintenance was more labor-intensive. Safety features on both elevators and escalator were not as abundant as they are today. Noise levels were higher. Ride quality was poor in comparison. Signal lights and gongs operated poorly when compared to modern fixtures. That's the *bad* news if your equipment was built in the 1970s or earlier. The *good* news is that the bulk of the elevator equipment can be reused, and the remaining equipment can be replaced (modernized, updated) in stages. Unfortunately the good news does not apply to escalators. The only good news about escalators is that they can be removed and replaced in less time, and installation labor is not as intensive, provided there is a passageway into the building for a preassembled escalator.

15.7.1 Why, When, and What to Modernize

Most of the reasons were listed above. "Keeping up with the Jones's" is another reason. Making your equipment more pleasing to the eye than the equipment in the office building across the street could attract more tenants. But if your equipment is over 20 years of age and has been subjected to heavy usage and/or abused, you may have more than one reason to modernize. Equipment manufactured in the 25 years from 1950 to 1975 contained hundreds of relays and similar devices. The life expectancy is 10 million operations; however, reliability studies will usually indicate a replacement at 2 to 3 million operations. Unfortunately most of these devices are "hard wired" and not readily replaceable, when compared to plug-in devices. Many failures can occur, making the new microprocessor control very attractive. Elevator controllers and energy-efficient drive units may be included in the first stage of modernization. Following close behind are cabs and car stations inside the cab, mostly for aesthetic reasons since the old cab architecture and pushbuttons may look aged and worn. Usually one elevator is modernized at a time, thereby minimizing disruptions in traffic flow. Doors, their operator, and the door safety equipment are next. Door panels can be optional. The cab superstructure, guiderails, counterweight, overhead driving machine, and sheaves usually need not be replaced and can last several decades. Most likely the motor-generator (old power source for the drive motor) will not be needed. A solid-state drive will replace it. So, as we stated above, the bulk of the elevator equipment will remain and the updated additions will make the riding public feel that they are using the latest, most advanced elevator system money can buy.

15.7.2   How to Select a Modernization Contractor

So, you've decided to modernize some of your equipment in stages, depending on the available budget each year, for the next several years. What do you buy first and what do you need most? We hate to keep harping on this, but contact your elevator consultant and have the firm survey your equipment. If you don't have a consultant, look in the yellow pages or the consultant directory in the *Elevator World Magazine* Source Directory (under the consultant tab). Have the consultant write the specifications for the modernization, assist you with contractor selection, and oversee the installation. All that! Now you know why we mentioned the consultant.

Consultant responsibilities for equipment modernization include the following:

- Evaluate the equipment in the building. The evaluation should include the equipment condition, current and projected reliability, ride quality, noise performance, and passenger-handling capability. All of the above should be compared with industry standards to highlight any deficiencies. Tests should include motor insulation breakdown, underground cylinder and piping pressure tests for leakage, and a complete structural examination of the escalator's internal parts including all step frames. The final report should indicate test results.

- Determine how many elevators can be modernized at a time. Based on the passenger-handling capabilities above, establish the minimum traffic-handling performance that would be tolerable.

- Establish the modernization stages, how much time each stage will take, which stage should be done first and why, and the projected costs for each stage.

- Write the specifications that will be submitted to a list of prequalified elevator contractors for competitive bidding. The specifications must include equipment changes necessary to meet current local code requirements for the alterations. They must also include performance assurances and the test methods that will be used to assure compliance. Usually there will be patch-up and other work associated with the modernization. The specifications must define what work is the responsibility of the elevator contractor and what other contractors will do. Interim maintenance required on the installed equipment must be included, along with a maintenance contract for the new and existing equipment.

- Rationalize any exceptions to the specifications and determine the impact of the changes.

- Oversee the installation of the new equipment by making periodic job inspections.

- Assure that specifications have been met for each stage of modernization. If not, make a list of deficiencies (punch list) and a time frame for correcting the deficiencies; then check and verify conformance.

- Assist building management in determining when final payments should be made for each stage of modernization.

The building manager's responsibilities prior to and during the modernization will include:

- Notify all tenants of the purpose and starting date.

- Explain that longer wait times for elevator response will occur and ask for their indulgence.

- Ask them to follow good elevatoring practices, e.g., walk up one floor and down two, do not detain the elevator door from closing, and stagger their working hours if possible.

- If the modernization elevator contractor is not the current maintenance contractor, consider selecting only the modernization contractor to perform maintenance on all equipment during all stages of modernization. Otherwise personnel issues may occur.

Bibliography

"Well It's About Time," *Elevator World Magazine,* 1997.
"What Callbacks Cost," *Elevator World Magazine,* 1997.

Water Treatment Services

Arthur J. Freedman

Arthur Freedman Associates, Inc.
East Stroudsburg, PA

Commercial facilities depend upon water systems for space heating and cooling, for heat removal and environmental control in database facilities, for humidity control in sensitive environments, and for domestic (potable) water service. These systems range from simple to very large and complex. The equipment may include many different combinations of cooling towers, boilers, chiller/evaporators, heat exchangers, direct expansion (DX) units, air-handling units, fan coils, and related equipment. The piping systems may include combinations of large-diameter main risers, smaller-diameter lateral distribution piping, and lines carrying water to and from individual heating and cooling units.

Water is a reactive substance. In passing through piping and equipment, it can cause corrosion of metals and/or it can lay down mineral scale deposits. Corrosion can lead to metal failures and leaks. Corrosion products and mineral scales form deposits that can reduce water flow in pipes and block condenser tubes, fan coils, and other heat exchange equipment. Microbiological fouling from bacteria and organic matter in the water can also lead to corrosion and to loss of heat transfer and water flow capacity.

Water treatment chemicals and services are used to help control these problems and maintain trouble-free operation of the facility heating and cooling systems. Water treatment may include various combinations of chemicals and services provided by a water treatment vendor, plus work done by facilities personnel. This work includes applying chemicals, controlling operating systems, chemical testing, and reporting of control and performance data.

This chapter briefly explains facility water systems and water-related operating problems. It then describes the process of obtaining contracted water treatment services and working with a water treatment supplier to operate and protect these systems.

16.1   Facility Water Systems

This section describes common facility water systems and the problems that can be encountered when operating these systems. Only water-side operations are discussed here. The mechanical aspects of chiller and boiler operations are described elsewhere.

16.1.1   Air-conditioning systems

Figure 16.1 is a schematic diagram of a typical simple building air-conditioning system, including a cooling tower, a chiller unit, air-handling units, and condenser water and chilled water circulating systems. Systems such as this one are sometimes referred to as heating, ventilating, and air-conditioning (HVAC) systems. Specific systems may be very complex, but in principle all are variations of the equipment and circulating water systems shown in Fig. 16.1. Following is a brief description of the operation of this typical HVAC system.

16.1.1.1   Chilled water. Building air is cooled by circulating the air over small fan coil units in individual rooms, or through large air-handling units that deliver cooled air to public spaces and to duct systems. Chilled water circulating in a closed loop (see Fig. 16.1) absorbs heat from the air passing over these fan coils and air-handling units.

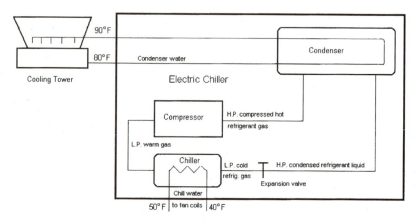

Figure 16.1   Schematic drawing of a simple building air-conditioning system.

The warmed water returns to the chiller machines, where the heat is removed by heat exchange with a refrigerant.

Chilled water systems may be small, supplying all or part of single buildings, or they may be very large, servicing many buildings in a campus facility. Sometimes a single large chilled water loop may be cooled by chillers in several buildings. In campus facilities or in high-rise buildings, secondary chilled water loops may cool specific buildings or tenant spaces. Secondary loops may be separated from the primary chilled water by plate-and-frame heat exchangers, or they may receive direct injection of primary chilled water as needed to maintain the required temperature in the secondary loop.

16.1.1.2 Condenser water and cooling towers. The chiller equipment in an HVAC system contains two shell-and-tube heat exchangers. The first, often called the *chiller,* transfers heat from the circulating chilled water to a refrigerant system. This can be either Freon gas in a compression chiller or lithium bromide solution in an absorption chiller.

The second heat exchanger in the chiller machine, often called the *condenser,* transfers heat from the refrigerant to an open circulating water system called the *condenser water* (see Fig. 16.1). The cooling tower in this system is the ultimate heat sink in an HVAC system. Heated condenser water falls through the cooling tower and is cooled by evaporation in a stream of air. In natural-draft cooling towers, the heat absorbed by the air creates a density difference sufficient to cause airflow through the tower. Large cooling towers in power stations and industrial plants often use natural-draft designs. More often in commercial facilities, induced- or forced-draft fans pull or push air through the cooling tower. References 1, 2, and 3 contain photographs and descriptions of different types of cooling towers.

16.1.1.3 Free cooling. Electric drive compression chillers, commonly used in large facilities, require large amounts of electric power to drive the compressors. In some parts of the country and at some seasons of the year, the condenser water, after passing through the cooling tower, becomes cold enough to be used directly for cooling the building air, so that there is no need for chilled water. Under these conditions, the chillers can be bypassed, and the cold condenser water can be sent directly through the chilled water pipes to the fan coils and air-handling units.

Bypassing the chillers in this way mixes the condenser water and the chilled water and leaves condenser water in the chilled water pipes when the system is returned to normal operating mode. Condenser water and chilled water chemical treatment programs are necessarily very different. Therefore, leaving condenser water in the chilled water system means that the chilled water system may be inadequately treated. This can create severe corrosion, deposition, and microbiological problems. To avoid these problems, the condenser and chilled water systems are sometimes separated by a plate-and-frame heat exchanger, in which the chilled water is cooled by, but separated from, the condenser water. This eliminates mixing of the system waters, albeit with some loss in system efficiency due to the heat exchanger.

This process is called "free cooling" because the energy cost associated with operating the chillers is eliminated. This is an attractive option, particularly for systems that can operate in free cooling mode for several months each year. However, the energy savings must be balanced against the increased costs of water treatment, the problems involved in providing effective chemical treatment to a mixed water system, and the potential for serious corrosion damage to the chilled water system.

For these reasons, use of a plate-and-frame heat exchanger to separate the condenser and chilled water systems may appear to be an attractive option. However, it is not always physically possible or cost-effective to retrofit a heat exchanger into an existing free cooling system. Many free cooling systems have been abandoned because a heat exchanger could not be installed and the costs of increased chemical treatment and repairs, plus the reduced reliability and availability of the system, outweighed the energy savings. Reference 4 provides further information on the advantages and disadvantages of free cooling systems.

16.1.2 Heating Systems

HVAC heating systems utilize circulating steam or hot water. Steam may be supplied by a local utility station or generated on-site in low- or high-pressure boilers. Some large campus facilities and high-rise buildings make their own steam and drive turbines to generate electric power. Steam extracted from the turbines is used for heating purposes. These are called *cogeneration systems.*

There are two basic designs for boilers in commercial facilities. In fire-tube designs, the hot gases from burning fuel pass through the boiler tubes, and the water to be heated or boiled to make steam surrounds the outsides of the tubes. These boilers are simple to build and operate and are commonly found in low-pressure systems (below about 300 psig). The tubes are usually horizontal, and the general appearance of the boiler is similar to that of a large shell-and-tube heat exchanger.

In water-tube boilers, the water circulates through tubes surrounding a firebox in which fuel is burned. Water-tube boilers may be more efficient in some low-pressure applications and are always used at higher pressures because of pressure-related design considerations. Many different configurations are used. Reference 5 discusses the principles of steam generation and describes many different designs of commercial and industrial boilers.

In some systems, hot water is circulated to heat the facility. So-called hot water boilers are not actually boilers. They are similar in appearance and operation to low-pressure fire-tube steaming boilers, except that no boiling occurs and no steam is generated. Closed hot-water systems are, in fact, more similar to closed chilled water loops than they are to boilers.

Many different piping configurations are used for heating purposes. In some cases, the same pipes and fan coils are used for both chilled and hot water. In other systems, piping for the cooling and heating systems may be different, but the same fan coils and air-handling units may be used. When planning a water treatment program, it is important to understand the operation of each system, including possibilities for mixing waters and for introducing air or losing water from otherwise "tight" systems.

16.1.3 Humidification systems

Hospitals, computer facilities, and other sensitive locations may require humidity control in defined parts of the facility. Humidity is controlled by injection of steam to the air ducts as required. Steam injection is controlled automatically by humidity sensors.

In some cases, facility boiler steam is used directly for humidification. This is the simplest and least expensive route to humidity control. However, this method limits both the type and amount of additives that can be used to protect the steam and condensate system from corrosion. Additives for use in humidification steam must meet the requirements of the Federal Food and Drug Administration (FDA). See Sec. 16.3.2 for more on this subject.

An alternative method for generating humidification steam is to use boiler steam through a heat exchanger to boil potable water or deionized water to make clean humidification steam. Small electrically heated steam generators can also be used in this way. Simple units that can be installed as part of the air duct system are available for this purpose. If potable water is used to generate humidification steam, provision must be made to remove the solids deposited from the evaporated water.

16.1.4   Potable water systems

Facilities may use a municipal water supply or may develop a private water supply, usually well water or river water. Large facilities may use municipal water for their potable systems, and a private system for nonpotable cooling, boiler, and process water.

16.1.4.1   Municipal water supplies.   In most cases, a municipal water plant can supply properly treated water for all of a facility's needs. This is the simplest, safest, and most reliable source of facility water. However, use of a municipal water supply does not guarantee a trouble-free system. A soft (low-hardness) corrosive water can corrode steel pipes and equipment, creating corrosion product (chip scale) problems and red water. Pipe failures can occur, first at threaded joints, welds, and other stressed areas, and then as pitting corrosion failures throughout the piping system. Soft municipal water supplies can also cause pitting corrosion of copper pipes. At the other extreme, hard water supplies can lay down calcium carbonate deposits in pipes and heat exchangers that can seriously impede water flow and heat transfer.

16.1.4.2   Private water supplies.   In large facilities, private water supplies may be cost-effective compared to municipal water. However, the apparent savings available from using a private supply must be offset by the costs of the required treatment and maintenance responsibilities.

Well water, if available in sufficient quantity and quality, can often be used for potable water simply with chlorination. In other cases, treatment to correct red water or black water problems and to control calcium carbonate scale may be required. Some very large facilities may find it cost-effective to use an available surface water as a potable supply. In such cases, additional processes such as clarification, filtration, softening, and color removal may be required. These processes require space, equipment, and personnel for continuous operation, and they create solid and liquid wastes for disposal. Such plants normally require state environmental discharge permits. Any facility considering development of a private water supply should contract with consultants and water treatment vendors to do feasibility studies, design the treatment plant, and estimate required operating personnel and costs.

16.1.5   Principles of water treatment

Water treatment programs, including combinations of physical and chemical methods, are used to control the water-related problems described above. This section is a brief introduction to the basic principles of facility water treatment. More information can be found in textbooks prepared by the major water treatment vendor companies (Refs. 1, 2, and 3). Reference 6 provides an excellent overview of water treatment in commercial buildings.

Water that evaporates in a cooling tower must be replaced by makeup water. Since the dissolved and suspended solids in the water do not evaporate, the circulating condenser water becomes more concentrated than the makeup water. The ratio of the concentration of dissolved solids in the circulating condenser water to the makeup water is called *cycles of concentration.*

Because of this increased level of dissolved solids, the concentrated condenser water may become more corrosive and/or more scale-forming than the makeup, depending upon the water composition. This can lead to corrosion of piping systems and scale formation in condenser tubes. For example, Fig. 16.2 is a photograph of corrosion product deposits in a condenser water pipe from a commercial office building, and Fig. 16.3 shows calcium carbonate scale coating the inner surface of a condenser tube.

Also, air drawn through the cooling tower brings with it soil, construction debris, and various kinds of organic and microbiological matter. In industrial facilities, oil and process contaminants can enter the cooling water. All of these materials can coat pipes and heat

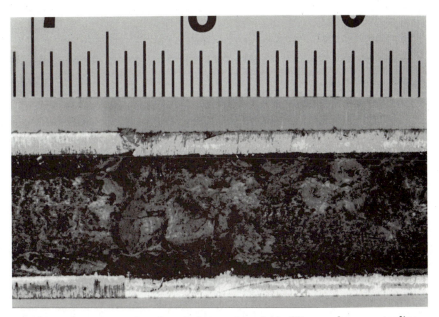

Figure 16.2   Corrosion product deposits in a commercial building condenser water line.

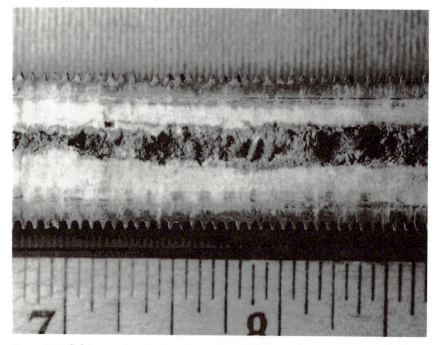

Figure 16.3   Calcium carbonate deposits coating the inner surface of a condenser tube.

transfer surfaces, restrict flow, and collect in the cooling tower basin. Deposits formed in this way become breeding grounds for organisms that can cause slime formation and under-deposit corrosion.

Closed chilled water and hot water systems do not concentrate by evaporation. However, corrosion and scaling problems still exist in closed systems. Any water lost from these systems must be replaced. Makeup water added to the system brings with it opportunities for contamination as described above. Expansion tanks on closed systems are often open to the atmosphere, providing an entrance point for oxygen, airborne debris, and bacteria. In industrial facilities, process contaminants can enter the systems. Hot water systems are especially prone to oxygen pitting corrosion. Under-deposit localized pitting, sometimes involving bacteria, is the most common form of corrosion in closed systems.

Steam generating boilers must be protected from many potential problems, including pitting corrosion by dissolved oxygen, scaling and overheating in the boiler tubes, and caustic corrosion caused by over-concentration of boiler solids beneath deposits.

There are three cardinal principles for successful water treatment in any facility. These are (1) circulate the water, (2) provide proper levels of appropriate water treatment, and (3) control the system within established operating parameters. Each of these principles is discussed briefly below.

16.1.5.1 Circulate the water. With few exceptions, systems with moving water will be in better condition than similar systems in which the water circulates only occasionally if at all. Moving water provides fresh supplies of treatment chemicals, carries away waste products, and helps to prevent hard deposits from forming in the system. Unfortunately, not all systems can be circulated continuously. Hot water systems are laid up in the summer, and some condenser water and chilled water systems in the winter. Sometimes these systems can be circulated intermittently without a heating or cooling load, but in other cases this is not possible. Dead legs never see circulation except when portions of the system are drained.

Solutions to circulation problems that prevent proper water treatment must be developed cooperatively by the water treatment vendor and the facility engineers. Possibilities include intermittent regular circulation, high-level chemical treatment with replenishment as needed, dry layup, and installation of bypass lines to make circulation possible. All of these and other possibilities should be considered. Systems that are simply closed up and ignored for several months will deteriorate and eventually cause serious and expensive problems.

16.1.5.2 Provide proper chemical treatment. In operating water systems, it is not practical to completely eliminate corrosion and mineral scale formation. However, these problems can be controlled at levels that do not affect system operations and that do not reduce the availability and life of the equipment.

The key to success is good quality water treatment. The nature and amount of water treatment that is required in any facility is site-specific. It depends upon the facility, the type of equipment involved, the heating and cooling loads, the nature and availability of the water supply, and applicable air and water discharge regulations. Water treatment can vary from simple chlorination of once-through cooling water to complex chemical treatment of high-cycle cooling towers and high-pressure boilers. Some systems can, in fact, operate without chemical treatment, but this is not generally the case.

Water treatment decisions should be made on the advice of experienced water treatment vendors and consultants. In developing their recommendations, vendors and consultants will consider the need to protect the systems against four general problems, while maintaining good operating efficiency. These problems are:

Corrosion   Damage to a system caused by reaction of both metals and non-metals with the water in the system. Corrosion is particularly serious when it occurs in crevices such as pipe threads, and beneath deposits of corrosion products and other materials in the system (see Fig. 16.2). This can lead to localized corrosion (pitting) failures. Corrosion is controlled by adjusting the alkalinity and calcium hardness of the water and by adding corrosion inhibitors such as phosphates, molybdates, zinc, and selected organic compounds.

Scaling   Deposition of minerals, such as calcium carbonate and calcium phosphate, from the water. Calcium carbonate and calcium phosphate, as well as some other minerals, become less soluble as the water temperature increases. These deposits therefore tend to appear first in condenser tubes and on other hot metal surfaces (see Fig. 16.3). Scaling can reduce heat transfer rates, restrict flow in condenser tubes and pipes, and lead to boiler tube failures from overheating. Mineral scale inhibitors include phosphonates and water-soluble polymers that delay precipitation of scaling compounds or disperse precipitated material so that it does not form hard deposits in the system.

General deposition   Formation of deposits from insoluble material in the circulating water. This can include dirt and debris blown into the cooling tower and suspended solids carried in with the makeup water. These deposits can restrict flow, bring in microbiological contamination, and provide sites for under-deposit localized corrosion. Polymers similar to those used for scale inhibition can disperse suspended matter in the water and help to prevent deposits from accumulating in the system.

Microbiological fouling   Deposits formed from microbiological material and metabolic products in cooling systems. Microbiological slimes in condensers can reduce heat transfer rates. Algae and bacteria can collect on cooling tower decks and fill. Metabolic products of bacterial growth that accumulate beneath deposits can be corrosive to metals. This phenomenon, called *microbiologically influenced corrosion* (MIC), occurs beneath corrosion products and general debris in the system as well as beneath microbiological deposits. MIC can lead to severe localized pitting failures in pipes and condenser tubes. Two general types of microbiocides are used: oxidizing microbiocides such as sodium hypochlorite, that directly attack and destroy cell walls, and various nonoxidizing, metabolic microbiocides that provide slower but sometimes longer-lasting kill. The microbiological control program must be compatible with all materials in the system and with other parts of the water treatment program.

All these problems can be effectively controlled by combinations of pretreatment of the makeup water and internal treatment of the circulating water. In open and closed cooling water systems, and in low-pressure boilers, most treatment is done internally, although the makeup water may be filtered or softened to remove excessive calcium hardness if necessary. High-pressure boilers require very pure water. Incoming water usually requires combinations of demineralization (to remove inorganic solids) and deaeration (to remove dissolved gases) in order to prepare suitable quality boiler feedwater. References 1 through 3, 7, and 8 contain more detailed information on chemical treatment of cooling and boiler water systems.

16.1.5.3 Control the systems.

In addition to providing good chemical treatment programs, it is important to control the dosage levels of chemicals and the operating parameters of the system. These can include, for example, the pH and conductivity of the water, required minimum flow rates, etc.

The water treatment vendor will establish minimum and maximum control ranges for each treatment chemical. Figure 16.4 is a typical control chart for a condenser water chemical treatment program. This chart shows actual measured corrosion inhibitor levels and the upper and lower control ranges established by the water treatment vendor. The control ranges are important. Exceeding the maximum limit, as in Fig. 16.4, will unnecessarily increase the cost of chemical treatment and may in some cases cause adverse chemical reactions to occur.

The minimum of the control range is the lowest level at which the chemical can be expected to perform properly. If a chemical dosage

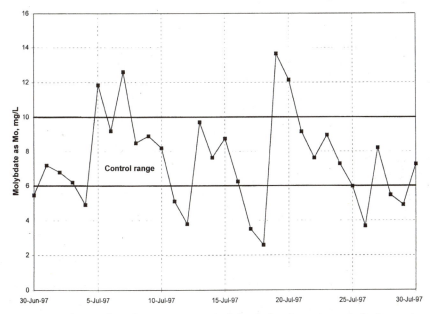

Figure 16.4 A typical condenser water control chart showing poor control of corrosion inhibitor level.

falls repeatedly below the lower limit, as in Fig. 16.4, protection may be lost. Corrosion that occurs under such circumstances cannot be reversed, and deposits that form may require off-line cleaning with strong chemicals for removal before chemical treatment can again be effective. Also, simply returning a chemical dosage to the control range may not be sufficient to reestablish protection. High-level treatment may be required to establish protective films that can then be maintained by dosages within the control range.

For these reasons, chemical feeding and control equipment should be an integral part of the total chemical treatment program. While manual control of water treatment systems is possible, some degree of automation provides improved reliability with fewer demands on personnel. Simple control equipment for condenser water systems includes timed additions of chemicals and bleed based on conductivity. More advanced technology can include chemical feed based on makeup water meter readings, control based on online analytical measurements, and precise-dosage chemical feed pumps.

Manual feed of chemicals as needed is often acceptable for closed cooling and heating systems that do not lose much water. The degree of automation and control required in chemical feed systems for steaming boilers depends upon the boiler pressure and the service involved. Site-specific recommendations should be made by water treatment vendors and consultants. Remote monitoring and control capabilities allow vendors and facility engineers to read data and make system adjustments from off-site locations. Feed and monitoring equipment should be selected for each facility, and perhaps for each system within a facility, to provide the necessary degree of control at that site.

Required chemical feed, monitoring, and control equipment, and the associated proprietary software programs, will be supplied by the water treatment vendor. This equipment and software may be purchased by the facility or it may remain the property of the vendor, on loan to the facility during the vendor's tenure as water treatment supplier. References 9, 10, and 11 provide more detailed information on control equipment for facility water systems. Reference 12 discusses the entire process of selecting, monitoring, and controlling water treatment programs.

16.2   Obtaining Water Treatment Services

This section deals with the types of water treatment services that are available, and with the process of selecting and installing a new water treatment vendor at a facility.

16.2.1   Contracted services vs. in-house water treatment

Water treatment vendors provide both the required chemical products and necessary service to apply the chemicals, monitor their performance, and report the results. Service levels vary from simple testing, monitoring of pumps, and inventory control to sophisticated equipment-oriented chemical addition, control, and reporting programs. Vendors also offer training programs for facilities personnel. The costs of these service programs are usually included in chemical prices but may, if required by the facility, be stated separately as part of a water treatment contract.

Large manufacturing plants and public utility stations sometimes find that it is cost-effective for them to take the service aspects of water treatment in-house. A facility may, for example, contract with a vendor to supply specified formulated water treatment products in bulk. Facility personnel then assume the entire job of feeding chemicals, controlling dosage and system parameters, maintaining equipment, monitoring performance, and making water treatment decisions.

This operation, while often quite successful, makes heavy demands for time and technical knowledge on the part of facility personnel. A consultant may be involved in the design and monitoring of the program, but the overall operating burden still falls on the facility. Because of the trend toward plant automation, and the accompanying reductions in plant personnel, most facilities, and particularly commercial facilities, have elected to leave water treatment service as a shared responsibility of the facility and the vendor. However, rather than leave service decisions to the vendors, larger facilities may now separate the costs of products and services, and specify the services to be performed.

16.2.2   Full-service water treatment

At the opposite extreme from total in-house water treatment, some facilities contract with a vendor for a "full-service" program. In this mode, the vendor supplies all products and does all the work. Facility personnel take no readings, do no testing, and make no adjustments to the system. In a typical full-service program, a vendor service person visits the facility weekly, biweekly, or monthly, depending upon the size of the system. He or she runs simple water tests, adjusts pumps or adds chemicals manually as needed, replenishes chemical inventory, and prepares a brief report. The vendor may also take quarterly water samples for laboratory analysis and do other routine monitoring. Between these visits, the water systems run unattended unless the vendor is called to respond to a problem.

Full-service contracts appeal mostly to facilities that do not have time or personnel to devote to water treatment operations. These may be small commercial buildings and manufacturing plants. The water systems in these facilities are usually straightforward: low-pressure steaming or hot water boilers and simple circulating condenser water and chilled water loops. Correspondingly, basic chemical treatment programs are used and feeding and control systems are elementary. High doses of chemicals are fed to help ensure that active levels in the water do not run below range between service visits. Performance monitoring tools such as corrosion coupons are not often used unless specifically requested by the facility.

Some large commercial facilities and manufacturing plants do write full-service water treatment contracts. In these cases, the vendor may use modern remote monitoring equipment to "watch" the systems from a central office. Also, at least one facility engineer will usually be qualified in water treatment and able to recognize and respond to unusual situations.

The major difficulty with full-service water treatment is that problems can arise and become serious before they are recognized and corrected. Algae blooms in cooling towers, microbiological fouling and MIC in both condenser water and chilled water systems, and dissolved oxygen corrosion in boilers and hot water systems are examples of problems that can go unnoticed until operations are impacted unless systems are inspected and tested regularly by trained facility and service personnel.

16.2.3   Support service water treatment

Support service is by far the most widely used form of vendor water treatment service in medium to large facilities. Support service involves joint responsibility for the facility water treatment operations by facility operating engineers and the vendor service representative. The vendor will supply chemicals and all feeding and monitoring equipment. This equipment can range from simple to highly complex and automated, as explained in Sec. 16.1.5.3. The vendor will visit the facility at a frequency ranging from weekly to monthly, depending upon the needs of the facility and the size and complexity of the water systems. Routine vendor service work is discussed in Secs. 16.4.1 and 16.4.2.

In a support service water treatment program, the facility operating engineers are responsible for the daily operation of the water systems. This includes regular chemical testing of the cooling water systems, plus testing each shift of the boiler systems if high-pressure boilers are involved. Daily service work by facility engineers is discussed in Sec. 16.4.1.

16.2.4   The role of consultants in contracted water treatment services

Experienced water treatment consultants are assuming an increasingly important role in modern facility water treatment service programs. As chemical programs have become more complex, feeding and control equipment more sophisticated, and service programs more elaborate, managers of large facilities have found it advantageous to have a technical person working with and representing them in dealing with the water treatment vendor. Consultants serve in several different ways:

- They prepare specifications and evaluate proposals, as explained in Sec. 16.3.

- They help to ensure that chemical programs are appropriate for the facility systems and that all feeding and monitoring equipment is properly installed and maintained.

- They review facility water treatment logs and vendor service reports and prepare periodic independent assessments of the status of each system. As part of this assessment, they perform independent water analyses and inspections of facility piping and equipment, and work cooperatively with the vendor to make necessary corrections.

- They assist the vendor in recommending needed changes and improvements to water treatment programs and to facility systems or operating procedures.

- They provide general technical advice to facility management and facility engineers concerning water-related issues as they arise.

The role of a water treatment consultant should be as a facilitator working cooperatively with both facilities management and the water treatment vendor to provide optimum protection for all water systems while at the same time maximizing system performance.

16.3   Installing a New Water Treatment Program

This section discusses the process of selecting a water treatment vendor and installing a new chemical treatment program.

16.3.1   Selecting a water treatment vendor

Installing a new water treatment program or replacing an old program involves as a first step selecting a water treatment vendor. Environmental discharge and air quality restrictions have narrowed the available choices for water treatment chemicals. As a result, most treatment programs now involve combinations of known corrosion inhibitors, scale and deposit control agents, and microbiocides. The opposite is true, however, of modern chemical feeding and control equipment and chemical delivery systems. The software associated with operating the systems and collecting and presenting performance data is also unique to each vendor. For these reasons, water treatment vendors are differentiated today more by their equipment options and the quality of their service work than by their chemical programs.

Most facilities select their water treatment vendors based on recommendations and personal contacts or on sales calls and proposals by the vendors. This process works well, and it has one major advantage. The most important part of any service program is the knowledge, experience, and dedication of the vendor personnel providing the service. Sales calls, interviews, and recommendations provide the opportunity to evaluate the companies and their people as well as the chemical programs and equipment.

Prospective vendors should ask for an opportunity to survey the facility, speak with knowledgeable employees, and study the history of water treatment in the facility, to help them prepare appropriate proposals. The resulting proposals provide insight into each vendor's understanding of the facility and its water treatment needs. It is important to remember that proposals must be individually adapted for each facility. A chemical and service program that works well at one building or manufacturing plant may not be suitable for another, similar facility close at hand. Differences in water quality, operating procedures, critical cooling and heating demands, and many other factors can combine to make apparently similar facilities quite different in terms of water treatment needs.

Cost should be only one important factor involved in vendor selection. Vendors will, of course, try to be competitive, but responsible vendors will base their proposals on the products and services they believe are needed for success in the facility. Large differences in the annual costs of proposals should be viewed with suspicion until the proposals have been technically evaluated and found to be sound.

16.3.2   Water treatment specifications

Larger facilities, and facilities that include critical cooling and heating needs or complex equipment such as high-pressure boilers have found that the use of bid specifications and formal requests for proposals (RFP) offers several advantages:

1. Specifications provide a controlled bid environment that puts technical proposals on an equalized basis for evaluation and may lead to more competitive pricing.

2. Specifications ensure that vendors will have available the critical facts about the facility that affect water treatment, so that proposals may be more realistic.

3. Specifications can define specific required services, with vendor options to offer additional services. This helps to limit service proposals to work that is really needed and is cost-effective at the facility.

16.3.2.1 Performance specifications. It is important to remember that water treatment specifications, and the resulting proposals, are different from purely mechanical specifications and proposals. Mechanical projects can be clearly defined in terms of what is needed, what materials are to be used, how the work is to be done, etc. This is not the case with water treatment. In spite of limitations on chemical programs that can be used, many options are available to the vendors. Each vendor has developed its own product line and application technology, and its own equipment and service programs that it trusts and uses effectively.

Forcing vendors to bid to a specifically defined chemical program makes evaluation of proposals simple. However, it severely limits each vendor's ability to offer the programs the vendor thinks will be most efficient and cost-effective in the facility. It also limits the facility's ability to obtain and use the best available technology that meets the facility's needs.

A better way is to set performance-oriented specifications and allow vendors to propose the programs that they believe will most cost-effectively meet these specifications. The basic parameters that determine the performance of facility cooling water treatment programs are:

- Corrosion rates on system metals
- Mineral scaling
- General deposition on surfaces
- Microbiological fouling

Table 16.1 is a typical performance specification for cooling water treatment that includes criteria for each of these parameters.

TABLE 16.1   A typical cooling water performance specification.

Parameter		Open Systems		Closed Systems
Corrosion on mild steel		Less than 2.0 mpy		Less than 1.0 mpy
Pitting attack on mild steel		None		None
Corrosion on copper alloys		Less than 0.2 mpy		Less than 0.1 mpy
Scaling and deposition		None		None
Microbiological fouling	1.	No visible deposits	1.	No visible deposits
	2.	No health hazards	2.	No health hazards
	3.	Total aerobic count	3.	Total aerobic count
		less than 10,000/ml		less than 1,000/ml

Performance specifications such as those in Table 16.1 make the job of evaluating vendor proposals more difficult. The process is further complicated by the fact that there are no standards governing the quality or performance of water treatment products. By contrast, standards for pipefitting and electrical materials and methods help to ensure that work done according to these standards will provide the expected service. That is not the case with water treatment work.

Differences in water quality, facility design, and service requirements make each facility water treatment program unique. Consider, for example, the water treatment needs of a steel mill in Indiana, a power plant in California, a pharmaceutical plant in North Carolina, a financial data center in New York, and a hospital in Texas. All of these facilities use water and need water treatment. However, their requirements differ widely. Specifications and performance requirements written for one of these facilities would have little value for another facility in the group.

Even within narrowly defined groups, such as commercial buildings in Manhattan or power plants on Lake Michigan, for example, water use and performance requirements vary. Some commercial buildings simply provide 5 days per week office space, while others contain financial centers that operate continuously, generate high heat loads, and require precise environmental control. Some power plants use once-through cooling water, while others use recirculating systems with several different configurations.

For all these reasons, vendor proposals must be tailored to the specifications and operating conditions at each facility. Water treatment programs are sometimes recommended simply because they are successful elsewhere. Such proposals should be rejected unless the vendor can demonstrate in detail why the proposal is appropriate at the facility in question.

By the same logic, chemical feeding methods and control equipment should be selected to match the needs of each facility. For example, a large power station operating with minimum personnel may require the highest possible level of automation in chemical feed and dosage control. At the other extreme, noncritical manufacturing facilities and commercial buildings may not need more than simple controls based on timers, water meters, and conductivity.

16.3.2.2 Facility technical information. A well-written specification will include the fundamental technical information about the facility that all vendors must have in order to prepare their proposals. This can include, for example, the number and type of water systems, volumes of water, special operating requirements, etc. However, this information will not be sufficient. Each vendor should be expected to survey the facility in order to thoroughly understand the water treatment requirements. Vendor surveys should include equipment inspections, water and deposit analyses, and reviews of historical water treatment records, among other work. The vendor's proposal should demonstrate an understanding of the facility water systems, current operating problems and needs for improvement. The proposal should show how the vendor has matched the proposed water treatment program to the specific operating requirements and performance needs of the facility.

16.3.2.3 Water treatment contracts. Support service water treatment contracts may be written on a "not-to-exceed" annual basis. The contract price may include both chemicals and service, or the costs of chemicals and service may be specified separately in the contract. The facility may order chemicals as needed, or the vendor may manage the inventory. This can be done either manually or by automated "keep-full" systems. "Not-to-exceed" contracts are usually based on an estimate of the facility demand for chemicals developed from historical use data, annual use of makeup water, number of facility operating days per year, etc. If facility demand cannot be reasonably predicted in this way, the contract may be open-ended and written in terms of cost per 1000 gallons of makeup water, per 100 megawatts of power generated, etc.

It is good practice to include required minimum vendor service work in the contract. The vendor may also want to list additional work that the vendor will do, plus work that will be done at additional cost as required. This helps to avoid later disagreements over contract provisions and service performance. Reference 13 provides additional information on the design of water treatment specifications and the evaluation of vendor proposals.

16.4   Managing a Facility Water Treatment Program

This section discusses cost-effective management of water treatment service programs, methods for evaluating program performance, and responses to problems that may arise. A brief discussion of water treatment technology and the problems that must be considered in managing water treatment programs is included in Sec. 16.1.5.

16.4.1   Service programs

16.4.1.1   Water treatment service team.   Once a bid has been approved or a vendor has been selected by another process, a conference should be scheduled to plan the water treatment operation. The purpose of this conference is to plan the service program as a cooperative venture involving both vendor and facility personnel. Serious problems have resulted, for example, from failure of facility personnel to respond to requests from the vendor service representative or from failure of the vendor to react to information on operating changes supplied by the facility.

These problems are the result of communication errors that can usually be avoided by developing a team approach to facility water treatment service. Members of this team should include individuals directly responsible for water treatment, their supervisors, facility management representatives, and, if appropriate, representatives of the owner or the management firm. For example, in a typical large commercial building or a campus facility, the water treatment team might include:

- The operating engineers directly responsible for water treatment
- The chief operating engineer
- The facility manager or assistant manager
- A water treatment consultant representing facility management
- The water treatment vendor representative(s) servicing the facility
- The local vendor district manager

In most cases, the facility manager or assistant manager will act as team leader. A water treatment team structured in this way should be accountable to facility management, to ensure that all facility water systems are properly and cost-effectively treated. This responsibility includes the domestic (potable) hot and cold water, nonpotable cooling and heating water, and steam systems. The team should meet monthly or quarterly, as required, to review service reports and operating logs, confirm the status of all systems, and take action on recommendations from the vendor and the operating engineers.

It should be expected that in spite of the best efforts of all concerned, water treatment related problems will occur. Pumps can fail, water can become contaminated, makeup water composition and service loads can change; these are just a few examples of problems that can appear without warning. An important function of the water treatment team should be to establish an atmosphere of mutual confidence and respect in which such problems can be discussed and resolved without recrimination. Quick action is always required to prevent simple problems from becoming serious and expensive. An operating engineer who does not report a problem for fear of reprisal and a vendor representative who is less than honest in service reports in order to avoid dealing with poor results are not serving the facility well. This problem is discussed further in Refs. 12 and 13.

16.4.1.2 Service work. There should be a clear division between the work done by operating engineers and vendor representatives. As discussed in Sec. 16.2.3, in a support service agreement, the facility operating engineers are responsible for periodic (preferably daily) water testing, data collection from online measurements, adjustments to pumps and control modules, etc. By agreement with the vendor, operating engineers may also remove and replace corrosion coupons, check inventory and place chemical orders, and do other ongoing water treatment work specific to each facility. The vendor representative provides the necessary training, analytical equipment, and other tools needed for this work. If properly trained, operating engineers can also recognize problems, such as cloudy or red water, out-of-range chemical readings, etc., and can alert the vendor and the water treatment team leader.

Vendor representatives, during regular service visits, review the facility water treatment logs and the operating logs if appropriate. They do independent water testing, including parameters that are not checked regularly by the operating engineers. They will calibrate meters, check inventory, and inspect monitoring equipment such as corrosion coupons and heat transfer test units.

During every service visit, the vendor representative should discuss water treatment conditions with the operating engineers and with facility management (usually the water treatment team leader) and should prepare a written service report. This report contains the results of the vendor's on-site tests, comments on the status of each system, and the vendor's specific recommendations for actions. These can include simple items such as adjustments to chemical feed rates, and more complex issues such as the need to repair or replace equipment, change chemicals or dosages, etc.

The vendor service reports become the basis for action by the operating engineers between service visits. As mentioned earlier in this section, it is critically important that the vendor's recommendations be objective and that the operating engineers respond quickly to these recommendations. The vendor must never gloss over or try to explain away poor results. Problems must be faced squarely by all concerned, not to assign responsibility but to correct the situation before it becomes serious. Similarly, vendor recommendations to repair, replace, or upgrade equipment must be taken seriously. As examples, loss of conductivity control in a cooling tower or loss of chemical feed to a boiler can lead to a need for repairs that will far outweigh the cost of the control equipment.

16.4.1.3 Documentation. Documentation is a vital part of the total service program. Documentation begins with the facility water treatment logs and the vendor service reports, discussed above. To these are added performance data, including corrosion coupon results, online deposit monitor data, microbiological tests, water and deposit analyses, and equipment inspections.

The vendor representative should summarize all of this information in a quarterly report to management or to the water treatment team. The quarterly report should also include a summary of all recommendations from the vendor and the operating engineers, including the actions taken on each recommendation. The quarterly report then becomes the basis for a review meeting with the water treatment team or with management. At this meeting, the status of each system is reported, recommendations are reviewed, and action plans for the next quarter are decided. This meeting helps to ensure that all problems are openly discussed and that well-considered decisions are made. This meeting also provides management with planning information for future repairs, capital improvements, and new water treatment–related equipment. Proper use of facility and vendor test data is discussed further later in this chapter.

In addition to quarterly reports, water treatment vendors may include in their service program an annual report that reviews both technical progress and the actual cost of the water treatment program vs. forecasts. These data are useful in contract negotiations and in planning future service programs.

16.4.1.4 Training. As explained above, the water treatment vendor is responsible for training the facility operating engineers to perform routine testing and equipment maintenance. The vendor should also provide sufficient general background in water chemistry and equipment operations to help the engineers understand the importance of their water treatment work and the need for maintaining careful, accurate records.

In addition to this basic training, vendors may offer specialized courses in safe handling of equipment and chemicals, pesticide applications, new water treatment technologies, etc. These courses may be offered on-site for large facilities or off-site in a classroom setting. Operating engineers should be encouraged to take advantage of as much training as practical, to improve their understanding of water treatment technology and their ability to recognize and respond to unusual situations as they occur.

16.4.2 Evaluating water treatment program performance

This section discusses the methods commonly used to evaluate the performance of facility water treatment programs.

16.4.2.1 Water and deposit analyses. Clear, foam-free water with all chemical tests within established control ranges is a reassuring sign that the water systems are in good condition. However, changes in water chemistry are often the first available indication of developing problems in both cooling and boiler water systems.

Operating engineers run periodic water analyses for parameters that are important in controlling the chemical treatment program. In an open condenser water system, these analyses might include, for example, pH, conductivity, biocide, and chemical treatment levels. Boiler water testing might also include silica, alkalinity, and hardness. Some of these data may be obtained by automatic online analytical equipment and corroborated by periodic tests.

Vendor service representatives repeat these tests and supplement them with chemical tests that indicate system performance. For example, high iron and copper levels in the water are signs of corrosion. Lower cycles of concentration based on calcium hardness and/or alkalinity compared to cycles based on conductivity or chloride may be a sign of calcium carbonate or calcium phosphate mineral scaling. Foam may indicate microbiological problems. High levels of turbidity and/or suspended solids should be analyzed to determine the source of the suspended matter.

The operating engineers' water treatment log data and vendor service report data should be plotted on control charts, such as the condenser water data shown in Fig. 16.4. Some vendors supply software that can plot these graphs directly as data are entered at a keyboard. The graphs clearly show how the test data relate to established ranges. In Fig. 16.4, the variability in molybdate level (a corrosion inhibitor) compared to the control range is too wide. The importance of maintaining chemical levels within control ranges is discussed in Sec. 16.1.5.3.

In addition to the regular service work, the vendor representative or a consultant should take quarterly water samples from all systems for complete laboratory analyses. If deposits or suspended solids are present, these also should be analyzed. For example, Table 16.2 shows typical laboratory analyses of a makeup water, a condenser water, and a condenser water deposit from a food manufacturing plant. Table 16.2 also shows cycles of concentration calculated by

TABLE 16.2   Typical laboratory water analyses of a makeup water, an open condenser water, and a condenser water deposit.

Parameter	As	Makeup Water mg/L	Process Water mg/L	Cycles	Cond.Deposit Wt. %
pH	pH	7.46	8.25		
Conductivity	umhos	87.1	730	8.4	
M Alkalinity	CaCO3	25.7	134	5.2	
Tot. suspended solids	mg/L	6.50	6.10		
Aluminum	Al	0.12	0.14		2.84
Calcium	CaCO3	60.1	295	4.9	3.51
Copper	Cu	0.14	0.19		
Iron	Fe	0.35	0.27		63.2
Lead	Pb	<0.01	<0.01		
Magnesium	CaCO3	8.16	64.5	7.9	1.29
Manganese	Mn	0.14	0.14		
Molybdenum	Mo	<0.01	6.37		
Potassium	K	0.58	4.80	8.3	
Sodium	Na	7.89	75.0	9.5	3.87
Strontium	Sr	0.03	0.18	6.0	
Zinc	Zn	<0.01	1.28		
Bromide	Br	<0.20	1.20		
Chloride	Cl	30.1	240	8.0	
Fluoride	F	0.93	7.08	7.6	
Nitrate	N	0.17	2.66		
Nitrite	N	<0.05	0.20		
Orthophosphate	PO4	0.05	0.21		
Total phosphate	PO4	0.17	1.42		
Silica	SiO2	3.60	28.4	7.9	12.6
Sulfate	SO4	1.54	12.4	8.1	2.61
Total aerobic count	CFU/ml		<10,000		>10,000,000
Loss on ignition @ 500 F					35.4
LSI @ 90 F	LSI	-0.9	+1.15		
Cation/anion ratio		0.95	0.95		

dividing the concentrations of various components in the condenser water by the concentrations in the makeup water. Abnormal or unexpected results from these analyses are highlighted.

Briefly, the data in Table 16.2 show that the condenser water treatment level is good and levels of iron and copper (corrosion products) are low. However, low cycles for calcium and alkalinity show that the water is precipitating calcium carbonate, as discussed above. Total microbiological counts in the condenser water are low. However, the deposit analysis shows the composition to be primarily iron oxide, with a very high organic content as indicated by the total organic carbon (TOC) level. The organic carbon could be, in this case, either process contamination or microbiological growth in the system piping.

This is one example of how analytical data can be used to detect system problems that would be missed by routine service work. In this case, further work will be needed to determine the source of the iron oxide and the nature of the organic matter, before corrective action can be taken. Early action, however, will help to prevent significant organic fouling and/or microbiological contamination in the system.

16.4.2.2 Corrosion monitoring. Corrosion coupons are by far the most common method for monitoring corrosion rates in cooling water systems. Coupons may simply be suspended in the cooling tower basin, but this is not good practice. Aeration and flow conditions in the basin are very different from those in the piping system, so that coupon corrosion rates measured in the basin are not representative of the system.

The best and most common practice is to install corrosion coupons in a rack. Figure 16.5 is a photograph of a typical coupon rack in service. The rack is made from 1-in-diameter mild steel pipe and has spaces for four coupon holders attached to pipe plugs. The rack must be installed where the pressure drop is sufficient to provide a velocity of at least 3 ft/s (8 gal/min through 1-in pipe).

Corrosion coupons and coupon holder/pipe plug assemblies are normally provided by the water treatment vendor. The vendor may also offer coupon racks fabricated from PVC. These can be used. However, mild steel coupon racks are preferred. Steel coupon racks provide a metal surface, readily available for inspection, that is as close as possible to the condition of the system piping. The rack should also include a spool piece, connected by unions, that can be removed periodically for inspection and then either returned to service or replaced. Finally, the rack should include a flowmeter and appropriate throttling, isolation, and drain valves.

At least one coupon rack should be installed on each condenser water system. In high-rise buildings, this rack should be near the low

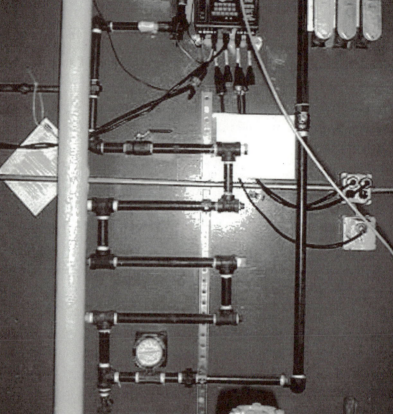

Figure 16.5 A typical corrosion coupon rack in service.

point of the system. In specific situations, it may also be wise to place coupon racks near the top of the system and on branch lines leading, for example, to tenant space. One rack should be installed on each closed cooling and heating loop.

The vendor will establish a routine for removing and replacing corrosion coupons. This work may be done either by the vendor or by facility personnel. The vendor will process the coupons and provide reports that include both a calculated corrosion rate and a brief

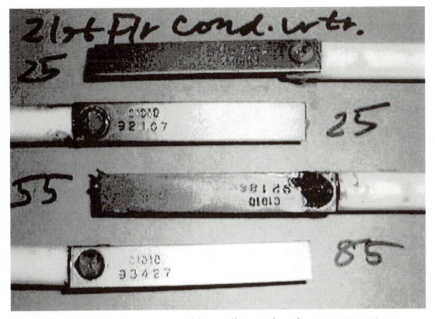

Figure 16.6 Corrosion coupons exposed in a well-treated condenser water system.

description of the condition of the coupon. It is good practice to photograph corrosion coupons as they are removed from the racks and while they are still wet. The photographs provide an ongoing record of the condition of each system. As examples, Figs. 16.6 and 16.7 show two sets of corrosion coupons as removed from condenser water racks in high-rise commercial buildings. The coupons in Fig. 16.6 are in excellent condition after exposure, while those in Fig. 16.7 clearly indicate the need for improved water treatment.

Corrosion coupons measure the average corrosion rate on the coupon over the period of exposure. Coupons should be exposed for at least 30 days to allow the metal surface to acclimate to the system. About 120 days is a reasonable maximum exposure time in condenser water. Longer exposures are acceptable in closed loops that show very low corrosion rates. The vendor will establish a coupon exposure routine to match conditions at the facility.

Online methods of measuring corrosion rates are also available. Polarization resistance devices apply a small potential difference across two electrodes exposed to the water system and translate the resulting current flowing between these electrodes into an "instantaneous" corrosion rate. Data may be taken manually or sent to a recorder for a continuous record. The Corrator (Ref. 14) is one commercial example of a polarization resistance device.

Figure 16.7 Poor corrosion coupon results from a condenser water system with inadequate water treatment.

Instantaneous polarization resistance data may not necessarily agree closely with average coupon corrosion rates. However, polarization resistance data are very useful for detecting changes in system conditions. For example, a polarization resistance graph will show an immediate change in corrosion rate due to a change in chemical treatment, long before this change will be recognized in coupon data. Figure 16.8 is a typical polarization resistance graph, showing response to temporary loss of chemical treatment.

The methods described above for monitoring corrosion in cooling water systems are not generally applicable to boiler water. Boilers are monitored for corrosion by chemical analyses of the water as described above, plus periodic inspections of the boiler tubes and drums. In high-pressure boilers making turbine-quality steam, the steam may be monitored for the presence of hydrogen gas, a product of the corrosion of iron in water at high temperatures.

16.4.2.3 Mineral scale and deposit monitoring. The most common mineral scales in cooling water systems are calcium carbonate and calcium phosphate. Both of these minerals are inversely soluble; that is, their solubilities decrease with increasing temperature. For that reason, these scales tend to appear first on heat transfer surfaces, where

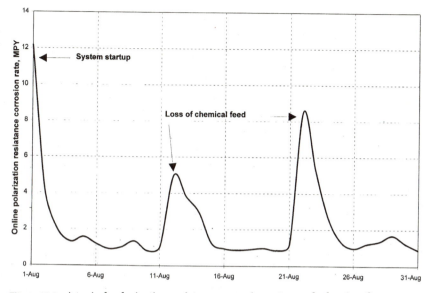

Figure 16.8 A typical polarization resistance corrosion rate graph showing the response to a loss of chemical feed.

the water temperature is highest (see Sec. 16.1.5). Figure 16.3 is a photograph of condenser tubes covered with calcium carbonate scale. Depending upon the system and the water supply, calcium phosphate, calcium sulfate, silica, and magnesium silicate can also become problems. All of these mineral scales can appear alone or mixed with iron and copper oxide corrosion products, plus general debris and microbiological matter.

In cooling water systems, online monitoring of heat transfer rates, combined with visual inspection of a test heat transfer surface, is a practical way to monitor deposit formation. In a typical deposit monitor, a sidestream from the circulating water is passed across the outside of a heated tube or through a heated block. This sidestream may be the same water used for the corrosion coupon rack. Scale deposits, and also general debris and microbiological deposits, cause changes in heat transfer rate that can be recorded online as a "fouling factor." In some devices, the heat transfer surface is visible, so that deposits can be observed as they form.

Figure 16.9 is a photograph of a typical commercial online scale and fouling measurement device. Water treatment vendors can supply similar equipment with various levels of instrumentation. Deposit monitors are commonly used in cooling systems where fouling of heat transfer surfaces is expected to be a problem. It is important that the

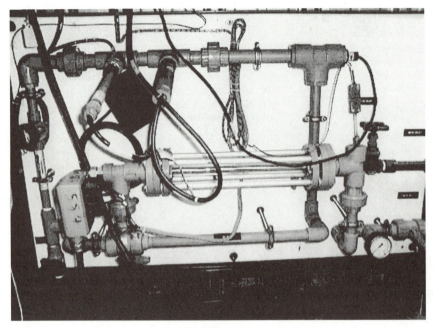

Figure 16.9   A typical commercial scale and fouling monitoring device.

heat transfer surface be representative of the system condenser tubes. If sections of actual condenser tubes are used, any corrosion that occurs on the heated tubes can also be monitored.

Deposit formation in boilers is difficult to monitor visually because of the pressure involved. Material balance calculations based on water analyses can provide an indication of problems. Boilers should be inspected at least once per year. Any indication of boiler tube deposits must be corrected immediately to prevent overheating of tubes and severe boiler damage.

16.4.2.4 Microbiological monitoring.   Cooling tower basins are a comfortable place for microorganisms to proliferate. Modern condenser water chemical treatment programs operate in the neutral or alkaline pH range. This pH range and the warm temperatures in the cooling tower basin are ideal for microbiological growth. Chemical treatment programs often contain phosphates, plus various organic compounds that can help to support growth. The air blowing through a cooling tower is a continuing source of both microorganisms and nutrients. For these reasons, microbiological control is an essential part of every open cooling water chemical treatment program. Closed cooling systems may also have microbiological problems if the makeup water is contaminated or if process leaks bring organic matter into the system.

Microbiological monitoring is a critical part of the water treatment service program. It is not practical to sterilize an open cooling tower system. It is therefore necessary to maintain microbiological activity at low levels that will not encourage rapid growth and slime formation in the system. Control levels can be set by specification (see Sec. 16.3.2) or by the vendor, based on the needs of each system. Monitoring should consist of several stages:

- Simple "dipstick" testing by facility operating engineers and by the vendor, for total aerobic bacteria.

- Quarterly laboratory tests of the circulating water and deposits from the cooling tower basin and the system piping, for both aerobic and anaerobic bacteria that could encourage MIC.

- Periodic testing of the tower water for pathogenic bacteria. The required frequency for this testing will depend upon location, opportunities for contamination, and exposure of facility personnel and the public to cooling tower water or to drift from the cooling tower.

- Inspections of fouling monitors, cooling tower basins and fill, and coupon racks for the presence of microbiological slimes.

16.4.2.5 Equipment inspections. The ultimate test of water treatment program performance, of course, is the condition of the equipment. Cooling towers, chillers, steam condensers, boilers, and other water-using equipment, including the system piping, should be inspected annually. Many different tools are available to aid in these inspections:

- Inspections of head boxes, tube sheets, boiler drums, etc., provide direct visual evidence of water treatment program performance. Photographs should be taken whenever possible, to provide an ongoing record of progress in keeping systems clean and corrosion under control. Thus Figs. 16.2 and 16.3 show corrosion product deposits in piping and calcium carbonate scale in a condenser tube. Figure 16.10 is a photograph of a condenser water tube sheet showing essentially clean conditions after a full year of service. Vendor photographs should be included in the vendor quarterly technical reports as described in Sec. 16.4.1.3.

- Fiberoptic equipment allows visual inspection of the internal surfaces of condenser tubes, boiler tubes, and pipes. Videotape records can be made of these inspections.

- Ultrasonic thickness (UT) probes can be used to measure pipe wall thicknesses online. UT measurements are especially useful in evaluating the condition of old piping systems to determine their gener-

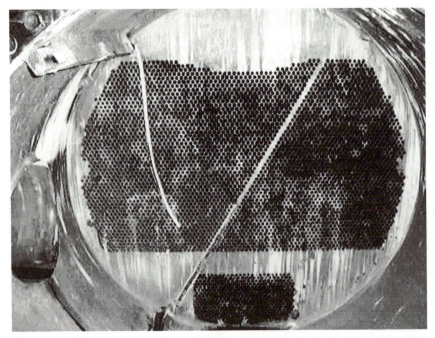

Figure 16.10 A condenser water tube sheet in a large commercial building, clean after 1 year service.

al condition. Water treatment vendors and consultants may offer both UT and fiberoptic inspection equipment as part of their service capabilities.

- Eddy-current inspection of condenser and boiler tubes can detect both internal and external damage. These inspections should be done annually or biannually as needed. Eddy-current testing is done by qualified contractors that provide the equipment and trained personnel. Comparisons of historical eddy-current reports provide a record of changing conditions in condensers over time.

- Various forms of radiography can be used to check welds and to evaluate the condition of piping beneath insulation. Qualified contractors provide the equipment and personnel for this work.

- Destructive examination of spool pieces and pipe sections in the laboratory can aid in determining the extent of damage and in predicting the projected remaining life of system equipment. Corrosion coupon rack spool pieces should be destructively examined on a regular basis. System piping should be destructively examined if other testing indicates severe pluggage or metal loss, if failures occur, or as a prerequisite for chemically cleaning a system (see Sec. 16.4.3).

16.4.3 Cleaning water system equipment and piping

16.4.3.1 Defining the problem. In spite of all best efforts, water systems sometimes become fouled with deposits. Deposits interfere with heat transfer and can restrict flow in piping and heat exchangers. In some cases, water systems that have reached this condition should be cleaned to maintain system performance and to protect the equipment. Also, old systems may require cleaning before a new chemical treatment program can be installed.

Cleaning of any water system is a major project that requires careful consideration and planning. The decision to clean a water system, and the selection of cleaning methods, should be made by the water treatment team (Sec. 16.4.1) or by facility management on the basis of recommendations from facility engineers and operators, vendor representatives, consultants, and cleaning contractors who are familiar with the system, the facility as a whole, and local safety and environmental regulations.

As a general rule, deposits that interfere with flow or heat transfer, or that create sites for under-deposit corrosion, should be removed. However, thin, hard deposits that protect the surface should be left in place. Removing protective deposits may, in fact, do more harm than good by exposing fresh, unpassivated metal to the system.

The first step in the cleaning decision process is therefore the destructive examination of representative pipe sections. This examination will help in determining whether the deposits are harmful or protective and whether or not sufficient wall thickness remains in the piping system to make cleaning worthwhile. Chemical analysis and physical characteristics of the deposits will aid in selecting the mechanical or chemical cleaning method to be used. Finally, from the amount of deposit in the pipe samples, the total deposit weight density in the system, in ounces per square foot, can be estimated. This will determine the quantity of cleaning chemicals needed.

The following general information on cleaning methods will aid in understanding the various processes available for cleaning of facility water systems.

16.4.3.2 Online chemical cleaning. Water treatment vendors may recommend cleaning a system online—that is, while the system is operating—using high concentrations of dispersant chemicals and microbiocides. Cleaning chemicals may be circulated in the system for a matter of days to weeks. The microbiocides help to kill and remove biofilms that may be bonding deposits to surfaces, and the dispersants then loosen and disperse these deposits into the water. In condenser water systems, the loosened deposits can settle in the cooling tower basin, or they can be filtered or removed with the blowdown.

In closed systems, some plan must be in place to remove loosened suspended matter that would otherwise settle elsewhere in the system. If the system can be shut down, it can be drained to remove these solids. Several fill and drain cycles may be necessary. An online "bleed and feed" process may also be used. This process is slow, wastes chemicals, and may be only partially effective in removing suspended solids. The best method is to use either a permanently installed or temporary sidestream filtration system to help remove suspended solids as they are formed.

Online dispersant cleaning is less expensive and simpler than any form of offline cleaning. For that reason, it is often tried as a first step. Online cleaning is helpful in removing loose deposits, particularly mud and silt, and microbiological deposits. However, this process is not very effective in removing mineral scales or hard corrosion products. Some water treatment vendors also offer stronger cleaning solutions that can be circulated online for several days and then must be drained to remove the cleaning solutions.

16.4.3.3 Offline mechanical cleaning. In some cooling systems, the main circulating pipes, particularly the vertical risers and large lateral lines, can be cleaned by a mechanical hydroblasting process. High-pressure water is forced through rotating nozzles that are drawn through the pipes. The resulting slurry of deposits and water is allowed to settle. In most cases, the solids must be carried away for disposal as a toxic waste. The water may be suitable for disposal in the sewers, or, depending upon the location, it also may be classified as a toxic waste and require off-site disposal.

Cleaning contractors have the equipment and technology to do this work. Mechanical cleaning is expensive, but it is simpler, faster, safer, and less costly than offline chemical cleaning. This process will remove all loose deposits, but it will not clean hard mineral scales or tightly adherent corrosion products. Nevertheless, mechanical cleaning may be sufficient. If not, it may at least greatly reduce the amount of deposit that must be chemically removed, thus shortening the time required and reducing the cost of offline chemical cleaning.

In some closed loop systems, it is possible to mechanically flush parts of the system and components such as air handlers and fan coils, before beginning either online or offline chemical cleaning. This work, although time-consuming and labor-intensive, makes the chemical cleaning work much simpler and more effective.

16.4.3.4 Offline chemical cleaning. The most difficult, most expensive, and most effective way to remove deposits from a facility water piping system is by offline chemical cleaning. This is a short-term, high-intensity process. The facility must be evacuated, at least for a weekend. Strong chemicals are brought in, heated and pumped through the system, and then drained and carried away for disposal by the contractor. In most cases, the first rinse water must also be disposed of off-site. The entire system being cleaned must be monitored continuously in order to find and immediately clamp any leaks that may develop. The cleaning contractor's personnel will test the cleaning solution frequently and determine when the various stages in the process begin and end.

In spite of these problems and the accompanying high cost, offline chemical cleaning is often the process of choice for restoring facility water system piping to working condition. The only viable alternative may be replacement of large sections of a piping system. This is especially true in facilities that house data processing centers and other operations that can only be shut down briefly, with long-term advance planning. The result of a well-done offline chemical cleaning should be piping that is clean essentially to bare metal or to thin, protective deposits, and is ready for passivation and treatment as a new system.

16.4.3.5 Planning the cleaning process. The following discussion applies equally well to cleaning boilers and open and closed cooling water systems. Any water system cleaning process, from use of a simple online dispersant to a complete offline chemical cleaning of a large system, should be carefully thought through, step by step, by the water treatment team with the addition of cleaning contractors and consultants. Every eventuality must be considered and appropriate responses must be planned. The result of this planning process should be a flowchart that covers the responsibilities of all parties concerned, from initial facility planning, through arrival of contractor personnel and chemicals on site, to immediate filling and chemical treatment of the system after the final rinse in order to provide corrosion protection to the freshly cleaned pipes.

The reason for this extraordinary concern is that any unplanned interruption in the cleaning process, particularly an offline chemical cleaning process, can be very expensive and potentially hazardous. For example, three problems that have occurred during chemical cleaning projects are sudden appearance of a major pipe failure, lack of circulation through small-diameter lines, and lack of sufficient chemicals on site to replenish chemicals depleted during the cleaning process. Each of these problems delayed a cleaning project beyond the available window of time, thus forcing the facility to cancel the cleaning project before it was completed. These problems could have been avoided by thorough inspection of the system before cleaning was attempted, and by careful planning and staging of the project.

These cautions are intended, not to discourage chemical cleaning, but to emphasize the importance of careful planning to ensure that the job is done right the first time. Many facility water systems have been successfully chemically cleaned, restoring full operating capability and extending their useful life.

References

1. "Principles of Industrial Water Treatment," Drew Industrial Division, Ashland Chemical Company, Boonton, NJ, Eleventh Edition, 1994.
2. "Betz Handbook of Industrial Water Conditioning," Betz Laboratories, Inc., Trevose, PA, Ninth Edition, 1991.
3. "The Nalco Water Handbook," F. N. Kemmer, ed., New York: McGraw-Hill, Second Edition, 1988.
4. K. A. Selby and A. J. Freedman: "Why Free Cooling Isn't Free," Paper 715 presented at Corrosion/98, National Association of Corrosion Engineers Annual Conference, San Diego, CA, March 1998.
5. "Boilers and Auxiliary Equipment—A Special Section," *Power,* June 1988.
6. R. W. Lane: "Control of Scale and Corrosion in Building Water Systems," New York: McGraw-Hill, 1993.
7. S. D. Strauss and P. R. Puckorius: "Cooling Water Treatment for Control of Scaling, Fouling and Corrosion—A Special Report," *Power,* June, 1984.
8. S. D. Strauss: "Water Treatment—A Special Report," *Power,* June, 1974.
9. A. J. Freedman and T. M. Laronge: "The Importance of Control in Operating Modern Cooling Water Systems," *Industrial Water Treatment,* July/August 1995.
10. O. Hollander and A. J. Freedman: "Monitoring and Control in Industrial Cooling Water Systems, Part 1: Cooling Water," *Industrial Water Treatment,* September/October 1996.
11. O. Hollander and A. J. Freedman: "Monitoring and Control in Industrial Cooling Water Systems, Part 2: Boiler Water," *Ultrapure Water,* February 1997.
12. A. J. Freedman and T. M. Laronge: "Fail-Safe Cooling Water Operations," Paper TP91-04, presented at the Cooling Tower Institute Annual Meeting, New Orleans, LA, February 6–8, 1991.
13. A. J. Freedman and T. M. Laronge: "Problems and Pitfalls in Water Treatment Specifications," Paper TP90-10, presented at the Cooling Tower Institute Annual Meeting, Houston, TX, February 5–7, 1990.
14. Rohrback Cosasco Systems, Inc., Santa Fe Springs, CA 90670.

17

Architectural, Structural, and Sustaining Maintenance and Repair for Structures

William G. Suter, Jr.
Director of Physical Plant Operations
American University
Washington, DC

17.1 Introduction

Effective management of architectural and structural maintenance and repair can only be achieved through a combination of resources. In architectural and structural maintenance and repair no level of hard work, expertise, technology, or money is sufficient without an effective plan. The plan must integrate elements that make up an interdependent maintenance management system. And it is management's responsibility to develop and continue to monitor this maintenance management system to assure that the changing needs of employees, facilities, and their occupants are addressed.

There are a number of decisions to be made about the various types and quantities of elements required in planning and developing a high-quality maintenance and repair system. For instance, the elements needed in a maintenance and repair system for a single-story elementary school are vastly different from those of a high-rise, hospital, or hotel, just as the maintenance and repair system for a hospital is different from that required of a shopping center. There is no one best system. However, high-quality systems share key elements, without which no system can be successful.

17.2   Information for Maintenance and Repair Systems and Work Control

All successful maintenance and repair systems will have a process for determining the needs of customers and structural and architectural components of the facility. The term "customers" in this text refers to all types of facility users. These users may be tenants who are residents, business owners, employees, shoppers, or visitors. The differences between systems for various types of customers are important but the basic elements remain the same. Regardless of the type of customer, the system must include an avenue by which the customers can communicate their needs to the facility department and put this information into meaningful direction for the personnel who must respond to customer needs. The system must also be able to deal with the ongoing preventive and predictive maintenance requirements of the facilities. The more integration in the scheduling and prioritization of these needs, the more successful the maintenance or repair system. In responding to requests from customers, the system must be able to differentiate between requests that require an emergency response from requests that can be handled routinely. The political reality is that the system must be established so that these more sensitive requests are identifiable. While life safety and structural integrity are obvious emergencies, some situations might not be. The system must be clear on emergency responses that require immediate attention. The distinction between emergency and nonemergency responses must be clear for the facility department in order to respond successfully to the customers' expectations. The maintenance and repair system must be developed to meet the needs of different customers. For instance, formal reception or formal meeting facilities that require impeccable appearance and routine availability can only exist if the maintenance and repair system is set up so that mechanics respond immediately to minor problems in these formal areas as though they were emergencies. It is also imperative to know that a "minor" problem in one area may be an emergency in another area with a different function. The system for collecting information about customer requirements is typically part of a larger "information system." This information system is used not only to collect data about customer requirements and preventive maintenance schedules but also to collect data related to facility component inventory, component condition, maintenance history, occupants, as well as safety and regulatory information. These systems require a considerable amount of work whether they are computerized or paper-based, but the value of the information justifies the effort during decision making. Any type of effort to organize information in a maintenance management system must involve knowledge about the facilities involved, and the system must be developed in a way to identify the various areas of the facility. The system may include an efficient way of identifying areas and room numbers so that workers will know where to perform work requests.

17.3 Human Resources

All successfully maintained facilities, regardless of function and customers, have people as the most essential and valuable element of the maintenance management system. Fancy computerized maintenance management systems are worthless without qualified people who understand how to use the systems as tools to serve customers. People are needed to monitor sophisticated systems to assure that they are maintained. It is your people who deal with customers one-on-one. Remember, it is not the supervisors, managers, or computers who satisfy customers. Front-line employees are the people who satisfy customers.

A number of important things must be considered when determining the knowledge, skills, and abilities that must be represented in the maintenance and repair staff. Technical aspects of the facilities must be taken into consideration so that the skills of the maintenance and repair staff can be matched. While most modern facilities have some computerized systems, they range from simple access control systems to complex energy-management systems such as those that operate HVAC equipment in response to weather conditions, prearranged schedules, facility use, and utility consumption. In older, less sophisticated facilities, technology may not require expert staff. Yet a facility of modest size and operation may require the expertise of several trades such as carpenters, custodians, electricians, painters, plumbers, and HVAC mechanics. There might be regulations that require credentials in order to operate and maintain a facility. For instance, facilities with sophisticated HVAC systems may require specially licensed operators. Many single-building facilities can be successfully maintained with just a few people who possess the requisite skills or the judgment to know when to use specialists or additional labor power when a job grows increasingly intricate or the volume of work rises. The sophistication of the maintenance management system must parallel the growth of buildings, facilities, and their intricacies.

When more than a couple of people are working together, leadership must be coordinated. There are basically two ways to organize the human resources involved in managing complex facilities. People can either be organized around their trade and expertise or they can be organized by the functions within the facility. Decisions about the type of organizational structure may not be as important for smaller, simpler facilities as they are for larger, more complex facilities. Traditionally organizations have been structured by categorizing construction trades where highly skilled foremen are responsible for planning, assigning, and supervising the work of their trade. This type of organization seems to be rooted in specialty trade unions with a focus on technical work quality as defined by traditional trade principles. While the quality of work is a critical element in the success of a facilities maintenance organization, it is not the only criterion by which the organization is judged. Since the mid-1970's there has been a growing interest in creating organizations that are more aligned with the needs of the facility's occupants. In theory, this is being accomplished through what have become known as *zone maintenance organizations.* This organizational structure emphasizes the maintenance management system's sensitivity and responsiveness to customer needs. If a zone system is selected for a facility, a decision must be made to either distribute or centralize the operation. One successful model of a distributed zone system involves the establishment of satellite shops. The satellite shop model tends to bring the workers closer to their customers, which leads to shorter response times. A centralized zone system can help achieve department unity and a sense of common purpose and interdependence that are more difficult to achieve in a distributed system. Perhaps these two examples of organizational structures are extremes, but it is important to note that neither approach is correct for all situations. Using either approach, the emphasis is still placed on work quality control, communication, accessibility, and customer satisfaction. All of these elements are essential for the success of any facility department system. Naturally, the system that best accommodates a facility's needs would be the one that strikes the right balance between the two approaches. One important point to recognize is that work quality is not necessarily sacrificed when there is an emphasis on customer satisfaction. Customers will be unhappy if they receive poor-quality work. Customers will also be unhappy if the technical quality of the work is good but those who deliver the work are insensitive to their needs. Figures 17.1 and 17.2 show the two types of facility organizations discussed above.

Larger facilities will quickly develop the need for additional types of expertise for the workforce. Facilities that have lawns and gardens require the support of groundskeepers. And owing to the varied tasks of grounds maintenance, there may be a need for varying levels of expertise in this field. Responsibility for grounds maintenance and the equipment used to attain professional results adds a level of complexity to the maintenance management system. Once the facility

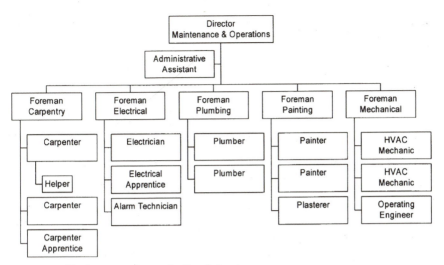

Figure 17.1 Trade-centered organizational structure.

manager determines the nature of the expertise needed in a field, a decision must be made to either hire individuals to perform the required services or outsource the services to a contractor who can produce at the required level of expertise.

The decision can be complex when choosing in-house services versus contracting services. Many of the issues discussed already will be involved in the decision. Decisions about delivering service, like all other decisions, will need to be revisited. Service quality is the ultimate criterion involved in deciding whether or not to outsource a service. Getting a service performed by a contractor for less may be attractive. But if the quality of work and customer satisfaction is sacrificed at the expense of saving money, then it is not a good decision. We will always be required to extract every ounce of value for the dollars we spend. However, we must avoid putting efficiency first and effectiveness second. Doing an effective job efficiently should be the goal of facility department organizations. Contractors may have narrow or wide areas of responsibility. If you need to clean windows occasionally, you may want to outsource this service rather than perform this job in-house. Many facility managers have discovered that housekeeping services can be effectively contracted and that there are a number of very good vendors able to provide a full range of cleaning services. Other contractors are able to provide complete facilities management services at competitive costs. No one option is best for all cases. In fact, the age and condition of the facilities, the customers' expectations, the expertise of facility managers, and the expertise in the local workforce must be taken into consideration. If the right

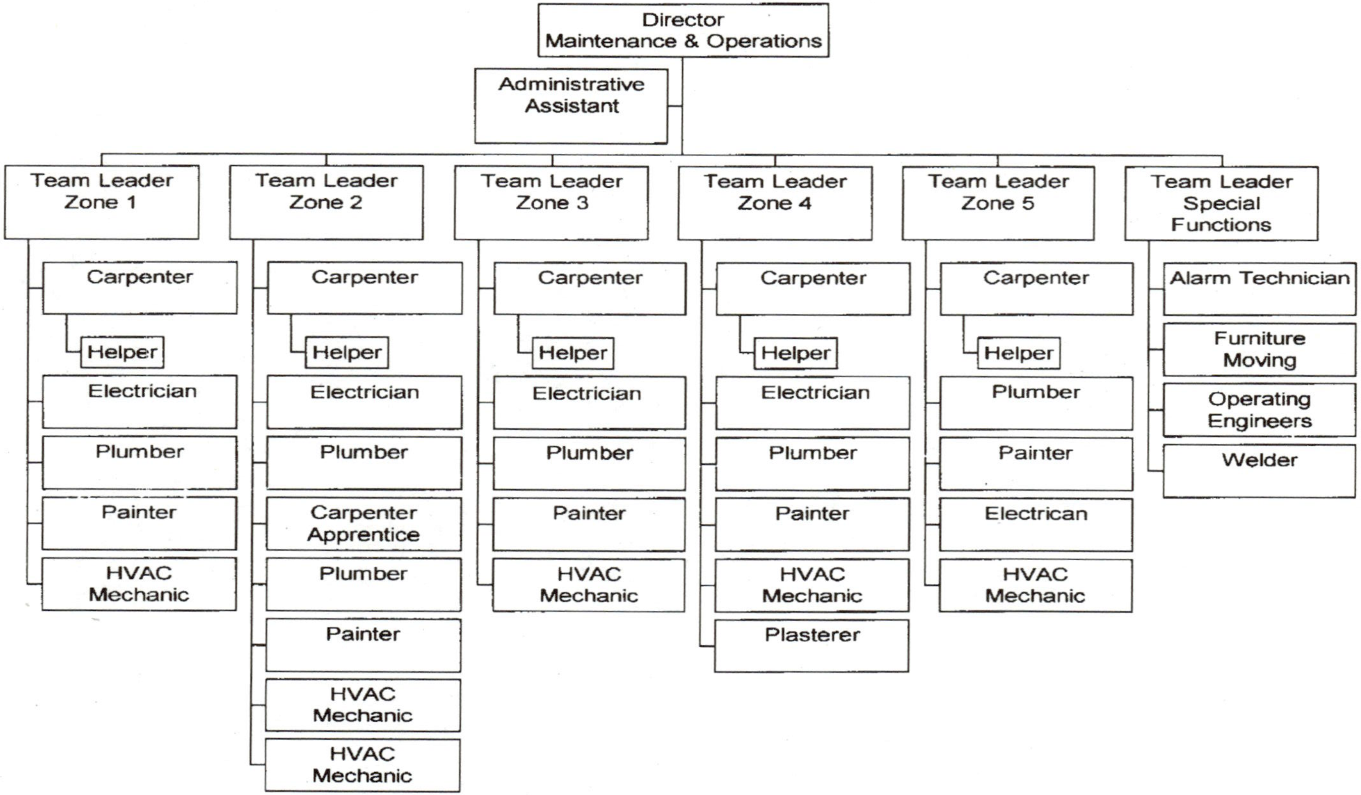

Figure 17.2 Customer-centered organizational structure.

information is available, choosing to use in-house or out-source resources is a difficult decision. In fact, if the decision ever appears easy, some important and relevant piece of information is probably not being considered.

17.4 Customer and Occupant Interaction

No matter the size of the facility, there is a need for both reactive and proactive methods of formal interaction with customers. In order to maintain good relations with customers, a facility department must be able to document facility problems and follow up with customers on the status of work in progress. And after the work is complete, determine whether or not expectations were met. There should also be a standard and routine form of formal communication between the facility department and customers to ascertain whether or not expectations for services are continually being met. Thus it is essential for a facility department to constantly adapt its services and improve its quality of work in order to meet the changing requirements of customers.

Being able to understand customer requirements plays an important role in prioritizing and scheduling work requests. Often those reporting deficiencies do not have sufficient knowledge to provide accurate information about what is wrong. They often do not know or care that a solution to their problem is complex, they just want it fixed. Understanding the functions represented by customers, knowing the nature of customers, and being familiar with the capacities of building systems are all vital to being able to meet customers' expectations. One of the most difficult realities is that many customers do not know what they expect from workers until workers fail to provide it. An effective facility department needs to have an understanding of its own capabilities so that, through the exchanges involved in routine business, they can help set customer expectations. Customers should be told when to expect a response to their requests and, if possible, who will respond and in what way. One example is responding to a temperature call in the mechanical equipment room. Even if a mechanic goes directly to the equipment and solves the problem, the customers' expectations may not be met until they see and talk to the mechanic who resolved the problem.

17.5 Scheduling

Scheduling repair and minor alteration work is a critical function that can be facilitated by having available data on the types of activities in the facility. Daily scheduling of minor repair work in a hotel, for example, involves having data on occupied rooms, the time rooms are occupied and checked out, and the time housekeepers clean the rooms. Long-range scheduling must take advantage of the natural cycles of facilities occupants. School-year cycles and summer tourist season are other examples that illustrate the information needed to make good scheduling decisions. It is always preferable to involve customers in scheduling decisions, as they know best when their function or operation should be interrupted by repair work. They also know exactly what parts of their operation are being interrupted by lack of repair work. Obviously, real or political emergencies must be handled before routine work. However, organizations must be careful in balancing emergency and routine requests. Organizations that do nothing but respond to emergencies will do nothing more than that, and the number of emergencies will continually rise. There are no facilities without preventive maintenance requirements. Preventive maintenance work is important but may not have the same feel of urgency as an emergency request. Every day of their career, facility managers must fight the battle of urgent versus important. Facility managers are in this business to help people. When they can respond effectively to a request, they feel the accomplishment of helping customers with a problem out of their control.

In the bigger picture, the highest-valued activity is implementing the preventive and predictive maintenance programs. These programs prevent the need for the customers' requests in the first place. Preventive maintenance programs must successfully compete with the urgent and often fun job of responding to customer requests. A "scheduler" in this context refers to anyone who schedules work or job requests. Most people in a facility department in some way play this role. Workers who receive multiple requests must schedule their time for each task. Naturally, the front-line supervisor will schedule a higher volume of work as well as more variety.

A facility department large enough to require a person to perform the primary role of scheduler may be involved in scheduling urgent repairs, major remodeling, and preventive maintenance. Everyone must schedule, and effective scheduling means having access to a wide range of information. The scheduler's task is to constantly strike the right balance between proactive (preventive) and reactive activities. On one hand, too much emphasis on proactivity will result in unhappy customers because customers require an effective response capability, regardless of the long-range effectiveness of proactivity. There will always be the need to respond well. On the other hand, too much emphasis on reactivity will result in the need for more and more reactive attention, as equipment and systems will deteriorate owing to lack of scheduled service. The scheduler must also know the capabilities of the workers performing the work. One worker might be able to handle multiple tasks and make good priority decisions while another might not. Dispatching workers to perform work is another function that requires good information. Not only does the dispatcher need to know the nature of the work (i.e., whether or not it is critical); they must also know the appropriate worker to dispatch to a particular job. This decision is made more complex because of the marginal quality of information given by customers and the high probability that whoever is selected to respond may be preoccupied on another assignment. Workers who are in direct contact with customers at the time of the work request need to help customer needs to extract the best possible information about the criticality of the deficiency so that they can communicate this to the worker or workers who must respond. The most frequently asked question by those responding to dispatch is "what emergency do we put off to respond to this call?" The more trades that are involved, the more complex the decision on how to route a work request. Each trade will eventually have to develop a decision tree approach to help those who are managing the work flow determine how best to route a call. Plumbing calls will have a very different decision process than those of locksmiths or electricians. The differences in the expertise of tradespersons will also make work routing and dispatch decisions more complex. Constant feedback is required between workers routing and dispatching work and those performing the work to continually improve the information on which the system depends.

Owing to the constant flow of requests from facility users and from ongoing preventive and predictive maintenance programs, scheduling of work must be done constantly. At any time of the day a request can be received and may change previously made scheduling decisions. You should not expect to get urgent or important work requests before routine requests. Therefore, a first-in first-out scheduling model will not be successful in emergency cases. Routine repairs may be handled using the first-in first-out schedule, but you must always respond to urgent and important work before responding to routine requests. Achieving the right balance between urgent and important response and routine response is important. A routine request that takes too long (from the customer's perspective) may possibly become an urgent or important request when the customer makes a call to complain. The scheduling system is part of the overall maintenance management system, and it must utilize existing resources to accomplish all types of work so as not to funnel all resources toward the work that is the most urgent and important at the moment.

17.6   Work Documentation

All work must be documented so that the expended flow of resources can be understood. Those who have to back-charge for their work need to know the period of time within which the work was performed, the labor rate, and costs of materials, parts, and supplies used. Preventive and predictive maintenance work must be documented so that the effectiveness of the programs can be measured and improvements made as needed. There is also a range of federal and municipal documentation required for personnel, work practices, hazardous substances, and even specific charges for work-related grants or special contracts.

17.7   Safety

In recent years the importance of safety in the workforce and for facility users has increased and has become the focus of intense federal and local regulation. The level of required training and other safety-related program activities is still increasing. Large and more complex organizations might designate someone to the safety programs, while smaller organizations might use contractors or even attempt to develop in-house expertise. The increasing changes in requirements make the growth of in-house expertise a very difficult proposition. An example is internal air quality (IAQ). With the advent of sophisticated HVAC systems came the need for buildings to be sealed from contact with outside air. Technology continues to improve and many of the earlier problems are now being addressed, but there remains a need for the capability of any maintenance organization to respond to IAQ complaints. The response typically involves air testing, HVAC system inspection, and communication with occupants to see how they are using the space and what type of equipment is being used. As new materials are incorporated into facility components and furniture, the IAQ issue becomes even more important.

17.8   Planning and Estimating

Facility departments must be able to predict the amount of labor and overall costs that are associated with different types of work. The ability to predict and prevent problems means establishing a system that will estimate the costs and the length of time a job will take, and which will involve scheduling and planning. There are several high quality estimating systems that can be used to formalize the estimating process. These systems require trained personnel who can integrate computer programming into the system to speed and organize the process so that it is easy to use.

17.9 Quality Control and Customer Satisfaction

Finally, facility departments must deal with the issue of quality control and customer satisfaction. The organization must know how well they are meeting customer expectations. And they must know something about the quality of work being performed by facility department workers. The organization that relies solely upon unsolicited customer feedback for work quality and customer satisfaction is courting disaster. Problems with work quality or work that does not meet customer expectations can quickly lead to customer dissatisfaction. The facility will also suffer and low-quality work will cause facility deterioration. It is always best for workers and leaders in facility departments to have firsthand knowledge about quality and satisfaction problems so that they can be proactive by making changes and communicating with customers rather than waiting for complaints. Even an effective response to customers' calls can control the situation even though real improvement has not taken place. Focusing continuously on work quality and customer satisfaction has the potential to improve the facilities and the satisfaction of those who use them. The ultimate goal of any facilities organization is to create an environment that promotes success and excellence. Whenever facility occupants are hindered or limited by problems in their surroundings, the facilities organization is not meeting their requirements. Successful facility department users create opportunities for themselves and for the organizations that interact with them. While the success of your customers might not guarantee your success, it cannot hurt, and not giving customers and facilities proper attention may just guarantee your failure.

Supplementary Practices and Procedures

Bernard T. Lewis

Facilities Management Consultant
Potomac, Maryland

The following practices and procedures are provided to assist the facility manager in achieving an efficient, cost-effective operation...without sacrificing high management, engineering, operation, and maintenance performance standards. These practices and procedures supplement the practices and procedures provided in the 17 chapters of the handbook.

Preliminary Practices

Identifying Problem Areas

Facility managers are being forced to find new ways and means of reducing their operations and maintenance costs, owing to increased pressure by top management for organization-wide cost reductions. There are only two avenues open to the facility manager in this respect: (1) squeeze more operations and maintenance effort from the current workforce or (2) obtain the same operations and maintenance effort from a reduced workforce.

How to pinpoint problem areas. On the surface, facility departments in most organizations appear to function properly until a facility-wide cost-reduction program discloses excessive operation and maintenance costs, and service complaints by department customers. Facility management controls, normally used—such as written work orders, management-approved funding of major repairs, work order schedule boards, cost accumulations by the computerized maintenance management system (CMMS), strict procurement procedures, etc.—will not pinpoint the specific areas that require close management attention. Where to look and what to do for *corrective actions?* This is the key to a successful facility management cost-reduction program.

Investigating operations and maintenance practices will normally reveal the following problem areas:

1. Budget standards are not being met.
2. Facility department customers complain of inadequate service. These complaints are directed at loss of production and payments for idle time, resulting from slow service and improper repairs; at failure to meet promised completion dates on major overhauls; at excessive repair parts costs; at outsourcing for some overhauls; and at a buildup of costly demands for equipment and systems replacement.
3. Industrial engineering work sampling surveys will disclose the amount of craftsperson idle time—and its causes. Normally the surveys will reveal 30 percent idle time—5 percent of it wholly unnecessary and 25 percent over which craftspeople have no control. Of the total idle time, 85 percent will be caused by unrealistic management policies and practices.

14 soft spots.   Based on the above indicated problem areas, 14 soft spots (vital functional areas), if placed under intensive study, and corrective action taken, will normally pay off as follows:

Fewer machine failures

Better management control of facilities workerpower

Performance of machine repairs within budgeted standards

Smarter decisions among supervisors, who now have more information and data than before

An increase of 50 percent in the number of major overhauls handled in the department (instead of being contracted to vendors)

A sharp drop-off in customer complaints

Smooth operations, resulting from positive direction and controls instead of a series of crises

A decrease in the cost of running the facility department.

1. *Loose Supervision*
Problem: Supervision spread too thin. Craftspeople lose time waiting for instructions or decisions. Supervisors lose time due to an inefficient telephone system and an excess of routine paperwork.

Employees receive no advice on job requirements, either before or during job performance.

Solution: Improve the telephone system by going to cell phones, one for each supervisor. Provide clerk(s) and a CMMS to handle routine paperwork. Provide supervisors access to electric one-person personnel carriers to make them mobile. In this manner, supervisors can control every maintenance or repair job of 2 hours or more. There will be time for supervisors to instruct on procedures and immediately reassign craftspeople, as they become available, to other jobs. With more direct supervision, maintenance and repair work will be more effective, and seldom needs to be done over. Craftsperson-controlled idle time and unavoidable craftsperson delays will be substantially reduced.

2. *Slow Service*

Problem: The method of distributing paychecks and providing check cashing service. Because lunch hour for various trades is not uniform, time is lost on many jobs requiring more than one trade.

Solution: Improve check distribution delivery coordination, and relocate and centralize check cashing services. A dollar savings will accrue plus various intangible but equally important benefits with improved employee morale and motivation.

3. *Poor Assignments*

Problem: Craftspeople were working without prints. This meant there was need for oral or sketched instructions for nearly every job. For this reason, craftspeople were not really controlled through the schedule board where written work orders were normally issued. Misunderstandings were common and delays were frequent when a foreman was not on the spot to assign the next job.

Solution: Assign each craftsperson a slot on the schedule board. Supervisors load each slot with several jobs, not just one. Provide written instructions, information on location of materials, supplies, and parts; job time estimates; and a sketch for each job. Craftspeople will no longer have to wait for the next job. Foremen will have more freedom to supervise jobs.

4. *Oiling Failures*

Problem: The problem of repetitive costly repairs that justify specific controls to reduce their frequency. Use of preventive and predictive maintenance is desirable from the cost standpoint.

Solution: For lubrication failures. Have oilers immediately notify a foreman if improperly serviced lube points turn up. Modify lube points to ease servicing. Keep sight gauges clean and in good order. Tag idle machines with a warning that lube service is required before starting the machine to prevent waste of oil and oiling labor in unnecessary applications.

Inspect each machine in operation at least once in each 5 weeks. Known problem equipment should be inspected as often as required. Oilers should be classified as repairers so quickie repairs can be made as observed. This approach, together with improved maintenance procedures, will avert machine damage due to lube failure. Continuity of lube service will be obtained in spite of personnel changes.

5. *Faulty Procedures*

Problem: Take scored ways, for example. They are usually caused by inoperative way wipers, which often were not reported until machine operation was affected. Contributing factors are use of air hoses to blow chips off fixtures (and into machine wear points) and inability to enforce routine machine care by operators.

Solution: Ensure that way wipers are checked at each preventive maintenance inspection. Repair those that are inoperative at once. Remove air hoses that cause machine damage by chip blowing.

6. *Quickie Repairs*

Problem: Five-minute jobs often take 30 minutes by the time make-ready, travel, delay, and put-away time is added.

Solution: If quickie repairs can be done in less than 30 minutes, they should be performed by either preventive maintenance or facility operations inspectors as discovered and charged to the cognizant patrol work order. If the repair time is estimated to be less than 30 minutes and the parts and tools are not available at the time, inspectors should note what is needed while on patrol, then take out the necessary tools and repair parts for use on their next job site visit. Performance of quickie repairs, while on patrol, will substantially reduce make-ready and put-away times for job performance at a later date.

7. *Uncontrolled Labor*

Problem: All the above steps can be taken to improve department efficiency. There is also need for a sound basis for better workerpower management control.

Solution: Budget controls, based on organization-wide productive workerpower estimated requirements, provide only a general measure of the projected facility department workload. These controls should be refined with use of a manning chart. This chart relates personnel needed for performing running repairs to productive workerpower. For example, in a production shop, the chart provides 1 repairer for each 50 operating machines; and 1 repair machinist for each 150 operating machines.

8. *Random Overhauls*

Problem and Solution: Overhauls should be approved by facility management after proper investigation. Job time estimated hours should be added to the department work backlog. Personnel performing overhauls should not be used for running repairs. Requests for overhauls should be disallowed if the capacity is not needed, if replacement is more economical, or if production method changes are projected. The facility manager should control overhaul performance, including setting of priorities.

9. *Unpredictable Work*

Problem: This type of work includes equipment breakdowns, experimental sample production, special orders, etc.—any work requirements that cannot be foreseen. These jobs are normally handled as backlog rather than running repairs.

Solution: The facility manager can have a good measure of the work ahead of the department and the workerpower required to perform the work. All scheduled work that is added is evaluated by management before it is performed.

10. *Costly Patchwork*

Problem: Equipment overhauls are usually based on need for production purposes. Some equipment can be patched at greater long-term cost effectiveness than performing a proper overhaul.

Solution: The facility manager's planning staff can pick up such cases from CMMS data or from craftsperson supervisors and initiate overhaul requests. As part of this program it is advisable to install centralized lube systems and limit switches where required. In such cases, savings can be realized in scrap production reductions, inspection time reductions, increased tool life, operator downtime, and other areas not shown as costs against a machine by standard accounting procedures.

11. *Loose Inventory*

Problem: Lack of formal provisions for storing and recording usable materials, parts, equipment, and supplies once issued, and not returned to stock, for formal inventory control procedures.

Solution: Develop and use procedures for a salvage crib to handle the above items until action can be taken to return to stock, to salvage for reuse, or to dispose of.

12. *Disorderly Buying*

Problem: Facility supervisors not assigned to the organization's purchasing and/or inventory control functions are responsible for ordering materials without sufficient cost data available and for determining which material to stock and whether to make or buy materials.

Materials orders are processed through the organization's general stores, which keep records only for its own purposes. Follow-up of an urgently needed line item required checking through general stores, receiving, and purchasing, an extremely time-consuming process.

Solution: Assign responsibility to the facility department for ordering, stocking, determining whether to make or buy, issuing and costing materials, and keying the data into the CMMS. Ideally, if the material workload warrants, create a facility department stockroom, located in the facility department area, run by the facility department following the organization's general storeroom procedures, and staffed by stock clerks assigned by the organization's general stores. However, the facility department stockroom should report to the facility manager for guidance and direction.

13. *Unrecorded Costs*

Problem: Material usage costs are not being recorded in the appropriate CMMS modules databases. Usage data was either not being keyed into the database or was being incorrectly keyed in.

Solution: Institute procedures to eliminate these happenings. Do not pursue minor database errors, but if the error is in a single expenditure transaction of more than $10, take corrective actions.

14. *Weak Control*

Problem: Production foremen sometimes would not release a machine that required repair but could still be operated to meet production schedules.

Solution: After instances of severe damage due to this practice, assign such repair or replace decisions to the facility manager. This approach is not necessarily part of the department's management control procedure, but each instance represents a real or potential source of excess cost to the organization. It is thus necessary for facility managers to have these transactions under their control.

Operations and maintenance costs vs. net profits. It is inconclusive to state that an improved CMMS will reduce breakdowns, cut labor costs, and check the expense of materials and parts replacements without having performed realistic cost evaluations over a long time period. It has been shown in past cost evaluations for some for-profit organizations that every dollar saved in efficient facility operations and maintenance work performance translates into 50 cents added to the organization net profits.

Problem: Organizations facing rising outlays for facilities operations and maintenance costs each year must find ways to keep these costs in line.

Solution: The following steps, when performed, will produce the desired cost savings results.

1. Assign facilities operations and maintenance functions to a specific department under a responsible facility manager.

2. Provide the facility manager the authority to back up his or her decisions.

3. Establish a computorized operations and maintenance management system that will generate and produce vital information and data records. This information and data will be needed at a later date to measure the success of the new program.

4. Establish clearly defined and complete work orders, work order scheduling, and job controls for these work categories: routine maintenance, preventive and predictive maintenance, scheduled overhauls, and construction alteration and improvement projects.

5. Calculate operations and maintenance costs and balance them against repair and lost production costs. Set work performance standards for craftspeople (both time and servicing methods), and ensure that supervisors and craftspeople are familiar with the work performance standards.

Cost-reduction areas. Management will never really know if facility cost-reduction programs are paying off until a *before and after* picture is obtained in three major cost-reduction areas. The answers to the following questions will provide the needed information and data:

Labor productivity. Is the same operations and maintenance workforce producing at least 20 percent more work?

Machine breakdowns. Have machine breakdowns been reduced by at least 20 percent?

Inventory control. Has the cost of running out of materials, supplies, and/or parts been reduced by at least 20 percent?

Engineering experience has shown that it is not an unreasonable expectation to achieve, with improved facility management procedures and methods and CMMS use, the following results:

1. Breakdowns reduced by 20 percent

2. Labor costs cut by 20 percent

3. The cost of running out of materials, supplies, and/or parts cut by 40 percent

Improving the organization structure. Facility departments are organized and operated to keep organization facilities in good operating condition. To achieve this goal requires a formal organizational plan, qualified people to staff the organization, and a good organizational location, in the facility, for providing support services. Organizationally speaking, however, what is good for one organization may not be good for another. A standard organization chart is only a guideline, in this instance. The basic principle to follow is to move judiciously into organizing or reorganizing a facility department. Ideas gained from "expert" sources may not apply in all cases.

Experience has shown that the following ground rules in facility department organization, staffing, and location are basic and have applicability in all organizations:

1. An organization needs a facility department based on the organization's size and needs to ensure availability for operational use of structures, systems, and equipment.

2. This is out of necessity, know-how, precedent, or because there is no other logical department to which these responsibilities may be assigned.

3. Dependence on this group has grown with the complexity of structures, systems, and equipment used in modern facilities.

4. Facility operations and maintenance functions exist today as a necessary facet of the whole organization's operations.

5. There is no formula for facility department space allocation other than to use common sense. Sufficient space and centralized locations may be key factors in the location equation.

Organizational Plans

A certain amount of formal organization is essential to gain the benefits of use of a cost-conscious CMMS. The organization designed must be flexible enough to change with the wide variety of work that is often outside the planned work schedule. *Never impose a good facility CMMS on a poorly designed organization.*

Reorganizations accompanied by necessary departmental relayouts inevitably produce the following results:

1. An improvement in the quality of work produced

2. Meeting of schedule commitments

3. Equipment downtime decreased

4. Reduction in overall operations and maintenance costs

Line and staff duties. The use of staff specialists in the facility department must be carefully considered. The need to separate responsibility for control of work input and planning, from control of work performance, requires staff specialists in facilities engineering, work planning and estimating, work order scheduling, etc. These specialists provide and prepare plans, procedures, and work performance standards and also furnish advisory services to line units. Supervision is always assigned to line officials. The sole responsibility for producing results is theirs also.

If there is no separation between line and staff functions, and the responsibility for execution of both line and staff functions rests with the same person(s); then, because of the press of normal business, the staff functions will always slide in favor of line functions. No supervisor can properly execute field supervisory duties if chained to a desk by paperwork.

For these reasons facility departments of sufficient size (medium or large) require assignment of a staff engineering and work control group reporting to the facility manager. Line personnel do not have the time to investigate and analyze the problems they are confronted with. This group should also question the entire facility operation—policies, organization, methods, skills and training, equipment design, etc., thus adding to the effectiveness of the departmental program.

Typical staff functions.   Functions that should be made the responsibility of the staff engineering and work control group, to provide relief to line supervisors, are:

1. *Facilities and tools*—determination of the best equipment and tools to be used and where they should be placed; location and use of such facilities as area shops, tool cribs, materials storerooms, and equipment storerooms.

2. *Systems and procedures*—introduction of operations and maintenance management systems to include writing the procedures, developing the necessary forms and reports, analyzing the reports, and conducting economic analysis of operations in such vital areas as lubrication, preventive and predictive maintenance, work order scheduling, work order planning and estimating, operational inspections, material handling, material ordering and storage, etc.

3. *Engineering support*—provision of design, development, drafting, specification writing, and engineering evaluations in such vital areas as electrical, and instrumentation, maintenance, repair, and construction; HVAC production, operation, and distribution; structural maintenance, repair and construction; mechanical maintenance, repair, and construction; and mobile equipment operation and maintenance.

4. *Planning support*—provision of job planning and estimating of labor and material requirements; job scheduling; reports preparation, including data collection in such areas as downtime, production efficiency, job costs, etc.; and material ordering and expediting.

5. *Cost-reduction support*—industrial-engineering-type studies designed to reduce costs and improve efficiency made in different subject areas, such as ratio delay studies, material substitutions, refuse collection, grounds maintenance, painting, job materials, custodial, etc.

Line and staff cooperation.   Conflicts arise between line and staff personnel when engineering studies made by staff personnel are arbitrarily rejected by line supervisors. This is costly in terms of wasted workerpower. Steps can be taken to minimize frictions by keeping line supervisors abreast of all developments in studies that affect them. Management should insist on solid and proved reasons for rejecting staff studies recommendations.

A method to ensure success in this regard is shown in the following example. A staff study to be worthwhile must show ways to be administered in the line organization. For example, a procedure for group replacement of light bulbs could be carried to various stages of completion before it is turned over to the line for execution. It could be as simple as a final report that analyzes the economics of the problem and recommends a replacement period (6 months, 1 year, etc.). But if the study goes only this far, execution probably will not go much further.

The final staff product should include a detailed procedure, i.e., a schedule of light bulbs replacement by buildings and a workerpower schedule that can be turned over to the scheduling group. If special bulbs are to be used, the staff engineer should determine quantities needed and arrange for delivery to stock. Proper storeroom code numbers can then be given to the line organization. The procedures should also include a report of work completed as compared to work scheduled.

Follow-up corrective action.   Before and after controls are essential to good management in any facility department. Planning support requires constant follow-up during the progress of a job and, if the allowed time has not been met, corrective action taken, after the job has been completed. This can only come about through accurate and timely reports preparation, analysis, and evaluation; and use of the factual data by facility department management and supervisory personnel.

Engineering and cost-reduction studies show that follow-up is extremely important to ensure that savings generated in one area is not dissipated in another area. To accomplish this objective, all cost-reduction projects should be summarized in a report to the facility manager presenting actual and estimated savings. With this data the facility manager can compare dollar savings to the number of craftspeople and then assign these craftspeople to other work or let attrition take its course.

The cognizant staff facilities engineer should place a "bring-up" on the study for follow-up after a reasonable time period. Normally after 1 years' operation, the facility engineer should review the study to see if it is working as planned or if further improvement is required.

Departmental size. *Maintenance Craftspeople*—one method for determining the number of direct personnel required, exclusive of supervision, indirect, and staff employees, is to base personnel requirements on today's replacement cost of systems, equipment, and buildings. Then find the percentage of this cost for maintenance labor that will be a realistic evaluation.

Use the following formula to consider the replacement cost of all systems, equipment, and buildings [current facility value (CFV), which is furnished by the accounting department]:

N = (replacement cost of systems and equipment in dollars X .11*
+ replacement cost of buildings in dollars X .023**) (.6***)/
average annual cost, in dollars, per maintenance craftsperson

N = number of maintenance craftspeople required for primary
maintenance functions

*.11 = 11 percent, which is the midpoint of the range of 7 to 15 percent of the replacement cost of systems and equipment, considered the appropriate range to cover any industry's needs. Selection of the correct percentage depends on conditions existing in the facility.

**.023 = 2.3 percent, which is the midpoint of the range of 1 to 3 percent of the replacement cost of buildings, considered the appropriate range to cover any industry's needs. Selection of the correct percentage depends on conditions existing in the facility.

***.6 = 60 percent, which is the high end of the range of 30 to 60 percent of the total maintenance cost, which normally represents the labor portion of the maintenance dollar to cover any industry's needs. Selection of the correct percentage depends on conditions existing in the facility.

N +.2N = total number of maintenance craftspeople required for both primary and secondary maintenance functions, plus an allowance for additional laborer-type personnel, as well as craftspeople required for alterations and improvements and minor new construction work.

.2N = 20 percent, which is the high end of the range of 10 to 20 percent of the total primary maintenance function workforce, which normally represents the additional labor required for secondary functions—laborers, and alterations and minor new construction craftspeople. Selection of the correct percentage depends on conditions existing in the facility.

These calculations will give the base target, or planning, figure that can be utilized in budget preparations or in decision making regarding the necessity of adding to or subtracting from the workforce.

Guideline calculations for determining proper staffing requirements are:

If $N + .2N$ is over 10 percent greater than the current staffing totals, then it is safe to assume that overstaffing exists and that maintenance resources are being dissipated in excess craftsperson lost time.

If $N + .2N$ is over 10 percent less than the current staffing totals, then it is safe to assume that there is understaffing and that a "breakdown" maintenance condition exists which is characterized by excessive overtime charges and the repair of an item only on breakdown.

Supervisory personnel. The number of shops and the number of personnel being directed will determine the levels and numbers of supervisors needed. This will automatically establish an organization chart in line with the principles of optimum span-of-control.

It is best to expand the span-of-control of supervisors to the maximum they can effectively handle, thus reducing the number of levels and the quantity of supervisors. The density of craftspersons per supervisor should run 1 to 10 to 15 at the first-line level, depending upon skills encountered and geographic dispersion of job sites.

In actual practice, this provides for closer relationships between supervisors and their employees. In the communication area, it results in closer coordination and more control since it shortens the length of communication lines. The major problem facing the facility manager at this point is when to delegate responsibility and when to retain it. Supervisors should look for opportunities to make maximum use of their lead craftspersons when delegating responsibility for various functional duties. Facility managers, on the other hand, should not add assistants unless such assistants will free them for more valuable duties.

In the case of either the first-line supervisor or the facility manager, the addition of assistants adds another layer of supervision and lengthens the lines of communication. The outstanding organization will always ask: "Is it really worthwhile for supervisors to increase their span-of-control, or would it be better for them to reduce it?"

Management by Objectives

Introduction. In the following sections, budgeting, costs, and related items are discussed. Step-by-step methods are presented for applying industrial engineering principles to a facility department to improve operational efficiency and effectiveness.

Definition of management by objectives. Almost every facility manager today has had academic exposure and field experience in the art of management. To be a good manager one must be able to *plan, organize, direct,* and *control,* and one must be able to deal with *people.*

Buildings, machinery, systems, and equipment need little or no direction or control, but people do. People want and need help in obtaining objectives. Thus the term "management by objectives" refers to people and dollars.

Planning, step 1, involves formulation. It answers the questions, What is to be done? How? When and who? Planning is a preliminary activity where one can formulate ideas, goals, and objectives. It is the first step a good manager must always take. Without good planning most goals and objectives are not met.

In the planning phase, the question must be asked: Who is involved in planning? The person involved, usually, was the "boss," the person in each organization who established goals and objectives for the organization. Today planning is a team effort, accomplished not just by one person but by the whole management team, each adding functional expertise to the overall plan. What is the purpose, the goal and objective? How is this accomplished? Planning involves "brainstorming" by the team. Written goals and objectives are the results of good planning.

Organizing is step 2: Organizing involves utilization of workers, activities, and skills. It also involves finding and delivering the right tools or equipment. An excellent tool for the application of step 2 is the organizational chart. An organizational chart should show the direct and indirect lines of responsibility and communications. Each job involved should have its place on the organization chart.

Definite lines of communications and responsibility are established. Who reports to whom? Who is responsible for what? The organizational chart should include every job within the activity, project, or department, from the highest to the lowest. Laborers, semiskilled and skilled, workers, supervisors, and managers should each have their place.

Each position designated on the organization chart requires an approved, published job description. Another good step is preparation of actual job assignment sheets. These are detailed, step-by-step, descriptions of daily duties for each job. You may also want to include approximate allowed standard work performance times for each job in each departmental function. Allowed work performance standard time data can be used in formulating a job assignment sheet with readjustments as necessary to include delays, breaks, lunch, etc. This is a useful tool for good organization design and fulfillment. You may want to include this procedure as the first implementing step.

Implementing, step 3, includes action tasks. Instruction, telling, ordering, and assigning are all included under implementation. A continuous training program for supervisors and workers alike, keeping everyone current on changes and project updating is important. A good manager must be a good teacher. New employees must be able to perform their jobs properly and as quickly as possible to save time and money. New personnel must see how their skills and talents fit into the overall goals and objectives picture. A good indoctrination program shows immediate results by helping employees reduce errors, accidents, frictions, tensions, frustrations, absences, and dissatisfactions.

Supervisors and older employees need periodical retraining. Each individual plays a key and unique role in the implementation program. Time needs to be spent in the classroom as well as in on-the-job training. All employees need constant stimulus.

Evaluation is step 4. Evaluation is the management method used for achieving complete overall control of work performance in the department. "Evaluation" means comparing, evaluating, and analyzing actual performance with standard performance for each task, work order, organizational unit, etc. Quantity and quality output is a method of measuring people and the work they accomplish.

To measure an individual's performance one must have incorporated, in the original planning phase, predetermined allowed work performance standards—both quantity and quality. Allowed time standards required to complete a given task or a group of tasks should be spelled out in the overall goals and objectives statement. Allowed time standards are now readily available for almost any procedure. Someone assigned the task of overhauling an engine should be able to complete the job within a certain number of allowed worker-hours, according to established allowed time standards. If the job is not completed within the standard time, management needs to know why. This is a very important part of the overall objective.

Quality is another part of the evaluation process. Good quality control procedures will eliminate rework, redoing what has already been done, and the expenditure of more time and money. Here again work performance quality standards, established by past performance and facilities engineering determinations, can be utilized as evidence of good quality or poor quality of work.

Practical application. It should be understood that applications of the management by objective theory and practices must follow a step-by-step procedure. But like all realistic endeavors, the procedures must be flexible enough to be practical. People change, things change, aims and goals change; so too management programs and interests are subject to change. When the goals and objectives are changed, the end result also changes. But the step-by-step procedure for attaining the result stays the same.

Long- and short-range objectives. Keep in mind, though, that management by objective theory can be applied to almost any given objective or group of objectives. However, every member of the organization team must be included.

Overall goals are established by coordinators. This group should always include the key company executive. This is the individual who establishes the overall organization goals and objectives, in concert with the team. A good group of facility department overall goals and objectives could be the following:

To provide better care and services to our customers through

- Better quality control
- Better performance

To execute a 5 percent overall cost reduction

To reduce employment turnover by 10 percent

These are goals, or objectives established by the organization's management team. The goals must be realistic and attainable. Never set goals that are impossible to accomplish. This will defeat the process attainments in the long run. Each member of the management team must keep the overall goals and objectives in mind at all times, during each step. Achievement of the goals and objectives, using the process, is the target. A definite time limit should be established for achieving the goal and objective. It could be a month, 6 months, 1 year, 5 years, or whatever is felt is a realistic time limit.

The time limit is another process step. There should be pauses to allow the team to evaluate the results. Most goals or objectives are best accomplished within a 1-year period, from January 1 to December 31 or July 1 to June 30. Anything less than a year should be classified as short-term objectives. Long-range objectives are those involving 1- to 5-year plans.

Test Case 1

The following is an example of a 1-year plan for implementing a preventive maintenance program for all facility physical assets to maximize facility efficiency and minimize overall maintenance costs. Here the overall objective is to comply with the administrative goals of:

Providing better care and services

Executing an overall cost reduction

The program planning. The goal is to develop and apply a preventive maintenance program for the entire facility. The total task will be divided into smaller, more easily handled elements. In this way, more immediate results will be obtained and can be evaluated. The following are brief descriptions of the project phases arranged in priority order:

Phase 1: Develop and implement a program to assure proper lubrication of machinery, systems, and equipment.

Phase 2: Develop a program for the adequate maintenance of clean rooms and dust-free areas.

Phase 3: Develop a program for the routine calibration of facility equipment, systems, and instrumentation.

Phase 4: Develop a program for the routine maintenance of process equipment.

Phase 5: Develop a program for the routine maintenance of machinery, systems, and equipment.

Phase 6: Develop a program for the routine maintenance of inspection equipment; i.e., Rockwell hardness testers, comparators, etc.

Phase 7: Develop a program for the routine maintenance of facility power and HVAC systems.

Phase 8: Develop a program for the routine maintenance of buildings and building installations.

Timing.   Complete within 1 year. It is recognized that all projects cannot be delayed to coincide with a schedule, so stopgap measures can be used until a total program is developed.

Organizing.   Each phase should be divided into organization steps. For example, the first phase might be divided as shown.

Phase 1: Develop and implement a program to assure the proper lubrication of machinery, systems, and equipment.

Determine lubrication requirements based on analysis of manufacturers' recommendations and local experience.

Investigate and evaluate lubrication methods and equipment for possible job improvements.

Evaluate lubrication control systems such as Texaco Stoploss, Mobil Spectrol, and other systems in use in the industry. Develop a work control system.

Evaluate methods for scheduling machine lubrication such as calendar period, machine run time, oil sampling, etc. Develop a system for work scheduling.

Prepare a written machine lubrication program including controls, methods, routings, and procedures, or manuals as required.

Estimated worker-hours to complete the phase: 172. Time span to complete the phase: 14.5 weeks.

Implementing. Each phase should be implemented separately. Each step should stand on its own merit. At this point actual performance can be measured and results indicated.

Phase 1: Lubrications of machinery, systems, and equipment. Periodic reports of progress.

Lubrication standards within the divisions are to be compatible with the corporate system.

Evaluation of lubrication requirements, and permanent lubrication processes, inspection of recommended checkpoints, and itemizing lubrication requirements for approximately 400 machines located in various divisions.

Permanent lubrication and inspection recommended checkpoints are for machines in the production building during the next month. Permanent maintenance lubrication record cares for machine tools in the R&D building are being developed at the present time. Balance of lubrication requirements for miscellaneous machine tools and equipment throughout the plant will follow.

Daily routing for oiling and equipment checks for machines in the production building within column area A is completed. Oiling routes have been established but because of various rearrangements need to be updated.

Development of decals to be placed at all lubrication points on machines denoting daily, weekly type of lubricant, etc., has been completed and request for quotes has been issued to purchasing.

Evaluation. Again, for each phase there should be periodic evaluation reports issued after a number of months have elapsed covering such items as shown:

Estimated completion of development and implementation of the entire preventive maintenance program detailed above is January 31.

Estimated total additional workerpower required to completely implement this program is:

1. Five additional maintenance mechanics

2. Two additional electricians

3. One additional oiler

4. One additional plumber

Two additional mechanics could be utilized now in area 8 to allow completion of preventive maintenance work on equipment in the production building, column area A, as scheduled. One of these mechanics could be made available from column area B after the machine tool rebuilder trainee is on-board. The other mechanic could be picked up as a replacement for the mechanic being terminated out of column area B. The charts showing progress in the various project areas listed above will be completed and ready for use by August.

As can be seen by the above information, things are not looking too good. Instead of saving money, the project is costing time and money with no payback. Again, project completion must be forecast. Common sense and experience tells us that in the long run, with proper preventive maintenance task applications, downtime for machinery, systems, and equipment can be reduced. This will undoubtedly save the organization time and money.

A final evaluation. By the end of a 14-month period all 8 phases in this sample case had been completed. A detailed analysis of the results was then undertaken. All aspects of the detailed analysis will not be covered, but a part will show a good example of what was accomplished:

Ten machines were selected for this study. A 2-year study of downtime for the 10 machines was made. It included breakdown repair time and preventive maintenance time. Total downtime was 25.5 percent. These results were recorded before the new preventive maintenance plan was put into effect.

Two years after the preventive maintenance plan was placed in effect, four more machines were added. Total downtime was now 10.8 percent.

Also, the maintenance staff was reduced from 17 to 12 workers. Total annual cost savings amounted to approximately $100,000.

Summary. This section presented general information on management by objectives theory, practices, and procedures. It should be repeated that this is an overall team effort, from top management down to individual workers who perform assigned tasks.

Planning begins at the administrative level. The establishment of goals and objectives at this level must be broad in scope and realistic in establishment. They must encompass the entire organization. Each member of the management team is then required to establish specific ways of accomplishing the overall goals or objectives. A good communication system up and down the chain of command is an absolute must. Each department and each division must be a part of the overall plan. Regular conferences need to be held and exact methods of reporting progress should be established on a daily, weekly, and monthly basis. Constant evaluation of quality and performance will show progress toward the goals and objectives. Department heads and supervisory levels must be free to change direction when needed, yet always keeping objectives in mind.

Training

Why a training budget is needed. Because technology is rapidly becoming more complex, the importance of training to better facility management cannot be overemphasized; everyone in industry and government is affected by technological advances. During the last 30 years, production in the United States has doubled, while the number of production workers has remained almost the same. During the same period, the number of facility department employees has increased by nearly 50 percent. These statistics reflect the fact that highly productive automated systems and equipment require fewer production personnel but *more maintenance* personnel. Because the end result of this trend is higher production and lower costs, it is certain to continue.

Often, semiskilled and unskilled people must be hired. To be effectively trained, they should be tested and placed in a program that will waste no time in preparing them for the highest level of skill they are able to achieve.

Because of the shortage of skilled craftspeople, a suitable training program is needed to fulfill present as well as future workerpower requirements. Training too frequently takes a back seat to other functions such as production or marketing. Such an attitude can hurt workerpower effectiveness as well as company profits.

Training is necessary to keep craft personnel and first-line supervisors abreast of technical advances continually made in facility systems, equipment, and processes and to constantly update the knowledge of the older, more experienced workers.

The spending of time, effort, and money on operations and maintenance training is justifiable for the following reasons:

Employees are unable to perform certain jobs.

They are unable to do the jobs well enough.

They are doing a job wrong.

With proper training, these conditions can be corrected and the organization will accrue either additional profits or additional production output.

Training can be a cure as well as a preventive to a job performance problem. Used as a preventive—before an employee starts to do a job wrong—training can save time and costly errors. Used as a cure—after mistakes become obvious—training can only get the job back to normal, not make up for the lost profits and time.

If mistakes are being made on the job or if the work is not being done fast enough, training can become a cure—provided that the deficiency was a lack of training but *not* a cure-all. First determine what the problem really is. Then take a comprehensive look at the task to find the weak spots and determine exactly what the job requires.

Some interesting things may appear when you approach the problem realistically. For example, a job may have been set up inefficiently. In this case, an analysis may show that it is impossible to perform the specified operations properly. Methods improvement is needed here rather than training.

Looking closely at a job may also reveal that work operations are inefficiently laid out. If the employees have been properly trained and the task still is not being performed acceptably, the problem could be caused by fatigue, poor attitudes, or some other seemingly unrelated reason which had not been recognized. In this case, job training is *not* the answer, although other training may be.

Training responsibility. Some facility managers may not think of training as an essential part of their jobs. Some may insist that the training department should handle all training. However, training is the responsibility of all facility managerial and supervisory personnel. It is essential that employees be trained in all aspects of their craft. Training should be an integral part of job enrichment within the department and company.

Training requires facility department management, including supervisors, to take a close look at the job and the employee. It is necessary to define what it is the employee should be doing and to determine whether he or she is doing it properly. It is equally important to determine why a particular employee is not performing as well as he or she can, so that the employee can be made aware of any deficiencies or problems.

The development and promotion of an effective training plan is especially dependent on a managerial climate that encourages employees to seek training. This climate can be created through sound managerial techniques. Management support must be effectively continued and not relegated to just lip service with no real support or follow-up.

In all situations, especially training, let your employees know that they are important members of your team and that their jobs are vital to the successful operation of the department. Make them aware that training not only gives them a chance to learn about new equipment and techniques but also can qualify them for more responsible positions. Training can also help you show your employees that you are interested in them personally. People need to feel important and wanted, individually as well as collectively. Nothing will do more for a worker's morale than your recognition of him or her as a person.

Determining training needs. Questions that may arise about your training needs and training programs follow:

What are the results of your present operations and maintenance training or retraining programs?

Do they accomplish their objectives?

Are employees able to perform specific job requirements after training has been completed?

Are the trainers able to relate textbook knowledge to the skills and know-how of practical "nuts and bolts" jobs?

A *training needs analysis* can be made and used as an appraisal technique by which a training supervisor can determine the present scope of individual skills and knowledge requirements. This will help answer the questions relating to what training is needed. Training needs may be determined by determining what is going on now and matching this against what should go on, now or in the future. The gap, if any, gives clues to the kind as well as the amount of training needed.

When analyzing a problem for training purposes, the thinking and suggestions of others can be helpful and can increase the value of the eventual solution. However, weigh all ideas carefully, for training may not be the best solution in a given situation, even though some of the people involved may feel it is.

Clues to training needs can also come from an analysis of the behavior of individuals or groups. Chronic absence, spoilage of work, carelessness, accidents, irritability, resistance, or resentment toward instruction are symptoms of conditions that may call for corrective action involving training. For example, a manager may need to be a better planner or communicator, or a group may need to know more about a policy. The department policies, or job standards, may not be clearly understood. In many cases, the standards for job performance can be written in clear terms, lending themselves to discrete measurement. The standards for management jobs are often expressed as the goals or objectives of the organizational unit for which a manager has responsibility.

The training needs of new workers, and the benefits present craftspeople can get from additional training, should be determined before a training program is developed. A training needs analysis can be of great value in making this determination.

The scope and level of knowledge a training program should have will be determined by the difficulty of the tasks to be learned. Training is most valid when the knowledge and skills taught are required and will be practiced on the job. The equipment and facilities to be maintained will help you determine the needs of your training program. Some basic steps for instituting a good training program are as follows:

Accurately determine job requirements.

Accurately determine skills requirements.

Ensure that training materials are at the trainees' level of comprehension.

Use training materials that can be easily adapted to the needs of the organization.

Maintain good supervision, personal contact, and follow-up with trainees.

Institute good environmental conditions that encourage learning.

A successful training program depends on the trainer's and the employees' knowledge of the job. Usually, the supervisor has the background to recognize the need for training and its value to the company while also understanding the capabilities and limitations of the workers. Often, only the supervisor can determine whether the workers will understand a subject that is taught a certain way.

The qualities of a good trainer are difficult to define; they may include such characteristics as the ability to like people, to be enthusiastic, and to speak distinctly. However, each of us need only think back to our schooldays to remember someone who violated all the rules and was still a good trainer.

A trainer is expected to produce to be acceptable, and the trainer must be held accountable.

Too often, employees are entrusted to an unqualified trainer, and then their performance on the job is evaluated as though they had had professional training.

Once the responsibility of training is accepted, the trainer should consider the best way to instruct the trainees. This involves learning about visual aids and communications and knowing how to generate involvement, motivation, interest, and enthusiasm.

Determining training objectives.   When training is needed, a *job analysis* can be quite valuable. A job analysis can help you determine exactly what constitutes a good job, what an employee should be able to do, what parts of the job are not being done properly, and where the training emphasis should be placed. A job analysis will also justify training because it reflects directly on any job deficiency. It can be used to help set realistic objectives for an employee's performance after training has been completed. You can then tell what that employee will be able to operate, maintain, and repair (specific equipment) within a certain time (repair certain equipment or locate defective components). An analysis will also reveal what the employee knows and can do before the training. This is essential, because the training should start where the trainees are, not where they may think they are or where they should be.

If the training starts beyond an employee's ability level, that individual has little chance or no chance of progressing to the goal you have set. The purpose of training is to change behavior or work habits. This means that the supervisor must know what behavior now exists and what behavior or change is desired. Training should not begin until this is determined. Training should be designed to fill the gap between where an employee is and that employee's predetermined goal after training. Every employee will not be at the same level and should not be expected to end up at the same place. This is why some training methods can be formal and other training methods informal or on-the-job procedures.

Studies have shown that while they are demonstrating what they have learned, people remember approximately 90 percent of what they say. But they will remember only 10 percent of what they have learned if they do not apply the information (by repeating it or using it) immediately or very soon after the training sessions. This means that the trainee must participate by saying and doing whatever he or she should have learned.

The objectives of an in-house operations and maintenance training program are to:

Train selected personnel in the knowledge and performance of tasks required for them to progress from trainee to craftsperson or operator.

Provide a means by which they can refresh or update present knowledge.

Provide steps for continued development of all operators and craftspeople.

Provide a climate that encourages individual effort so that knowledge can be most effectively gained.

Provide sufficient flexibility so that training will not interfere with department work needs.

Provide planned work assignments that meet experience capabilities (on-the-job training).

Provide scheduled participation and study time in the training program.

Provide help and support for overall trainee achievement.

Factors important to trainee achievement follow:

A skills outline and a knowledge reference list should be developed and used to help trainees direct their efforts to becoming competent craftspeople and operators. A skills outline can also serve as a guide for the trainee to determine the knowledge and skills he or she must improve. This can also help motivate a trainee to have a stronger desire for improvement.

Specialized training conducted by factory representatives or suppliers can be used very effectively to complement your training program. Supplemental reference texts and manuals may be made available to provide information on specific equipment operation and repair and maintenance instructions. Manufacturers' instruction manuals on special equipment should be available and kept updated. Training films or other visual aids are helpful and can be shown during supervised study sessions as needs and availability warrant. Information sheets or pamphlets should be written and distributed to strengthen knowledge and provide specific instruction about the operation, maintenance, or repair of facility equipment and systems most applicable to the trainees' job responsibilities.

Scheduling training. Training in the facility department should be scheduled at the same time a new machine, new employee, or new job assignment is planned. Any new person working on an existing machine or assignment will need some training. An experienced employee on a new machine or new job will require training in his or her new responsibilities and environment. Training is simply insurance against making errors at the start and can save costly downtime and repairs.

When new equipment is installed, employees need to know its operations and maintenance requirements. The complexity of the equipment and the number of people to be trained will obviously determine the extent of preparation, cost, and mode of training. The quality of a training program should never suffer because of relatively simple operations or too few trainees. The machine, the company, and the technicians deserve adequate training if they are to perform effectively. "Telling" and "training" are, unfortunately, often considered synonymous. This can cause a hit-or-miss training approach, which is costly and will result in inferior results. Assess the problem situation and at the same time seek out ideas for solutions that can be achieved through training.

You may find three levels of employees involved in the problem, and interviews should be conducted with the more knowledgeable and experienced individuals from all three levels. The personnel involved follow:

Those who supervise the people for whom the training is being scheduled

Those for whom the training is being scheduled

Those supervised by the people for whom the training is being programmed

The amount of training necessary depends on the present level of craft and operator skills in the various shops.

Sophisticated equipment requires updated operations, repair, and maintenance techniques. This means educating existing personnel in advanced methods of operations, maintenance, and mechanical repairs.

Specific items that need to be determined before a training schedule can be established are:

The training group size

The frequency of the training sessions

The length of the training sessions

The time of day for the training

The place scheduled for training sessions

Group size.   How many trainees will be enrolled in the total program? How many separate sessions will be required to train all the trainees enrolled? Usually 12 to 18 trainees is considered ideal for each training session.

Frequency of training.   How often are the instruction sessions to be held? Experience has shown that too much training in too short a time is not effective. Additional time is often required of the trainee outside working hours to complete homework assignments. With this in mind, it is a good idea to schedule at least one but probably no more than two sessions per week.

Length of sessions.   The exact schedule depends on local conditions and requirements for getting the job done. When training periods are too short, there tends to be a lack of continuity with too much time lost in starting and stopping the session. On the other hand, the interest span of adults who have been out of school for some time is quite short. If the sessions run for too long a time, trainees become bored and much of the training effectiveness is lost.

For most industrial operations and maintenance training situations, the schedule should probably fall within the range of one to two sessions per week for 1 to 2 hours each session. Work not completed during the allotted time period should be assigned as homework to be turned in at the next training session.

The time training is conducted. As near as possible, the sessions should always be conducted during the same time of day except when emergencies disrupt the schedule. The actual time of day is not too critical, except that the hour before lunch or the first hour in the morning are probably best from the standpoint of the trainee's alertness. The period immediately after lunch is probably the worst. Sometimes, training sessions are scheduled during shift changes. Their employees from the first shift stay extra time and employees from the second shift come in early. With this system, both the company and the employees contribute some time to the training program.

Training location. The trainees should know the *training place,* where and exactly when the program is intended to take place, what is expected to be accomplished, and the date when that particular segment of training is to start and stop.

When developing a training program for new equipment, each facility should determine the details of its own program. However, these basic rules should be followed:

Plans for training should be developed when plans for acquiring the new equipment are made.

Money for training should be included in the cost of installing and converting to new equipment.

Time for training should be allocated when personnel requirements for the new equipment are determined.

Responsibility for training should be specified when other responsibilities pertaining to the equipment are assigned.

Accountability for training should be just as strict as the accountability for equipment operation.

When these rules are followed, training for the operation and maintenance of any new equipment will not present any serious problem. If the rules are not followed, the training will be mediocre.

Costs and the training budget. When organization funds are expended for training, you must know why it is needed. Just as expenditures for mechanical equipment must be justified, so must expenditures for training the workforce be justified.

The use of properly administered pretraining tests can reduce training costs and the time needed to complete training. Trainees can be given pretests to determine the subject areas they know best as well as the extent of their present knowledge.

The training budget must provide sufficient flexibility to allow you to carry out your responsibilities for training. Training is not a science. The approach to budgeting should allow for both known and unknown factors in the daily activities of your department. Thus you will be able to predict with accuracy the number of trainees for any given training program and the resulting costs for each trainee. However, if you decide to change from one type of training to another one heavily supported by visual aids, for example, there should be sufficient room in the budget to provide for the expense of preparing the visual aids without causing a financial crisis.

An accurate forecast of training expense can be made only if there are records from previous years or programs to support decisions made for the coming period. If there are no such records, or if the entire budgeting procedure is new, there never will be a better opportunity to fashion a system for your training budget. Such typical questions as these should be examined slowly and carefully:

Why is training needed?

What are the objectives?

Who will be trained?

When should training take place?

Where should training be done?

How will training be conducted (media)?

Standard budget periods are usually established by the organization. Conformance with the organization's financial structure is essential, whether it follows a quarterly, semiannual, annual, or special pattern. Determining this framework before analyzing training budget requirements will place the training program in proper perspective. Budgeting for training is sound financial practice for the facility manager. Budgets should include funding for special training such as seminars and vendor schools or other off-site training projects.

Sometimes a functional budget, rather than a project budget, will meet your needs for operating your training program more effectively. A functional budget is most applicable when the department is organized on a functional basis. Thus there might be sections devoted to education, training aids, and special training. Each of the functional sections would have budgets for all activities. A budget of this type has some advantages in that it helps decentralize control and audit.

By division of the budget into small units, it also helps localize any problem areas.

Wherever possible, however, it is to the facility manager's credit if every training dollar is followed to its destination. Training is a business now, and as such it takes on the responsibility of supporting itself. Salaries and other costs—administrative, supervisory, training, and clerical—must not be economically out of line with the results of training. Trainees trained in productive assignments must justify the costs. Training films produced or purchased must be evaluated for definite results in terms of better service and increased skills. Training must provide the instruction and knowledge necessary to help employees do the job. Granted, these are intangible areas, but their relation to an organization's profit picture or fiscal soundness is obvious.

Determining training effectiveness.   To check overall effectiveness, periodically make spot checks of the training program to assure that satisfactory progress is being made. Spot-check the work performance of trainees to assure that knowledge gained in the training program is being applied to the job. It is a good idea occasionally to survey the shop supervisors, the trainers, and the trainees themselves to determine whether the training is producing satisfactory results and to spot any trouble areas that may be developing.

It is more difficult to measure learning than it is to measure reaction to a training program. A great deal of work is required in planning the evaluation procedure, in analyzing the data obtained, and in interpreting the results. Wherever possible, it is suggested that training supervisors devise their own methods and techniques. It is relatively easy to plan classroom demonstrations and presentations to measure learning where the program is aimed at the teaching of skills. Where principles and facts are the objectives of the training program, it is advisable to use a written test. Where suitable standardized tests can be found, it is easier to use them. In many programs, however, it is not possible to find a standardized test. Then the trainer must resort to skill and ingenuity in devising a measuring instrument.

A *performance rating sheet* can be used as an appraisal tool to evaluate individual performance with respect to the quantity and quality of work. This management tool can also be used to determine growth. As workers become more skilled and knowledgeable, any change in performance can be recognized and evaluated.

From an evaluation standpoint, it is best to evaluate training programs directly in terms of results desired. There are, however, so many complicating factors that it is extremely difficult, if not impossible, to evaluate certain kinds of programs in terms of results. Therefore, it is recommended that training be evaluated in terms of reaction, learning, and behavior.

This can be accompanied by a job analysis, or write-up, to help determine the degree of skill needed for a particular job. Progress reports on the effectiveness of classroom training and facility exposure can be used to evaluate training techniques.

A training course should be tailored to specific requirements by using special training materials. This would apply to such courses as blueprint reading, equipment operation and repair, and electricity. If a vocational school is near your facility, try to arrange a training program in cooperation with the school. Upon completion of the training program, each trainee should receive recognition in the form of a certificate. This will encourage the trainees to continue their training.

Customer service schools, although somewhat restricted to special crafts or skills, are also helpful for specialized equipment training.

The trainees should be given jobs that have the greatest learning potential. Be sure they are assigned a variety of facility-wide work. Nothing can hurt an employee's initiative more quickly than to be stalled on one type of job in one department because he or she does this job best or the department has a heavy workload.

Workers who are not interested in advancing will usually not apply themselves very effectively in a training program. Occasionally, however, a demonstrated lack of interest may be a sign of poor motivation. The employee may not see any advantage in giving up time for study. There are also some employees who prefer to pass up the opportunity to advance to higher-level jobs because they would rather be "one of the gang" or do not want the added responsibility.

Summary. When an employee does not perform properly but knows how, skills training is not likely to help. A communications problem as to exactly what is to be done could exist between the employee and the supervisor.

Always keep in mind that skill is developed through concentrated training. Training means telling, showing, and supervising. The trainer must have the desire to teach and supervise a trainee. You may want to start the trainee's on-the-job training by assigning him or her to work with the most skilled people in the crew.

All classroom practice, programmed instruction, text, and equipment manual instructions should be supplemented as much as possible with actual hands-on demonstrations for the most effective results.

Equipment models should be used to allow trainees to practice, under the supervision of an experienced trainer, what has been learned. For example, a model could be used to illustrate a coupling alignment problem. A motor coupled to a pump mounted on an adjustable base will permit the trainee to actually align the drive and use such tools as feeler gauges, a dial indicator, and a micrometer. This situation covers text study concerned not only with coupling alignment but with mechanical drives and measuring tools as well.

With a proper balance of study and in-house rotated work experience, trainees should soon become proficient in their particular craft fields. Equally important, the chances are good that they will develop morale and a sense of responsibility. The achievement of such a combination of skill and proper attitude can soon repay your training investment. Improved performance on the job should be the prime objective of any training program.

Staff Development

This section will do four things: (1) assess the "state of the art" in management development as it applies to facility management, (2) identify some of the critical elements in the management climate that must be generated in an organization to make possible the development of its technical people, (3) describe the major barriers or blocks that prevent the development of employees and managers in the organization, and (4) present an administrative blueprint or model for an experiential approach to staff development within the facility department.

The state of the art.   Facility managers who look to the current literature on management development for answers to questions on the development of people are apt to become both disillusioned and confused. They will be disillusioned because they will find little theoretical support for evaluating programs of selection, placement, training, progression, career utilization, termination, or retirement. They may be confused because there is so little consistency in theory or practice among the writers on the subject.

There is accumulating evidence that the greatest waste of human resources lies in the unused potential of technical and other professional employees. Performance of these people is, of course, quite satisfactory when compared with production worker efficiency levels of traditional organizations. However, when we realize what could be their contribution to the organization when highly motivated and given stimulating intellectual leadership which their special abilities and professional backgrounds make possible, the opportunities are exciting!

These technical personnel hold tremendous potential for desperately needed creativity and resourcefulness in the *idea generation* needs of the organization. However, the price tag on the creative utilization of these people is a challenging developmental climate far more sensitive to the requirements of the individual than industry or government has had any experience with in the traditional organization. It would be difficult to overemphasize the highly strategic value of this select group. The industrial world has a vital equity in their contribution. Yet, from the point of view of on-the-job development, they have been the most neglected people in business, industry, and government.

Most companies, even though research in the behavioral sciences has gone on for years, stress the management of their financial and physical resources and pay only sporadic attention to the human side. A survey of the voluminous literature describing industry's struggle to learn how to manage the development of its people reveals very little that facility managers can put to work in their own situations. They will find hypotheses and theories still frozen in the jargon of the behavioral sciences; they will find case studies of companies who have rushed into print to tell the stories of their bright and shining programs before they have been evaluated or tested.

Need to separate the wheat from the chaff. However, if we take care to separate the wheat from the chaff, we will be able to find some useful information that will give us some guidance in designing a development program good enough to do the job intended. With considerable agreement, the literature records the belief that the development of personnel can take place only in a management climate that is strongly supportive of such development.

Before we can understand the nature of a climate which makes possible the development of people, we need a workable knowledge of what we mean by "development." We will need to discriminate between "development" and "growth," "education" "training," and "experience"—all key terms that come into the picture when a development program is being explained. All these terms have a place in the developmental processes. All too often, they are used interchangeably in describing phases of the learning processes that take place in development. This is misleading. It encourages limited approaches to so-called development programs that fall short of the comprehensive thinking needed to achieve real developmental goals.

Need for a common understanding of major concepts. Though any number of activities have a place in a development program, such as *appraisals of performance, human relations seminars, sensitivity training, confidence leadership training, simulation, coaching, and* *rotation,* they are at best only a small part of a comprehensive development program. One of the most costly mistakes an organization can make is to set up an educational or training program and then depend on it to take care of the development of its people. Let us reexamine some of the major terms we find in the language of development and see if we can define them more usefully for the purposes of this section.

Development. This is the overall term that should be used to describe the ongoing processes throughout the employee's lifetime. It includes living a life as well as making a living. It includes the growing-up stages from infancy to adulthood. It includes the maturity goals of the adult—from where the person is right now to where he or she will be in the future. It has to do with ongoing experience, what it does to and for the individual. More specifically, from the facility manager's point of view, it has to do with a person's career in the company. This includes one's capability, competence, job progress, values and ambitions, and potential for grasping new opportunities and taking responsibilities. It concerns one's long-term ability to adjust to the needs of the organization and continue to make the contribution the department requires. And also it concerns the employee's ability to realize his or her own career expectations as a member of the department staff.

Thus we see that development is the comprehensive concept. All other terms that we use are more limited, having to do with specific aspects of development as a learning process.

Growth. In this section we will use the term "growth" as a step in development at any point in time. We will apply the concept of physical growth, so relevant to childhood, to mental growth in the areas of personality, maturity, intellectuality, decision making, and perspective. In general this might be summarized as the translation of the individual's potential into improved performance, the development of a specific ability from a person's general bank of capability.

Education. We will use this term in the traditional sense of schooling. In the minds of most people education is equated to the acquisition of knowledge. It has become the major avenue of learning. For better or worse, our society has institutionalized education and has depended on it to be the primary change agent in equipping children to move into the world of adults. In actual practice, education is mostly thought of as the presentation or discussion of subject matter in formal classroom situations. Educational activities are generally conducted at the verbal level of theory, principles, subject matter, or knowledge. Too often the result is weak, vicarious experience—experience ahead of the gut experience it substitutes for. In most schools the students get knowledge long before they will have occasion to use it. In the meantime they may forget much of it or it may become obsolete. The same thing is true of management seminars, plus the added factor that the language of the seminar is seldom the language of the job situation where it is supposed to be applied.

A classroom with an instructor and a textbook is a good place to *start* a learning process that may lead to development, but it is a poor place to *end* it. To put the matter more bluntly, an organization that expects to have its people develop by sending them away to schools and seminars must inevitably be disappointed. Such educational digressions are not likely to turn poor employees into good ones or to turn even the most promising recruits into good managers. It is not that education has failed us—just that we have blindly expected education to take over the whole developmental process of which it is only a small part. This mistake has taken place because we know a lot more about education than we do about development.

Training. In its relation to education, most people think of training as a more "plebeian" process than education. Universities have high status, while training institutions have much less. Education has to do with theory; training has to do with practice. Falsely perceived, education has to do with the "head" while training has to do merely with the "hands." Training concerns itself with improving skills, habits, and routines. Industry and government uses it as the mechanism to improve performance. In fact, training—with its wide variety of sophisticated techniques (role playing, case study, audiovisuals, simulation, to mention a few of the common ones)—has been so useful in doing the job of improving performance that companies are expecting it to do more than it is capable of doing. Like education, training can do only a part of the development job. Calling a training program a development program does not make it one. Experience in this regard supports these conclusions: the kind of supervisory leadership a person works under which will importantly affect his or her training techniques are at their best when jobs are relatively stable, know-how is constant, and the status quo can be maintained. However, a word of caution is in order. In a rapidly changing technology, training as we now conduct it may bring undesirable results, such as resistance to change.

Experience. The role of experience in the developmental sequence has usually been misunderstood in industry. Consequently it has been greatly misused. In the narrow sense, experience is simply doing a job. An old German adage had it that "you can steal a trade with your eyes." Literally, the job being done is what an observer can see the worker doing. Industrial engineering has tended to measure experience by the amount of work being done accurately in a given length of time. The most frequently heard statement an individual makes about his or her experience is "I have had x years of experience on this job." For our purpose in this section, we need to redefine "experience" in terms of its role in the learning processes that contribute to development. In this broader sense, real experience is what the individual learns from doing a job.

Really, what do people learn? Do they learn to just get by, to ignore or get away with mistakes, to hate their jobs, to resent their bosses, dislike their organizations, and perhaps a host of other negative reactions? Or do they learn to improve their knowledge and skills, to use their jobs as stepping-stones to better ones, to feel the importance of what they are doing, to appreciate their own contributions to the organization's success? In short, are these people learning the things that make them outstanding employees and also qualify them for the advancement of their careers?

This perception of experience places it as the last link in a learning chain that finally results in development. We might not be far wrong if we thought of education as *eating,* training as *digestion,* and experience as *assimilating.* At the point of assimilation, food becomes new protoplasm. In this chain, what began as new knowledge and information ends as new understandings, skills, attitudes, and abilities. The individual has grown in competence and has taken another step in his or her ongoing development.

All too often, organizations have had good educational programs, good training programs, and good systems for administering experience without ever seeing their relation to each other or to the development of their personnel. Through a lack of management of these programs as interrelated parts of a whole, an organization may well be spending the money and the time needed for development and get no development!

The behavioral scientists have a new term to describe the results of this developmental process. It is "behavior modification." The organization's management climate must be good enough to produce favorable behavior modification in individual employees sufficient to motivate them to achieve their realistic life goals and to make their work contributions productive and profitable throughout their working careers.

The management climate.   The top management team in an organization must realize that the development of brain power in the technical and managerial levels of the company is a *must.* There can be no compromise with complacency. Development cannot, at this late date, be thought of by the top management team as a luxury, a fringe activity, an on-again-off-again proposition. Nor can it be thought of as a program that can be initiated at the top and then turned over to lower groups in the chain of command to implement. The responsibility for creating and maintaining a management climate strongly supportive of employee development is the top management team's continuing and never-to-be-relinquished *first* order of business. The future competitiveness and profitability of the organization depend on it.

Responsibility of top management. A top management climate study in a middle-sized company showed strong evidence that when a company's top management group consciously organized its developmental resources into a systematic plan for the development and training of its personnel, (1) it measurably increased the number of its promotable managers in the developmental progress; (2) an employee-centered executive climate was more conducive to growth and development of managers at lower levels than a work-centered executive climate; (3) a climate that provided opportunities to utilize newly developed skills was more conducive to continuing growth than a static climate not providing such utilization; (4) managers whose subordinates are developing will themselves be more apt to develop; and (5) the perception of a subordinate as a high performer on his or her present job will increase that subordinate's likelihood of receiving a promotion.

The study further indicated that more acceleration in the development of managers could be obtained: as the organization (1) obtained greater integration of its staff support and line training activities; (2) educated both the individual manager and the chain of leadership above that manager to realize that development is a team effort with reciprocal relationships; and (3) brought more clearly into focus throughout the management structure an awareness that the goals and objectives of the individual for his or her own development and progress are mutually supporting and compatible with the goals and objectives of the organization.

The most significant change top management can make in its understanding of employees is the realization that they must be managed in terms of their *potential* as well as their achievement. For most executives, this calls for new insights into the wellsprings of human behavior. It calls for a deeper understanding of human nature than management has previously felt to be necessary in order to get the work out.

Even a brief review of some of the characteristics of a development climate will show the central role of potential in the growth processes. The following list highlights some of the characteristics of the growth climate as perceived by supervisors from middle-sized organizations:

The most important characteristics of the growth climate. The company is a profitable company, consistently earning a high rate of return on investment.

The company is an expanding company, constantly increasing its share of the business in the industry.

There is a systematic planned approach to the development of people (not hit or miss).

People who do outstanding work are identified, are religiously looked over, and are considered for appropriate promotional vacancies.

There is a fundamental faith that the potential of people exceeds their achievement by enough margin to make their training and development a profitable undertaking.

The subordinate is being supervised by a boss who is also growing in maturity and considers a supervisor's major responsibility to be that of getting the job done through people who can grow.

The accountability of managers is such that they will not be promoted themselves unless they are successfully training the personnel under them to meet the demands of group maintenance, including the demand for their own replacement.

The basic learning experiences are provided through the work experience by means of good delegation, guided experience, and interpreted experience.

- High demands are placed on people, and much is expected of them.

- Supervision capitalizes on the mistakes of subordinates and uses them as growth points.

- Security is earned through competence rather than seniority.

- Leadership puts a very high premium on behavior that shows originality, resourcefulness, creativity, innovation, suggestions, and uninhibited comment.

- Training is primarily beamed on job improvement and preparation for advancement.

- The subordinate is given a maximum opportunity to share in decision making, planning, and problem solving.

- Individuals know where they stand—how they are doing—and what else is required of them to progress.

- The individual is not exploited or deprived of the rewards and recognition of increased productivity.

It should be noted that the list taken as a whole reveals a strong supervisory faith that employees grow and mature through their work experience.

Need for organizational flexibility. Organizational psychology is just beginning to probe the effect of organizational structure on the performance and contribution of employees. As the work of the world becomes more sophisticated—more mental and less physical in nature—older, rigid patterns of structure have become a brake on employee development. An age-old question reasserts itself: Was humanity made for the Sabbath or was the Sabbath made for humanity? Should an organization set up a logical or "ideal" organization and then seek readymade people to fill its blocks? Or should an organization find good workers and build an organization around them, a structure flexible enough to support their growing abilities? For any number of compelling reasons, the answer must be the latter. The traditional organization is bogged down with an inherited line-staff organization heavily practiced in downward communication, pressure, and control. It has paid scant attention to upward communication except during crises, and then only until things got back to normal. New ideas have usually come from the outside. The feeling seems to be that it is better not to have any ideas coming from the ranks than to have to cope with an embarrassing one once in a while! This traditional pattern of organization was good enough to get the work out when that work was engineered to be routine, repetitive, and geared to the mass production of interchangeable parts. With automation taking over more of the jobs of this dimension, management must find more flexible ways to organize people into specialized work teams.

The development of technical people can best take place within an organizational structure that will fully support and stimulate competence, opportunity, innovation, experiment—in short, a rich learning environment. People who are growing will always break out of organization blocks and stretch the limits of job descriptions. A flexible organization will help them do it.

Key role of the supervisor for staff development. The role of the supervisor takes on new dimensions when his or her organization begins to take on the responsibility for the development of its technical and professional personnel. In general, growing professional and technical employees place higher demands on their organization and expect more from their leadership than do nonprofessional employees. This is especially true of facility department employees. They also expect more specialized services and support for their endeavors. These increased demands and higher expectations may take shape in ways such as the following:

Much more of the communication between the supervisor and professional and technical employee must be at the intellectual level rather than at the administrative level.

The level of motivation needs to be much higher to achieve creativity and *intrinsic* job satisfaction than would be necessary for productivity in more routine types of work.

- Activities must be perceived at meaningful and reasonable levels in terms of goals and purposes. Professional and technical employees expect competence in leadership, competence well above the level of conclusion and drift. They will not accept with docility, improvisation and drift.

- The supervisor must take the initiative in meeting the needs of professional and technical employees. For example, a competent employee with a salary raise long overdue would be more apt to quit on a pretext and get another job than he or she would be to ask for a salary increase.

- Professional and technical employees expect to have their own views and insights fully considered in any evaluation or review of their work performance.

- The administration of the professional and technical work group must be service-oriented. The supervisor is expected to provide the resources and tools needed for competence at the expert level. Conversely, the supervisor is expected to remove any administrative or red-tape barriers that get in the way of top-notch performance.

- The professional and technical employee expects that the standards and ground rules that bind the work group members together and control the intergroup relationships in the organization will be compatible with professional ethics and moral values.

- The professional and technical employee is much more highly sensitive to concepts of fairness, mutual sacrifice, objectivity, respect for personal privacy, status, authority of knowledge and expertise, etc.

- Professional and technical employees tend to find their security in their competence and contribution rather than in loyalty or seniority. This is especially true in their earlier years of service.

- Professional employees are really not different from less educated employees in the kinds of qualities they possess. The difference lies rather in the degree to which these qualities are disciplined or sensitized in a person's behavior. Long years of academic and professional schooling have emphasized objectivity, the scientific approach to problem solving, and the importance of logic and reasoning. Such training tends to make people distrust their feelings somewhat and to repress any expression of emotionality. All too often, professionals come to their first job with little training in self-understanding or in the skills of interpersonal communications. Nor is such a person apt to be highly experienced in the working relationships needed to produce teamwork and cooperation in sophisticated group endeavors.

In the light of these observations, it is evident that the supervisor of technical and professional employees will need to develop some special abilities and skills. For a start, each supervisor should:

- Learn to understand the nature of motivation, appreciate its influence on behavior of the individual, and become skillful in its applications on the job.

- Learn an acceptable way to evaluate and feed back performance results as a basis for stimulating the development of the employee.

- Learn how to use on-the-job training to ensure competence, confidence, and acceptance throughout the work team.

- Be able to get employees to place high demands upon themselves.

- Learn the career expectations of employees and know how to supervise them both for present competence and for preparation for the next jobs in their career progression.

- Know how to accelerate the experience of the employees of high potential.

- Find new and better ways to secure teamwork and cooperation through mutual-support relationships.

- Be able to lead employees to work out developmental work programs for themselves.

- Be able to build and maintain a flexible, dynamic work group organization able to cope with the changing demands of the company.

While these nine supervisory growth areas are by no means the whole story, they do provide a place to start development.

Goals of staff development. There is clearly much greater definitiveness in management's ability to manage its physical and financial resources than in its ability to manage its human resources. In the management of equipment, systems, machinery, materials, and money, management has learned how to set minimum standards, establish break-even points of costs and profits, and measure quantitatively the results obtained. The movement in management toward "management by objectives" and "results management" is evidence of rapidly growing skills in these fairly concrete management responsibilities.

When it comes to the management of human resources, these newly designed systems are less applicable. Setting standards for *professional* services is much more difficult than setting standards for *making products*. Whenever management has been unable to determine and measure results, it does the next best thing: it evaluates methods, procedures, or techniques. In this retreat, management is really saying, "If I can't measure the results of your work, I'll make sure you are kept doing it the way I want you to do it." When an engineer leaves the drawing board and looks out the window, is he or she wool-gathering, reliving last night's party, or putting new ideas into perspective? Management will never know for sure! Since the development of people is so complex and abstract, it is without doubt the least managed of all managerial responsibilities.

Barriers to development.   Few people question the importance of development for individuals in all their life roles. The individual perceives this as *self-development*. Personal development is each person's own responsibility, and anyone would rightly resent the attempts of others to take over that responsibility. The individual senses that such manipulative practices, engaged in under the guise of "personnel management," are a threat to his or her self-respect.

Few managers question the need of the organization to have in its employ a staff of developing people able to fill the jobs vital to the continuing profitability of the organization. However, in their eagerness to have key people ready for use right now, managers have allowed themselves to become manipulative—moving people around like checkers on a checkerboard, arbitrarily assigning them to jobs, ignoring the wishes or needs of those most immediately affected by these administrative moves. Essentially, the employee is expected to believe that "father knows best" and that what is best for the organization is best for the employee. This may be, in part, the reason why organizations have produced so few examples of successful development programs and why they have produced so many technical and professional people who turn into "deadwood" long before their normal retirement years.

Much of the answer to lack of development lies in the negative elements in the management climate that act as blocks or barriers to development. Following are six major blocks that prevent the individual from realizing his or her potential. It has been found that these blocks are present to an unhealthy degree in most organizations. They are caused both by growth problems in the individual and by restrictive forces in the organization. All six can be significantly reduced by a comprehensive development program.

1. Underutilization. This barrier takes on many forms. It may be a by-product of overselection. Incompetent supervisors may be holding people back. Management may be limiting its vision of employees to their achievement, ignoring their potential. An underutilized employee soon learns to cover up with busy work, routine activities, or work at the lowest level.

2. Chronic frustration. Unrelieved tensions, low threshold of capability, meaningless work activities, state of confusion or disorganization—all these take a heavy toll in lowered motivation and eventually result in chronic frustration. Unhappily from the point of view of management, a good employee who is frustrated behaves like a poor employee. For all practical purposes, while such a person is in a state of frustration he or she is not likely to make progress in development.

3. Low standards of performance. Where performance standards are set by the manager without the participation and acceptance of the employee, they are usually minimum standards—much lower than the employees would set for themselves were they motivated to do so. Low standards take the challenge out of work. They make it difficult for the employees to test their strengths or become aware of their potential.

4. Obsolescence. Technical change is an ever-present challenge to the employee to keep updated. The older employees feel the pressure to keep their skills and abilities competitive with those of the young graduates entering the department. Two decades ago, 7 or 8 years of experience on the same job was thought of as evidence of superior know-how; today, it is apt to be thought of as evidence of obsolescence. Sophisticated know-how is forcing more employees to become highly specialized in areas that can make them victims of technological change. Under these conditions, our older and more experienced employees are increasingly likely to "become obsolete." The threat is one of not only technical obsolescence but of mental obsolescence as well. The problem has become so acute in some companies that retirement plans, originally designed to hold people to age 65, are more and more being used to cope with middle-age obsolescence through early retirement.

5. Professional myopia. With the trend toward more intensive specialization, the specialist or expert is often confined to a narrow work niche. Such a person is always working with small parts of tasks and rarely gets to see the task as a whole, suffering from a narrow time frame and myopic vision. Without prompt recognition, this predicament can lead a person to lose the broad perspective that a generalist (and almost by definition, a manager) must have to further a career. We are all familiar with the specialist who has become imprisoned by the wrong kind of experience or too much of the same kind.

6. Career drift. It is painlessly easy for some employees to slip comfortably into the status quo and become passive or procrastinating about their career aspirations. Under the guise of cooperation, they can let their company place them or move them around in terms of company expediency, only to wake up too late and discover that they have drifted through the best years of their lives in terms of career development and qualifying themselves for greater opportunities. They slide through their twenties, their thirties, and their early forties still waiting for something good to happen to them.

The above six barriers are fatal to development. In the good old days when workerpower was cheap and expendable, an organization could muddle through with fairly high levels of block in its people. In the future, an organization's survival may depend on its batting average in learning how to reduce these important causes of poor work performance and costly waste of employee potential.

Administrative action needed to motivate development. The goal of development is the more efficient utilization of people to make them more productive and more likely to realize their own growth potential. Both employees and the organization gain when development takes place. *Sterile* management development programs seem to forget this goal; they are sterile because they make activities—brought into the company in the name of development—ends *in themselves.* They ignore a most basic principle of human nature—the intrinsic motivation for development is more rewarding utilization. The following administrative action pattern is a psychological approach designed to "build in" organizational *readiness* for development. The administrative action taken by top management puts high demands on the staff to move up to the competence needed for developmental implementation. Thus the organization is quickly motivated to learn.

Model of administrative action needed to set the goals for optimum personnel utilization

Step 1. Adopt and begin administering a policy of *promotion from within* that will provide maximum opportunity for employees to take increased responsibility and qualify for advancement within the organization.

Step 2. Set up internal selection procedures (*A, B,* and *C*) that will make sure all employees who are qualified and available for advancement get looked over and are not overlooked. This calls for the development and maintenance of:

A. Organizational replacement charts to forecast present and future workerpower needs

B. Inventory of personnel in terms of developmental status—readiness and availability for filling organizational vacancies and new positions

C. Accelerated *individual development programs* for high-potential candidates in critical replacement areas

Step 3. Review supervisory evaluations of employee performance; identify outstanding performance and evidence of growth potential.

Step 4. Require periodic supervisory appraisal and evaluation of employee's work performance and developmental progress.

Step 5. Require, at the work-group level, a developmental work program for employees, designed and carried out jointly by the supervisor and the employee, to achieve present job competence with supportive job training *plus* developmental experiences prerequisite to furthering career values and opportunities.

Step 6. Provide the on-the-job training resources needed for implementing the individual employees' developmental work programs.

Step 7. Provide employee orientation to get people off to a good start on their new jobs, help them become aware of their own potential for growth and development, and acquaint them with the opportunities available to qualified employees. In short, equip them to be *active participants* in their own education, training, growth experiences, and development.

The reader will note that each step in this sequence becomes an objective for the preceding step and puts pressure on the prerequisite function to achieve that objective. The developmental learning that the individual engages in to support each administrative action step becomes *immediately* useful and meaningful to him or her.

Summary. This section explained the nature of development, what it should be and what it should not be, and provided an overall blueprint of what has to happen in order for a facility department to have a comprehensive management development program. Obviously, it is not possible within the scope of this section to present the "how to" side of the picture. However, most organizations have within their own staffs the personnel capable of making staff development work.

Materials Management

Parts, Supplies, and Materials—Storage, Location, and Distribution

The storage, location, and distribution of parts, supplies, and materials is unique to each facility's requirements. Many factors must be considered to develop the most economical and practical methods. These factors—for determination of the resources necessary for proper storage, location, and distribution of parts, supplies, and materials—are discussed in the following paragraphs.

Storage space.   Storage space is defined as that space where parts, supplies, and materials are held for future use.

Storage categories.   The first thing that must be determined is the amount and type of storage space necessary. Some of the categories that must be considered in determining storage space requirements are the following:

Physical.   *An inventory listing of the parts, supplies, and materials to be stored.* This should include a 5- to 10-year forecast of future inventory requirements for determining and justifying future storage space.

The mean number of individual items stored. Determined by adding half the order point to the order quantity for fast-moving materials. For slow-moving items, such as spare parts, add the order point to the order quantity.

Three-dimensional space requirements of items to be stored. The cost of storage space, on a square foot basis, usually contributes more to storage cost than any other factor. The storage of materials, parts, and supplies must be considered on a volume basis in order to reduce square foot costs.

Weight of each item. Stock materials, parts, and supplies double, triple, etc., floor loads proportionally. Storage space bearing loads must be known to properly design storage facilities.

Environmental.   Dry storage. Those items that must be protected against the elements and require storage in an enclosed building.

Moisture protection. Items that must be protected individually, or in an assembled state, such as bearings on a shelf; or for equipment and items that must be protected in bulk form, such as cement and some types of coated welding rods.

Shelf life. Epoxy paints, plastics, and O rings must be considered against aging factors such as sunlight, high temperatures, and time.

Freeze protection. Liquid items stored in breakable containers. Special protection for stored equipment items must be considered.

Hazardous materials

Inflammables. Are the parts and materials combustible? Can they be protected by normal sprinkler systems, or must they have special storage facilities?

Explosives. Types and volume of explosives. Must they have electrostatic protection, special bunkers, etc.?

Toxic substances. Are materials toxic? Are special clothing or masks required to be worn when handling?

Security

Minimum. Items that can be left in open storage for routine use. Usually consists of parts being manufactured, bulk materials, bar stock and plate, and small expenses items stored in free bins (low-unit-cost items with high-volume usage). Protection is normally provided by guards at gate entrances.

Normal. Items that can be pilfered due to in-plant use during off hours for personal use. Normally protected by locking of storerooms during off hours and prohibiting of entry by unauthorized persons. Many items in this category are protected by inventory records control.

Maximum. Valuable articles, usually small with a high resalable value or attractive for home use; and *insurance items.* Insurance items are items that must be on hand, available for immediate emergency use 365 days a year, 24 hours a day, regardless of the inventory control system's activity criteria. Usually protected by locked doors or cabinets within the storage area with only authorized persons permitted to open. Additional maximum security can be provided by safes or bank vault storage.

Storage space requirements.

Storage space requirements calculations, for figuring the volume and type of storage space necessary, should consider the following information and data items:

Stock number is the assigned inventory number of the item being considered; usually consisting of a six- to nine-digit number with certain digits identifying the family or class of the item it is identified with.

Order point is the bin balance level of stock inventory items designated to set the time at which an order must be placed to replenish the stock.

Order quantity is the number of stock items that must be ordered and placed to replenish the inventory item's bin balance level.

Activity is the number of times, and the quantity of inventory stock items issued at each issue time, in a given time period.

Mean storage is the number of items stored determined by adding half the order point quantity to the order quantity for fast-moving items, and by adding the order point quantity to the order quantity for slow-moving items such as spare parts.

Unit of storage is the quantity involved for packaging and issuing, i.e., each foot, pound, gross, 100 pounds, gallon, etc.

Height occupied by the stock item in the storage unit.

Width occupied by the stock item in the storage unit.

Length occupied by the stock item in the storage unit.

Weight of the stock item in the storage unit.

Type of storage nomenclature is designated by the following terms:

Bins	Racks
Shelves	Pallets
Drawers	Skids
Boxes	Open

Size and type of bin, shelf, drawers, racks, pallets, or skids required. Usually specified by a coding system designating the space required for the type of designated storage. Storage types and sizes come in numerous shapes and designs. Therefore, some one or two manufacturers of equipment should be decided upon, and their equipment coded accordingly for storeroom space planning use.

Environmental designates the type of storage necessary for protection against the elements.

Hazard designates special precautionary matters of handling and storage necessary for designated items.

Security designates for the type of security necessary to protect items for both future availability and individual items costs.

At this time, two important factors must be considered. First, the cost of tabulating the above indicated 15 data items is high. It's estimated that an average of 15 minutes is necessary for determining the required data for one stock item. If 8000 items are in inventory, then approximately one worker-year of work is required. In many facilities workerpower is too scarce for such work of a temporary nature.

In lieu of analyzing every item in the inventory, a statistical analysis can be made using valid statistical methods, based on normal sample sizes, assuming the probable error can be tolerated. If the probable error cannot be tolerated, then a larger sample or total sample must be made.

Functional space. In addition to the above, within the storage area, additional space must be provided for:

Railroad rights of way	Disbursements
Truck turning radius	Shipping
Unloading ramps	Offices
Receiving	Restrooms
Inspections	Janitor's closet

In many storerooms, it is found that the functional space requirements often exceed the storage space requirements.

Location of the storeroom. Work sampling studies normally reveal that craftspeople traveling for job materials acquisitions, from the shops or from job sites, can spend from 5 to 20 percent of their job times, depending upon a number of factors. One of the most important of these factors is the storeroom location. The closer the storeroom is to the center of activity of craftspeople using materials, supplies, and parts, the less time is spent in traveling in this regard.

If the parameters are variable, that is, if one is free to locate the storeroom where practical, then the location is usually adjacent to or within the facility department shops area. In turn, the department's area shops are located within the center of production facilities in order to keep traveling distances to a minimum.

One of the best practices is to have a centralized organization storeroom adjacent to the facility department shop area, with a telephone ordering service for delivery of parts, supplies, and materials beyond the shop limits. This practice allows craftspeople to remain at the job sites for work continuance while additional tools and materials are being delivered to them.

Storage facilities (storerooms) are ideally located within the facility department's assigned spaces. Storeroom design criteria should include the following space planning factors:

Rack storage for bar stock and pipe are located adjacent to the machine and pipe shops, respectively.

The toolroom (tool crib) is located within the machine shop area for tool disbursement and, for convenience, adjacent to the storeroom.

Both toolroom and storeroom have accessibility for disbursements external to the shops to eliminate traffic within the shop spaces and for external deliveries.

Internal and external delivery windows and the truck pickup area are located in the same vicinity so they can be manned by the same personnel.

Materials are received at one end of the storeroom and disbursed at the other end.

There are many parameters that determine storage facilities location. All fixed parameters should be determined before a storage room layout is made. In a "grass roots" facility, the parameters are variable except that they may be determined by policy or cost. In an existing facility, the fixed parameters may be:

Storeroom location

Storage space

Building designs

Material handling facilities

Storeroom accessibility

Parts, supplies, and materials.   A survey should be made of the inventory list to determine which items should be isolated and not kept in the regular storeroom space. Then a determination should be made as to where to store the following items.

Free bin are fast-moving, low-unit-cost items, stocked in open bins (lazy susans) near their point of usage. The bins are stocked from the storeroom at regular intervals and expensed to an overhead account. Free bins are normally located in shop areas and are accessible only to craftspeople during normal working hours.

Machine tools and regular tools are normally stored within the toolroom (tool crib) area. Tools and tool parts are delivered directly to the toolroom for storage and use.

Insurance items consist of large, bulky, capitalized items, spare parts with long delivery dates for replacement, or items that have to be on hand for use 365 days of the year. The items normally involve special approval for stocking and are relatively expensive. Also, the inventory control system's activity requirements do not apply to these items.

Environmental and hazardous materials are isolated and stored according to their particular requirements. Where possible, their location is within or near the storeroom proper.

Remainder of the items are stored in the storeroom after determining and isolating all storage items that are not stored in the storeroom proper.

Aisle space must be provided for the mechanical equipment which will handle the materials. The type and size of the handling equipment needed are determined. Then the proper aisle space must be provided for width and turning radius. Central aisles should run the length of the storage area, with direct access from the unloading dock to the disbursement areas. Forklift trucks require an aisle of approximately 10 ft for passing. Narrower aisles of approximately 4 ft should be provided for stacker cranes and other one-way traffic. Turning room should be provided at each end of one-way aisles. For hand truck and carts, a space of 3 ft should be provided between rows.

Item location. Once the space necessary for storage within the storeroom is determined, the next concern is where to locate particular items. The three factors that determine where items are located are:

Family storage—storage of items of a particular classification. These items may be classed according to stock catalog numbers, parts for a major item of equipment, names of parts for a particular type of equipment (such as that all spare pump impellers be stored together), or functions (such as that all safety equipment or clothing, storage by shop or foreman, etc., be together).

Activity storage—locating items within the storeroom according to usage. The faster the turnover of an item, the closer it should be located to the disbursement area. Slow-moving items should be located away from the disbursement area.

Size storage—includes the total volume required for the mean number to be stored as well as the individual item. If weight is a critical factor, it should be considered also on an item-by-item basis.

A combination of these three factors should provide the best solution for most facility department storerooms. Where family storage is specified—as for safety items, light bulbs, clothing, etc.—the family storage should be located according to activity and size. Light, bulky items should be stored in the tops of bays or overhead. Heavy items should be located near the floor for ease of handling. Easy-handling items should be stored in the middle of the bays.

All items handled by a storeroom should have a stores location code assigned that identifies where the item may be found. The stores location code gives every single storage space, big or small, a unique designation. Each item within the storage system, usually identified by a stores catalog number, is also assigned a stores location code.

With this code, or a similar one, each bin has its own address in a logical sequence throughout the storeroom. This allows an individual or a computer to prepare "pick lists" that will enable stockclerks to fill an order without wasted movement. The code also helps those who are unfamiliar with the storeroom to find items in an emergency.

Distribution. On any job where materials are required, the materials must be requisitioned from the storeroom, held for a job, and distributed to the various job sites for usage. Various techniques for use to provide these services follow:

Requisitioning of materials is handled by some form of written order known by numerous names that vary according to the organization and the purpose the order serves. The most commonly known are "material requisition," "stores requisition," and "bill of materials." All material disbursing requires a written order for control purposes that has been properly approved and dated by an authorized signature.

Requisitions may be received manually from the requisitioner at the disbursing center, by mail, by telephone, or through any type of transmitting device (such as computer printout, FAX transmission, or written transmission). When verbal or visual requisitions are received, storeroom clerks must convert the orders into written requisitions.

Disbursing of materials.   Upon receipt of the requisition, action must be taken depending upon its nature. The requisition is usually used as the "pick list." Normally the order is filled and handled as follows:

Counter requisitions are filled and delivered to the mechanic who has waited at the counter for the requisition to be filled.

Phone orders can be filled and held for the requisitioners to pick them up, or they can be delivered by a route truck for routine deliveries or by a "hot shot" truck for emergency deliveries.

Requisitions for materials with a future delivery date can have the materials in the bins held by a "hold tag," or special "hold bins" can be designated for a particular order within the storeroom near the disbursing area. This provides a means for making sure the materials will be available for the job the day it is scheduled. A release order is necessary to have the material delivered to the job site.

On larger jobs, special areas near the job site may be designated as "hold" areas for accumulation of materials. These areas are usually called "end use staging areas."

Routine materials delivery should be normally handled by route trucks, which deliver materials to designated stations throughout the facility. Emergency delivery items should be handled by "hot shot" trucks which deliver only those special items called for. Special delivery items of a bulky or heavy nature usually require additional handling facilities. These items are delivered directly to the job site. One copy of the requisition should normally be a hardback. This copy should be attached to the material or container for delivery to job sites.

Route trucks normally deliver materials from station to station throughout the facility. A copy of the shop work order should be used to authorize movement of materials between truck stations and movement back to the shops for repairs.

Locked boxes can be provided at delivery stations for pilferage reduction. Only the foremen and the truck drivers should have keys for opening the boxes.

Important and key practices for improvement of materials delivery follow:

Storeroom clerks usage—Where "free bins" are designated within work areas, storeroom clerks should be made responsible for restocking the bins as required. Where material catalogs are provided in the work areas, the storeroom clerks should be responsible for keeping the catalogs posted and up to date. This requirement should not be assigned to a craftsperson or a foreman. The same philosophy should apply to obtaining materials from the stockroom.

Requisition procedures—Counter requisitions should be discouraged. Telephone requisitions and "hot shot" delivery keeps craftspeople at job sites and provides greater craftsperson work utilization.

Relationship with Purchasing

Since providing service is a prime facility manager's function, it is their responsibility to be in a position to offer such services as quickly as possible and at the lowest cost possible. All such services should be accomplished within the bounds of the organization's current policies and procedures. To meet this objective the facility manager will have to use all the organization's resources, including centralized purchasing. The organization's purchasing manager provides specialized aid and assistance to all departments, including the facilities department, in receiving the best values, timely deliveries, and proper specifications for materials ordered and resources expended. It is important therefore that the facility manager who is tasked to maintain facilities operating at minimum cost relate to the purchasing function at the highest possible level.

Ally in purchasing.   Unfortunately this is not always the case. The facility manager, in some organizations, is not aware of the benefits to be derived through a good relationship with the purchasing function, i.e., the purchasing manager. The facility manager, normally deeply engrossed in facility operations and maintenance problems, sometimes fails to appreciate the fact that the purchasing manager can be an ally. On the other hand, the facility manager's relationship with the purchasing manager should be one of mutual trust and understanding, with respect for each other's skills and a desire to work together voluntarily in the best interests of the organization. One has to recognize the fact that facilities engineers, by nature, are individuals who are strong of character, who consciously or unconsciously demonstrate that they usually expect to be the deciding factor in any transaction involving them. Instead of downgrading the purchasing function, the facility manager, as a professional, should strive to establish and maintain a friendly and cordial association with the purchasing manager as another professional, with mutual respect for each other's responsibilities.

Small facility departments.   Facility managers in small departments accustomed to doing their own buying may find it initially difficult to work with centralized purchasing. They are usually in good control of their buying procedures, probably have excellent supply sources, and in general feel no need to have a centralized purchasing organization informing them of their needs. They maintain their own storerooms, approve payment for all invoices, and essentially manage good operations for their organizations. Then why break this combination up? One good reason could be that the organization may wish to grow. As it grows, maintenance and operations will have to grow; and as it grows, it becomes more unwieldy to operate. Thus keeping the operation "under one hat" begins to be a problem.

Facility managers then have to depend on other departmental staff members to assume some of their responsibilities. When the facility manager begins to lose close control over all buying functions, problems begin to develop not only in the procurement of materials and supplies but in other areas as well. A quick solution would be to transfer all buying responsibilities. It is advisable for facility managers to meet with purchasing and go over their requirements. Provide the opportunity for purchasing to:

Plan volume buying

Assist in inventory control

Research new items on the market

Process and clear invoices

In general, perform the organization function in which they are specialists

The responsibility for buying should be that of the buyers in purchasing, not facilities department staff members. Once the buying responsibility has been transferred, the facility manager should help purchasing by offering ideas and the fruits of experience.

Medium-sized facility departments.   The facility manager, in a medium-sized operation, should strive constantly to improve relations with purchasing. Both functions, however, should operate as an integral part of the organization, helping to meet the organization's mission, with the same objective in mind: *Provide the right materials, to the right places, at the right times, at the lowest overall optimum costs.* The purchasing department should be kept informed, by the facility manager, of annual requirements so deliveries can be properly scheduled. In turn, the facility manager should explore "open end" contracts with the purchasing manager. The facility manager, in assisting purchasing, should:

Standardize specifications

Stay alert to possible materials scarcity

Keep abreast of market conditions

Provide technical assistance, as required

Allow purchasing to perform its function to the best of its ability in its area of responsibilities.

Large-sized facility departments. In this area the facility manager and the purchasing manager should develop the highest type of interrelationship possible. The work crossing the facility manager's desk is so complex and of such a nature and volume that accomplishing procurement actions without the purchasing department's assistance is almost unthinkable. Organization policy normally dictates procedures in this regard. To work within this framework will require that all buying be accomplished through purchasing, in any case. The facility manager should take advantage of purchasing's technical expertise by allowing purchasing to:

Locate sources of supply for specialty items

Contact testing laboratories, as required, to assist with complex problems solutions

Work out, jointly, legal problems in connection with buying

Develop standard specifications to cover repeated buying problems

Assist in managing stock inventory

Assist in developing vendor revolving accounts for facility department

These actions can be of valuable assistance to facility managers, thus freeing them to perform other duties.

Comparison of basic objectives of facility managers and purchasing. In general, the key and prime goal and objective of the facility manager is to provide quality, responsive service at the lowest overall total cost. The key and prime goal and objective of the purchasing manager is to obtain maximum value, at least overall costs with proper quality and timely deliveries, for the organization's resource dollar expenditures. It appears then that both are striving for the same objectives. In the area of service and support, both have a common objective, i.e., to improve upon these goals and objectives. Both functions require an effective planning system with good controls, systems that will further relationships with other organization divisions through proper communications.

Other goals and objectives shared by both functions are to avoid waste and duplication, to optimize systems and procedures, to use to advantage outsourcing to fulfill the organization's needs for goods and services, and—finally—to require the full cooperation of all departments to the end that the organization may obtain maximum advantage of the services offered by the facility and purchasing departments.

Improving upon buying for maintenance, repair, and operations (MRO). Proper and timely acquisition of MRO materials and supplies is usually neglected in most purchasing organizations owing to the low processing priorities assigned. What are MRO materials and supplies? MRO materials and supplies are products that are used in the organization's production process but do not become a part of the end product.

Maintenance supplies. Maintenance supply items are those items required to keep the physical plant and grounds in good condition, well lighted, easy to clean, attractive, and pleasant to employees. Such items include paint, lumber, hardware, electrical supplies, HVAC preventive maintenance supplies, custodial supplies, etc.

Operating supplies. Operating supplies are those supplies used in processing or making an end product that do not become part of the product. Such items include lubricating oils; cutting oils; processing materials for painting, plastering, and heat treating; perishable tooling; and abrasives. Office equipment, stationery supplies, printing, forms, HVAC operating supplies, etc., also fall into this category.

Repair parts. Repair parts are replacement parts required to repair lubricating equipment, machine tools, test equipment, HVAC systems and equipment, motors, and other capital equipment in the course of normal wear, tear, or breakdown. Hence sources for major repair parts are more limited than either maintenance or operating supplies. The original equipment manufacturer (OEM) is usually the only dependable source. But, not infrequently some replacement parts must to be duplicated by a vendor for lack of OEM availability.

From the above it can be noted that there is a great variety of materials, supplies, and parts carried under MRO. These items differ not only as to what they are but as to where they come from, how they are priced, and—most importantly—how much lead time is necessary. A repair part for a vital piece of production or HVAC equipment can cause costly downtime. Hence these items need to be on hand and available for use when required.

Major areas of real assistance to the facility manager by purchasing. Key questions regarding areas where purchasing can provide real assistance to the facility manager follow:

- Will involving purchasing in the acquisition process upset the facility manager's usual close control of materials buying in such areas as quality, quantity, cost, and delivery?
- Who will control storage and inventory control procedures?
- What computerization use will be specified in the areas of hardware, software, etc.?
- What about the third party effect on a contractual service?

Following are answers to such questions of concern on a variety of subjects involving the facility manager's relations with purchasing.

Standardization. Too many facility managers wish to "go it alone"—in lieu of coordinating with purchasing on the problem—when it comes to standardization of parts and supplies. Normally it is felt that they may lose control of quantity and availability, particularly if they have good supply sources. What is not realized is that purchasing can, after reviewing usage, combine or consolidate the facility manager's requirements with other similar organization departmental requirements, incorporate them into a single specification, and place an order of considerable quantity, which because of its considerable volume will show an appreciable cost savings. The factor of cost saving alone should make facility managers consider parts standardization. Also, there will be the advantages of better delivery and reduced inventory needs.

What are some of the objections that the facility manager has toward standardization other than losing a good supply source or giving up a supply item with a name brand? Is it that the quality of an item might be reduced? If so, a good specification will eliminate the poorer product. Purchasing will support quality and cost optimization and welcome a change in the specification that will ensure that the facility manager receives the needed product. Usually quality suffers if specifications are poorly written or maintained in a general format. If there is a quality question regarding a specific item, the facility manager should feel free to request a report from an independent testing laboratory covering the material. Also, the facility manager has the option of turning down, based on a field test, what seems to be a poor material. Take these questions regarding paint as an example: (1) Does it cover adequately? (2) Does it apply smoothly?

Even so, generally speaking, a well-written specification will bring to the facility the product that is wanted; and if the specification is used as a standard, there are the added benefits—in cost savings, better delivery, and fewer items to be carried in stock—that accrue when purchasing buys the product.

Reducing paperwork costs. Reducing paperwork costs is a major goal in obtaining less costly facility department overhead operations. Many of the items purchased are small by volume and relatively inexpensive on a per item basis. Processing these materials requisitions, and follow-on purchase orders, and finally their payments based on established procedures, costs more, in some instances, than the acquired materials themselves. The facility manager should work with purchasing to develop a simplified system and procedures for low-valued items. The facility manager's objective should, in this regard, be to reduce to as few procedural steps as possible the purchase of an item. It is useful to develop special small-order techniques, i.e., use of petty cash, direct purchases, consignment buying, stockless buying, costless buying, and other methods that will be described later. Also, it should be noted that every time a formal purchase order is bypassed, other paperwork such as receiving reports, invoices, and checks are eliminated. The overall results are a net cost savings to the organization.

Possibly, a forms change could assist in making "rush" or emergency pickups. Even so, the facility manager should process all such actions through purchasing so they have an overview of what is needed. This provides purchasing the opportunity to devise the best procurement system possible that will satisfy the organization's requirements at the lowest possible overall cost.

Some buying techniques

Blanket order. Since buying for facility operations and maintenance needs differs from buying for the normal organization production needs, the facility manager should explore with purchasing several general buying approaches. One such approach to be explored is use of blanket orders. This buying method is probably the most widely used today. It involves negotiating an agreed-to price for special items which are needed repetitively throughout the fiscal year. Or it may be an agreement to supply all a buyer's needs for agreed-upon prices. The facility manager should:

Review all problem areas

List all items that relate to such a piece of equipment with reference to a supplier's catalog or some other form of description

Show quantity needed

List the usual supply sources

Forward the data to purchasing for action

Purchasing then should:

Analyze the data

Review operations and maintenance needs

Review data for suitability

Find a supply source (one who will deliver if possible)

Place an order

Immediate savings will be shown in the elimination of paperwork alone, and in the time required to locate an item when needed.

Direct purchases.   The facility manager should arrange with purchasing to be able to direct procure an item up to a prearranged price of say $100 or $200, whenever necessary. This approach can be used when emergencies arise. A direct purchase order is written to cover the emergency need.

When costs of an item(s) exceed the prearranged figure, the facility manager should arrange with purchasing to be able to call a buyer over the telephone for clearance to buy. The buyer then assigns a purchase order number for the facility manager, thus clearing the way to make the transaction. Many purchases of sizable amounts can be made through this controlled process. Why use this approach? Organization policy does not normally permit the purchase of items directly from a source without purchasing being involved. In general, the facility manager should use purchasing to assist with emergencies. Once purchasing is made aware of a problem, they normally grant the facility manager all the freedom needed to complete the job. This approach prevents having to justify the actions when the invoices start arriving in accounting. An axiom worth repeating, in this regard, is "resolving an emergency is fine, but resolving it through established procedures is better." This is the mark of a good facility manager.

Revolving account for inventory.   Another technique is use of a revolving inventory account for purchased stock items. Purchasing sets up a special account to permit the procurement of sizable quantities of materials against this account for stocking. This allows the advantage of lower per item costs due to significant quantity purchases. The material is not charged against a facility department account until it is withdrawn for use by authorized employees.

Consignment buying. This method permits the maintenance of a high stock level of items, in the storeroom, which are paid for only after they are put to use. Only special materials fall into this class, i.e., electric motor belts which come in hundreds of sizes and types. The vendor maintains predetermined stock levels of the items, which are distributed by the facility's storeroom clerks. The clerks reorder the items as needed. They are also responsible for furnishing the vendor with accurate usage reports. The vendor then bills after receiving the usage reports.

The advantages that accrue when using this technique are

Ready stock of materials is always available

Tremendous reduction in paperwork

Reasonably priced items on the shelf

Expectations of further price reductions if the system works to the vendor's satisfaction

Stockless buying. The same procedure is used as discussed above except that the vendor does all the work, including maintaining the items in the vendor's stockroom inventory. The material is made available and is delivered to the facility upon receipt of an authorized request.

Catalogs. Some large purchasing departments publish, periodically, catalogs which show materials available for use. Facility department staff can refer to the catalog when preparing materials requisitions, citing the catalog item number for identification. Delivery is usually prompt, since the majority of items shown are available in the central warehouse. This approach mitigates against long delivery times and also reduces paperwork.

Petty cash. This approach is normally used in buying low-cost, hard-to-procure items. Using petty cash has been found to be the most advantageous method in this instance. The facility manager's office is usually held responsible for this fund and its periodic audit. This is a minor inconvenience for the derived advantages from its use.

Cashless buying. This involves use of a purchase order, issued by the buyer, with a check attached for completion by the supplier upon delivery of the order.

Using an integrated management information system (MIS) as a control device or buying aid. Organizations having in use an integrated MIS should ensure that the system is used for purchasing and inventory control for both production and facility operations and maintenance materials and supplies. The facility manager should consult with the purchasing manager to determine, before using the MIS, the impact on the department of answers to the following questions:

What information and data would have to be made available for MIS use?

What are the overall requirements to become an MIS user?

What reports would the MIS have to generate for the facility manager's use?

What would be the staff requirements impact on using the MIS—both data input and reports output use—evaluation and corrective actions as needed?

Is the facility department staff's competence and reliability such that the MIS data input is both accurate and timely? *Remember, the information received from the MIS is only as accurate as the data put into the system.*

The advantages to the facility engineer for using an MIS that is closely controlled and has accurate results for facility procurement and inventory control are relief from responsibilities for inventory control system usage, reorder points determinations, purchase order issuance, and lead time determinations. MIS use also reduces considerably the attendant paperwork and record keeping involved in purchasing actions.

Contractual services. A much needed area for purchasing support is the function of contracting for facilities operations and maintenance services. Repetitious heating plant repairs, such as boiler tube replacements and circulatory pump installations, are an example. Roofing installations and/or repairs, exterior painting, exterior structural maintenance, janitorial services, and waterproofing also are in this category. These contracts are those that cannot be completed because of lack of in-house staff, i.e., skills missing and/or skills nonavailability.

The procedural steps to be used follow:

The facility department prepares a technical work specification describing, in detail, the work tasks to be contracted.

The purchasing department would then:

■ Provide instructions to bidders, general conditions, etc., *normally called "boiler plate."*

- Ensure that the contract is legally sufficient.

- Advertise for bids.

- Accept and review bids for necessary negotiations.

- Award contract.

Some disadvantages found in working with purchasing feedback. From the above, it can be deduced that the facility manager should develop a high level of respect for the purchasing manager's knowledge and judgment concerning purchasing and inventory control procedures. In turn, the purchasing manager should develop a high level of respect for the facility manager's knowledge and judgment concerning facilities operations and maintenance work performance.

Even so, certain pitfalls will become apparent once purchasing has total responsibility for procurement of facilities operations and maintenance materials and for contractual services. One of the most frustrating problems is the complete lack of feedback. Every facility manager wants and needs to know the status of each order at all times. Why the lengthy delays? What is the source of supply? Even the answers to these questions are hard to come by at times.

In any takeover by purchasing of facility department procurements, the need for constant feedback must be emphasized. The facility manager must be:

Kept abreast of the progress status of all items on order in a real-time environment

Informed of the exact status on projected deliveries

Informed as to which firms have the orders

Constant feedback, at all times, is the important thing.

Understaffing of the purchasing department. Understaffing of the purchasing department is another disadvantage the facility manager has to be made aware of. Lengthy delays in processing facility department materials requisitions develop when this occurs. A facility manager confronted with this situation should immediately make the purchasing manager aware of the problem. The purchasing manager has to be made aware of the delay problems and how they are affecting facilities operations and maintenance work performance as well as the organization mission performance. Usually the purchasing manager has to be convinced of the need for corrective action. How to do this?

Document the problems discovered. Take the following actions:

Document everything that even approaches being a problem when purchasing is involved.

Let the purchasing manager be made aware of the concerns by use of telephone calls and/or letters as needed.

Give purchasing managers an opportunity to make the necessary changes. Once they are made aware of the problems, it is assumed that corrective actions will follow.

Budgets and Costs

Identification, Preparation, and Administration of the Facility Department Budget

Introduction. It is important when discussing the terms "facilities cost" or "facilities budget" to define exactly what one means by them. Many times, individuals discussing and comparing costs for their respective organizations find after lengthy discussion that they are talking about a term that needs further definition. What is sometimes termed facilities cost, and included in the budget, is not necessarily the same in all facilities. Therefore, the terms "facilities budget" and "facilities cost" will be discussed in the context of today's organization layouts. The differences between various cost accounting methods as used by many organizations are discussed in detail following.

Costs, in today's environment, are defined as costs incurred to keep the facility operational, in other words, the cost of workerpower, supporting machinery and equipment, repair and replacement parts, building materials, and daily operating supplies required to operate and maintain the facility on a daily basis. All these, totaled, make up facility cost. These are the costs which must be taken into consideration when discussing the term.

Facility administrative costs are often included in the overall cost under the item *supervision.* Sometimes HVAC costs, as well as their usage, are included in the originally stated budget. Some facilities include housekeeping, safety, and security; others include none of the above. When discussing the term "facility cost" or preparing the facility budget, be specific and define the term first.

Most accounting departments today use six major categories to define costs for their respective facility organizations. These six categories are (1) equipment, (2) supplies, (3) labor (direct and indirect), (4) outside services, (5) department overhead, and (6) organization overhead. There are many variations to this, as there are with almost anything. Some organizations may combine department and organization overhead, while others choose to break them down further with a series of number identifications. Every organization usually has individual facility cost accounting methods peculiar to its individual operation, but the above are the basics.

Equipment. Items purchased as units from a manufacturer or distributor, or items manufactured in facility shops as complete assemblies for support of the facility function within the plant are usually defined as *equipment cost.* Cost of the equipment may range from a $45 amperage meter to a $5500 roof moisture detector or a $10,000 computer used by the facility in troubleshooting and diagnosing operating problems on numerically controlled machine tools.

Maintenance performed in the facility depends upon two major factors: (1) qualifications and expertise of the skilled craftspeople and (2) the equipment and tools the craftspeople have with which to perform their assigned work. Craftspeople cannot be expected to perform quality work in support of modern production machine tools and sophisticated production processes unless they have the equipment and tools to diagnose and troubleshoot the problems. Once they have defined a problem, they also need, in many cases, special tools to assist them in performing their jobs efficiently and at minimum cost. These special tools include equipment needed to do the job. Costs of such items are termed *equipment cost.* New equipment such as portable tools, pumps, hoist lifts, and other related items (such as pullers required by craftspeople to support continuous and efficient daily facility operations) are also termed *equipment cost.*

Supplies. Items carried on the shelf either in facility shops or in facility storerooms to support continuous, efficient, and reliable daily operations of production machinery, and facilities equipment, are defined for cost accounting purposes as *supplies.* This cost area includes such material as mechanical packing and seals, pipe, fittings, valves, electrical wiring devices, V-belts, lubricants, filter media, paint and coatings, water-conditioning chemicals, insulation, and many other items used daily in support of facility operations.

Special machine repair parts are maintained in facility storerooms and are controlled on a minimum-maximum basis based on usage. Items such as bearings, power transmission equipment, motors, and special individual machine components (such as a machine spindle head or a special shaft or gear head) are all part of facility supplies as defined by cost accounting. Maintenance supplies are items determined by maintenance supervision (foremen, planners, scheduler, craftspeople, etc.) as being essential to continuous and efficient facility operations.

Operations and maintenance (O and M) supervisors determine what items are to be maintained in the storerooms. They also determine the specification by brand name or approved equal, along with minimum quantity, as well as maximum quantity, to be maintained in stores. On certain items which have heavy usage, the control level is based on the amount withdrawn from stores within a given period of time. Purchasing may set up an annual contract and have a specified quantity of items delivered to stores each month, based on the anticipated usage. The original requests are determined by operations and maintenance supervisors.

Individual items maintained in stores are usually assigned a material code number by inventory control planners for cost control and ease of identification and usage.

An important point to consider when discussing facility budgets and costs is how much of facility department resources are devoted to materials, parts, and supplies. Another important consideration, and an area each facility manager should monitor, is overstocking of materials, parts, supplies, and provisions as well as removing obsolete materials, parts, and supplies from inventory. A review of items in the storerooms should be made annually to determine usage for the 12-month period. Items that have low usage (the demand annual quantity number is determined by accounting and the facility department staff based on projected carrying charges) should be removed from inventory and the materials codes canceled.

When machine tools are removed from service, all supporting spares should also be removed. This does two things: it reduces inventory costs and keeps storerooms clear of obsolete stock. This, a common oversight in many organizations, results in excess dollars tied up in inventory that may lead to space and material handling problems.

Not all items that do not show usage for a year can be included in this category and automatically removed from inventory. Some of the items in inventory are classified as *insurance items*. These items are available at the facility site for immediate use, at any time, should a failure occur that would shut the facility down or shut down a critical piece of machinery or equipment. Often transformers, power cables, or starter motors are included in the category. At any rate, a close scrutiny should be made of the items to be removed from inventory to ensure that no *insurance items* are included.

Labor.   Craft time charged against a specific work order is defined as *labor cost*. Labor costs are normally considered to be the wage rate plus fringe benefit cost times hours spent by a craftsperson in performance of work order assignments. Today, with the increased emphasis on more efficient utilization of laborpower, facilities, and equipment, labor costs are being more accurately defined. In most facility departments, especially where work performance allowed time standards are employed, the charges distinguish between direct and indirect labor cost.

Many factors make up the total labor cost on a job: planning, estimating, and scheduling; reviewing prints and job communications; preparation and put away of materials, tools, and equipment; use of special tools and/or equipment; travel to and from the job site; delays encountered; and work execution all go to make up the labor cost.

When discussing labor costs, one cannot talk only about labor, lumping all the other facets that support actual work performance under the heading of labor cost, and be completely accurate. What you are really talking about is direct and indirect labor cost. Direct labor costs are those costs incurred in actually performing the work. Indirect labor costs are those costs incurred in preparation for the work to be performed. Travel to and from the job site, requisitioning parts from stores, and reviewing job plans are but a few of the functions that make up indirect labor cost.

Labor costs, incurred by operations and maintenance functions, are usually charged to a predetermined organization order number which denotes a particular production operation. The charge is then billed against the production department operating costs. Labor charges that cannot be charged to a production-supporting organization order are charged to the facility department's overhead account.

Outside services.   Costs incurred under the heading of *outside services* (outsourcing) are in the form of labor and technical expertise obtained from a contractor, consultant, or supplier. Many facility managers supplement their in-house workforce with outside services for various reasons. One main reason for the use of outside services is to assist the existing staff in peak workload periods, enabling facility management to maintain a more stable in-house workforce.

Another facility management will use outside services is to employ special talent or services to meet an infrequent need. Sometimes special skills or tools and equipment are needed for a particular job that does not warrant an in-house investment in staff skills or equipment. Engineering consultant services and fees are included in this cost area.

Many times, especially in modern industrial facilities, temporary employment of outside labor is needed to supplement the in-house workforce. Use of outside labor is often necessary in the performance of highly technical work, such as compressor rebuilding or performance analysis or efficiency checks of compressors or boilers (steam analysis).

Outside services are also often employed to perform large-scale facility repair such as roof overlays, painting, new floor coverings, and other large-scale jobs that are infrequent and upsetting to the existing work schedule. Other areas where outside services are often employed is in the performance of specialty work such as cleaning air-handling systems and removal of waste material, both solid and liquid. All these costs are often lumped together and termed operations and maintenance costs.

In most facilities, the facility department has responsibility to maintain and operate the facility efficiently. The department's training, skills, and expertise, and the tools and equipment they have to work with, are geared toward daily routine operations and maintenance operations. When infrequent needs arise for special skill or work performance, as mentioned above, facility management often turns to outside sources for performance of the work.

Often it is more economical to contract special work to outside services (for reasons previously discussed) than to try to do it in house with the existing labor force.

There are no cut-and-dried guidelines for determining when work should be contracted to outside services. However, there are a couple of areas that are usually good indicators. One of them is the department's work backlog. When backlog hours reach a predetermined level, this indicates some work that needs to be done may have to be postponed. The other is criticalness of need for work to be completed; i.e., the work performance has to be accomplished immediately or in the very near future. When either of these conditions prevail, the use of contracted services has to be considered.

Department overhead.   All facility department operational expenses that cannot be charged directly to a specific work assignment (organization work order) are charged to *department overhead*. Usually the expense of supervision above the foreman level, and facilities engineering assistance, are charged as department overhead. In every facility budget there is a certain amount of cost incurred that stems from the fact that the department exists. These costs are usually budgeted for as department overhead operating costs.

Maintenance and depreciation of machine tools used by the department and also cost of supplies such as hand tools, drills, files, gloves, and nut-and-bolt items used in daily support of the operations and maintenance workforce all are budgeted as *department overhead*. In addition to materials and supplies, other items including vacations, sick leave, training, absenteeism for any reason, and labor discussions are also considered part of department overhead operating expense. Heating, steam, cooling, production of compressed air, water service, and other utility costs incurred for facility operations are also termed department overhead. These costs are usually totaled at year's end and distributed proportionally to production operations based on square footage of occupancy. In other facilities, department overhead costs are charged totally as an operations and maintenance cost and are not proportionally charged against production operations.

The difference in accounting methods—that is, the various ways in which departmental costs may be carried on the books—reemphasizes the importance of defining the term "facilities cost."

Organization overhead. Organization administrative expenses that must be shared by all departments, including the facility department, are termed *organization overhead.* Top-management salaries as well as the cost of the human resources department, purchasing, and other similar service organizations and also other expenses necessary for organization operation that cannot be conveniently allocated to a specific department or activity, are all assigned to the organization's overhead cost. These costs are divided among the organization's operating departments.

Costs charged to the overall facilities operations and maintenance costs are classified as part of the organization's overhead cost. Similarly, all these overall operations and maintenance costs constitute the total facilities cost which also is the total facilities department budget.

Preparing the budget. Every organization has individual budget preparation procedures, peculiar to itself, that are to be used to prepare departmental budget estimates. There are no cut-and-dried methods of preparing the departmental budget, mainly because of the various levels at which operations and maintenance is carried out in individual facilities.

Preparation of the proposed facility departmental budget should start at the foreman level. After all, the foreman is the one whose "head is on the block" to ensure continued and efficient operation of production and facilities equipment, not the purchasing or organization engineering department.

If failure of a critical piece of production equipment occurs, management does not call organization engineering to get the equipment returned to service. Rather, they call the facility department foreman either directly or indirectly. The foreman is the one responsible, through direction of the assigned workforce, to repair and return the equipment or facility to service. It is also the foreman's responsibility to ensure that repair and replacement parts are available for immediate use in the facility department storeroom. The foreman should be recognized as an important cog in the industrial management wheel when the forecast of anticipated future expenditures is being prepared.

All operations and maintenance foremen should be required to submit through the chain of command proposed expenditures for their areas of responsibility. This input should be only one of many required by management in preparing the departmental budget. Only individuals directly responsible for continued efficient operation of facilities, and the equipment thereof, are cognizant of individual equipment or facility needs. They have the cost records, through planned operation and maintenance activities, and also firsthand information, of current conditions that might warrant heavy expenditures at any given time.

Other inputs required, in addition to the estimated laborpower (labor cost) to prepare budget figures include cost of inventory materials, parts, daily supplies; utilities; and supporting services. Fringe benefits, vacations, holidays, sick leave, absenteeism, training, and even long-term laborpower demands should be part of the total figure.

Once all preliminary estimates of facilities O and M costs are developed, they are submitted through the chain of command for review and consideration. The proposed budget expenditure is either approved as submitted or returned with specific instructions for budget reduction. Often large expenditures such as resurfacing a large roof area or resurfacing a manufacturing floor area can be spread out over a period of years and the job done in phases to prevent large expenditures in a short period of time. In most budget control procedures, any large maintenance or repair expenditures for building or facilities equipment can be handled this way. Usually, production machinery and equipment is not included except in extremely large process facilities where a particular modernization or renovation may take place. Other methods of identifying and handling facilities operations and maintenance costs that are often used in preparing departmental budgets could include those described below.

Maintenance of facilities services. This cost accounting area includes all costs, both labor and materials, incurred in the maintenance of facilities services. Facilities services are defined as those services required to maintain and operate the plant on a daily basis. Maintenance of heating, ventilation, and air conditioning (HVAC) services, waste handling and disposal, etc., and housekeeping may or may not be included under facilities services.

Maintenance of production and support equipment. Material and labor costs incurred by the facility department in support of daily production operations of production and support equipment are charged to maintenance of production and support equipment. Daily operating supplies such as lubricants, coolants, and repair parts; preventive and predictive maintenance tasks; and all other such costs incurred are included in this cost area.

Operation of facilities services and utilities. Labor, materials, parts, and operating supplies costs incurred in the production and control of compressed air, steam, air conditioning, ventilation, and other utility services such as electricity and water, sewage and waste control are termed facility services and utility costs.

Usually, only charges incurred at the point of generation or control are included in this account. Once the distribution service lines leave the four walls of the power plant (room), the charges then become part of operations and maintenance of facility services. Again, this distinction between operation of facility services and utilities, and maintenance of facility services, is an individual organization definition. However, a distinction is usually made through use of a series of organization work or charge order numbers.

Vacations, holidays, sick days, etc. Days of missed work, and holiday and overtime pay, are also part of the facility department budget and must be considered when preparing the estimated cost of facilities operations. Because of the character of operations and maintenance work, often the only time a piece of critical production equipment is available for necessary maintenance and service is on a holiday or weekend. This premium time for work performance costs extra dollars and must be taken into consideration in planning the facility department budget. If these expenditures are not planned for and they do occur, they usually upset preplanned budget figures at the year's end. Then adjustments are hard to make, especially when an overexpenditure is involved.

Building and grounds maintenance. Costs incurred for labor, material, parts, supplies, and equipment to perform necessary repair and maintenance work on building structures, inside and out, such as roof repair, painting, floor repair, air distribution and handling systems, etc., or on the facility grounds, such as parking lot repair, fence work, grass mowing, etc., are termed *building and grounds maintenance costs.* Labor, material, parts, supplies, and/or equipment costs are distinguished by separate predetermined organization charge or work order numbers.

Other operations and maintenance costs. Other individually distinguished operations and maintenance cost areas may include transportation, portable tool repairs, furniture and office equipment repairs, office and shops rearrangements, safety program activities, facility security tasks, etc. As previously mentioned, these costs are not always deemed operations and maintenance costs. However, if they are included in operations and maintenance, they play an important role in developing the facility budget. No two facilities are the same in this respect. Facilities differ in their budget preparation methods and cost accounting procedures. However, most organizations use one of the two formats previously discussed as a base, incorporating their own needs and detail into the methods of preparation and control to achieve individual goals and objectives.

Use of organization work order numbers. When the organization work order numbering system is used, usually a basic work order number is established. In a number such as F.O. 0801956000, for example, the first three digits show that it is a maintenance work order number. The second three digits usually represent the specific shop within the facility department, such as 195 for electricians, 192 for welders, 191 for carpenters, etc. The third group, in four digits, is used for specific work areas. For example, all 2000-series numbers could be for a specific area, or building, within the facility. Individual jobs are assigned specific work order numbers, and work is then charged to these work order numbers. The use of these numbers can be as creative as the organization chooses to make it. The more specific the breakdown in numbering, the more specific the cost detail. Sometimes the last group of numerals is required to be a five- or six-digit number because of the organization's size and the complexity of its activities.

As previously stated, no two facility department budgets are the same. However, listed below is what a typical facility department budget would look like in a typical organization that is average to large in size. These percentages include both labor and material cost.

Budget item	Percent of budget
Maintenance of facility services	8.7
Maintenance of production machinery and equipment	23.1
Laboratory equipment maintenance	4.0
Operation of facility services and HVAC	22.3
Vacations, holidays, etc.	4.1
Buildings and grounds maintenance	0.5
Other functions	37.3

Control of facilities costs and the budget. Today increased emphasis is being placed on the facility manager to ensure efficient facility maintenance and operations. The managers of modern facilities require maximum return on dollars invested in all facilities. For this very reason, use of improved work methods and procedures, time standards, work sampling techniques, and quality control measures along with automation devices have become a way of life for facilities operations and maintenance.

Production control practitioners study methods for improving process and production procedures, looking for ways to improve effectiveness and utilization of laborpower, facilities, and equipment. Their ultimate objective is to increase productivity and minimize cost.

In organizations today, many of these same production control techniques have been carried over and used in facility department functions. Organization managers are no longer concerned only that a large piece of production or support equipment is producing an end product. They are concerned about unscheduled machine and equipment downtime, repetitive repairs, improved quality control measures, and more effective utilization of laborpower, facilities, and equipment.

Many of the same work measurement principles and techniques previously applied to production processes by industrial engineers are now being applied to facility operations and maintenance practices. The objective of applying work analysis and performance data is to improve the effectiveness and utilization of assigned laborpower.

Labor costs are further defined through work analysis. Work order schedulers and planner and estimators have been added in many facility departments to more effectively execute the operations and maintenance function, at minimum cost, with no organization mission degradation.

This has the final objective of obtaining maximum return on dollars invested. Sophisticated work measurement programs, materials standards, and improved work methods and procedures have given the facility organization today a new engineering profile and, in many places, a new organizational structure. To help facility managers implement their changing role, a new title and job function has been added in many organizations, that of the facilities engineer.

In facilities operations and maintenance today, there is an increased demand for:

Organization profitability

Maximum return on dollars invested

Engineering reliability in facility operations

Buying proved and performance tested supplies, equipment, and materials

Improved procurement and inventory control procedures

Efficient utilization of laborpower and facilities

Improved overall organization performance

How has facility operations and maintenance procedures and processes changed and what effect does this have on preparation of the facility budget? What made this change take place? Here are but a few of the reasons:

- New operating guidelines by federal, state, and local authorities (government regulations as dictated by OSHA and ADA)

- New and improved machinery, systems, and equipment operating at higher speeds and temperatures, more specifically, the numerically controlled machine operating on as many as 12 different axes and controlled by tapes and computers

- Greater need for higher-skilled craftspeople to troubleshoot, inspect, and maintain new high-rolling production machine tools with their electromechanical control valves and solid state and integrated circuit control

- Greater emphasis on cost reduction, improved operation, and more efficient utilization of existing laborpower, facilities, and equipment

- Shortage of trained and skilled laborpower for facilities operations and maintenance procedures because of the changing profile

- More sophisticated production methods and procedures

Facility operations and maintenance today is certainly no longer the "broom and mop" operation as thought of for so many years. Such an operation will not suffice in industrial plants or office or residence complexes where facility operations and maintenance has the distinct responsibility of maintaining several thousand square feet of building facilities under roof as well as the complex and sophisticated supporting equipment, tools, and systems.

Control of facilities operations and maintenance costs today starts with organization management. Efficient operation of production equipment and facilities is only as good as the programs implemented by top management.

Direction given by those responsible for developing and approving facility operations and maintenance plays a big part in determining the facility operations and maintenance budget. Often a particular organization may be preventive and predictive maintenance oriented, while another may have a run-till-breakdown operating philosophy. Certainly planning of the facility budget would be more realistic in the first organization than in the second.

There are no cut-and-dried rules for budget preparation or budget control. Each organization, however, must determine goals and objectives for its facility department and then ensure that all concerned in the chain of command know, understand, and will endeavor to attain these goals and objectives. From this basis, it is relatively easy to estimate the proposed facility department operations and maintenance costs and to provide the controls to ensure that these goals and objectives are met.

Capital Investment Analysis

Introduction. In the field of facilities operation and maintenance, certain fundamental tools and definitions are required to provide the facility manager with an understanding of basic economics, accounting, and finance. This section contains an overview of economics, accounting, and financial analysis as they apply to capital budgeting and in particular an approach to analyzing the costs of improving a facility's working conditions. Each facility manager should understand this approach before attempting to make simple economic comparisons of alternative investments. Improved or newly acquired facilities should be justified.

Depreciation. Deterioration of equipment and facilities takes place over time. This should be reflected in an economic study for capital investment. The deterioration, or lessening in value, is called depreciation. Factors influencing depreciation are

Life of the organization

Life of equipment

Inadequacy of equipment

Obsolescence of equipment

Requirements of law

Depreciation may not be equal to the actual physical lessening of value, since it is merely an accounting for this lessening in value and does not necessarily represent a fixed sum of money set aside. The objective of depreciation accounting is to amortize the invested capital funds. In the perfect situation, the value of the equipment, plus the accumulation in a depreciation fund, would be equal to the original investment.

Normally depreciation is taken into account yearly in accordance with a predetermined function. The service life, salvage value, and attendant mathematical function serve to determine the yearly depreciation cost. The usual depreciation methods are listed below.

Straight line

Declining balance

Sum of the digits

Sinking fund

Appraisal or book value

Unit method

Comparison of alternatives.   A rational decision concerning investment of capital usually requires a comparison of several courses of action. The difference between possible alternatives is expressed in monetary terms. Organization policy should stipulate that new or improved facilities may be justified solely on the basis of economic (monetary) benefits, requiring an economic study of various alternatives for all projects over an assigned total cost.

It is important that the alternatives be clearly defined and that all appropriate alternatives be evaluated. In many situations, decisions are made by default; that is, not all the alternatives are recognized and a decision is based upon a poor alternative. In other situations, no consideration is given to improving existing conditions. For example, a work simplification study might show reduced costs as an alternative to procuring new mechanization.

Comparisons are made by considering the alternatives in pairs. That is, a base alternative (alternative A) is a make-do alternative. Each of the other alternatives is compared to alternative A. It is necessary to develop calculations that will produce comparable figures for the alternatives. The common calculation methods follow:

Equivalent annual cost

Present worth

Rate of return

When it is possible to determine only one of the alternatives, each method of comparison will give the same result. However, other matters (minimum rate of return, for example) may add advantages or disadvantages to each method.

Equivalent annual cost.   In this method, a nonuniform series of expenditures is converted to an equivalent uniform annual cost over the life of the alternative. This term is sometimes shortened to "annual cost."

Generally, operating costs are stated as annual figures or can be estimated as average overall costs. The only nonuniform costs are investment costs. It is necessary, then, to convert the investment cost to an annual cost of capital recovery. Convention dictates that expenses or savings that occur during the year are treated as if they occurred at the year's end. This convention is almost universally used in economic studies. The use of capital recovery factors accounts for the "time value" of money. Sometimes it is useful to merely tabulate the annual cash flows of one alternative as compared with another. However, a comparison of this type is not useful for choosing between alternatives, since it does not take into account the time value of money.

Once the rate of interest is decided upon, the equivalent annual cost provides a satisfactory and simple method of comparing alternatives.

Present worth. The present worth method converts cash flows to an equivalent single figure at the start of the project life. It also depends upon deciding upon a standard interest rate. In one sense, present worth is the opposite of annual cost, since it converts annual figures to an equivalent single amount at zero date. In fact, present worth can be converted to equivalent annual cost by multiplying by the capital recovery factor. Calculations can be based upon either differential cash flows or annual expenditures.

Rate of return. Discounted cash flow rate of return is used when it is desired to determine the prospective interest rate of return for an investment. This is a trial-and-error method. It is the primary method used in large organizations.

Salvage value. Salvage value is the net amount of money that is expected to be realized from an equipment item when it is sold at the end of its useful life. If the economic study is for a shorter period than the expected life, residual values at the end of the economic study should be used instead of salvage values. Salvage values are normally determined as follows:

In equivalent annual cost studies, salvage value is subtracted from investment cost, but an annual simple interest charge is added to value.

In the present worth method, salvage value is converted to an amount which will yield the desired salvage value at the end of the investment life.

In cash flows, salvage value occurs as a positive value (receipt) at the end of the investment life.

Economic studies. The purpose of an economic study is to decide on a selection between quantified alternatives based on money and time.

The economic analysis involves an input of some type of investment over a period of time. This investment results in an output of an income over the same time frame. In order for the investment to be a measured success, the ratio of output to input must equal or be in excess of unity.

In comparing alternatives in an economic study, the goal is to produce the greatest output for a given input. Profit is the result of two components in the investment mode, investment and income. Investment involves the outlay for performance of an activity. Income is the return derived from the investment. The difference between income and investment is profit, or in the case where the outlay exceeds income, a loss results.

Consider the following alternatives A, B, and C.

Alternative	Investment	Income	Profit (Loss)
A	$500	$600	$100
B	$1000	$1100	$100
C	$2500	$2400	($100)

In terms of profit alternatives, A and B are equal, with C having a $100 loss.

Another measure of profitability is the ratio of income and investment. Alternatives A and B have a ratio greater than 1, which is favorable when compared to alternative C, in which investment exceeds income, resulting in a ratio of less than 1.

Alternative	Investment	Income	Investment income ratio
A	$500	$600	1.2
B	$1000	$1100	1.1
C	$2500	$2400	0.9

Income after deduction of investment is called net income. Income as a return on investment before deduction of operating expenses is gross income. Operating expenses deducted from gross income are operating costs, depreciation, interest, and taxes. The following is a mathematics model for income and investment:

N = net income

G = gross income

O = operating costs

D = depreciation

I = interest

T = taxes

where

$$
\underset{\text{output}}{\underset{N}{\text{Profit}}} \quad = \quad \underset{\text{input}}{\underset{G}{\text{Income}}} \quad - \quad \underset{}{\underset{(O + D + I + T)}{\text{Investment}}}
$$

Investment: Input investment includes the following under the general categories of costs:

Operating costs

Direct labor

- Direct material and services
- Indirect labor
- Indirect material and service

Depreciation

- Equipment and systems
- Buildings
- Land

Interest

Investment in depreciable items funds borrowed from operations

Taxes

Inventory

Sales

Net earnings

Income.   Income output includes such items as products, monetary return, physical plant, concepts, plans, and any act or object quantifiable in terms of output productivity in monetary terms.

Plan for economic analysis.   In an economic analysis, engineering and economic considerations are joined. In order to have a sound analysis, a logical sequence must be followed. As a guide to this sound analysis, a four-step plan should be followed:

Creative step

Definitive step

Conversion step

Decision step

Each of these steps will now be reviewed, in detail:

- *Creative Step*   The creative step involves searching for facts or new combinations of facts. Some facts may be gathered through combination of previously known data, or in other cases facts may be gathered through new research into a problem. The creative step results in finding an opening through a barrier of economic and physical limitations.

  Facts and economics combined with innovation are the prime ingredients needed to create a feasible solution for a problem. The creativity step can and usually does involve more than one solution to choose from.

- *Definitive Step*   The creative analysis step revolves around general concepts. For the purpose of evaluating alternatives, the broad concepts must be converted into detailed alternatives that can be quantified for future evaluation. Some alternatives can be eliminated on the basis of being beyond the scope of the current area of activity. Constraints and limitations narrow down and eliminate some alternative in this step of analysis.

- *Conversion Step*   Alternatives can be compared only if they have the same common denominator. This common denominator is normally expressed in some sort of monetary terms for economic analysis. The alternatives must first be converted into receipts and disbursements with a time base, and second be reduced to a common monetary base with a time frame. Items not definable in quantitative terms must be qualitatively described for evaluation. Common facts required in the conversion step are present cash outlay, capital recovery period, flexibility of proposal to future changes, finance, and effect on employees.

- *Decision Step*   In order to arrive at a decision, the previously developed alternatives must be reviewed along with the alternative of making no decision or maintaining the status quo. Quantitative and qualitative data are reviewed and a quantitative profit is arrived at with regard to a decision on an alternative. If facts are missing or unavailable, judgment must be used. The ultimate decision on an alternative is measured in quantitative and qualitative terms, and judgment is added where facts are missing. In the case of an economic decision, the ultimate decision reduces to what is the most economic or cost-effective alternative.

Estimates in economic analysis.   Income or a return on an investment is the normal measure used in economic analysis. Income is defined as

$$N = G - (O + D + I + T)$$

The outcome can, with judgment, be estimated.

Estimates of income should be based on objective information as much as possible. If income must be estimated, casual factors having a bearing on results should be estimated separately rather than as a whole.

Operating expense should be estimated in the same manner as income, using related parts to make up the total expense.

Depreciation is estimated by using four separate estimates, installed cost, service life, salvage value, and depreciation schedule.

Interest is measured as an expense using the current rate. The rate is determined in case of borrowing money at the current lending rate or, in the case of investment, the rate that would be obtained if the money were invested rather than used as capital money.

Tax rates vary but are normally considered on the basis of percentage of the investment.

Capital investment. Every organization has limited resources to fulfill its mission. Every organization must generate funds necessary to expand or improve its services or products. This must be accomplished with the realization that unlimited resources are not available to do everything that management might desire. This section has as its objectives the general criteria for facility investment analysis. Many current economic analysis textbooks are available to provide the theory and mathematics for implementing detailed cost analyses.

From a budgeting viewpoint, a capital investment is one in which organization resources are committed for expenditure that will, in a future planned period of time, return the invested monies and profits.

Many industry groups and various trade associations will make data available for comparison purposes. Other sources of ratio information as well as definition of the accounting terminology, and capital management concepts, are available from published texts and various cognizant organization publications.

Criteria and strategy for capital investment. Once policy for resource allocation has been established in terms of budget level and the accompanying procedures have been developed, then, within constraints of capital availability, several basic types of projects must be developed. These may then qualify for budgeting and, ultimately, commitment of capital funds. Such projects would be:

- Facilities, systems, and equipment which will improve service and return at least a 10 percent or an assigned rate of return over 10 years or less on a discounted cash flow basis

- Facilities, systems, and equipment essential for security

- Facilities, systems, and equipment essential to provide adequate capacity to operate even if an acceptable rate of return is not achieved

- Facilities that will provide an adequate and appropriate safe working environment for all employees

In other words, the project must pay back the original investment, plus an additional amount equal to the amount which could be earned by investing it at 10 percent interest, or other assigned factor, for the life of the project in 10 years, or whichever is shorter. This is considered a very reasonable expectation in today's security market, especially when the *risk* of new projects is considered. A general basis of justifying capital expenditures can be established with three general criteria:

Working environment

Capacity of facility to perform its mission

Economic opportunity

Capital expenditures that are a result of any of the three listed criteria require justification. The type of justification required will differ in each case.

Working environment.   A basic general objective, derived from policy and procedures as a strategy to invest funds, may be necessary to provide an adequate and appropriate working environment. Each facility manager has the prime responsibility to provide and maintain at a fully functional quality level the physical conditions surrounding all employees within that manager's jurisdiction. A priority and starting point to improve working conditions in the case of multiplant facilities is to direct resources to those facilities containing the greatest number of employees, since the available resources should be used accordingly. Benefits should be provided for all employees, as the resources permit. In general, upgrading facilities factors should include but not be limited to the following:

- Fire safety and environmental standards

- Building useful life

- Employee benefits per dollar of investment

- Mission of facility

For interim repair costs considerations, prior to replacement, four alternatives to be considered are these:

- Make do (meet all standards)
- Upgrade within policy and authority
- Upgrade beyond control limits
- Replacement

Selection of alternatives. A standard practice should be developed for implementing an economic analysis if facility improvements are justified due to environmental factors. Policy should support the view that under normal conditions the most cost-effective correction should be selected. Other alternatives, if selected, must be fully justified.

Capacity of facility to perform its mission. The capacity of a facility to perform its mission must be considered in decisions to improve, modify, extend, or replace these facilities. Always consider economic analysis in seeking realistic alternatives for meeting capacity requirements. The most cost-effective alternative will normally derive from sound management principles; where other alternatives are chosen, they must be fully justified.

Economic opportunity. Improved or newly acquired facilities may be justified solely on the basis of economic benefits to be gained.

Economic analysis of capital expenditures. The purpose of economic analysis is to determine which alternative is most economical when considering capital investments that are justified on the basis of working environment or facility capacity.

If the case is one of pure economic opportunity, the economic analysis serves to determine whether the investment is economically justified.

Methods and criteria of economic analysis follow:

- A primary method of economic analysis is the discounted cash flow rate of return.
- The organization's capital investment policies and procedures should stipulate that improved or newly acquired facilities may be justified solely on the basis of the economic benefits to be gained. In addition, this policy requires an economic analysis to determine the most cost-effective of the alternative means of correcting an environmental or capacity deficiency. These specifications should apply to all analyses of proposed requirements over a management-designated total cost. An individual project or procurement over assigned limitation of total cost may be considered for

approval based on economic factors exclusively if it is estimated to produce a benefit-cost ratio in excess of 1.0 when benefits and costs are discounted at 10 percent to determine their present value. Individual projects or procurement under the assigned limits of total cost require no formal economic analysis. Any investment over the maximum assigned may be considered for approval based upon economic factors exclusively only if it is estimated to produce a discounted cash flow rate of return of at least 10 percent.

- Studies normally commence with the date of decision and cover a 10-year operating span or the useful life of the investment, whichever is shorter. If an exception is warranted, it must be fully explained in the project analysis. A typical exception would be a lease agreement with option to purchase other than in the tenth year. Studies should consider a time span appropriately selected to fit the significant factors pertinent to the project, such as life of investment, period of occupancy, or purchase option dates.

- The base study should be presented based upon actual anticipated dollar amounts of cash flow. For each year under consideration, the best available projection should be used for the values of such elements as land, buildings, equipment, taxes, wages, rents, and contract services. A constant dollar study, based upon current costs, may also be presented if desired, but recommendations should be based upon the base study described above.

- The basis for each decision and assumption must be included in the analysis. The economic and decision analysis, as well as the assumptions, will be validated by the organization's comptroller or the comptroller's designee prior to submission to the approval authority.

Sources of information and assistance. Data sources for justification of economic analyses can be categorized as both internal and external. They follow:

Internal sources include:

Previous economic studies

- Current manning levels as shown on organization charts
- Current space usage as derived from CAD drawings
- Volume projections derived from sales and production
- Current financial reports
- Approved organization and staffing matrices
- Projected pay grades derived from human resources department

External sources of data follow:

Contractors

Sales and marketing representatives

City and regional planning boards (useful for obtaining projected growth figures)

Justification and requirements. The study report should contain the following information:

- Problem definition, and either the reasons for an economic analysis or conversely the reasons for not presenting an economic analysis.

- Qualification and quantification of the need to be satisfied by the recommended course of action.

- An analysis showing that present conditions must be corrected in order to meet present or projected sales and/or production requirements. Not necessary when investment is based strictly on economic reasons (opportunity).

- Economic analysis including, whenever applicable, the following:

  Projected workload level on future use to which the proposed alternative will be subjected.
  Description of alternatives presented in order of increasing initial investment.
  For constant dollar analysis, land and building valuation.

- For anticipated dollar analysis, land value may be escalated at a compound rate of 6 percent, and future buildings value may be increased at a compound annual rate of 3.5 percent while simultaneously declining elliptically at 50 percent.

- Cash flow analysis for each alternative and cash flow comparisons between pairs of alternatives.

Economic evaluation and intangibles. The discounted cash flow analysis of investment alternatives gives a basis for comparing alternatives on an economic scale. However, alternatives may differ on more than just an economic basis. For example, two alternative methods of air conditioning may be under consideration. Alternative A involves individual units placed in windows, while alternative B involves one large, centralized air-conditioning unit. Although the economic analysis may show that A is slightly more attractive than B, consideration should be given to intangible benefits of having one single unit with less chance of failure than several small units. In addition, the single large unit may be quieter and less unsightly than the small window units. Notice that the economic analysis should take into consideration the cost of maintenance of the units in each alternative, but it cannot consider the inconvenience of failure and repair.

Economic analysis does provide a means of evaluating intangible benefits. If alternative A is superior, both economically and intangibly, then the decision is clear. However, if the intangibles of alternative B are more attractive than the intangibles of alternative A, the decision is more complex. An economic approach can be taken, however, by asking whether the difference between alternative A and alternative B should be paid in order to pick up the intangible advantages of B.

Again, if alternative A on a present worth basis is valued at $5000 less than alternative B, it must be asked if whether the intangible benefits of B are worth $5000 over the life of the alternatives. If the answer is yes, then B may be preferred to A, although on a pure economic basis the decision would be reversed.

There is an inherent danger in this type of analysis, however, as it always permits noneconomic elements to suddenly take on economic status. It is essential that *all* intangible benefits of each alternative be enumerated before such analysis is undertaken.

Validation.   The purpose of the validation study follows:

To assure full compliance with organization policies and procedures

To assure the integrity of all economic and arithmetic calculations, and all facts and figures

To assure the validity of all reasoning, rationale, logic, and resulting conclusions

To assure accuracy of all statements of historical fact and future prognosis

To assure that no pertinent information is omitted

To assure that all reasonable alternatives have been given adequate consideration

To assure that all alternatives presented fulfill the operational, safety, and environmental requirements of the organization

The responsibility for validation rests with the comptroller, or the comptroller's designee, at each designated level of authority

The validation process is initiated during the development of the economic analysis and continues through final approval. The validation of an analysis should take place after the analyst developing the analysis has compiled all the justifications, assumptions, and economic data, but prior to completion of the economic analysis.

The validation report will be composed of at least:

The name of the manager or engineer who developed the study

The findings, and validation of the findings, on each of the points mentioned in the specifications for validation

A statement of omissions, reasons for the omissions, and impact of omissions on the analysis (i.e., missing alternatives)

Safeguarding the Facility

Applied Biology

Introduction. Deterioration is considered to be the process of making a material unfit for its intended use. Biodeterioration is caused by the direct or indirect action of biological organisms. Direct action indicates the actual involvement of the organism in material deterioration. An example of direct action is the removal of wood fibers from a timber by the feeding of termites. Indirect action indicates deterioration by an environment which has been physically or chemically changed by biological organisms. An example of indirect biodeterioration is the removal of hydrogen ions from metal surfaces by certain species of bacteria. The microgalvanic cells which result increase the rate of corrosion.

Method of presentation. Included will be an outline of materials subject to biodeterioration, a review of the organisms and mechanisms contributing to the problem, an evaluation of the economics of biodeterioration, the safe control of common biological problems, and a discussion of methods for determining laborpower requirements to correct the problems.

Biodeterioration. Biodeterioration causes billions of dollars worth of damage annually to facilities in the United States. Awareness of its economic importance seems to be limited to biologists or entomologists working in the field of deterioration, and personnel of various organizations adversely affected by the biological environment. Here only those materials used by facility departments in erecting or altering structures or performing operations and maintenance of equipment, systems, or structures will be discussed. The scope will be further limited by selecting typical materials for detailed examination. The purpose is to point out how biological organisms contribute to deterioration.

Type of facilities materials affected by biodeterioration. The bald statement, "No material, whether natural or man-made, is immune to biodeterioration" is open to serious question. However, the only materials known to be immune are too new to have been fully studied. Such a situation is not surprising. The "function" of many bioorganisms is to reduce complex compounds, which are useless in nature, to their simple, useful chemical components. Some of the new polymers were thought to be biologically inert. However, each has, in time, shown some capacity to support bioorganisms. Teflon, for example, has been reported as supporting microbial growth (although the microbes were found to be obtaining support from plasticizers). Some materials subject to deterioration follow:

Wood and paper are subject to deterioration by slime molds, cellulolytic and lignolytic fungi, mollusks, isopods, insects, birds, and mammals.

Metals can be corroded by bacteria, fungi, and a variety of marine organisms. In addition, some of the softer metals, such as lead, are subject to direct attack by a number of insects and mammals, as well as indirect attack by birds.

Rock, long thought to be immune to biological attack, can be deteriorated by rock-boring mollusks.

Masonry may also be destroyed by rock-boring mollusks. Birds, insects, vines, and plants contribute to the deterioration of cement and mortar.

Paints can be discolored by bacteria, blistered by fungi, and destroyed by vines, insects, birds, and rodents.

Plastics and polymers can be dissolved by bacteria and fungi and can be destroyed by mollusks, insects, and birds.

Glass can be dissolved by bacteria and can be etched by fungi and marine fouling organisms.

The above list presents a general view of the scope of biodeterioration. It serves to emphasize the value of accepting three facts as basic engineering axioms:

1. Any structure or mechanism must be placed in a preexisting biological environment.

2. No known material has the capacity of adapting to changes in its environment.

3. All biological organisms have an inherent capacity to adapt to changes in their environment.

Bioorganisms causing biodeterioration. The *general biological requirements* for growth and reproduction are food, living space, and creature comfort. The satisfaction of each of these requirements is in some way responsible for biodeterioration.

1. All living organisms require nitrogen, oxygen, carbon, and hydrogen in some form. Some organisms may have specific requirements for metals and other chemicals. Most, if not all, organisms obtain the chemicals essential for life through the reaction of enzymes and their environment. Enzymes are protein molecules secreted by the organism. Although the mechanism of enzyme action is not completely understood, they may be considered catalysts or even chemical templates. Their function is to separate essential chemical molecules from the more complex molecules found in the environment. Enzymes are considered quite specific in their action. That is, a given enzyme will break down a given chemical because their chemical structures or shapes coincide, permitting the enzyme to split the complex chemical into smaller components. Sometimes two or more enzymes must act in series or parallel to complete the breakdown of certain chemicals. When this phenomenon occurs, the enzymes are called *coenzymes*. Once the chemicals are broken down by the enzymes, they must be brought into the organism to be converted into protein and energy. Normally, the chemicals are brought in by osmotic pressure. Osmotic pressure is the pressure built up by a chemical on one side of a semipermeable membrane. As the concentration of the chemical increases, the osmotic pressure increases. If the membrane is permeable to the chemical, a flow through the membrane will begin from the side of high concentration to the side of low concentration until a balance is achieved. The basic unit of life is the cell. The walls which surround living cells are semipermeable membranes. The above non sequitur has been used to emphasize the need for enzymatic action in living organisms. The enzyme function is to make chemicals which will pass through the semipermeable membranes.

2. Living space, although not as important in biodeterioration as food requirements, is responsible for considerable destruction. Damage to wood by carpenter ants, pholads, and teredinids is the direct result of living-space requirements. The problems occur because the damaging organism needs a space in which to live, an access to the space in which it lives, or material to furnish the space in which it lives. The carpenter ant, the wood-boring mollusks of the teredinid family, and the rock-boring pholads all establish a shelter in materials used by humans. Rats and mice damage wood and other soft materials to establish runways between their living and food-supplying areas. They also destroy the same materials to build nests.

3. Creature comfort, the least important factor, does cause some damage to materials. Certain microorganisms thrive better in acidic or basic environments. Their metabolic processes tend to make the environment more favorable to their survival. As they spread from an area containing a surplus of water, they carry the water with them. The water-bearing tubes they possess enable them to alter an otherwise unfavorable environment into one in which they can thrive. *Microbiological organisms,* such as bacteria, molds, and fungi, are normally too small to be seen with the unaided eye. They are responsible for the expenditure of millions of dollars and thousands of labor-hours annually. The money and laborpower mentioned are spent to determine ways of combating microbiological deterioration, not to correct the damage caused.

So far, three general ways of microbial deterioration have been found:

1. They weaken a material by using part of it as food.

2. They secrete or excrete chemicals which weaken materials.

3. They precondition a material or its environment so that other organisms or mechanisms can carry on deterioration.

In addition microbes may deteriorate materials by their simple presence. The discussion of microbes will be limited to the growth requirements of bacteria and fungi, the mechanisms of bacterial corrosion, the effects of bacteria and fungi on paints and films, and the destruction of wood by fungi.

Biodeterioration is a potent factor in determining the useful life of engineering and construction materials. At this point, one wonders about the cost of equipment or structures. Is the cost that of design and construction? Or is it the cost of design, construction, operations, and maintenance for the designed life? From the facility department's point of view, operations and maintenance costs for the designed life would seem to be a pertinent part of the total cost. If so, it would follow that in determining the lowest overall total cost of a structure consideration must be given to the effects of biodeterioration at the design stage, since biodeterioration is an important factor in the useful life of materials.

Common biological problems and their control. Rodents, birds, and insects are the animals most often causing problems in facility buildings. Weeds, vines, and wood-rotting fungi are the common plant problems. Each group mentioned is a pest inasmuch as it interferes with organization operations, destroys or downgrades raw materials or finished products, creates safety hazards in the working environment, or poses a health threat to personnel. Control in each instance will depend upon the particular pest and the environment in which the organization is located. The life history of the pest, the frequency of its introduction, and sanitation are all factors in planning a control program.

A well-integrated program consisting of inspections and physical, chemical, or biological control measures can prevent or reduce biological problems. Once the pest populations have been reduced to acceptable proportions, a preventive program will minimize recurrence.

Rodents.   Many facilities are located on the outskirts of densely populated areas. As a result, they are continually subjected to invasions by rodents. Rodent pests found within buildings are the Norway rat, roof rat, house mouse, and field mouse. Their seriousness is often underestimated in facility buildings because it is felt that they can do little damage. However, they can cause fires, disrupt communications, and weaken structures. They can cause rat-bite fever and are implicated in the transmission of marine typhus, plague, hemorrhagic jaundice, and salmonellosis.

Rodents invade structures for shelter or food. Rodent inspections should be conducted periodically so that rodent presence can be detected before a buildup of large populations can take place. Rats and mice are positively thigmotropic or stereotropic. They tend to travel along solid objects. They prefer walls and tunnels to open spaces. Therefore, inspection for rodent signs should concentrate on peripheral areas: along walls or under pallets, stacks of stored material, machinery, and equipment. Ideal runways and nesting areas are provided by utility tunnels and sewers.

Fresh droppings are an indication of active rodent infestation. They will be black, shiny, and have the consistency of putty. Old droppings will vary from black to gray and be brittle.

Rodent tracks are also possible clues to runway locations. If dust has accumulated in likely areas, it should be inspected for tracks. If there is no dust, smooth patches of flour (tracking patches) may be used to detect runways. Examination of the tracking patch or dust should be made with a flashlight held near and almost parallel to the floor. In this position the light will cast shadows, making tracks more visible. Rub marks are often found along rodent runways. The marks are made by natural oils and dirt clining to the rodent hair. They rub off on materials frequently passed by the rodents. They generally indicate a long-standing infestation. If the marks are fresh, they will be soft and greasy. Old marks will flake if scratched.

Once the presence of rodents is verified, entry points should be sought. Entry points for mice may be as small as $\frac{1}{4}$ inch. For rats an opening 2 inches in diameter is large enough for entry. Common entry points are oversized holes for utility lines through foundations or floors, settling cracks or holes in the foundation, crawl spaces, and open or ill-fitting doors and windows. Chimneys, skylights, and openings under eaves have been used for entry. It should be noted that if the head of a rat or mouse can go through an opening, the entire body can follow. The ground around the outside perimeter of buildings should be inspected for rat burrows or holes. If holes are found, they should be covered with dirt and reexamined the next day. Holes found to be open are active. Crawl spaces should also be examined for burrows. In crawl spaces, attention should be paid to support columns as well as the foundation perimeter. Once entry points have been located and blocked with cement, sheet metal, or hardware cloth, a rodent-reduction program can be initiated by trapping, poisoning, or (in the case of mice) use of insecticidal dusts.

Trapping may be time-consuming, as the traps must be examined daily and reset as necessary. For small infestations, traps could be the most economic control method. Contrary to popular belief, rodents will not avoid traps with human odor. Nor will they avoid traps contaminated with blood of previously caught rodents. However, they will avoid any new object placed in their environment. Unset, unbaited traps should be placed in the runways 2 or 3 days before the trapping program starts, to overcome this fear of strange objects. Traps should be placed against the wall so that they will snap toward the wall. In the case of roof rats, traps can be nailed to joists. If baits are used, peanut butter and cookies can be as effective as meat or cheese. However, baits are not necessary. A four-way trap with an expanded trigger will catch as many rodents as a baited trap. The trigger can be expanded by firmly attaching a piece of cardboard or sheet metal to it. Care should be taken to make sure that the expansion does not interfere with the locking or snapping action of the trap. The expanded trigger will release the snap when touched by a rodent crossing the trap.

Poisoning is the best method of rodent reduction for medium and large infestations. Some highly toxic chemicals have been recommended. They have the advantage of killing rodents immediately. However, they have two disadvantages. One is secondary poisoning. That is, an animal eating a rodent which has been killed by a toxicant will also die. Secondary poisoning can result in expensive lawsuits. Another disadvantage is that rats eating a sublethal dose will associate pain with the bait and refuse to eat what is offered. This is known as "bait shyness." Bait shyness will destroy the effectiveness of a poisoning program. Anticoagulant baits at 0.025 percent in cornmeal are recommended because the risk of secondary poisoning is minimal; and because the effects on the rodents are painless, no bait shyness develops. However, rodents must consume anticoagulant baits for a period of from 5 to 15 days before they will be killed. At times rodents will ignore the cornmeal bait. This refusal is generally because of food preference rather than bait shyness. The bait can be made more attractive by adding a little sugar, honey, vegetable oil, or peanut butter.

Construction, or demolition, near facility buildings may result in the movement of large numbers of rats. Permanent bait stations outside buildings will attract them and reduce the probability of rat invasion. In colder areas, field mice will tend to move toward the shelter of buildings as winter approaches. Permanent bait stations should be established in the fall to control the mice before they can enter the building. Bait stations can be set along fence lines or building perimeters.

A bait station may be any object, open at both ends, in which bait can be placed. A section of pipe, a small rectangular wooden box, or any other object which will protect the bait from the weather and permit rodent access may be used as a bait station. Bait stations, once established, should be inspected once or twice a week to ensure an ample supply of bait.

Rodent populations are generally as large as the food supply will support. In areas where rodents have free access, rodent populations reduced by trapping or baiting will rise to the original level after these control measures stop. Therefore, sanitation is an important aspect of rodent control. All food and garbage should be made inaccessible to rodents. Food should be stored in rodentproof containers. Spilled foods should be cleaned up promptly and stored in metal garbage cans with tight-fitting lids.

Further information on rodent control may be obtained from one of the following sources:

U.S. Department of Health, Education and Welfare Public Health Service Communicable Disease Center, Atlanta, GA

U.S. Department of the Interior, Fish and Wildlife Service Bureau of Sports, Fisheries and Wildlife

Any state or local health department

Birds. Birds present a legal problem, as well as production, operations, and maintenance problems, in and around buildings. They are often protected by international treaty or federal and state laws. Most laws contain escape clauses which permit the control of protected species when an economic or health hazard can be traced to them. Therefore, thorough consultation with states' fish and wildlife, natural resources, or game commissions is advised before any bird-management program is undertaken. Generally speaking, pigeons, starlings, and English sparrows are not protected by law. These three are the most common problem birds. However, permits may be required for their control.

Bird problems are created by their nests and their droppings. Bird nests create fire hazards. Birds have been observed carrying lighted cigarette butts to their nests. They have been known to build their nests in such a way as to cause motors to overheat. Bird nests attract carpet beetles, which are serious pests of animal products such as silk, feathers, furs, hides, bristles, and leather.

Bird droppings cause safety hazards, since they are very slippery when wet, especially on metal surfaces. Droppings also cause health hazards. Histoplasmosis-, cryptococcosis-, and salmonellosis-causing organisms have all been found in bird droppings. Production, operations, and maintenance problems develop in areas heavily contaminated with bird droppings. Contaminated material takes longer to move because it must be cleaned before it can be used or shipped. Contaminated material also deteriorates faster. Painting costs are considerably higher when bird droppings must be removed prior to painting.

It may seem foolish to recommend a bird survey, since everyone can see where the birds are. The real key to bird management is knowing not just where the birds are but how they arrived. Each building presents a different environment and requires knowledgeable observations to determine local flight patterns. Control measures, whether trapping, repelling, or baiting, must be planned according to the species and habits of the particular group involved. Observations should begin in the early morning before the birds leave their overnight roosts. The behavior patterns to be noted are the areas chosen for overnight roosts, the areas chosen for nesting, secondary roosts, or perches as they leave the building, and feeding areas. Evening flight patterns are also important. After several observations, key control points may be noted, that is, roosting or perching areas used by most birds as they work their way to or from their night roosts. Control efforts should concentrate on those points.

The most effective bird control measure is exclusion. See to it that all windows and doors are kept closed or screened. Block off or build up all roosting and nesting areas. The roosting areas can be blocked by screening with ½-inch-mesh hardware cloth. Ledges may be built up with smooth wood or masonry so that they have a slope of 45 degrees or more. Birdproofing of large facility buildings is often impractical. Overhead cranes prevent the screening off of nesting sites, exposed girders offer a multitude of roosting and nesting sites, and doors are constantly open for the receipt and shipment of material. Such problems can be solved only in the design stage. Girders can be hidden. Shipping areas can be designed so that they do not open directly into plants and warehouses.

One method of bird management is to trap the problem birds and remove them far from their roosting areas or dispose of them humanely. The size and design of the trap will vary with the location. The species of bird to be controlled will also affect trap selection. Information on bird traps is available from the U.S. Fish and Wildlife Service, Division of Wildlife Services, Agricultural Experiment Station, Purdue University, Lafayette, IN.

Many devices have been proposed to scare birds. Temporary effects have been achieved by stuffed owls, noise makers, revolving lights, and supersonic sound. None has had long-lasting success. As soon as the birds become accustomed to the device and learn that it does not cause pain, they return. Starlings can be driven from roosting areas by high-fidelity recordings of their attention and alarm calls. The results can be improved with sporadic use of bird shot. The shot helps teach the starlings that the calls are not all false alarms. An electronic-sound producer has been claimed to drive birds from large areas. However, it is not known if the effect will be long-lasting. It is suspected that the judicious use of bird shot in conjunction with the sound will increase its effectiveness. Another proposed scare device is a system of vibrating wires installed along roosting areas. The theory is that the birds will find the vibrations uncomfortable and will not land.

Sanitation is important in bird management. Food is a major attractant for birds. They will often feed on lunch scraps or be fed by personnel. Bird feeding should be prohibited. Lunch scraps should be placed in covered containers.

One other physical method of bird control is the removal of eggs and nests. In order for this method to be successful, all nests and eggs must be removed at 2-week intervals until the birds become discouraged and leave.

Chemical repellents have been used to reduce bird populations within buildings or to move outdoor populations. They are effective as long as they maintain the stickiness or slipperiness upon which the repellent action depends. With time, in an area of heavy dust, they often become covered with soot, dust, or other airborne material and lose effectiveness.

A number of roost poisons have been used to reduce bird populations. Some have been applied directly to the roosts. Others have been placed in artificial perches with wicks to draw the poison to the area contacted by the bird's feet on landing. The roost poisons are effective but have drawbacks. The poisons used can harm humans and will kill any bird landing on them. Their use should be restricted to indoor areas where personnel will not come in contact with them and where nontarget birds will not be killed by accident.

Several bait poisons have been used to reduce bird populations. These poisons are generally mixed with corn or the seed upon which the birds will feed. They are either exposed where the birds normally feed or in an area where the birds have been taught to feed by "pre-baiting." The major disadvantage of these poisons is that they sometimes kill nontarget birds and result in a flurry of unfavorable publicity. Pigeons and sparrows roosting indoors can be controlled with 0.6 percent strychnine on an appropriate-sized grain. There is a secondary-poisoning hazard, and its use should be limited. Starlicide has been used for the control of starlings at their feeding grounds. Unfortunately, birds other than starlings have been killed by this method. Avitrol has been used for the control of sea gulls at airports and of pigeons indoors. In order to control birds indoors, bait stations (platforms firmly attached or suspended from rafters and beams) should be established. The bait stations should be high enough not to interfere with organization operations and should have a lip to prevent the bait from spilling. Unpoisoned grain of the appropriate size should be exposed on the platforms. The birds may avoid the bait for a period of time. It may take 7 to 10 days before the birds recognize the bait stations as a source of food. Bait should be replenished as it is consumed (once a day at the start). When most of the birds have accepted the baiting area, treated bait may be exposed.

Strychnine is a fast-acting poison. Birds that eat a sublethal dose feel a gripping pain. Since they are near the bait when the pain occurs, they will avoid that area in the future. This reaction is known as "site shyness." It differs from the bait shyness in rodents in that the birds will continue to eat the bait if it is placed a short distance from the original feeding area. This reaction is used to make Avitrol effective. Avitrol is a slow-acting chemical. It takes up to 16 hours before its effects are manifested. At this time the birds will be in their roosting areas. Because of the instinct to avoid places associated with pain, they will move. Company literature claims that few birds are killed (the old and the weak). However, up to 10 percent kill may be expected. Dead birds should be picked up and incinerated or buried. These disposal methods reduce adverse reaction.

Outdoor baiting of pigeons and starlings, when permissible by state law, can be accomplished at known feeding sites. Outdoor baiting should be scheduled only at those seasons of the year when migratory and nontarget birds are absent from the feeding sites. In the north, the best time for pest-bird control is winter, when snow covers the ground and natural food sources are scarce. Dead birds should be collected and disposed of discreetly.

Plants.   Any plant that grows where it is not wanted is a weed. Sometimes weeds create fire or security hazards. Plants growing around oil tanks or along fence lines are examples of this type of hazard. Sometimes weeds interfere with normal maintenance procedures, like plants growing along power-distribution lines or around utility poles or runway lights. At other times weeds detract from the natural or planned aesthetics of an area. Lawns do not appear as neat when dandelions, plantains, and crabgrass appear in patchy spots. The forgoing situations can all be corrected.

Herbicides may be classified by the kinds of plants they affect as selective or nonselective. The selective herbicides at the recommended rates will kill one or a number of plant species while not affecting others. Herbicides are also classified by how they affect the individual plant as contact or translocated herbicides. Contact chemicals kill that portion of the plant they contact. Translocated chemicals are absorbed by one portion of the plant and move to another through the plant's circulatory system to kill the plant. Some herbicides, when applied to the soil, kill all plants and are called "soil sterilants." Soil sterilants may be contact or translocated chemicals.

Care must be taken in the application of all herbicides. Overdosing with a translocated chemical on the leaves of a plant may kill them too fast for it to be absorbed. As a result the roots, which are really the objective of the application, will be left intact for later growth. This is particularly true in poison ivy control. Overdosing may make a selective herbicide nonselective. Excessive dandelion control may kill an entire lawn. Careless application may result in valuable plantings being lost. Herbicides should never be placed within the trip line of a desirable plant which is susceptible.

Timing is also important in herbicide applications. Preemergence applications, as implied, are made before the weeds emerge from the soil. The objective is to kill the weeds as they begin to sprout. Postemergence applications are made after the weeds emerge. Here the objective is to place the chemical on the leaves while the plant is growing vigorously, making translocation easier. There is also a preplanting treatment which is made before desirable species are planted. Here the object is to kill the weed roots and have the chemical deteriorate before the desired plants begin to root. Nonselective herbicides, designed to kill all vegetables, may generally be applied at any time. Some herbicides have a very short time period within which they can be effectively applied. Be certain that label directions are followed in respect to tuning as well as to dosage rate. The variables in climate, soil, water conditions, and plant varieties are such that no specific recommendations can be made here. Current information can best be obtained from the local county extension office of the state department of agriculture or from a local university.

Fungus attack on wood has been described above. Damage can be minimized during the design stage by the use of treated wood and by adherence to practices which reduce moisture within the structure. With all the biodeterioration wood encounters, one might think it was a poor choice as a construction material. However, wood preservation prevents these problems. The treatment of wood with insecticidal-fungicidal chemicals isolates the wood from insects and fungi and makes it immune to biodeterioration, as long as the treatment barrier is unbroken. Many seem to think that wood "just naturally deteriorates." While this statement is true in the technical sense that natural forces are at work in wood deterioration, it is false in the implied sense of inevitability. Wood deterioration can be prevented or stowed to one-fifth to one-tenth the normal rate. This means that a piece of lumber which must be replaced every 5 years because of exposure conditions may be replaced at 25- to 50-year intervals. If one considers the cost of labor, the total cost of untreated wood is many times greater than the cost of treated wood, even if a 100 percent premium is paid for treatment.

The techniques for treating wood are varied. Although the chemicals are relatively standard, the solvents used and the methods of treatment will differ according to the degree of protection needed or desired. Wood may be pressure-treated, dip-treated, or surface-treated. The choice of the method will depend upon whether or not the wood is already in place, the degree of exposure to moisture, the climate of the area where the wood is to be used, and the function the wood is to perform in the structure. Wood in contact with soil, masonry, or metal will need greater protection than wood which is completely protected from the weather and condensation. Wood used in the warmer southern states will need greater protection than wood in northern states. Supporting members of a structure will need more protection than incidental wood.

Pressure-treated wood offers the greatest protection for the cost of any wood-treatment method. There are two basic methods of pressure treatment, the full-cell and the empty-cell methods. In the full-cell method, the wood is placed in a treating cylinder and a vacuum is drawn. The preservative solution is introduced into the cylinder, and then pressure is exerted to force the preservative into the wood. Wood treated by the full-cell method has a tendency to bleed and should not be used where appearance is a factor or where it will be behind dry walls. Nails driven into full-cell-treated wood will form a channel through which the preservative solution will bleed. The result will be a wall that is stained in the areas of the nail heads. Nothing can stop the bleeding except the reduction of pressure within the wood itself. This process may last several months. Painting will have no effect on the bleeding. The full-cell method is best for wood that is exposed to marine and severe leaching conditions.

In the empty-cell method, the wood in the cylinder is subjected to pressure before the preservative is introduced. Then even greater pressure is exerted. The result is that, when the pressure is released, most of the preservative solution is forced from the wood cells, leaving a coating of protective chemical on the cell walls. The empty-cell method is best for conditions where there will be moderate leaching conditions and where appearance and paintability are factors.

In the solvent-recovery process the wood is steamed after treatment to remove excess solvent and reduce bleeding. In the liquefied-petro-leum-gas process, the preservative chemical is dissolved in a liquefied gas and no solvent is left in the wood. This process gives the cleanest wood.

There are three general classes of chemicals used in wood treatment: creosote solutions, oil-borne solutions, and water-borne salts. Each class has its limits and values. All except the liquefied-petroleum-gas process will leave the wood discolored. Most, except those using cre-osote, can be treated to leave the wood paintable. Each will vary in its resistance to leaching. Leaching is a tendency of the chemical to migrate from the wood when exposed to moisture. The rate of leaching is in direct proportion to the solubility of the preservative chemical in water. This would seem to indicate that all water-borne salts have a high leaching rate. However, some of the salts become bound to the wood after the water has evaporated, and current research indicates that they will not leach under severe exposure conditions.

Coal tar–creosote, creosote–coal tar, and creosote-petroleum solu-tions are the best preservatives known. But they can be used only where appearance and paintability are not factors. They have a ten-dency to bleed and should not be used inside a building. Their most common use has been for piles in waterfront structures or as building supports, utility poles, posts, and railroad ties.

The oil-borne preservatives give excellent protection. They are resistant to leaching and can be made water-repellent and paintable. Treatment with oil-borne preservatives must be specific for the intended use. Wood subject to wetting by rain should have water repellency added. The water-borne salts are clean. If a moisture con-tent of 19 percent or less is desirable, the wood can be dried after treatment. Water repellency cannot be added to the water-borne-salt treatments.

All these treatments are recommended in pounds per cubic foot. The amount of chemical added to the wood will vary with the intend-ed use. It is recommended that standards established by the American Wood Preservers Institute be followed. These standards include a quality-control program that enables the buyer to inspect the wood after delivery and to reject nonconforming pieces at the seller's expense. The standards also permit incising so that the weight requirements may be more easily met. Incising is slicing the wood parallel to the fiber direction to a predetermined depth at predetermined intervals. If wood is to be painted, the buyer should specify that incising on one side only is permitted. If water repellent is required, it must be specified. If paintability is required, it also must be specified.

The problem of what treatment to specify for various areas of a structure is not too difficult to solve. Pressure treatment should be used for:

1. Wood framing, woodwork, and plywood up to and including the subflooring of the first level in structures having crawl spaces.

2. Lumber and plywood used on the exterior of buildings and exposed to rain wetting such as fascia boards, soffits, columns, posts, balconies, porches, steps, platforms, railings, or members in contact with masonry or metal.

3. Framing, sheathing, and other wood used in cold-storage room walls, floors, and ceilings.

4. Wood-framing sheathing or trim used in the walls and floors of shower rooms, including openings from shower to dressing rooms; wood sills, soles, plates, furring, and nailers that are set into or in contact with masonry or concrete.

5. Other areas of potentially high moisture within any particular structure. Any wood exposed to rain wetting, whether or not it is painted, should be given at least a treatment with a preservative which has a water repellent added.

Once the wood is treated, the integrity of the treatment must be maintained during construction. If treated wood is cut, worked, or drilled after treatment, the newly exposed areas should be well brush-coated with at least two brush coats in which the preservative solution is flooded onto the wood surface. The chemical should be the same as used in treatment, and each coat should be absorbed by the wood but not dry before the second application is made.

Precautions to be observed in using pesticides. All chemicals recommended for the control of biological pests are biological poisons. They will kill or damage all living things. The danger increases with the concentration of the chemical being used. No one should be permitted to use concentrated pesticides unless they know the hazards involved not only to themselves but to people in the area they are treating, to fish in waters where the chemicals run off, to wildlife in the area being treated, and to plants adjacent to the treated area. The pest-control technician should also be thoroughly familiar with the applicable state regulations and should be licensed if state law requires licensing. If federal licensing is required, the pest-control technician should be thoroughly familiar with the applicable federal regulations and should also have a federal license. Information regarding the toxicity of each chemical is available from the manufacturer or from the state department of agriculture. First-aid information, procedures, and antidotes are available from the same sources.

Anyone who is mixing or applying pesticides should be required to wear an appropriate respirator, solvent-resistant gloves, and coveralls. If they are applying pesticides higher than the level of their shoulder, a hat and goggles should also be required. The person who is applying pesticides is constantly exposed to fumes and vapors, which are absorbed in small amounts through the skin, mucous membranes, mouth, and lungs. They need the protection to prevent the development of chronic poisoning.

In addition, an area for storage of toxic chemicals should always be under lock and key. Other important safety requirements are a sink with a surface for mixing chemicals and an emergency shower in case of accidental spills.

All personnel who mix or apply pesticides should be given an annual physical examination. Kidney- and liver-function tests should be part of the examination. In addition, those who handle organic phosphates or carbamates should be given a blood cholinesterase test at least quarterly. Any person found to have a low blood cholinesterase should not be permitted to handle organic phosphates and carbamates until the level returns to normal. At the time they return to work, application techniques and attention to safety precautions should be thoroughly checked. A person who cannot or will not adhere to safety requirements should be removed from the job.

As can be seen from the above, training is an essential part of any pest-control program. People must be able to read and understand what they are reading. They must be able to manipulate mathematical formulas. They must be observant and must be able to change plans without notice. Large organizations might consider the possibility of hiring a professional entomologist or applied biologist who would be responsible for training personnel and guiding the pest control program for all organization plants.

Pest control costs.   The cost of any pest control program will depend upon the facility areas and materials to be protected, the pest(s) involved, the equipment needed, and the knowledge and skills of the personnel who do the work. A rough estimate of the time required for preventive pest control within buildings is 1 person-year for each 1 million square feet of building floor space in the northern states. The time increases as you move south to 1 person-year for each 2 million square feet of floor space. A more accurate estimate may be made by using the following criteria, which are expressed in person-hours of productive labor per 1000 square feet of floor space per year:

Office, industrial, and miscellaneous areas	0.7
Warehouses with susceptible items	1.3
Living areas	1.7
Hospital, laboratory, and clinic areas	4.4
Food-handling and -serving areas	5.0

The figures quoted are based on the time required for a continual preventive program. They include a 10 percent lost-time factor. Since too many variables are involved, there are no reliable figures for generalizing pest-control person-hours for outdoor work. Each area must be analyzed on an individual basis.

Once the total workload has been determined, a decision must be made as to whether contract or in-house personnel should do the work. This decision has to be made after reviewing and evaluating all costs, management involvement, and mission accomplishment tasks involved; i.e., pest control contractor availability, reliability, and cost; facility department involvement with pest control contractor work performance, monitoring, and reporting; pest control program involvement and coordination with the organization department's work performance. One problem peculiar to pest-control contracts is that there is no way to determine what has been done after the fact. Many feel that a lack of complaints is evidence of work being done. However, complaints may be low while a problem is building up or getting out of hand. Therefore, it is always advisable to have a facility department representative who is knowledgeable in pest control procedures accompany contractors all the time they are on the premises. Very low bids for pest-control contracts must be scrutinized carefully. They generally indicate:

1. The contractor does not know much about pest-control operations.

2. The contractor has no intention of following the specifications.

3. The contractor is planning on change order extras after contract award.

Either such bids should be rejected or all work done under them should be closely supervised. Preferably, they should be rejected.

Against the contract costs one must compare the probable in-house cost to contract cost considering the in-house productive worker-hour cost, including all overhead costs, to contract productive worker-hour labor cost, including all overhead, G and A costs, and profit. The initial cost for equipment and materials may vary from several hundred to several thousand dollars. The larger figure is applicable if considerable outdoor work using heavy equipment is anticipated.

For indoor work the following equipment is required:

1-gallon compressed-air sprayer with spare	
Fan and pin-stream nozzles	2
Repair kit for the sprayer	1
Bulb dusters	2
Electric aerosol generator	1
Small compression pump	1
Respirators (per person)	1
Respirator cartridges	90
Goggles (per person)	1
Coveralls (per person)	3
Solvent-resistant gloves (pairs per person)	2

Summary.   The biota, or living organisms in the human environment, can cause considerable damage to humans and their structures. This damage can be avoided or minimized by manipulating the environment. Sanitation and exclusion are, in the long run, the most effective methods, since they cause the least environmental discord. Chemical barriers, properly applied, can reduce deterioration without disrupting the environment and without hazard to wildlife. The safe application of chemicals requires knowledge and skill. Knowledge applied to chemical management of the environment includes knowledge of the life cycle of the target organisms, the effect of chemicals on nontarget organisms, drainage flow from application sites, chemical application and disposal methods; and first aid and safety precautions. Skill includes skill in the selection of chemicals, equipment, and safety devices; in the methods of applying chemicals; in detecting the presence of unwanted organisms; and in equipment use and maintenance.

The use of untrained or poorly trained personnel in the control of pest organisms is dangerous and irresponsible. If problem conditions require company personnel, they should be thoroughly trained before they are given an opportunity to apply chemicals. If state laws require licensing of pesticide applicators, they should be licensed. Large organizations should consider the utilization of a full-time applied biologist, or entomologist, to train pesticide applicators, to establish company policy, and to ascertain continued adherence to safety procedures.

Corrosion Protection

Introduction. Corrosion protection of buildings, bridges, highways, and other structures can be divided into many classes. All these structures contain any or all of the many different materials of construction. They contain steel, nonferrous metals, wood, and masonry. The protection of each of these is different, in both application and choice of materials.

Corrosion is defined as the reaction of the substrate with the surroundings.

Erosion is defined as the wearing of the surface due to the environment.

Steel. Steel, of course, probably gives the biggest problems in protection; however, steel is subject to various types of corrosion which are much different from each other. Steel is used for structural members, in which it is subject only to the corrosion of the atmosphere, which may or may not contain contaminants from products manufactured in neighboring facilities.

Steel is used in the construction of storage tanks for many products, but here the inside has entirely different requirements from the outside. Steel includes pipes which also are differently protected inside and outside; and of course, the motors, machinery supports, and so on.

Steel protection methods. Steel is protected in two ways. It is protected by coating the steel with inhibitive pigments which pacify the action of the steel, such as red lead and the various chromates, and cathodically by the use of zinc and other metals connected to the steel either by separate cathodes or by coating directly on the steel.

Preparation of the surface. Regardless of the method of protection of the steel, the metal must first be exposed to the protection, as the protection is designed to protect the metal and not to protect the coatings which inherently form on the steel such as mill scale, rust, and dirt. The plain steel must be exposed to the protective chemicals. This steel should have a pattern on it which permits mechanical as well as chemical adhesion to the protective coatings.

Unquestionably the most satisfactory way of preparing the steel for coatings is sandblasting, and in most cases, considering the present price of labor, sandblasting is the lowest-cost as well as the best.

Many methods are listed and are commonly used for surface preparation. They are solvent cleaning, hand cleaning, power-tool cleaning, flame cleaning, blast cleaning of various types (white metal, commercial, brush-off, etc.), pickling, and combinations of weathering with cleaning.

Some of these are not usually applicable to maintenance painting unless they can be applied to the steel before the steel is erected. It is important to review these cleaning methods in order:

Solvent cleaning, of course, removes only those materials which are soluble in the solvents chosen. This method is usually a preclean method rather than a cleaning method. It involves primarily the removal of oils, greases, and lubricants from rolling compounds and in most cases should be done prior to any other cleaning method.

Hand cleaning is usually considered to be simply the vigorous application of wire brushes, sandpaper, or emery cloth to the surface of the steel, which removes only loose surface contamination. It is impractical to attempt to achieve a truly clean surface by hand-cleaning methods alone.

Power tools are various tools used to reduce labor costs slightly and achieve slightly better end results. They are power-driven wire brushes or scrapers which consist sometimes of impact needle devices which do a much better job only because they require less physical effort.

Flame cleaning is a very common method of cleaning, primarily in Europe. Flame cleaning, properly performed in accordance with the detailed instructions and specifications of the Steel Structures Painting Council and in accordance with these specifications, does remove most of the mill side and previous coatings from the metal. It cannot be used in flammable atmospheres.

Pickling, or chemical cleaning, is extremely effective. In effect, it is similar to high-speed weathering, but it must be performed under very carefully controlled conditions. Pickling performed on an erected structure is seldom satisfactory. Work is continuously being done on this method, however, and some newer materials may also do a rather good pickling job, cold and in place.

Weathering and cleaning are described as a separate method by the Steel Structures Painting Council. Their reasoning is that rust is relatively easier to remove than is the mill scale. If the mill scale is allowed to weather well, it can be removed nearly completely. It is then much easier to sandblast, hand-clean, power-tool, or flame-clean this weathered scale and the balance of the dirt from the surface.

Sandblasting. Blast cleaning is split into several different types. The so-called brush-off cleaning is, as the name implies, simply a case of rather quickly brushing the surface with a stream of air-supported sand. It removes no scale, but it does remove loose dust, loose paints, and some of the dirt.

Commercial sandblasting is similar except that it is much more thoroughly done. It is done to the point where only a haze of reddish discoloration of the steel is permitted to remain. The bulk of the previous coatings and all mill scale are removed by this technique if properly done.

White-metal sandblasting is the ultimate in sandblast cleaning. This removes all contamination from the surface. Ideally, the metal is so clean after white-metal cleaning that passing the sandblast gun over the surface again does not show an additional pattern.

The most objectionable coating on the uncleaned steel is the mill scale. No matter how it is covered, eventually under some exposure conditions, the mill scale will come off, taking with it any coating that is placed over it. Mill scale, in addition, is extremely hard and adherent when fresh but is chemically an entirely different composition from the metal itself with a different electrical potential. This combination is what makes it both so difficult to remove and so inadvisable to coat over. Flame cleaning and sandblasting and weathering remove the mill scale completely. Power tool cleaning is too expensive to use in large areas. On certain limited areas, dependent largely upon the ability and the honesty of the operator, it can do an excellent job on confined areas, particularly when the mill scale has previously been softened or allowed to lose its adhesion by weathering.

After any method of cleaning, the coating must be applied before the surface has become recontaminated.

Steel tank interiors.   On the interior of a tank or on steel surfaces subject to continuous immersion in any fluid, the only acceptable method of cleaning is white metal sandblasting, or what is sometimes called "near white metal." This also removes all contaminants except that it leaves the surface such that you can see the path of the sandblaster. Nevertheless it has removed all real contaminants such as mill scale, paint, and rust, as well as greases and dirt, and has exposed essentially pure iron or steel.

Primers for steel.   After the steel is properly prepared, the next item to be considered is, of course, the primer. The primer has two basic functions. First of all, and possibly the most important, is adhesion to the substrate, but secondarily, it is to inhibit corrosion of the iron and prevent undercutting from scratches or from pinholes which invariably occur in the coating. This sharply reduces the tendency to blister.

Before the structure is erected, while the steel can still be handled as simple pieces, the most effective primer in many areas is galvanizing. Here the zinc is applied to the steel surface after very careful chemical cleaning, pickling, under highly controlled conditions of time and temperature. This, of course, is the oldest method of applying zinc to the surface.

A newer method of applying zinc, which is applicable to a structure already in existence, is metallic zinc spraying. This requires complete and careful preparation of the surface. The application should be made immediately on pure iron metal obtained by white or near-white sandblasting. Even a slight delay between cleaning and painting is serious. This is done by hot spraying the zinc by passing either zinc wire or dust through a gun which heats it to the point where it will fuse to the metal. Zinc coating is also done by applying the dust using one of several types of binders.

Zinc-rich coatings. Much has been written about the advantages and disadvantages of so-called organic and inorganic binders, but in both cases, the purpose of the binder is simply that of holding the zinc dust to the metal. The zinc must be used in high enough quantity so that it forms an electrically conductive coating on the metal. This effectively galvanizes on the job. The advantages or disadvantages it offers over hot-dip galvanize are purely the relative economics of the individual job and the relative ease of galvanizing and sandblasting and painting.

The oldest inorganic binder for the application of the zinc dust is sodium silicate. It is performed by applying the zinc dust with sodium silicate (water glass) and then "curing" the system by an after application of a material (usually an acid) which converts this sodium silicate to silicic acid and, of course, a salt which will wash off in rain or can be washed off with water. Because this was a two-step operation, even though extremely successful, much research and development work has been performed to find a binder for the zinc which will give similar properties and not require this "curing" after step.

This has been accomplished very successfully with lithium silicate, potassium silicate, ethyl silicate, isopropyl silicate, ammonium silicate, and many other chemicals. Yet all these materials in the final analysis lay a coating of zinc dust well adhered to the properly sandblasted surface with silicic acid, and the percentage of silicic acid needed to adhere the dust to the surface is so small that it permits the dust to produce a conductive film.

In all the above cases, you end up with a coating of the zinc on the steel, highly resistant to organic solvents and to heat. Silicic acid is hurt by neither heat nor most organic solvents. It is a coating which, of course, is not resistant to acids and alkalies because the zinc itself is not resistant to acids and alkalies.

At the same time as the work being performed on the application of zinc without a topcoat for curing, many organizations worked on the application of zinc using more nearly conventional organic carrier vehicles. But they still ended up in a similar place in that the ratio of zinc to the binder must be high enough in concentration so that a conductive surface is formed and held in place by these organic binders.

The organic binders, of course, have the disadvantage of being more subject to degradation by heat and therefore less heat-resistant, and (depending on the choice of organic binders) they may be much less resistant to some organic solvents. Being organic in nature, they can be formulated to have much greater flexibility and are somewhat less critical to the surface preparation, although their application on other than a pure-iron surface is not usually successful.

The main characteristics needed of the organic binder are that it be alkali-resistant and that it have excellent adhesion to the surface. When zinc protects steel sacrificially, it forms an alkaline residue. This alkali is not strong enough to damage the silicic acid but is strong enough to damage any organic binder subject to saponification. The binders used with the organic zinc are therefore alkali-proof or -resistant materials like chlorinated rubber, styrene-butadiene, vinyl, vinyl ethers, epoxy esters, acrylics, and epoxies with various types of curing agents such as polyamide, amido amines, and amine adducts.

The particular choice of organic or inorganic binders depends on the end use, the case of application, and the proper surface preparation necessary for the particular job.

Organic binders would never be chosen for an application requiring extremely high temperatures or immersion in many organic solvents, and here the inorganic binder becomes unique. On the other hand, where certain amounts of contamination cannot possibly be avoided between the preparation of the steel and the topcoat, it is found that the organic binders are much more tolerant of these small discrepancies in surface preparation and are therefore easier and cheaper at the present time to apply.

All these binders are in the neighborhood of 10 to 15 percent of the final film, so it can easily be seen that the cost of the coating per mil is that primarily of the zinc metal and not that of the binder. The initial cost of the material itself should not affect the choice, but the differences in cost of application may be important.

Passivating steel primers. The other method of protecting steel is passivation of the steel to corrosion. It is well known that certain ions form protective layers on steel and drastically reduce the rate of its reactivity with oxygen.

The most common of these passivators are red lead and zinc chromate. More recently, strontium chromate, basic lead silicochromate, and certain molybdenum salts have been substituted, in many cases with great improvement in results. For example, the basic lead silicochromate has more reactive material per pound of pigment than the pure red lead. Therefore, a successful paint may be made with basic lead silicochromate weighing only 12 to 15 lb/gal, whereas a top-quality inhibitive primer with red lead probably weighs 18 to 25 lb/gal. It is much easier on the painter to carry a bucket or handle a brushful of the lighter material, the lighter material containing just as much protection. On a spray job, it is much easier for the pumps to lift the relatively light lead silicochromate than the relatively heavy pure red lead when spraying at higher levels than the spray pot.

Currently, the binders of all the above are always organic; so they can be chosen to be unaffected or relatively unaffected by acids and alkalies, while the zinc of zinc-rich primers is soluble in both. The various pigments can be selected to be much less affected than zinc metal to acids and alkalies, but because of the vehicles, they are not as resistant to some organic solvents, and great care must be taken to choose a binder which is resistant to the organic solvent or chemical which may be involved in the particular application. Fortunately, today, organic binders are available which are resistant to a large variety of solvents, and it is usually possible to pick a binder which will withstand the exposure required.

Other primers.   The next most common choice for pigmentation of lower initial cost primers are the iron oxides. These, however, rely primarily on the binders themselves for the adhesion and the protection, receiving little, if any, help from the pigment. The pigments are, however, very opaque, cheap, and easy to use, and they are therefore frequently used for a fast, quick job. They are often used in combination with the inhibitive-type pigments to assist in their coverage and make the application easier.

The primers, other than the zinc-rich, may include as vehicles any or all of the vehicles used in the organic zinc-rich—the epoxy esters, the vinyl ethers, the acrylic copolymers, styrene-butadiene, chlorinated rubber, epoxy polyamide, epoxy with other curing agents, other types of vinyls, and possibly many others.

With the other pigments, in addition to any of the above binders, it is possible to use:

Oil—linseed, fish, tung, or synthetic

Alkyd—short or long oil with any of the oils above

Varnishes—long or short oils

Phenolics—with or without the oil

In addition, organic primers usually contain various amounts of other pigments commonly called "extenders" or "inerts." The choice of the words "extenders" and "inerts" is unfortunate in that it gives the impression to the reader of the label that they are put in for the sole purpose of lowering the cost of the product. Admittedly, they do lower the cost, but the real reason for putting them in is to increase the adhesion of the subsequent coat by controlling the gloss and to make application easier, along with delaying settling both during application and in storage before the material is prepared for use on the job.

Finish coats. The final coat, which may be more than one coat, is chosen with special care. It should have the following characteristics:

It must be compatible with previous coats.

It should have a slightly different color from the intermediate coat to make application and inspection simpler.

It must be resistant to the environment and have the color and color retention, gloss and gloss retention needed in the condition of the exposure expected.

The prime difference in formation between a high gloss, a medium gloss, a semigloss, a satin finish, and a flat is the relative percentage by volume of pigments and vehicle. The higher the ratio of nonvolatile vehicle to pigment, the higher the gloss and the more impervious the film, and the more you are relying on the properties of the vehicle to resist the environment. To summarize:

Surface preparation permits the adhesion and chemical properties of the primer to be used to the best advantage.

The primer protects the steel.

The intermediate coat protects the primer.

The topcoat resists the environment.

Although the last two can be combined, and occasionally the first two are combined, best results require the use of all types of materials, and the choice becomes that of the system, not the item, which is best for the environment chosen.

Low initial cost coatings. Where it is desired to simply "hide" the surface at minimum cost, aluminum paints are frequently used. They do not really protect the metal, but they make it look good. They have the highest "hiding" per gallon.

Actually, they do protect the metal to a certain extent in that aluminum pigments leaf and have a tendency to keep out moisture, one of the three critical factors necessary for corrosion.

Aluminum pigments are sometimes combined with certain of the chromates to get both the inhibitive effect of the chromate and the leafing effect of the aluminum, giving some extremely good protection for surfaces that cannot be properly cleaned and prepared for the reception of the best coating system.

Aluminum. After steel, probably the most common metal is aluminum. In most cases, aluminum is almost self-protecting, but this varies with the environment, and aluminum becomes unsightly in many locations. Aluminum protects itself by the formation of an oxide coating which is so tightly adherent and so transparent that it is usually sufficient to keep the aluminum attractive.

Where this coating is not sufficient, and for an increased inherent beauty, chemical treatment such as anodizing is sometimes used. This is really a formation of an oxide coating under controlled conditions which is therefore much stronger and tougher than the natural oxide and can be colored in many metallic colors.

Occasionally, however, because of decoration and sometimes protection, it becomes desirable to paint aluminum. Here, as with steel, the most important item is the surface preparation. Although it is impractical to remove all the oxide coating which naturally protects the metal, the surface preparation after removing previous coatings consists usually of chemical preparation, the most common being TT-C-15328 if opaque colors are to be used.

Certain epoxies and urethanes can be used directly on aluminum or are used with the same pretreatment primer. The TT-C-15328 consists of an alcohol-soluble type of vinyl plus phosphoric acid plus zinc chromate, and these materials react with one another and with the aluminum on which it is placed to form a compound that, when properly applied, can be removed from the aluminum only mechanically. It forms a complex containing aluminum, chromium, and phosphate.

In order to get this complex to form, the material must, in addition to being put on clean aluminum only, be put on in a film thickness so low that the reaction can take place throughout the coating. Around 2/10 mil is the usual maximum thickness of this coating. If it is put on too thick, it simply has no adhesion. Other zinc chromate coatings can be and are used on aluminum, but this is the most commonly used pretreatment. It is also probably the most effective, and probably the most difficult to apply.

Magnesium. Magnesium is rather rarely used as a construction material. It has properties similar to aluminum in its ability to "protect itself," and when painting is desired, it is performed with techniques similar to those used on aluminum.

Copper. Copper and bronzes are usually used in locations where the lack of corrosion is desired. In some cases, however, the black or green formed by copper corrosion is desired. Copper and bronzes, of course, must be carefully protected and must not be used in atmospheres which contain substantial amounts of sulfur or sulfides in the air such as near sewage-disposal plants. Copper will react with sulfide to form an extremely brittle compound. Where it is desired to protect the copper, here again cleaning is the prime consideration. Epoxies and urethanes are being used and certain silicones are being designed which will go over this properly cleaned and prepared copper surface to prevent tarnishing.

Wood. Another common construction material is wood. Wood is one of the most frequently overlooked materials of corrosion-resistant construction. Wood withstands a wide variety of chemicals better than most metals.

The greatest difficulty in coating wood, however, is moisture. Moisture has a tendency to accumulate in the tissues of the wood and migrate through the wood to take off the coating on the other side. Coating wood therefore requires protection of both sides as well as the ends. It is important that both sides as well as the ends be protected and that there is no possibility for moisture to get through the wood from the other side.

A penetrating sealer is frequently used first on both sides and especially on the side from which the moisture can enter, which must be especially well protected.

It is important also, of course, that the topcoat material be compatible with the primer, that it does not lift, that it is flexible, and that it has a coefficient of expansion similar to that of the wood and the primer.

Masonry. Possibly the most common material of construction is masonry. Masonry, contrary to first thought, is severely affected by corrosion and by erosion, as well as by chipping and cracking.

Masonry consists, in addition to concrete, of brick, concrete block, tile, and most other nonmetallic surfaces.

Glazed tile has, at first thought and appearance, a sanitary, easily maintained, and corrosion-resistant surface. Careful examination of the tile wall shows that the tile is separated by cement or plaster grout joints, and it is usually adhered to the substructure with the same or similar cement or grout. This cement or grout between the blocks has no more resistance, and in fact frequently has less resistance, than even poured cement. From a bactericidal standpoint, it is a porous surface which readily absorbs both food for bacteria and the bacteria itself, and yet does not readily absorb sanitizing solutions.

From a chemical standpoint, the fairly resistant cement is normally modified by a much less resistant lime in order to get adhesion to the tile. Therefore, the ordinary tile surface is no more sanitary or chemical-resistant than the grouting and adhesive used.

Mortars.   Chemical mortars of various types have been introduced to the market for use in this construction, and a choice of chemical mortar may be made depending on the proposed end use of the surface. The various types of mortars follow.

Furane mortars are frequently used and, being resinous in character, are resistant to large lists of chemicals. They are quite impervious to absorption which will lead to future attack by bacteria.

Polyester mortars have certain resistance properties which make them desirable under many conditions, the largest failing being that of relatively poor adhesion to other than specially treated tile.

Epoxy mortars have the advantage of exceedingly good adhesion to the block; their adhesion actually exceeds the strength of the tile itself. They have a nonporous character which prevents absorption of contaminants and have a high degree of chemical resistance which could be modified by choice of curing agents to resist those solvents and chemicals desired.

Tile surfaces are used on interior and exterior walls. On interior walls, the contamination is dependent on the environment, and on exterior walls the absorbency of water by cementitious materials can and does lead eventually to failure by freeze-thaw. Tile can be made exceedingly resistant to freeze-thaw, but the mortar must be carefully protected.

Concrete.   Concrete is subject to attacks by acids, alkalies, and many salts, rarely by organic solvents.

Cement is also subject to chipping, erosion, and cracking.

Cement has high strength when monolithic, but it has the poor property of having exceedingly poor adhesion to itself in repairs. Concretes can be made to have very high compressive strengths. Compressive strengths of 5000 lb/in^2 and more are possible, but the elongation of cement is virtually zero. The tensile strength of concrete is normally figured at 10 percent (but actually runs less than 10 percent) of the compressive-strength figure. On floors especially, this combination of properties makes its use common but repair difficult.

Where concrete has proved to be a reasonably satisfactory material with its advantage of low initial cost, new cement can be made to adhere to old concrete by using an epoxy interface bond. This epoxy interface bond acts to adhere the new concrete to the old.

The epoxy bond acts to a more limited extent as a membrane coating to reduce the penetration of newly spilled liquids from the top through the new concrete and into the old concrete. It also tends to prevent migration upward of corrosive liquids from the old concrete into the new.

Repair of concrete. Where concrete has proved to be a poor construction material for the floor or machine base, for example, where it has shown severe erosion or corrosion from spillage of chemicals or solutions, this concrete should be either replaced or protected. Protection can take many forms.

If the original surface has broken off to a thickness of 1½ inch or more, it is frequently economical to use the above-mentioned interface adhesive, pour concrete to level, and then top the new concrete with a resistant material to withstand the erosive and corrosive effects which caused the trouble, preventing or delaying recurrences.

Another method is the use of modifiers in the new cement topping, usually consisting of latices of various types. These have several beneficial properties depending on the choice of latex which is used and the method by which it is used. Most latices, acrylics, polyvinyl acetate, and styrene-butadienes increase the tensile strength of the overlaid concrete by small margins and, when dry, slightly increase the compressive strengths and improve the abrasion resistance. Some of them also improve the resistance of the concrete to acids and alkalies. They sharply increase the adhesion of the new concrete to the old concrete. Generally they give about the same strength under average curing conditions as cement does under ideal conditions, plus far greater adhesion.

Epoxy polyamide latices and other epoxy latices can be used to modify the concrete in place of the thermoplastic latices, giving the same properties mentioned above plus the advantage that these properties do not deteriorate when wet. They are the same when wet as they are when dry. The thermoplastic latex modifications of concrete lose a large proportion of their improvements when they are continuously wet. The epoxy-modified concretes have almost the same compressive and tensile properties and abrasion resistance when wet as when dry and have about the same improvements on adhesion and other properties depending on the choice of epoxy and curing agent as the thermoplastic latices first mentioned.

Operations Research—Techniques

Introduction. Operations research methodology has had a substantial impact on facilities management and engineering over the past 30 years. With its impressive list of successful applications, this scientific discipline will undoubtedly continue to increase in importance in such activities as facilities operations and maintenance. In this section, the field of operations research will be described and examined, the use of mathematical models for solving facilities management and engineering problems will be considered, the areas of facilities management and engineering in which operations research techniques have been applied will be illustrated, and the problems encountered in implementing operations research techniques in facility management will be discussed.

Operations research defined.   Operations research is a term that has come into wide acceptance in the past 30 years. From its origins in the early 1940s it has developed into a burgeoning field with professional societies, journals, consultants, and departments and degree programs in colleges and universities. New work and significant claims for the merits of operations research continue to appear daily. Yet, despite this growth and general acceptance, operations research remains almost as difficult to define as it was 20 or 30 years ago.

Operations research has had almost as many definitions as it has had practitioners. Indeed, it is practically an annual occurrence for the retiring presidents of the professional societies to give their personal definition of operations research in their presidential addresses.

Within this diversity of opinion, two main categories of definitions emerge. Those in the first tend to stress the philosophical approach used in operations research. One definition normally stated is: Operations research is the application of scientific methods, techniques, and tools to problems involving the operations of systems so as to provide those in control of the operations with optimum solutions to the problems. Another definition is: Operations research is the application of the scientific method to the study of the operations of large complex organizations or activities. While operations research work does fall within the confines of the above descriptions, they nonetheless fail to distinguish operations research adequately from other fields. The above statements are no less true of industrial engineering or even cost accounting. Thus one must look beyond this first category of definitions.

The second category of definitions tends to list the many quantitative techniques that have become associated with operations research. Thus many define operations research as consisting of linear programming, inventory theory, queuing theory, replacement theory, simulation, etc. While operations research undoubtedly does use all the above techniques, it is too limiting to define it in terms of them. To define operations research in such a manner is similar to defining the field of medicine in terms of the collection of drugs which doctors use to cure their patients. If there is no more unity to the field of operations research than the latest collection of techniques that can be culled from the literature, then there is no particular merit in even giving it a name.

It is quite apparent that precision of definition is difficult to achieve while still including all the activities encompassed by operations research. However, if a short, simple definition is insufficient to describe operations research fully, perhaps a more complete understanding can be obtained by identifying its major characteristics. This will better allow one to sense the dimensions of the field of operations research and appreciate its general magnitude and direction. A consideration of the key words typically used in defining operations research will give a fuller appreciation of its true meaning.

Use of computers. Many of the mathematical models encountered in operations research work involve overwhelmingly large amounts of data. Others require that long and complex manipulation of data be performed. Thus many problems are soluble only with the aid of high-speed computational aids. Use of computers has become closely associated with operations research. Indeed, many of the techniques and tools of analysis of operations research would not be usable without the computer except at an unrealistically small scale. Computers have the important characteristics of speed of calculation, flexibility, and the performance of simple decision-making activities that are invaluable in experimenting with business-decision models.

Through simulation and other approximation techniques, problems have been solved with the aid of the computer that previously defied mathematical treatment. The computer has also provided a means for solving those problems which have long been quantifiable but computationally too complex or too time-consuming for human calculating effort. Organizations are daily solving problems which would have taken months to solve manually. Production scheduling, maintenance work order scheduling, blending, investing, procurement, and other activities are being reevaluated daily on the basis of current information. Previously such decisions were being made weekly, monthly, or even annually. By the time the decisions were reached, the information on which they were made was often obsolete.

Applications of operations research techniques in facilities management— Introduction. Facilities management and engineering is one of the most difficult industrial operations to manage and control owing to the varied nature of the work and the scattered locations at which it is performed. Because of this, operations research has been comparatively slow to develop in this area. Many of the early problems solved with the use of operations research involved production scheduling or inventory management. The immediate cost savings resulting from such applications were greater than could be achieved in the facility department function.

For example, earlier it had been found that while 42.6 percent of observed organizations had operations research departments, only 11 percent of the firms used operations research in facilities management, engineering, operations, and maintenance; and just 10 percent used it in equipment replacement studies. However, the use of operations research techniques in facilities management and engineering has more than doubled in the past 20 years. In a follow-up survey, 24 percent of the organizations reported using operation research in facilities maintenance and 20 percent in replacement analysis. Today, operations research tools and techniques not only are available to aid the facility in many managerial decisions, but they are also widely used.

In the following discussion various applications of operations research in facilities management and engineering will be described. Emphasis will be on the application rather than the tools of analysis. The techniques of operations research such as linear programming, queuing theory, and critical path analysis mentioned in these applications are covered in greater detail in various discipline textbooks.

Materials handling application. A case study concerning the design of a materials handling system illustrates clearly the importance of empirical observation of all the facts before problem solutions are suggested.

A firm manufacturing hot-water heaters was engaged in a continuing program of cost reduction. Many of the proposals under the program involved changing some portion of the materials handling equipment and procedures.

The existing system utilized fork trucks to move sheet steel through receiving, stores, and various blanking and forming press operations. The heaters were then moved onto a conveyor system for glass coating, welding, and assembly. Finished, crated heaters also were handled by fork truck. In addition, substantial transporting of in-process material from machines to storage and back was done by fork truck. Finally, the trucks assisted in moving and installing heavy dies for the machines.

Of particular interest to management was a proposal to extend the conveyor system back to include the initial fabrication operations. This addition promised a substantial immediate savings. Management decided, however, that by installing the conveyor, they would be beginning what could turn out to be a lengthy piece-by-piece improvement process. Even though each change individually might reduce operating costs, the overall design might be one not nearly as efficient as what might be obtained by considering all the materials handling requirements and problems of the entire plant at once.

Therefore, before designing improvements in any specific portions of the system, management decided to gather information on precisely what materials handling service was required in the facility. To obtain this information, they started by simply observing what was being done by the present system. For this step, they used the familiar industrial engineering tool of work sampling. A work sampling study consists of a large number of random instantaneous observations of the process under study, taking note of the condition of the object under study and classifying it into predetermined categories of activity pertinent to the particular work situation.

The operations of both the fork trucks and the operators were studied. Results from only the former study will be considered here. After 1427 observations of the six fork trucks, the following percentages of time in various activities were found:

Category	Percent of time
Move with load	13.4
Move, no load	15.7
Adjust stacked material	3.1
Move with die	0.4
Assist in installing or removing die	0.9
Not in use, no work available	46.1
Not in use, operator absent	9.4
Not in use, maintenance, adjustment, or inspection	6.3
Truck not in area being studied	4.7
Total	100.0

Further results showed that in only 0.2 percent of the rounds of observation were all fork trucks in use. In only 1.3 percent of the rounds were as many as four trucks in use at the same time.

The category "operator absent" accounted for 9.4 percent of the time, while personal time allowed operators was 5 percent. Actually, the operators realized what management did not, that too many fork trucks were assigned to the job. Also, maintenance was supposed to be performed at night. Only emergency repairs should have been made during the periods studied, yet the trucks were idle for maintenance 6.3 percent of the time.

As a result of the information obtained by the work sampling study, management was able to take action to reduce the cost of materials handling substantially using the existing system. The conveyor addition, which previously looked so attractive, was no longer economically justified.

Specific actions taken by management were:

Reduction of fork trucks from six to four

Institution of a new system of fork truck scheduling to reduce travel and waiting time

Enforcement of the maintenance rules which reduced maintenance time during the shift almost to zero

In addition, further improvements resulted from the study of machine-operator activities. In this application, the main tool of analysis, work sampling, is certainly not new or even specifically associated with operations research. Of specific interest from the point of view of operations research, however, is the approach taken to study this problem. It would have been quite typical of many managements simply to order the conveyor equipment based on the expected cost reductions. Management decided instead to seek an optimal material handling system for the entire facility and to define the materials handling requirements precisely before any changes were instituted.

Planned replacement. A pulp and paper company was using a fleet of 49 fork trucks of 4 different makes and different models for materials handling. Maintenance costs were very high, average $1.49 per hour of running time, a figure about twice as high as the industry average. Moreover, they had increased by 26.6 percent in 1 year to total $136,000 for the year. In addition, the cost of carrying an inventory of spare parts was $46,000 annually.

In order to reduce these costs, the firm considered the three following alternatives:

Replace the fork trucks, one at a time, with new ones, beginning with the unit with the greatest maintenance cost.

Recondition the fork trucks with a major overhaul, again starting with the unit with the greatest maintenance cost. The cost of such an overhaul was 40 percent of the cost of a new truck and resulted in maintenance costs one-third higher than for a new vehicle.

Replace the entire fleet at once by a new fleet consisting of fork trucks of two models from one manufacturer at a cost of $600,000.

By the use of a total cost minimization model, it was found that the third alternative was superior. The expected yearly savings were $50,000 on maintenance and $25,000 on spare parts inventory. Other intangible improvements cited by the organization were easy financing (since all purchases were from one manufacturer), reduced fuel cost, reduced driver fatigue, improved morale, easier training of both drivers and maintenance employees (because of the standardized fork trucks), and a higher moving capacity per fork truck.

Planned preventive maintenance. A large steel mill with high maintenance labor costs undertook an extensive conventional industrial engineering methods improvement program at the end of which maintenance labor costs were found to be $60 per $1000 worth of output. To reduce costs still further, the organization attempted to apply operations research models to improve the scheduling of the planned maintenance activities. A research team was formed which consisted of organization personnel from both the systems engineering and the facilities engineering departments, and also consultants from outside the organization.

The team developed mathematical models for maintenance work order scheduling and also designed a computer data-processing system to provide the necessary information to implement the models. As a result, optimal preventive maintenance frequencies were calculated, and a maintenance schedule developed based on actual machine running time. In addition, the data processing presented information on the comparison of actual preventive maintenance work order performance to scheduled work order performance, comparison of expenditures to budget, a complete history of every piece of equipment, and comparative information on machine running time as a function of the frequency and level of planned preventive maintenance activities.

The reported savings in the first year were $2.50 per $1000 of output for a total of $124,087. Other improvements were the possibility of quickly identifying the source of high maintenance costs; improved production scheduling; the provision of information to aid in the evaluation of the profitability of production equipment; help in maintenance prevention through the provision of information for better tool design; and additional information for assisting in the replacement evaluation and equipment-selection decisions.

Group relamping. A common task falling under the supervision of the facility manager is the maintenance and replacement of auxiliary facilities such as lamps. Some alternative policies which might be considered for lamp replacement are:

Replace lamps when they fail.

Replace lamps after they have reached a certain age.

Replace all lamps at periodic intervals.

Operations research models have been developed to determine the costs of operating under each of the above policies. The best policy will depend on the costs of the lamps, the cost of replacement, and the life-expectancy curve for the lamps.

Several organizations, which had been using a policy of replacing lamps when they failed without ever evaluating the cost of such a policy, found substantial savings were possible by employing the group relamping policy. The following improvements were reported:

Group relamping cost $3000 annually compared with $5200 for replacing only failed lamps.

Group relamping resulted in a net annual savings of $91 per fixture.

Group relamping reduced cost from $18,400 to $13,612.

Group relamping reduced costs by 33 percent.

Relamping costs were cut to one-tenth of the previous cost by group relamping.

In addition, the overall average lighting level in offices and shops was increased in some of the facilities.

Other applications.   In addition to the several case histories presented, there have been many other applications in facility management including:

Project planning and control (installation, modification, rearrangement, and removal of large production facilities)

Preventive and predictive maintenance decisions (specification of requirements, determination of frequencies, scheduling and routing, and deferring preventive maintenance)

Overhaul scheduling

Materials, supplies, and spare parts control (calculating reorder point, safety stock, and economic order quantity)

Crew-size determination

Investment and replacement decisions (for entire piece of equipment and for parts)

Reliability and maintenance prevention

Corrective maintenance

Maintenance shop scheduling

Assignment of crews in various facility areas

Subcontracting (outsourcing) determinations

Wage incentive plans

Staffing of a toolcrib

Make or buy spare parts

Diagnosis of complex systems failures

Placement and layout of the various operations and maintenance shops

Tools and special equipment acquisition evaluations

Limitations of operations research.   Operations research, just like any other management tool, is no substitute for good management. The facility manager must still decide what to investigate, what to do about the factors that cannot be measured, and how to interpret the results of operations research. Operations research performance requires capable, well-qualified, and experienced personnel in the field to support and aid the facility manager. Operations research studies frequently require a great deal of time and are very costly. Attempting shortcuts by superficially examining the problems, using inappropriate models or inaccurate data, may produce results that if applied will be far more costly than simply using the dictates of management's subjective judgment and experience.

The results certainly cannot be guaranteed. The potential benefits, however, may be enormous and well worth all the time, effort, and expense required to undertake the operations research approach.

The future outlook.   One of the most startling developments in the area of management in recent years is the penetrating and use of operations research studies and of electronic computers into almost every phase of human activity. Progress and improvements in operations research methodology to date has been continuously increasing, and the prospects are even brighter. Operations research is a science directed toward those tasks which human beings have not yet delegated to machines. It is a science of decision and its application. For this reason, it is difficult to assess the future growth of operations research in general or in the area of facility management and engineering. However, some indications point to a possible increase of use of operations research techniques in industry, business, education, and government. Some indicators follow:

First, we are witnesses to a dramatic growth in the computer industry.

Second, the properties of the operations research models are improving continuously. New techniques and applications are being developed in industry, business, education, and government.

Third, the level of education and knowledge of the personnel involved in operations research activities is continuously growing.

Finally, the need for operations research is growing because of the increasing competition and constant demand to increase our standard of living.

The growing need for use of operations research techniques on one side and the growing availability on the other side, coupled with the increased educational level, tougher competition, and cheaper and faster data-processing facilities, lead one to believe that the rate of using operations research techniques is accelerating and that use of operations research techniques will successfully meet the managerial requirements of the next millennium.

Glossary

ADA signage Signage required by the American Disabilities Act to aid the handicapped.

alert levels In the predictive maintenance program, alert levels must be established for each designated parameter measured to provide a flag or trigger for taking necessary additional corrective action(s).

AMMS (advanced maintenance management system) A PC/windows-based management tool for use as the department's MIS.

annual PM schedule An annual preventive maintenance (PM) work schedule is developed for equipment, systems, and assets to encompass intensive work performance tasks that will not impede production or mission performance.

Askarel (PCBs) Askarel is a generic term for the family of polychlorinated biphenyls (PCB). It is marketed under various trade names.

batteries Used in electrical switchgear applications, in uninterruptable power supply systems, and for emergency lighting.

benchmarking Used to improve a department's performance by comparing its practices to those of other similar-sized facility departments.

building operational plan Published plans for each facility building describing the required processes and procedures necessary to ensure safe, efficient operations and maintenance actions.

building tours Formally planned and scheduled building tours to confirm the effectiveness and efficiency of building equipment and systems, data collection, interaction with building users, and structural deficiency assessments.

cable and wire Conductors found throughout the power and control systems of all plants, facilities, and distribution systems.

CAFM (computer-aided facility management software) Software that provides the facility manager with "tracking" capability on what is going on in the department plus use as a planning and cost-saving tool.

callback The term used to indicate an unscheduled visit by a service technician.

CAMMS (computer assisted maintenance management system) A management tool for use as the department's MIS.

chilled water Provides cooled building air by circulating the air over small fan coil units in individual rooms, or through large air-handling units, that deliver cooled air to public spaces and to duct systems.

CMMS A computerized maintenance management system used by the facility manager as a tool for planning maintenance and repairs and keeping records using various cognizant modules such as work order control, preventive and predictive maintenance, inventory control, etc.

communications protocol A protocol developed and used for each communication systems level.

conditions assessments These define the needs of installed systems and equipment and a baseline of operating conditions, i.e, remaining useful life, level of maintenance achieved, and operating characteristics.

constant-volume air-handling unit (CAV) Consists of a supply fan, a cooling coil, a preheat coil, outside air and return air dampers, a return air fan, and exhaust dampers.

contract administration The facility department's outsourcing functions involved in contracts bidding and awarding, monitoring and inspection, acceptance, payment, and financial and production reporting.

contract services The providing, by contract, of general facility operations and maintenance services.

cooling tower fans These fans, single or multiple speed, are used to control condenser water temperature.

corrective maintenance The repair or replacement of structures, equipment, or systems that have failed.

corrosion coupons The most common method for monitoring corrosion rates in cooling water systems.

corrosion Damage to a system caused by reaction of both metals and nonmetals with water in the system.

criteria The facility manager's goals and objectives established for use in performance measurement.

customer service The service provided when the facility department meets and exceeds customers expectations of service.

customers All types of facility users who receive support from the facility department.

data transmission devices These devices (modems, transmitters, active hubs, etc.) are deployed in support of the remote control system. They should not be used for voice transmissions or other computer system data communications services.

demand control ventilation (DCV) A method of modulating fresh air intake to a facility in response to the level of carbon dioxide present in the building's return air system.

design assessments These assessments determine whether current electrical and HVAC needs are met, including current ventilation codes and standards to achieve enhanced indoor air quality.

designing healthy buildings   The issue of SBS (sick building syndrome) must be considered in the design and construction stages of a building's life. This approach recognizes the benefits of prevention over cure in treating building systems ailments.

direct digital control systems   With the advent of microprocessor technology and sophisticated programming techniques, these have become the current industry standard.

disconnect switches   These switches are used to connect or disconnect circuits or apparatus, and for isolating equipment or circuits.

duty cycling   Used because of spare capacity built into systems, making it possible to turn the system off for a period of time and the turn it back on again, and still maintain prescribed environmental space temperatures.

economizer operation   This operation can be implemented in many HVAC functions based on building design. It can be either air side or water side.

eddy-current inspection   Used to detect internal and external damage of condenser and boiler tubes.

electrical maintenance program   This program identifies the various types of equipment to be serviced, its characteristics, and the effecting conditions of its use and environment; and all other needs are identified.

electronic watch tours   The Energy Management Control Program may use the remote control and monitoring system to perform engineering watch tours electronically (as required by local codes) of the connected equipment.

emergency response plan   This plan defines the duties and responsibilities, identifies various emergencies that may occur, explains what to do in the emergency situation, and details procedures for resolving most conceivable emergencies.

empower   The delegation by the facility manager to both individuals, and value improvement teams, of both authority and responsibility to effect changes in processes, procedures, and methods, without prior approval.

energy conservation   Energy conservation is achieved when temperature controls are set to maintain an indoor environment consistent with general industry standards and practice during normal facility work hours.

energy management control system (EMCS)   A computer-based energy and temperature control system used to administer building operations, featuring a microprocessor-based remote field panel capable of standalone operation.

Energy Management Program   A program that regulates the HVAC systems to maintain tenant comfort conditions within facility spaces while at the same time managing energy costs.

engineering evaluations   The cost-benefit studies, developed in quantitative terms, that specify the feasibility and costs of each of the possible alternative courses of action for facility changes.

engineering input   The information and data obtained from the original equipment, systems, and building design and specifications, and from cognizant key operations personnel.

entrapments   Entrapments on escalators occur if passengers stand around or near the edges of the steps.

EPS (engineered performance standards)   The allowed job time, in manhours, developed by industrial engineering techniques for a qualified worker, working with average skill and effort under good supervision, to accomplish a defined amount of work.

equipment and systems logs   These logs, manual or computerized, provide the quantitative tools for technical, administrative, and management personnel to evaluate equipment and systems programs performance.

equipment inventory   The tabulation and recording of all pertinent and cognizant information and data for building systems, equipment, and controls; i.e., operational data, location, PM requirements, maintenance schedules, etc.

equipment shutdown   These are procedures designed to ensure that the systems and equipment are properly secured and that the desired mode of operation for the ensuing startup functions can be achieved.

equipment startup   These procedures for equipment and systems vary according to external environmental conditions and average building space temperatures.

estimate   Allowed job time standards developed by supervisors and/or planner-estimators, after a job site visit, based on judgement, performance standards and work experiences, for cost estimating purposes.

estimating   The development and establishment of materials, supplies, equipment, and worker-hour budgets for maintenance or repair work performance.

evaluation   The appraisal of craftspersons and supervisors' performance using a CMMS for reporting production and financial information and data.

facilitator   An individual who has detailed TQM knowledge of measurement processes and team facilitation skills—such as how to orchestrate and guide a team—as well as negotiation and conflict resolution skills.

facilities inspection and maintenance program (FIMP)   A set of programs designed to aid the facility manager in operating and maintaining the facility within budget and facility needs.

facilities management function   The organizational department that applies technical and management effort in the planning, design, and operational life of a facility.

facility engineer   A graduate engineer of any discipline, employed on a full-time basis, to provide needed facility technical services.

facility management   The effort required to provide complete operations and maintenance service support so that a facility (buildings, equipment, machinery, systems, and grounds) may operate at an optimum lowest overall total cost.

facility master plan   This establishes the facility capacity and a logical strategy for the development of a site.

facility occupant support plan   This plan ensures the provision of necessary operations and maintenance support in a caring manner that exceeds customers expectations.

facility planning studies   These studies provide cost estimates for site development, construction, operations, maintenance, support equipment, and other applicable items.

fiberoptic equipment   Equipment that allows visual inspection of internal surfaces of various mechanical equipment items.

field inspection procedures   Guides used by inspection personnel showing what and when to inspect; and how to service and maintain the equipment and systems.

find/fix   Programs that use scheduled teams of craftspersons to systematically inspect for and then perform contingency maintenance and repairs on a particular system or equipment, or a complete building.

fire protection   A developed fire protection protocol, unique for each building, describing the procedures to be followed in case of fire in the building.

firefighter's service   A test of the system that controls the elevators firefighting operations.

free cooling   A process used to obtain free cooling by eliminating the energy costs associated with operating the chillers.

full-service water treatment   Treatment services where the vendor supplies all the products and does all the work.

fuses   Either of the current-limiting or noncurrent-limiting type. Fuses are available in an array of voltage and current ratings.

general cleaning   The types of cleaning needed for the facility; i.e., tenant requirements, tenants uncommon needs, and budget guidelines.

ground fault circuit interrupters (GFCI)   Personnel protective devices intending to prevent electrocution when a person's body or other object bridges between an energized 115/230 V conductor and ground.

ground fault protection   Protection designed to sense abnormal current flow between energized conductors and ground, and to operate to disconnect the circuit for the purpose of limiting equipment damage, without concern for electrocution.

grounded   Being connected to earth or some conducting body that serves in place of earth.

group cleaning and/or relamping   Performed when facility areas indicate that lumen degradation is at a level that hinders the safe and secure usage of the space.

historical standards   Job time standards developed from historical work order actual times and procedures which include all the foibles, problems, and inconsistencies unique to the facility department.

house calls   The requests from tenants to perform numerous small tasks necessary to perform or meet mission operational requirements.

human relations   The art of dealing with department personnel when it is necessary to take corrective actions due to poor or improper work performance.

humidification systems   Systems used to control humidity by injection of steam to the air ducts, as required.

IAQ (indoor air quality)   This describes the pollution measured in indoor (in buildings) environments.

indefinite quantity contracts   Contracts that allow the facility manager to order an indefinite service or supply quantity within a fixed dollar limit.

infrared (IR) thermograph   The monitoring of equipment temperatures using infrared measurement instrumentation to provide a predictive maintenance tool.

insulating liquids   Various liquids used for cooling and insulating purposes.

insulation resistance measurements   These measurements are obtained from field tests performed on most but not all low, medium, and high-voltage apparatus.

integration   Although not actually a tool, integration of all predictive maintenance technologies is a powerful diagnostic function. Also, integrating the predictive maintenance program with all other maintenance functions is a desirable goal.

ISO 9000   An international quality standard used in a manner similar to TQM procedures.

labor standard hour   The amount of work that a mechanic can perform in one hour considering these time factors: EPS developed work, travel, preparation, and delays.

labor standard time   The EPS developed work order allowed standard time expressed in worker-hours.

landscaping services   Professional services that give properties curb appeal and enhance the appearance of buildings.

LAN (local area network)   This links (networks) PCs for the sharing of information and data.

lighting controls   These controls, either a standalone package or integrated into a building management system, are utilized to turn lights on and off during occupied and unoccupied time periods.

load shedding   A technique used to reduce the peak electrical demand charges from the utility company.

low-voltage air circuit breakers   600-V class air circuit breakers not housed in an insulating case.

low-voltage molded (insulated) case circuit breakers Molded circuit breakers are commonly used circuit breakers found in electrical panels in residences, commercial buildings, and institutional grade facilities.

machinery history A record of all problems, breakdowns, and repairs, in detail, concerning all individual facility equipment and systems items.

maintenance All work related to the economical preservation of structures, equipment, and systems at a level satisfactory to perform their designated functions.

maintenance file cards A manual maintenance card system using multi 3 × 5 or larger hardstock cards for each equipment and system item recording all the same information and data as in the CMMS's PM module database.

maintenance history The record maintained in the CMMS, in detail, of what maintenance action has been performed.

maintenance log The manual or computerized source data accumulation that provides a chart for the technician's maintenance work performance.

maintenance management manual A manual developed to describe the policies and procedures, in writing, to be used in planning and performing the facility maintenance effort.

maintenance outplanning To ensure that planned maintenance effort is conducted efficiently, it is necessary to review projected maintenance on a regular, annual basis.

maintenance system administrator An expert in maintaining equipment and systems. This person is responsible for ensuring that the preventive and predictive maintenance programs are working properly in all aspects.

major repairs The work required to restore a seriously deteriorated or broken-down structure, equipment, or system to a state of usability for its designated function.

mean time between failures (MTBF) These are analyses in a mathematical/statistical procedure used to determine the mean time between equipment and systems breakdowns.

medium-voltage breakers Three-pole devices that interrupt the circuit with contacts in air, oil, or vacuum mediums.

methods engineering (ME) Time standards developed by industrial engineers using the methods of time and motion study, work sampling, or predetermined time standards.

microbiological fouling Deposits formed from microbiological material and metabolic products in cooling systems.

microbiological monitoring The system used to maintain microbiological activity at low levels so as not to encourage rapid growth and slime formation in the open cooling tower system.

mineral scale and deposit monitoring In cooling water systems, online monitoring of heat transfer rates, combined with visual inspection of a heat transfer surface, is a practical way to monitor deposit formation.

monitoring frequency In the predictive maintenance program, after selecting the appropriate monitoring technologies needed for each equipment and system item, the next step is to identify and select the proper monitoring frequency.

motion analysis of circuit breakers These tests measure the time it takes the contacts to part after the trip coil has been energized or to make contact after the close coil has been energized.

motor control centers An assembly of motor control units (starters) built by modular design in order to centralize the motor control units' performance.

ODBC (open data base connectivity) The information systems ability to transfer information between disparate databases.

offline chemical cleaning The most expensive and effective way of removing deposits from a facility water piping system using strong chemicals.

offline mechanical cleaning The cleaning of cooling systems main circulating pipes performed by a mechanical hydroblasting process.

oil analysis Development and implementation of a formal lubricant analysis program to identify lubricant condition and equipment wear.

oil circuit breakers (and switches) These devices have their main and arcing contacts immersed in a tank of insulating mineral oil.

one hundred percent outside air systems These systems use complete outside air with no return air mixing.

online chemical cleaning Cleaning a system online, while operational, using high concentrations of dispersant chemicals and microbiocides.

operating watch tours These tours ensure that operations of the mechanical and electrical systems are in accordance with the building operating plan (BOP).

organization statement The facility manager's reflection of the way he or she views the facility department's organization design, structure, and staffing.

outsourcing Contracting with vendors to expand the facility department's capabilities and resources without expanding its workforce.

overall total cost The cost features that incorporate initial costs, life cycle maintenance and repair costs, and salvage costs.

planner-estimator An experienced and highly competent individual who can produce work orders that are accurate, complete, and easily understood by facility department personnel.

planning All administrative and technical effort necessary to ensure that each craftsperson and laborer will be working at maximum efficiency on a necessary task.

plans and specifications The facility engineer's technical drawings and detailed job specifications

PM "templates" A preventive maintenance template is the complete description of a specific PM procedure, including the equipment or sytems, interval or performance frequency, and required resources such as parts and labor, and other detail.

pneumatic control systems   Control systems that are utilized in larger buildings where long-term stable control systems and distribution of power via pneumatic air tubing is an economic alternative.

position descriptions   Documents created for each staff member that specifies individual work functions needed to execute assigned specific duties.

potable water systems   Systems that supply water either from a municipal or private water supply.

power factor and dielectric-loss measurements   Nondestructive tests which detect the change in the measurable characteristics of insulation.

power factor   The power factor of an insulation is the cosine of the angle between the charging current vector and the impressed voltage vector.

preconstruction conference   A conference called by the facility manager to define the limits of authority of the facility department's inspection staff and to alert the contractor to the degree of inspection tolerance to be expected.

predictive maintenance   Predictive maintenance requirements are identified and performed when empirical data that are collected and reviewed indicate that maintenance is required.

predictive process   The predictive maintenance program goal is to provide the information and data needed to make decisions on the need for maintenance action, or no action, without sacrificing reliability.

preventive maintenance program (PMP)   A program that, when properly conducted, will reduce operating costs, aid organization effectiveness, aid safety practices, and assure equipment and systems performance.

preventive maintenance tasks   Task assignments which can be applied to any structure, equipment, or system which could benefit from routine inspection, repair, or renewing.

preventive maintenance   The prebreakdown work performed on equipment and systems to eliminate failures and/or breakdowns, or to keep such failures and/or breakdowns within predetermined economic limits.

process improvement team (PIT)   The group formed to investigate problems, find root causes, and improve processes and systems that exhibit wasted effort, money, and time.

protective relays   These relays detect defective lines or apparatus or other power system conditions of an abnormal or dangerous nature, initiating appropriate control circuit action.

QCP (quality control plan)   An important management tool used to ensure maintenance of high quality standards in a sustained departmental quality control effort.

radiography   A system used to check welds and to evaluate the condition of piping beneath insulation.

reasonable expectancy (RE) Time standards, developed by industrial engineers, based on several macro observations. RE represents a reasonable, observable amount of work performed in a given time period.

reengineering The process of "wiping the slate clean" and recreating all or part of the facility department's organization.

refrigerant management A plan that describes procedures for tracking refrigerant usage in accordance with EPA requirements.

remote control operating Remote control operating for HVAC facilities is the fundamental ability of an operating engineer, from a remote location, to change the operating state of a piece of equipment or system.

requirements contracts Contracts that the facility manager has with one contractor who provides all the services and supplies on call.

reset programs These programs make it possible to reset air-handling unit temperatures to reduce energy consumption while maintaining required space specified temperatures.

rotating machines Electrical machines, motors, and generators that are the workhorse of modern society in plants and facilities.

routine maintenance The day-to-day upkeep of structures, equipment, and systems to ensure their capabilities to perform their designated functions.

scaling The deposition of minerals from the water supply.

scheduling (of facility department work) The actions taken to constantly strike the right balance between proactive (preventive) and reactive activities. The allocation of labor, materials, and equipment, at specific times and locations, for maintenance and repair work performance.

service orders Small, service-type maintenance jobs that require immediate attention and cannot be deferred.

sick building syndrome (SBS) A syndrome that is multifactorial in origin with no specific causes. Also, it is not a pathological illness, but rather a compilation of many vague symptoms.

standing operating orders (SOO) Standing work orders where the specific work and the manpower requirements are relatively constant and predictable.

step voltage regulators These regulators increase or decrease line voltage, usually by plus or minus 10 percent.

supervised communication formats Any and all telecommunication services where the availability of the communications path is supervised by the remote control and monitoring system.

support service water treatment Treatment services that are the performance responsibilities, jointly, of both the facility operating engineers and the vendor service representative.

surge arresters (protectors) Devices used to protect all systems and equipment from line voltage that exceeds a predetermined line voltage level.

switchgear   The general term applies to power switching and interrupting apparatus, typically consisting of switches or circuit breakers and associated protection and control equipment.

task assignment sheets   Sheets developed to describe tasks to be performed on building tours.

time-based maintenance   A fixed amount of time between scheduled maintenance visits.

time-of-day programming   The preplanned start and stop of equipment and systems based on the time of day rather than existing conditions.

TQM (Total Quality Management)   A management tool that fosters a continuous improvement mind-set in the facility department management team.

transformers   Devices that convert or transform electrical energy from one voltage or current level to another level, with relatively insignificant energy loss.

ultrasonic thickness probes   Probes used to measure pipe wall thicknesses online.

ultrasound   The use of ultrasound monitoring and measuring equipment to detect and zero in on particular equipment faults.

use-based maintenance   Scheduled examinations and periodic maintenance tasks performance dependent on events happenings; i.e., number of operating hours, number of stops, etc.

value engineering (VE)   A methodology for obtaining the best value for the lowest cost throughout the life cycle of the facility. It is more properly called *value analysis.*

variable-air-volume air-handling unit (VAV)   A variable air volume unit is configured in a similar manner to a constant air volume unit with the addition of supply and return volume varying devices and airflow measuring stations.

vertical transportation equipment   Equipment that consistently delivers people and freight safely and efficiently in buildings in a vertical mode.

vibration monitoring   The use of specialized testing equipment to determine the overall levels of vibration considered "acceptable" for each facility equipment or system item.

WAN (wide-area network)   A network consisting of mini- or mainframe computers acting in a client-server relationship for the sharing of information and data.

waste management program   A program that addresses all issues regarding the collection and disposal of trash and waste, including recyclable and non-recyclable trash and waste.

water and deposit analyses   These analyses are conducted by operating engineers, periodically, to analyze parameters that are important in controlling the chemical treatment program.

water systems   Systems used for space heating and cooling, heat removal and environmental control, humidity control, and domestic (potable) water service.

water treatment performance specifications   These specifications set basic parameters for cooling water treatment programs in the areas of corrosion rates on system metals, mineral scaling, general deposition, and biological fouling.

water treatment service work   The daily periodic work performed, by agreement on responsibilities, between the facility operating engineers and vendor representatives.

wheelchair lifts   Lifts established for use by physically disabled persons.

winding turns ratio measurements   Measurements obtained from tests that measure the ratio between windings of current or voltage transformers.

work control center   The central point where all work requirements are funneled. The work is then coordinated, planned, costed, scheduled, and results are measured.

work control cycle   Work control methods that allow work requirements to be identified, screened and evaluated, planned and scheduled, checked and inspected, closed out and cost accounted, results recorded, analyzed, and measured, and feedback delivered.

work order close outs   Work orders are closed out when the prescribed work has been completed, inspected, and accepted for quality, quantity, and performance, cost accounting has been performed including annotation on work orders and posting of labor, mate.

work priority system   A system used to prioritize work performance after work requests are received, categorized, classified, and approved for subsequent performance.

Index

ABOUT THE EDITOR

Bernard T. Lewis, an independent Facilities Management Consultant, has a Bachelor of Science degree in mechanical/civil engineering from the U.S. Military Academy, a Master of Arts degree in mathematics from Columbia University, and a Doctorate of Business Administration degree from Pactific Western University. He is a Registered Professional Engineer and a Certified Plant Engineer. He has been engaged in facilities engineering and management for over 35 years at the U.S. Army, American Machine and Foundry, Western Electric Company, the U.S. Navy, and on various public and private sector consulting projects. He has published 19 books covering facilities maintenance engineering and management including this handbook.